CIVIL ENGINEERING DRAWING AND DESIGN

Other Engineering Books by the Author

Engineering Drawing

Geometrical Drawing for Beginners

Engineering Mechanics and Strength of Materials

A Handbook of Mathematical Tables and Formulae

Materials of Construction

A Dictionary of Civil Engineering

Operation & Maintenance of Sewage Treatment Plants

CIVIL ENGINEERING DRAWING AND DESIGN

(A Comprehensive Guide for Civil Engineering Students)

Second Edition

by

D.N. GHOSE, MCE, FIE, FIWWA, FIPHE

CBSPD

CBS Publishers & Distributors Pvt Ltd

New Delhi • Bengaluru • Chennai • Kochi • Kolkata • Lucknow• Mumbai
Hyderabad • Jharkhand • Nagpur • Patna • Pune • Uttarakhand

Civil Engineering Drawing and Design

ISBN: 978-81-239-1809-9

Copyright © Author

Second Published: 1987
Reprint: 1998, 2000, 2003, 2005, 2007

Second Edition: 2010
Reprint: 2011, 2012, 2014, 2015, 2024

Published by Satish Kumar Jain and produced by Varun Jain for

CBS Publishers & Distributors Pvt Ltd

4819/XI Prahlad Street, 24 Ansari Road, Daryaganj, New Delhi 110 002, India
Ph: 011-23289259, 23266861   Website: www.cbspd.com
                                        e-mail: delhi@cbspd.com

Corporate Office: 204 FIE, Industrial Area, Patparganj, Delhi 110 092
Ph: 011-4934 4934   Fax: 011-4934 4935   e-mail: publishing@cbspd.com
                                                        publicity@cbspd.com

Branches

- **Bengaluru:** Seema House 2975, 17th Cross, KR Road, Banasankari 2nd Stage, Bengaluru 560 070, Karnataka, India
  Ph: +91-80-26771678/79   Fax: +91-80-26771680   e-mail: bangalore@cbspd.com
- **Chennai:** 7, Subbaraya Street, Shenoy Nagar, Chennai 600 030, Tamil Nadu, India
  Ph: +91-44-26680620, 26681266   Fax: +91-44-42032115   e-mail: chennai@cbspd.com
- **Kochi:** 42/1325, 1326, Power House Road, Opp KSEB, Power House, Ernakulam 682 018, India
  Ph: +91-484-4059061–65   Fax: +91-484-4059065   e-mail: kochi@cbspd.com
- **Kolkata:** 147, Hind Ceramics Compound, 1st Floor, Nilgunj Road, Belghoria, Kolkata 700 056, West Bengal, India
  Ph: +91-9096713055/56   e-mail: kolkata@cbspd.com
- **Lucknow:** Basement, Khushnuma Complex, 7-Meerabai Marg (behind Jawahar Bhawan), Lucknow 226 001, India
  Ph: +91-522-4000032   e-mail: tiwari.lucknow@cbspd.com
- **Mumbai:** PWD Shed. Gala no. 25/26, Ramchandra Bhatt Marg, Next to JJ Hospital Gate no. 2, Opp. Union Bank of India Noorbaug Mumbai 400 009, Maharashtra, India
  Ph: +91-22-66661880/89   e-mail: mumbai@cbspd.com

Representatives

- **Hyderabad** 0-9885175004
- **Jharkhand** 0-9811541605
- **Nagpur** 0-8692091830
- **Patna** 0-9334159340
- **Pune** 0-9664372571
- **Uttarakhand** 0-9716462459

Printed at Sanjay Printer, Sahibabad, UP. India

Preface to the Second Edition

It is the pleasure of all to see the popularity of the book earned during the past years. Further, the contents of the book have been thoroughly revised in this edition. No stone has been kept unturned to make the book need-based.

I hope that this volume of "Civil Engineering Drawing and Design" will be useful to the students, practising engineers and contractors in the field of civil engineering.

Lastly, I express my fervent thanks to the publisher Mr. Satish K. Jain and his helping hands.

D.N. Ghose

Preface to the First Edition

If is needless to mention that *'Drawing is the language of Engineers'*. An engineer for better expression of his ideas should have therefore thorough knowledge and conception in drawing.

At present there is hardly any book on Civil Engineering Drawing covering all the essential topics of civil engineering. Thus, difficulty arises with 'what more to teach and how to learn more'. It is to fufil that want that this book has been brought out.

While preparing this volume, every effort has been made to present the matter in a lucide style through self explanatory sketches. It covers the syllabi of civil engineering diploma and draughtsmanship courses of all the State Boards for Engineering and Technical Education and Industrial Training Institutes of all the states of India. This book will also be helpful to the students of civil Engineering Degree courses as well as practising Engineers and contractors.

It is expected that my reader friends will find no difficulty in going through this volume. Should the readers find out any mistake, I would appreciate having those brought to my notice. Any constructive suggestion for future development of this book would be highly appreciated.

D.N. Ghose

Contents

About the Author

Shri D.N. Ghose, born in 1938, is a well-known author of engineering books. His first book was published in 1956. Three of his books were selected by the Government of West Bengal (1956—1960) for Adult Education. In 1972, National Council of Educational Research and Training (NCERT), New Delhi, awarded Certificate of Merit for his manuscript "Purano Sei Diner Katha" which ranked second in 17th All India Children's Literature Competition.

Shri Ghose obtained Bachelors Degree (1963) and Master Degree (1969) in Civil Engineering from Jadavpur University. In early 1959, he started his career as a teacher and afterwards held the posts of Head of the Department of Civil Engineering and officiating Principal of Don Bosco Polytechnic, Calcutta.

In 1973, Shri Ghose as a Divisional Sub-committee Member of Bengal Centre of the Institution of Engineers, India, organised Refresher Course on Public Health Engineering. In the same year, he conducted Unitary Summer School on Public Health Engineering under the sponsorship of Indian Society for Technical Education (ISTE). He was also Examiner of various quailfying examinations in the field of Civil Engineering.

In late 1973, Shri Ghose associated himself with M/s Tata Engineering and Locomotive Company (TELCO). He was responsible for operation, control and maintenance of water treatment and sewage treatment plants and town water supply of TELCO. In early 1977, he joined Calcutta Metropolitan Development Authrity (CMDA) and visited U.S.A. with a work programme of CMDA on computer aided analysis of the water distribution network of the city of Calcutta. During his stay in CMDA, he was engaged in planning and development work of sewerage, drainage and solid waste management for the CMD Area. In late 1980 Sri Ghose joined a consulting firm.

Presently, Shri Ghose is working as a Consultant in tne field of Environmental Engineering. He has prepared quite a good number of Project Reports based on his field investigations, study findings and analysis.

1

Symbols, Sign Conventions, and Dimensioning

Symbols and Sign Conventions

There are various sign conventions and graphical symbols to represent the materials of construction, sanitary installations and their fittings and fixtures, electrical installations, rolled sections, pipeline fittings, etc. etc. These are chiefly employed in maps and layouts. Conventional symbols vary from place to place. Usually sign conventions and graphical symbols used in a map, layout or drawing are explained in legend or described in form of a note in the drawing sheet. A few symbols commonly used are presented in Figs. 1·1 to 1·7.

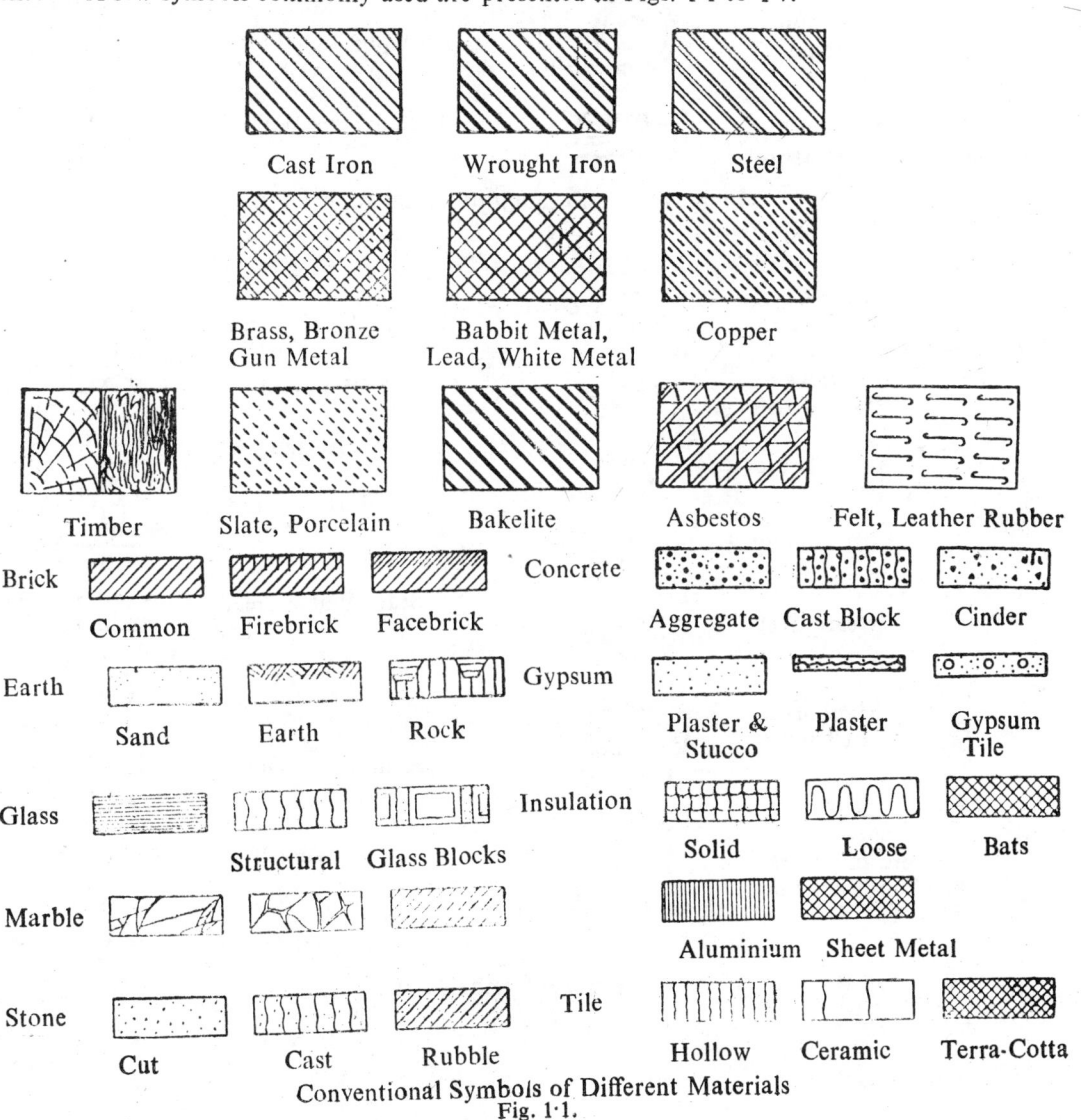

Conventional Symbols of Different Materials
Fig. 1·1.

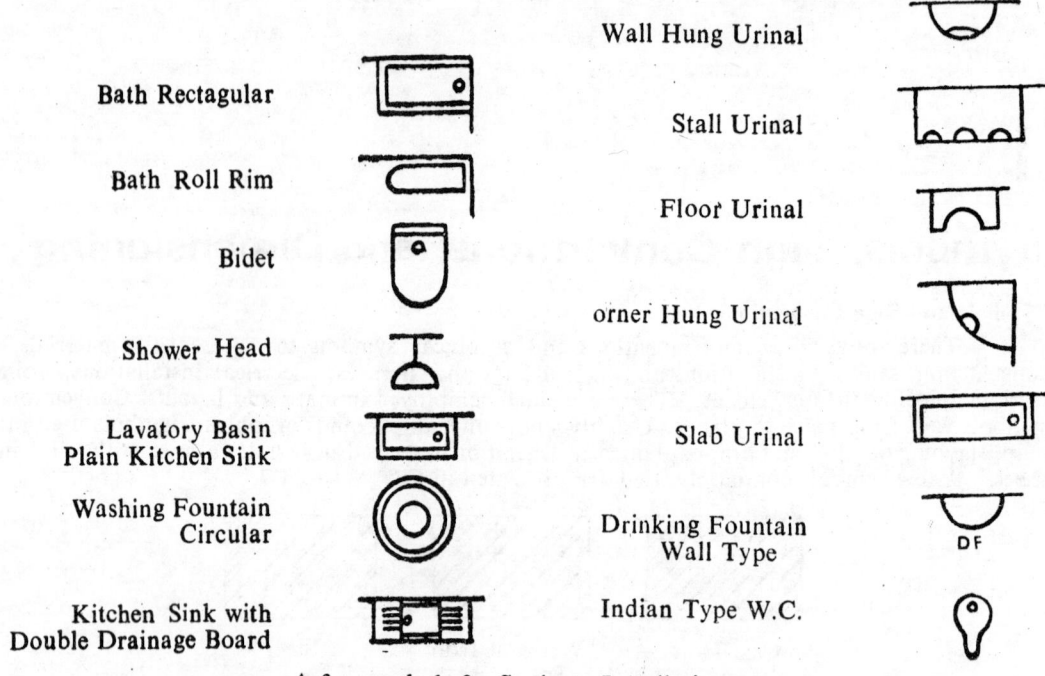

Wall Hung Urinal

Bath Rectagular

Stall Urinal

Bath Roll Rim

Floor Urinal

Bidet

orner Hung Urinal

Shower Head

Lavatory Basin
Plain Kitchen Sink

Slab Urinal

Washing Fountain
Circular

Drinking Fountain
Wall Type

Kitchen Sink with
Double Drainage Board

Indian Type W.C.

A few symbols for Sanitary Installation :
Fig 1·2.

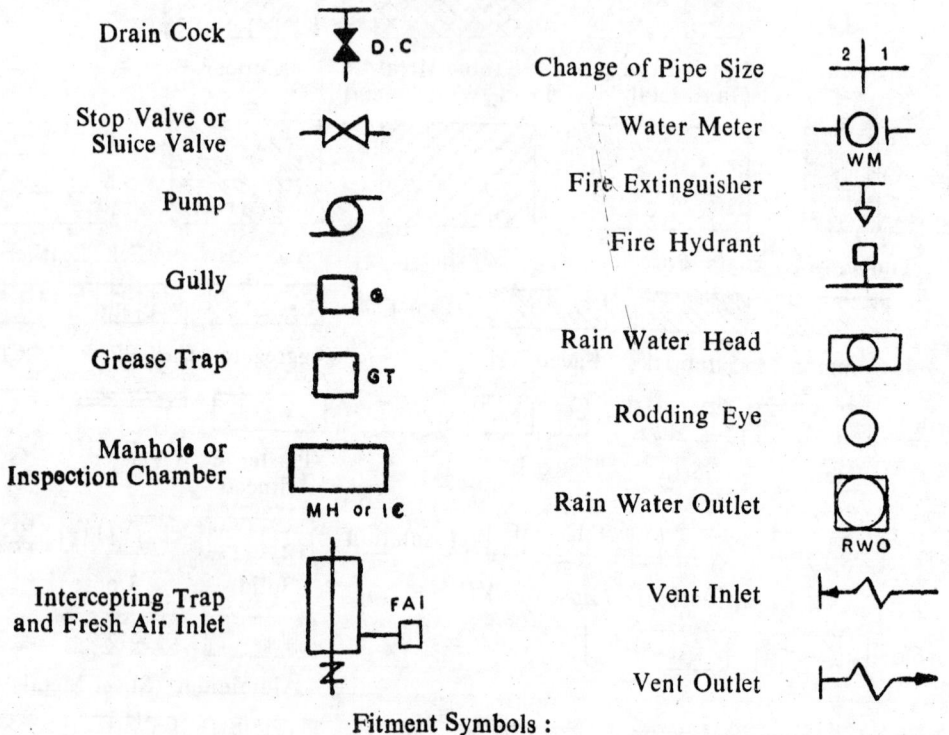

Drain Cock

Change of Pipe Size

Stop Valve or
Sluice Valve

Water Meter

Pump

Fire Extinguisher

Fire Hydrant

Gully

Grease Trap

Rain Water Head

Rodding Eye

Manhole or
Inspection Chamber

Rain Water Outlet

Intercepting Trap
and Fresh Air Inlet

Vent Inlet

Vent Outlet

Fitment Symbols :
Fig. 1.3.

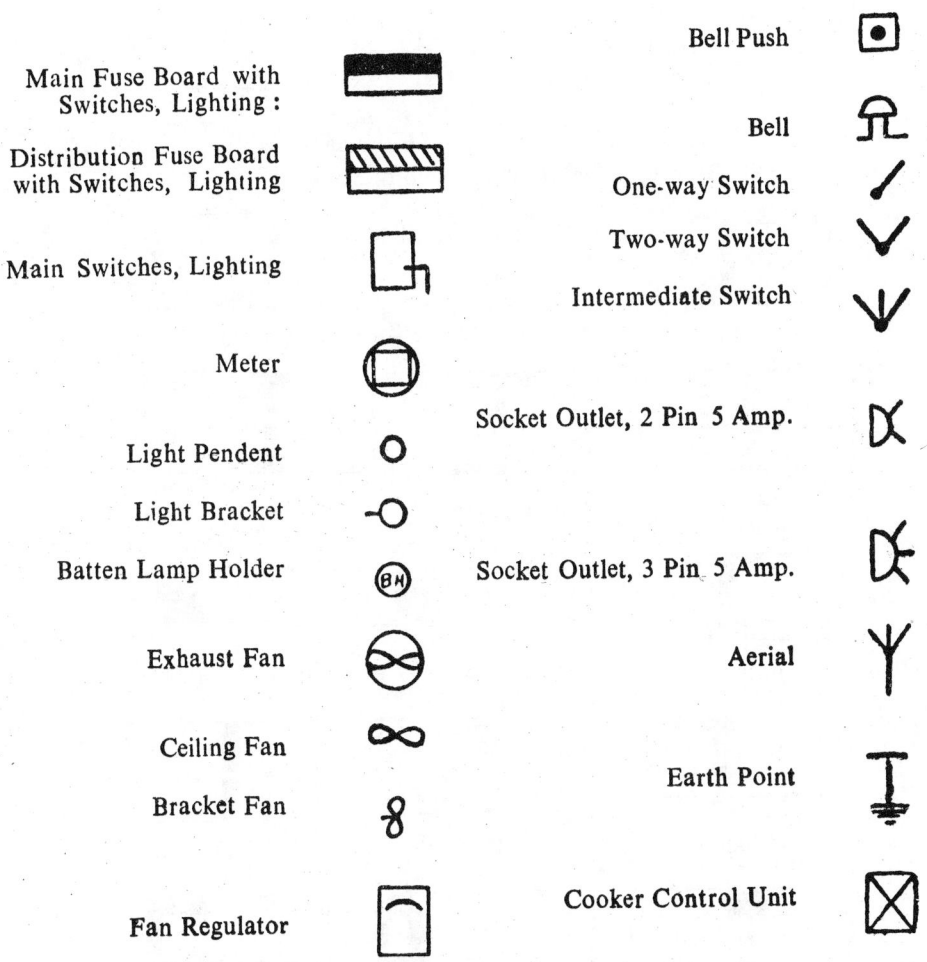

Main Fuse Board with Switches, Lighting :

Distribution Fuse Board with Switches, Lighting

Main Switches, Lighting

Meter

Light Pendent

Light Bracket

Batten Lamp Holder

Exhaust Fan

Ceiling Fan

Bracket Fan

Fan Regulator

Bell Push

Bell

One-way Switch

Two-way Switch

Intermediate Switch

Socket Outlet, 2 Pin 5 Amp.

Socket Outlet, 3 Pin 5 Amp.

Aerial

Earth Point

Cooker Control Unit

A few symbols for Electrical Installations Required in Building Wiring.

Fig. 1·4.

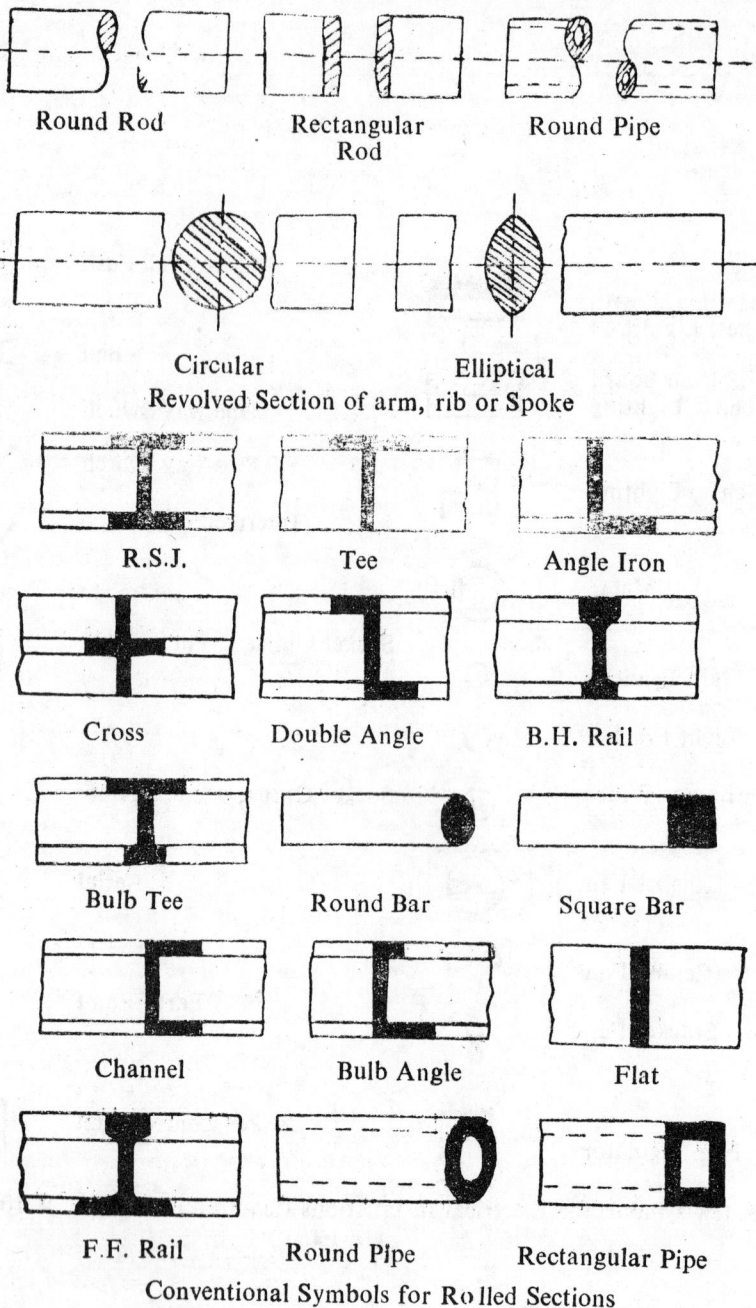

Round Rod   Rectangular   Round Pipe
                      Rod

Circular   Elliptical
Revolved Section of arm, rib or Spoke

R.S.J.   Tee   Angle Iron

Cross   Double Angle   B.H. Rail

Bulb Tee   Round Bar   Square Bar

Channel   Bulb Angle   Flat

F.F. Rail   Round Pipe   Rectangular Pipe

Conventional Symbols for Rolled Sections

Fig. 1·5.

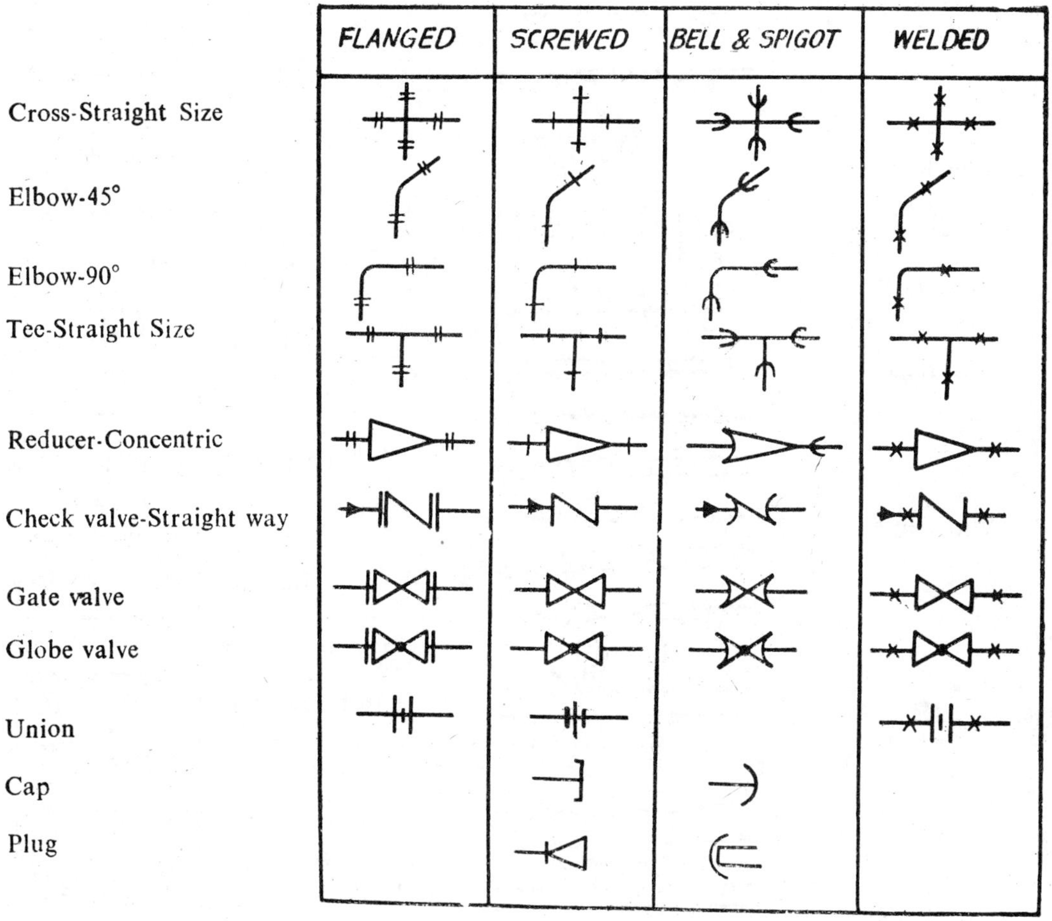

	FLANGED	SCREWED	BELL & SPIGOT	WELDED
Cross-Straight Size				
Elbow-45°				
Elbow-90°				
Tee-Straight Size				
Reducer-Concentric				
Check valve-Straight way				
Gate valve				
Globe valve				
Union				
Cap				
Plug				

Graphical Symbols of Pipe Fittings
Fig. 1·6.

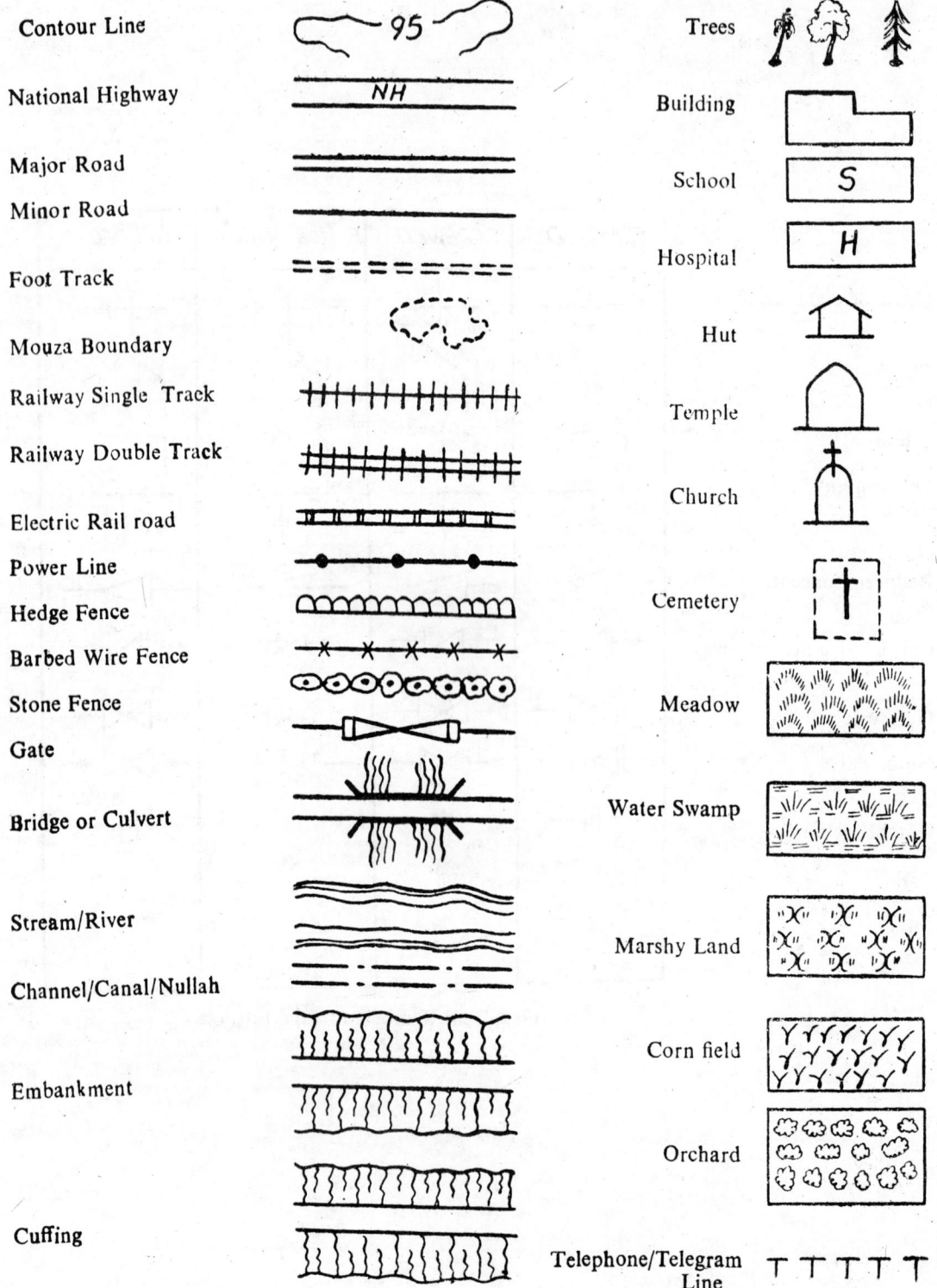

Fig. 1·7. Conventional symbols used in Survey Maps

Dimensioning

Any engineering sketch or drawing needs dimensioning, otherwise the size of the object and its components cannot be read. Any part may be dimensioned easily and systematically by dividing it into simple geometric solids. Each part should be dimensioned to indicate its size and its relative location from a centreline or baseline.

A dimension line is a light-weight line terminated at each end by an arrowhead. A numerical value written along the dimension line specifies the measurement. Under no circumstances should the line pass through the numerals.

Extension lines are light lines extending from a view to indicate the extent of measurement. These lines start 1·5 to 2 mm. from the view and extend 3 to 4 mm. beyond the dimension line.

Dimensions should be placed where they are expected by the reader. Major dimension. A drawing should appear on the principal view of the object. A drawing should not be crowded with too much of dimensions placed haphazardly. Dimension lines should be given at least 12 mm away from the object and the spacing between two parallel dimension lines should be 10 mm. minimum. Arrow head should be thin and proportionate and it should touch the extension line. Fig. 1·8 shows poor practice in dimensioning. The correct way of dimensioning is shown in Fig. 1·9.

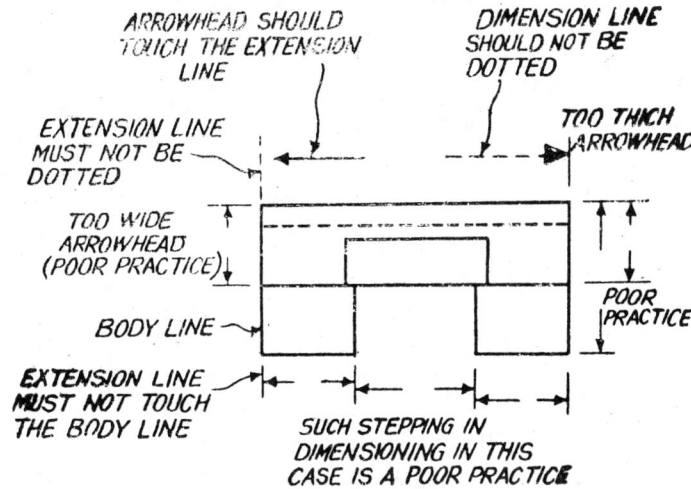

Poor Practice in Dimensioning
Fig. 1·8.

A few more points are to be remembered while dimensioning a drawing. These are :

(1) Never use the object line or centre line as the dimension line.

(2) Avoid crossing of two dimension lines, if possible.

(3) Keep parallel dimensions equally spaced.

(4) Extend centre line to act as an extension line, where needed.

(5) Arrange a series of dimensions in a continuous line as shown in Fig. 1·9.

(6) Place all the dimensions along the direction of the dimension line such that they are readable from bottom and right side of the drawing. Alternatively, follow unidirectional method of placing dimensions as shown in Fig. 1·9.

(7) Put dimensions outside the view excepting unavoidable situation.

(8) Place longer dimensions outside shorter one so that extension lines need notcross the dimension lines.

(9) In parallel dimension lines write the figures staggered.

(10) Put overall dimensions outside the view.

(11) Do not repeat any dimension unnecessarily.

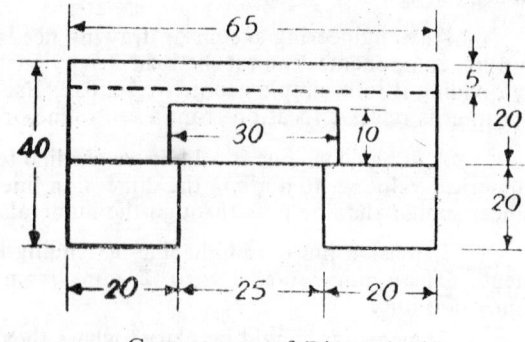

Correct way of Dimensioning
Fig. 1 9.

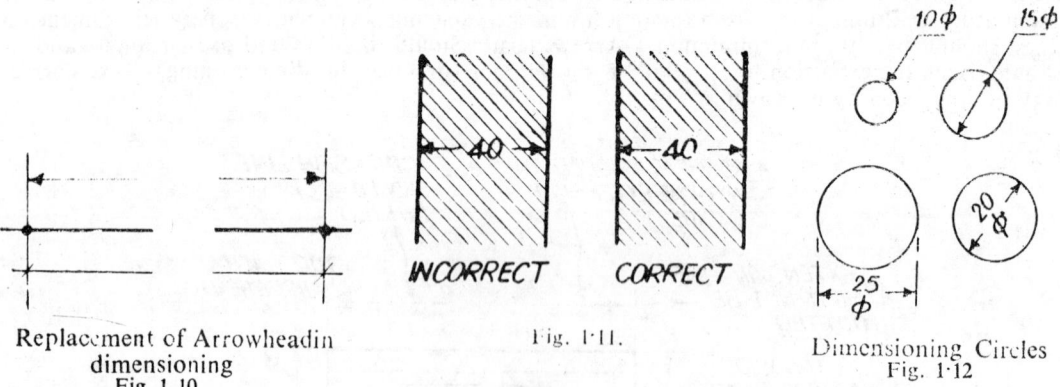

Replacement of Arrowheadin
dimensioning
Fig. 1 10.

Fig. 1·11.

Dimensioning Circles
Fig. 1·12

(12) When a dimension line appears on a hatched area show it in a partly unhatched area such that it can be easily read. See Fig. 1'11.

(13) Put dimension of a circle by showing the diameter and never the radius as shown in Fig. 1'12.

Fig. 1'13 shows haphzard and systematic way of dimensioning.

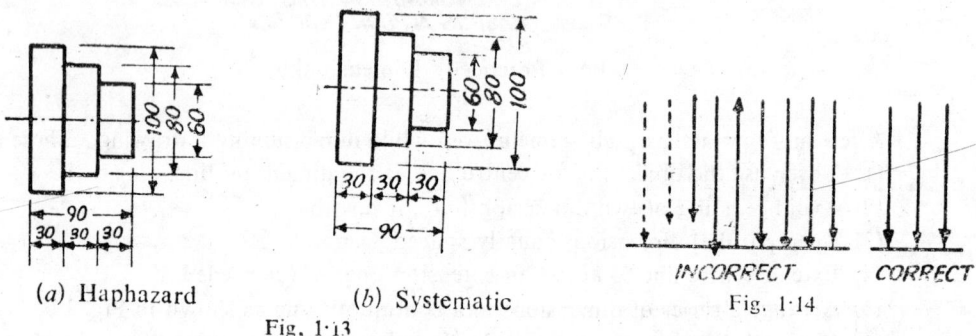

(a) Haphazard   (b) Systematic   Fig. 1·14
Fig. 1·13

Fig. 1'14 shows incorrect and correct ways of putting arrowheads and placing the dimension line.

2

Lettering Styles

Lettering styles are often neglected in engineering drawing. It should be remembered that careless lettering may result in rejects. The character of lettering imparts good effect to a drawing. Therefore, inscriptions accompanying a drawing should be made in clear and pleasing letters and figures.

There are various styles of lettering, each of which is appropriate for a particular purpose. The uniformity in height, width, inclination, spacing and depth (thickness and blackening) of line is essential for good writing. In composition *i e.*, in combining letters into words and words into a sentence, the spaces for the various combinations of letters are to be arranged in such a manner that the areas appear to be equal. Letters in words are not placed at equal distances from each other, but are spaced so that the areas of blank spaces between the letter are about equal and the letters appear to be spaced uniformly. Thus, two adjacent straight sided letters would be spaced much further apart than two curved letters. But in general, the letters should be kept fairly close together to form a word.

When the adjacent letters have straight sides, the area is obtained by keeping the distance between the letters slightly greater than one-half the height of a letter. The space between two words should be equal to or greater than the height of a letter but not more than twice the height. The space between two sentences should be somewhat greater. The space between two consecutive lines of lettering should vary from $\frac{1}{2}$ to $1\frac{1}{2}$ times the height of capitals.

Capital and small letters may be vertical or inclined, two-stroke or single-stroke and thick lined or thin ones. Two-stroke or single stroke capital vertical letters are commonly used in writing the title of a drawing. Commercial Gothic and modern Roman letters are used

[Method of writing two-stroke vertical capital letters on a graph paper. On formation of letters, graph markings are to be erased out. The letters are formed with the help of set squares. The corners may be rounded with the help of bow compass instead of making them kevelled by using set squares].

Fig. 2·1.

for their purpose In two-stroke lettering, the letters are drawn first in outline and then filled in with ink. Single-stroke lower-case or small letters, either vertical or inclined are commonly used on maps and drawings. These letters are particularly suitable for long notes and atstements. Single stroke lower case letters can be written faster than capitals The construction of inclined lower-case letters is based on straight line and ellipse.

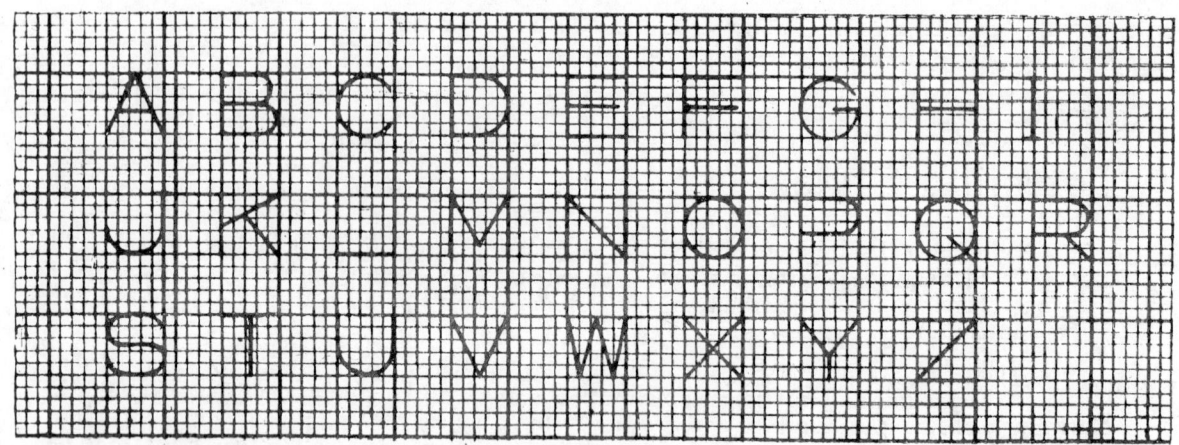

Single Stroke Vertical Capital Letters written on Graph paper using Set Squares and Bow compass.
Fig. 2 2

English Gothic Capital Letters (Small tyye)
Fig. 2·3.

English Gothic Small Letters
Fig 2·4.

Lettering Stencil for Capital Letters
Fig. 2·5.

Lettering Stencil for Small Letters
Fig. 2·6.

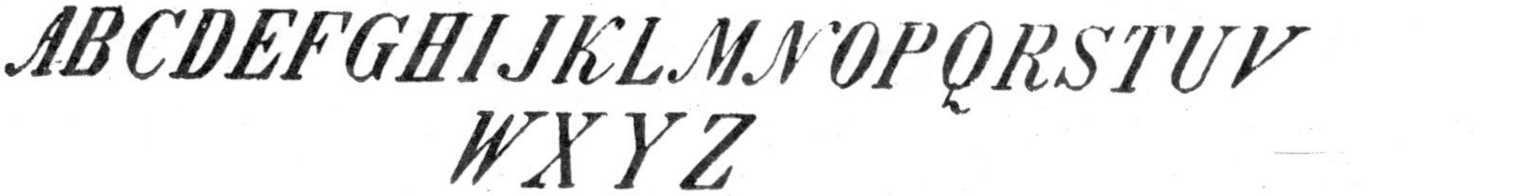

ABCDEFGHIJKLMNOPQRSTUV WXYZ

Inclined Capital Letters
Fig. 2·7.

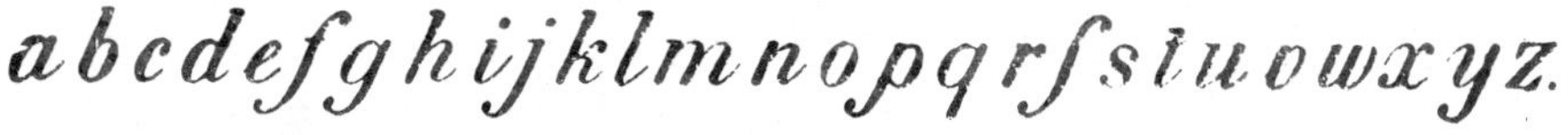

abcdefghijklmnopqrſstuvwxyz.

Inclined Small Letters (Modern Roman)
Fig. 2 8.

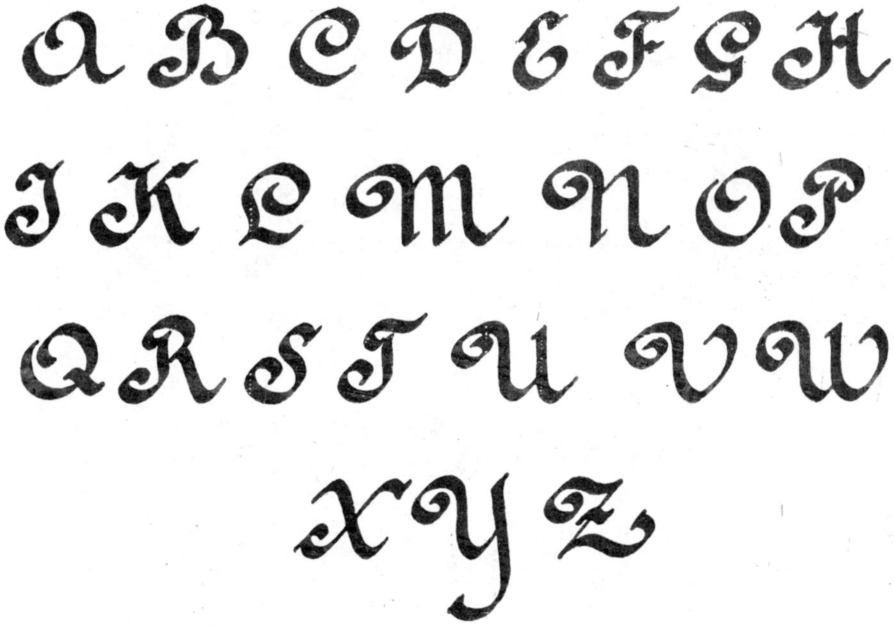

A B C D E F G H
I K L M N O P
Q R S T U V W
X Y Z

Old English Small Letters used in Stone carving
Fig. 2·9.

A B C D E F G H I J K
L M N O P Q R S T
U V W X Y Z

Ornamental Lettering
Fig. 2·10.

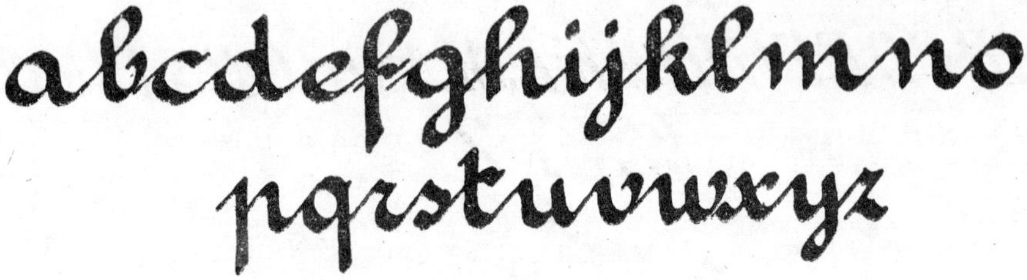

abcdefghijklmno
pqrstuvwxyz

Small letters used in Stone carving
Fig. 2·11.

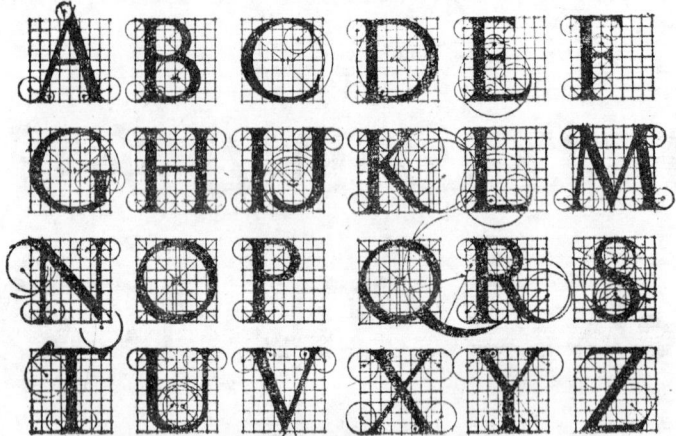

Classic Renaissance Letters
Fig 2 12

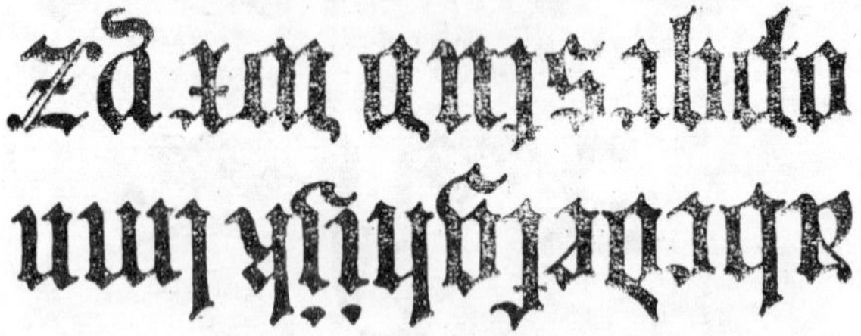

Italian Black Letters
Fig. 2·13

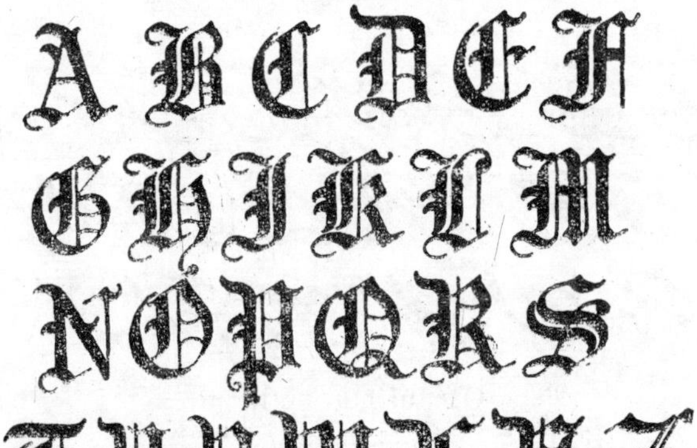

English Gothic Capital letters (Text)
Fig. 2·14.

abcdefgh ijklmn
opqrs suvwryz

English Gothic Small Letters
Fig. 2·15

A B C D
E F G H
I K L M
N O Q P
R S T V
W X Y Z

Italian Renaissance Letters
Fig. 2·16.

A B C D E
F G H I J
K L M N
O P Q R
S T U V
W X Y Z

Classic Roman Letters
Fig. 2·17.

Uncial Gothic Capital Letters
Fig. 2·18.

DRAWING IS THE LANGUAGE OF ENGINEERS

DRAWING IS THE LANGUAGE OF ENGINEERS

DRAWING IS THE LANGUAGE OF ENGINEERS

DR AWING IS THE LANGUAGE OF ENGINEERS

DRAWING IS THE LANGUAGE OF ENGINEERS

DRAWING IS THE LANGUAGE OF ENGINEERS

Use of various types of Lettering in Engineering Drawing
Fig. 2·19.

SEWERAGE

SEWERAGE

Use of cut Stencil Letters
Fig. 2·20.

3

Isometric Projection

In engineering drawing orthogonal views *i e.*, plan, elevation and end view are shown. With complicated drawings, the workmen find difficulty in understanding the shape and size of the object. It is therefore needed in such cases to provide with an auxilliary view, preferably a pictorial view of the object. Most commonly, isometric views or oblique views are sketched by engineers and draftsmen to convey to the workmen the shape and size of an object.

Isometric projection is a form of axonometric projection giving the pictorial effect of an object. Theoretically, axonometric projection may be treated as a form of orthographic projection. In an axonometric projection, the object is turned conveniently such that the three faces come in one plane. See Fig. 3·1.

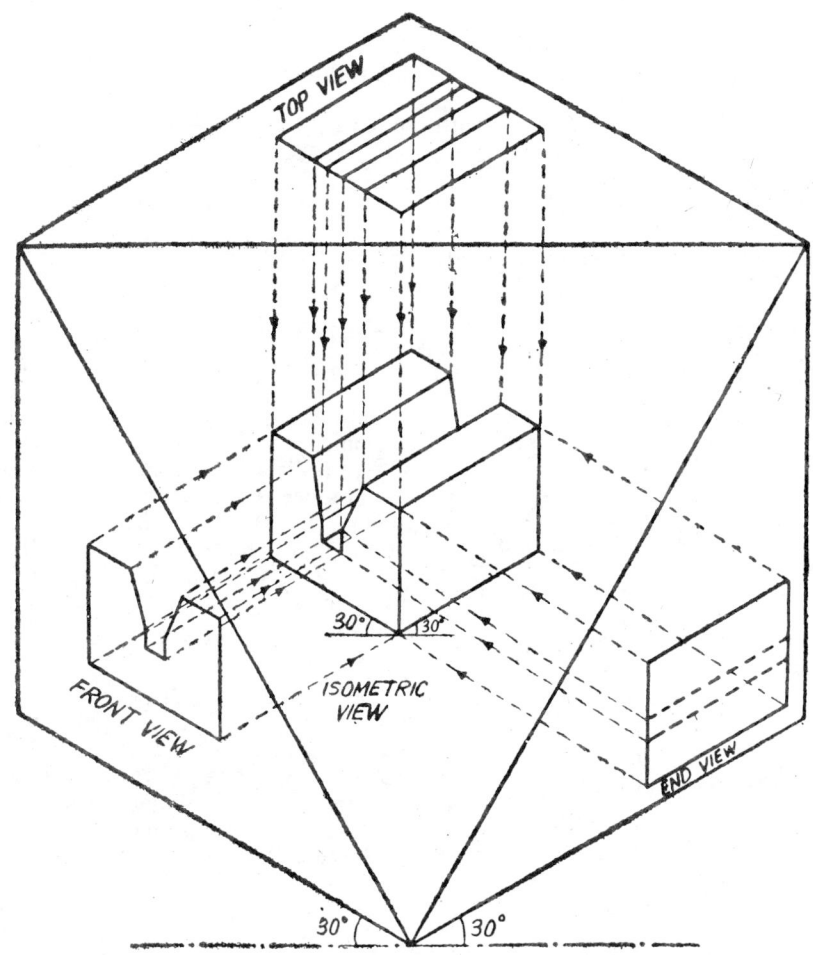

Isometric Projection from Orthographic views

Fig. 3·1.

The term *'Isometric'* is derived from two Greek words *'Iso'* and *'Metric'*, meaning *'equal measures'*. Isometric drawing is therefore based on equal measurements on three axes, called isometric axes as shown in Fig. 3·2.

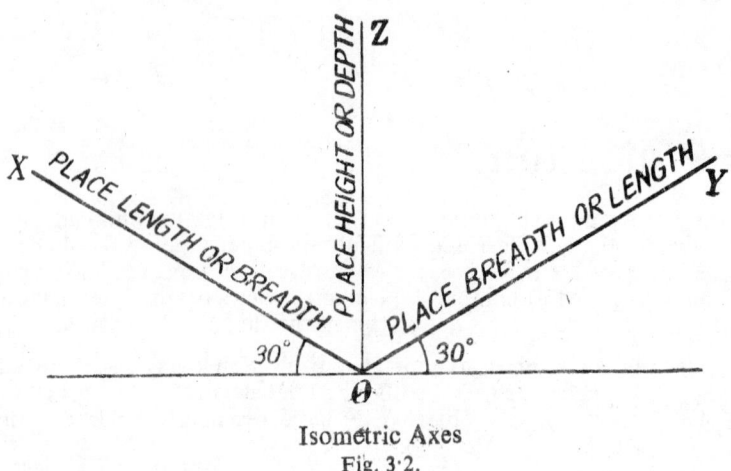

Isometric Axes
Fig. 3·2.

To draw an isometric view one should remember the following :

(1) Three faces make equal angles with the plane of projection.

(2) Line of sight is perpendicular to the plane of projection.

(3) All vertical lines of the object remain vertical in isometric view.

(4) Horizontal lines of the object are rotated through an angle of 30° to the horizontal.

Isometric Scale :

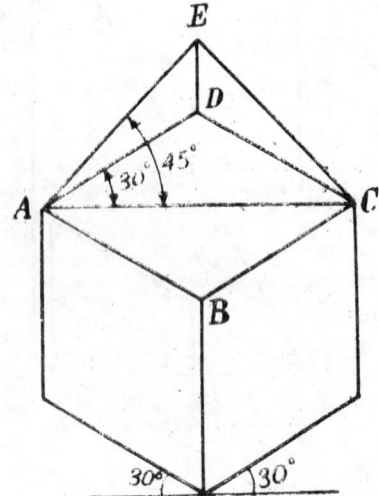

Fig. 3·3.

In making an isometric drawing it is very convenient to use an ordinary scale and mark off equal distances along the three axes. But, this practice is not correct as regard the size of the object. In a full size drawing the object would appear larger than it actually is, in the ratio $\sqrt{3}$ to $\sqrt{2}$. It is therefore preferred to use an isometric scale while drawing isometric views.

The principle of drawing an isometric scale is very simple. Fig. 3·3 shows the isometric view of a cube. If AE and CE be drawn at 45° to the diagonal AC, then the triangle AEC represents the true shape of the triangle ADC and AE and CE represents the actual length of AD and CD.

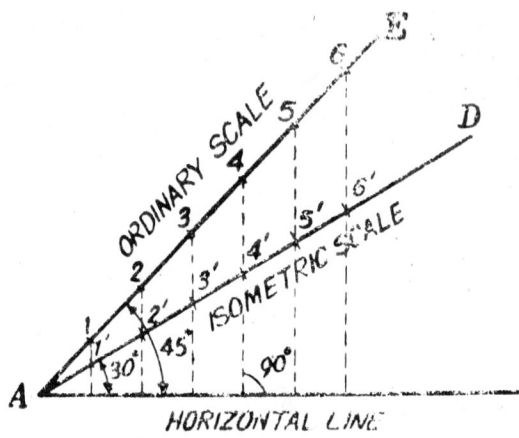

Based on this principle isometric scale is constructed. First, draw the ordinary scale AE placing it at 45° to the horizontal Draw a second line AD at 30° to the horizontal as shown. Then, draw normals on the horizontal line from the ordinary scale graduation dividing AD in a similar way. AD is the isometric scale. See Fig 3'4.

Fig. 3·4.

Problem 1. *Draw the isometric view of a cube from its orthographic views. Assume any dimension of the cube.*

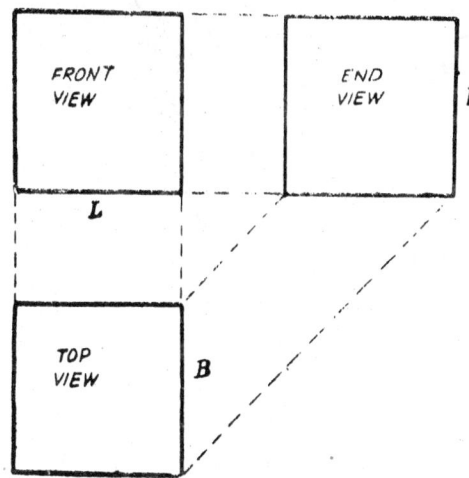

Fig. 3·5 shows three views of a cube in orthographic projection. The first step to draw its isometric view is shown in Fig. 3·6. Draw the three axes and cut the measurements L, B and H as shown.

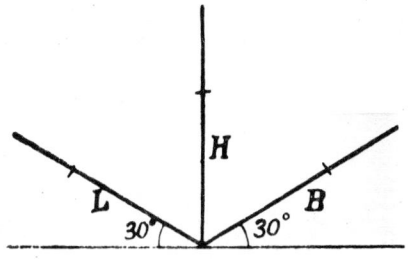

Fig. 3·5.   Fig. 3·6.

Then, draw the lines one by one as marked numerically in Fig. 3·7. and obtain the isometric view of the cube as shown in Fig. 3·8.

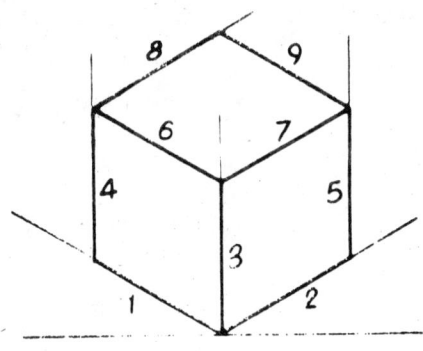

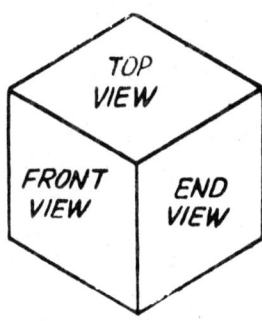

Fig. 3·7.   Fig. 3·8.

Problem 2. *Plan and elevation of an object are shown in Fig. 3·9. Draw its isometric view.*

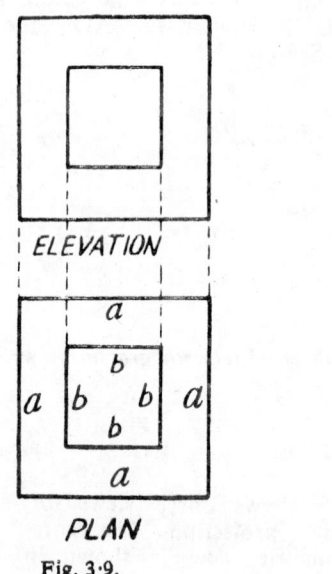

PLAN

Fig. 3·9.

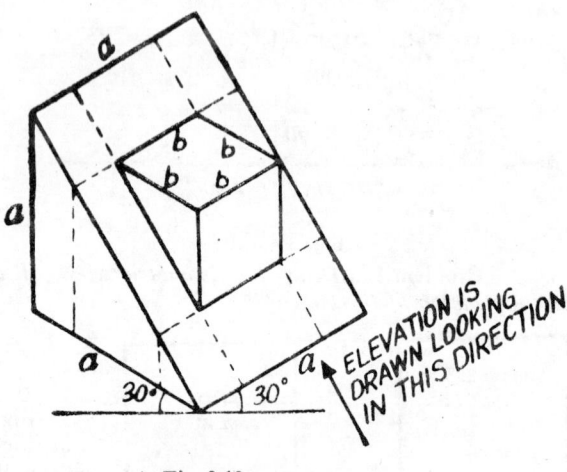

ELEVATION IS DRAWN LOOKING IN THIS DIRECTION

Fig. 3·10.

Fig. 3·10 shows the isometric view of the object. Alternatively, the object may be like one as shown in Fig. 3·11. In such cases, a third view is needed to define the object.

LOOK ALONG THIS DIRECTION TO DRAW THE ELEVATION

Isometric view
Fig. 3·11.

Problem 3. *Three views of an object are shown in Fig. 3·12. Draw its isometric view keeping the front view to the right.*

Place the left end view on the left-hand isometric axis. Project from its corners making parallel to the right-hand isometric axis Cut the lengths by obtaining measurements either from top view or from front view. Draw vertical lines as shown and complete the view as shown in Fig. 3·13.

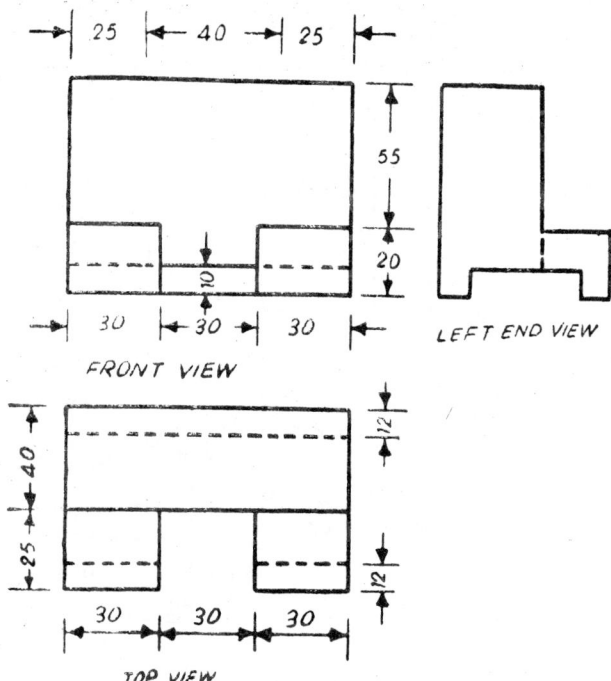

FRONT VIEW

LEFT END VIEW

TOP VIEW

Fig. 3·12.

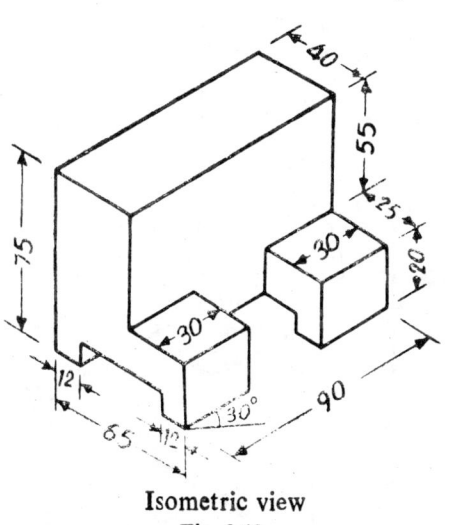

Isometric view
Fig. 3·13.

Problem 4. *The front view and left end view of an object are shown in Fig. 3·14. It is required to draw its isometric view.*

The object consists of three rectangular prisms and one hollow cylinder.

Fig. 3·15 shows the required isometric view first, draw the isometric view of the rectangular prisms and then place the hollow cylinder as shown.

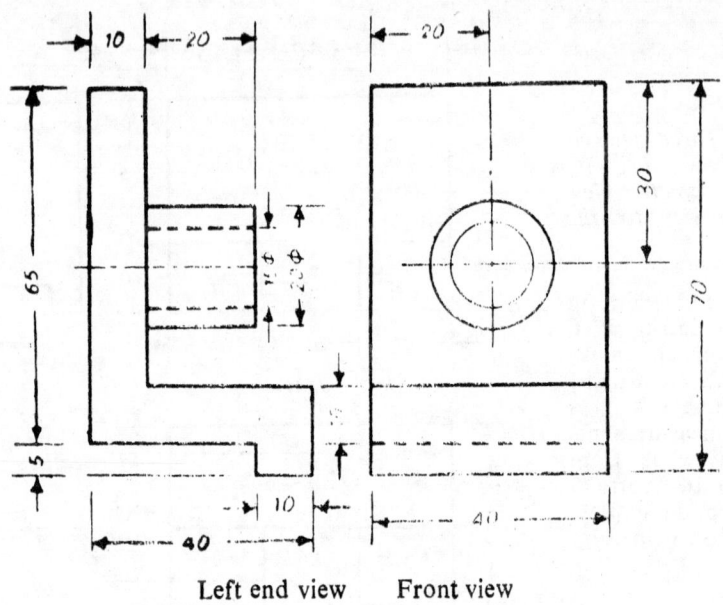

Left end view   Front view

Fig. 3·14.

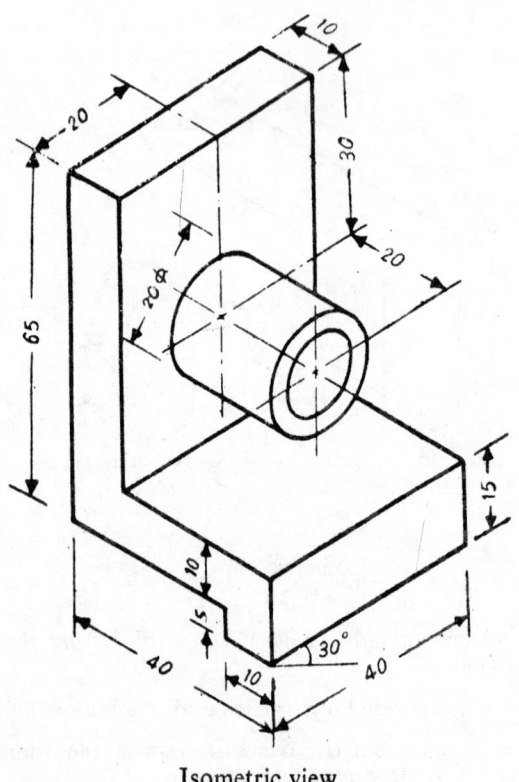

Isometric view
Fig. 3·15.

Problem 5. *Two views of an object are shown in orthographic projection. Draw its isometric view.*

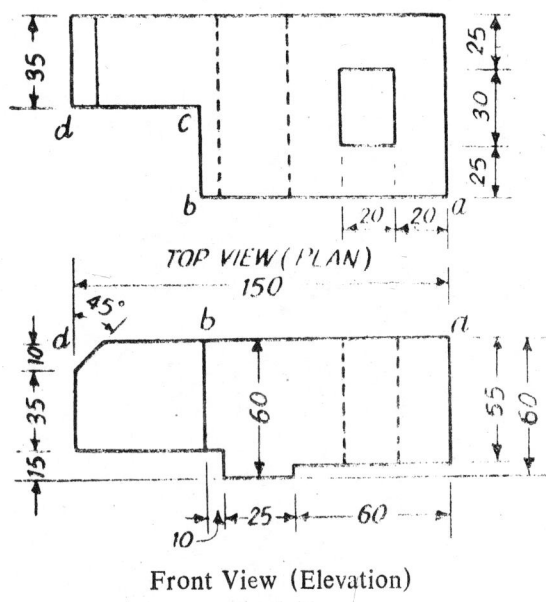

TOP VIEW (PLAN)

Front View (Elevation)
Fig. 3·16.

First, draw *a* to *b* portion of the elevation on the right-hand isometric axis and place *bc* (from plan) parallel to the left-hand isometric axis. Then, draw *b* to *d* portion of the elevation parallel to right-hand isometric axis. Chamfer the end and project all other lines. Lastly, place rectangular hole by the taking measurements from the plan and complete the figure with dimensions as shown in Fig. 3·17.

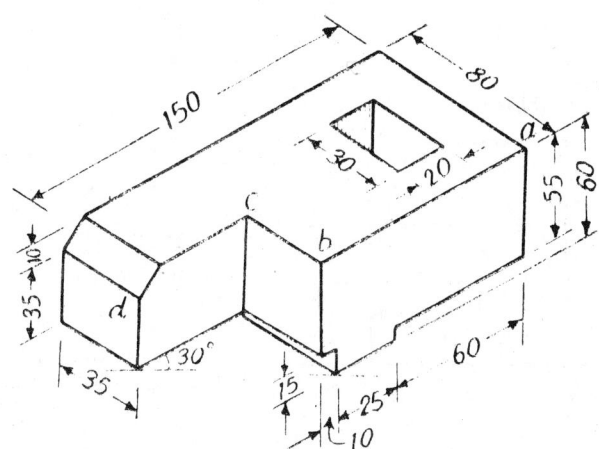

Isometric View
Fig. 3·17.

Problem 6. *Draw front view and top view of the model from its isometric view shown in Fig. 3·18.*

Place yourself in the direction of the arrow shown and draw the front view. In this view, you shall get cross (×) marked enclosures. Now, look from top. You shall get tick (√) marked enclosures. Project from the front view already drawn, complete the figure and put dimensions as shown in Fig. 3·19.

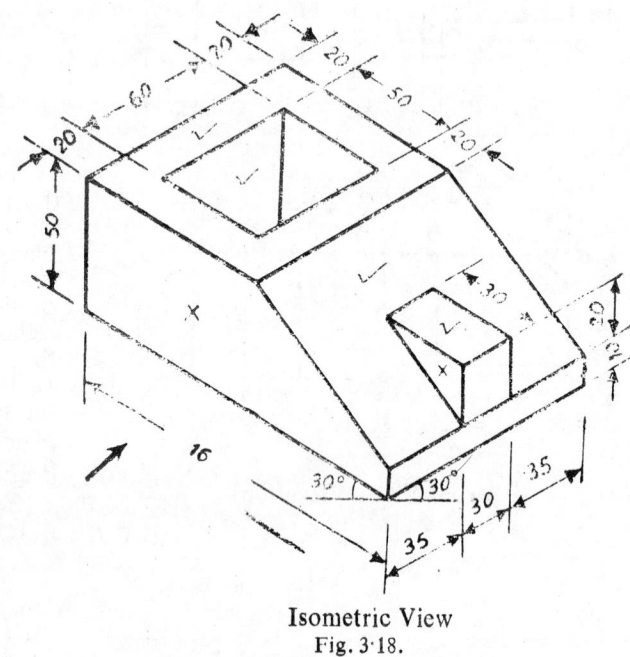

Isometric View
Fig. 3·18.

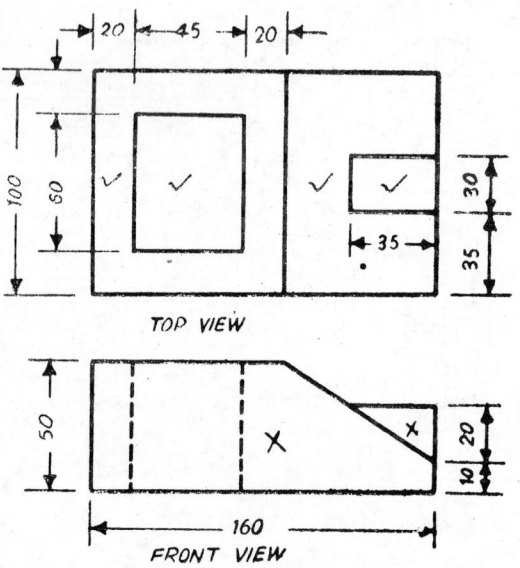

TOP VIEW

FRONT VIEW

Fig. 3·19,

Problem 7. *The orthographic views of a memorial tomb are shown in Fig. 3·20.*

Draw the base as shown and place the top. See Fig. 3·21.

Problem 8. *Draw the isometric veiw of the model from its top view and front view shown in Fig. 3'22.*

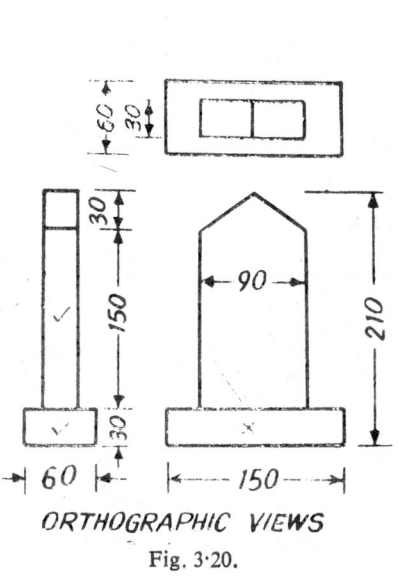

ORTHOGRAPHIC VIEWS

Fig. 3·20.

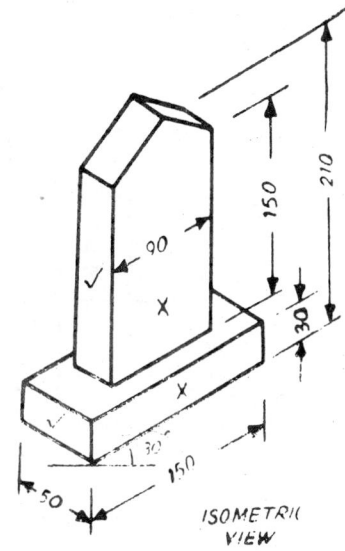

ISOMETRIC VIEW

Fig. 3·21.

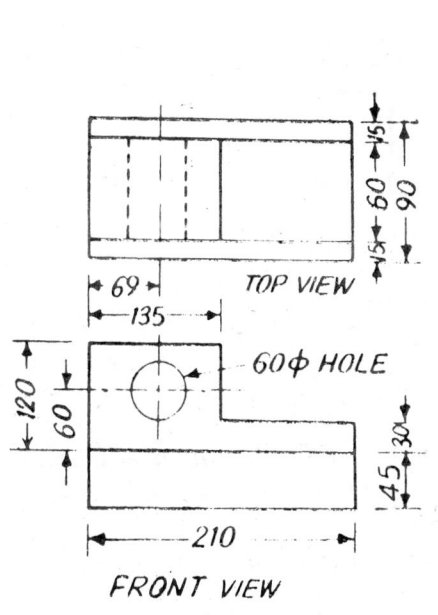

TOP VIEW

60φ HOLE

FRONT VIEW

Fig. 3·22.

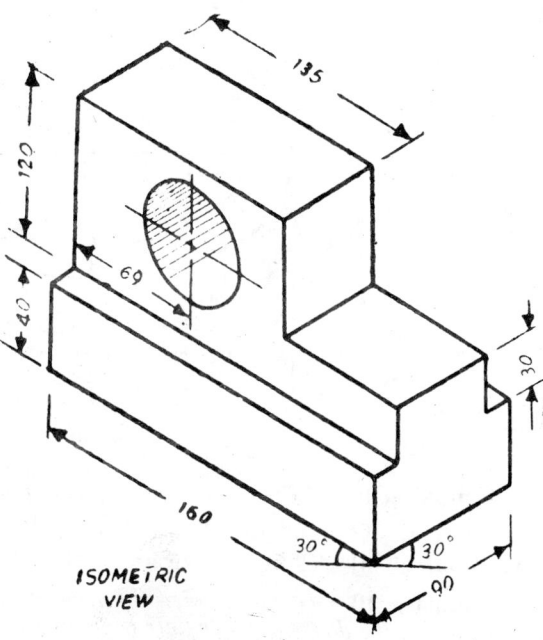

ISOMETRIC VIEW

Fig. 3·23.

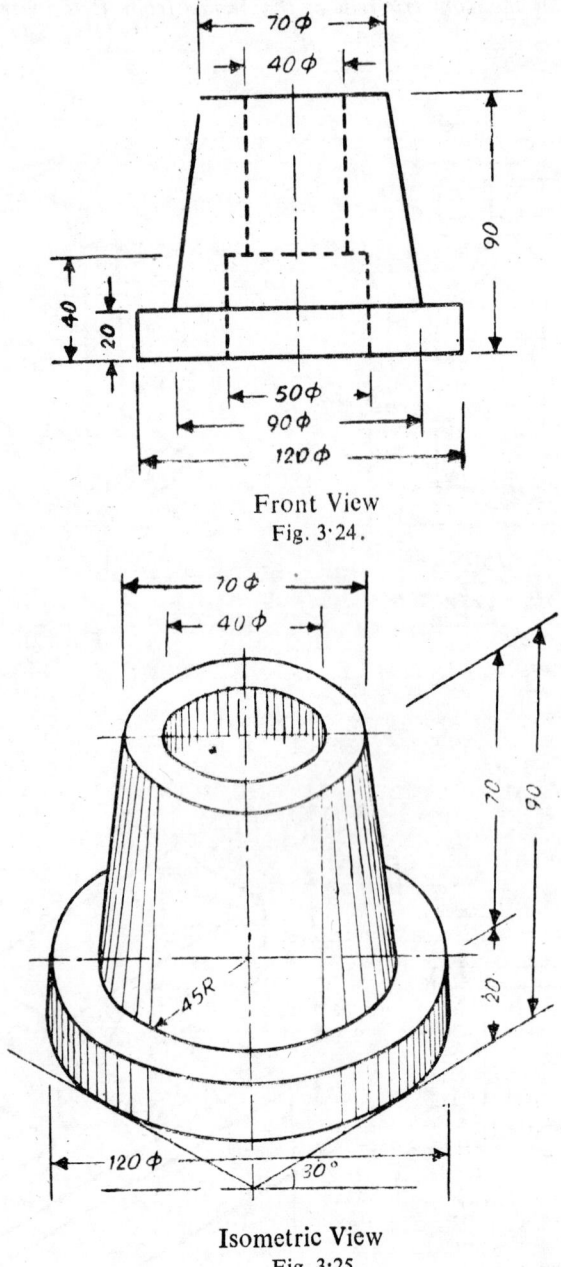

Front View
Fig. 3·24.

Isometric View
Fig. 3·25.

Problem 9. *The front view of an object is shown in Fig. 3 24. Draw its isometric view.*

In plan of the object all are circles. The circles will take form of ellipses in isometric view. Construct the isometric view as shown in Fig. 3 25.

Problem 10. *Isometric view of a machine part is shown in Fig. 3·26. Draw its front view looking in the direction of the arrow and its corresponding top view.*

The required views are shown in Fig. 3·27. Draw the elevation first and then project from it to get the view.

Problem 11. *Draw the isometric view of a model from its orthographic views shown in Fig. 3·28.*

Place the front view on the right-hand isometric axis as shown. Take measurements from top view and complete as shown in Fig. 3·29.

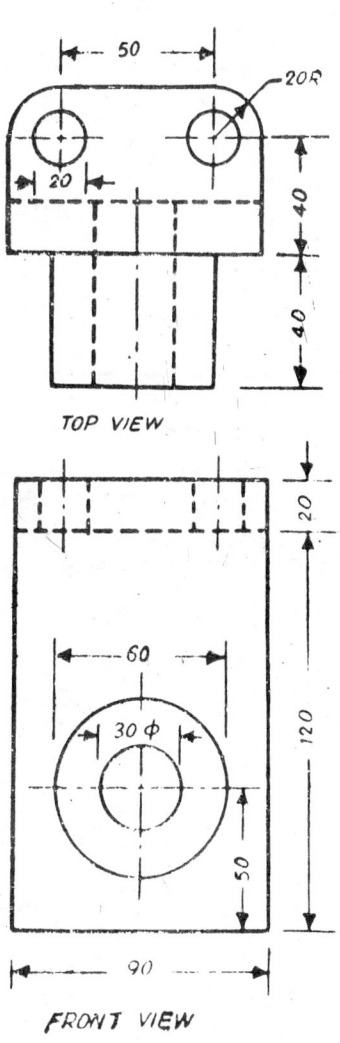

TOP VIEW

FRONT VIEW

Fig. 3·27.

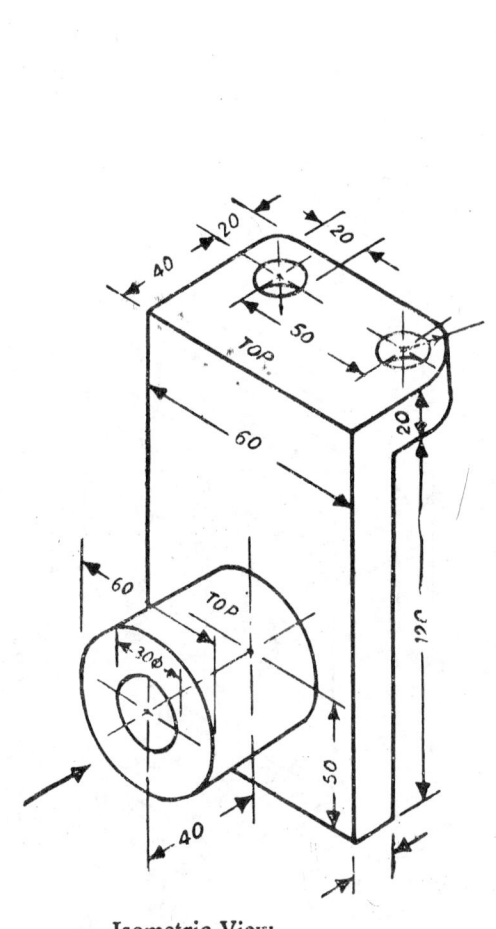

Isometric View

Fig. 3·26.

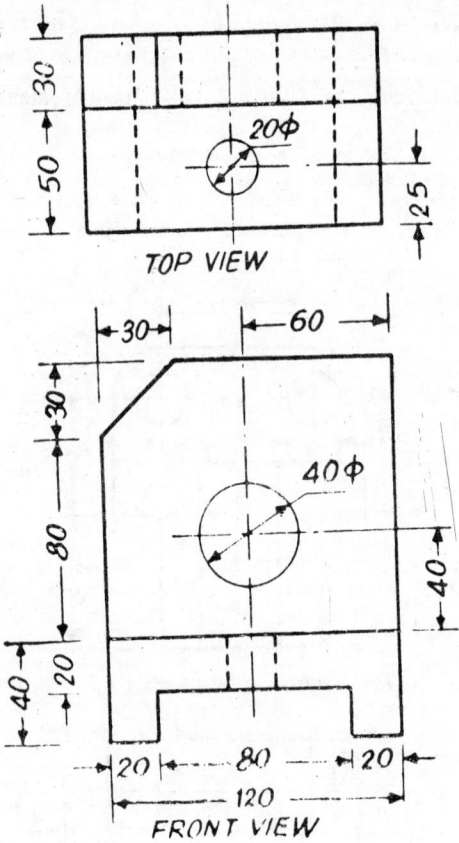

TOP VIEW

FRONT VIEW

Fig. 3·28.

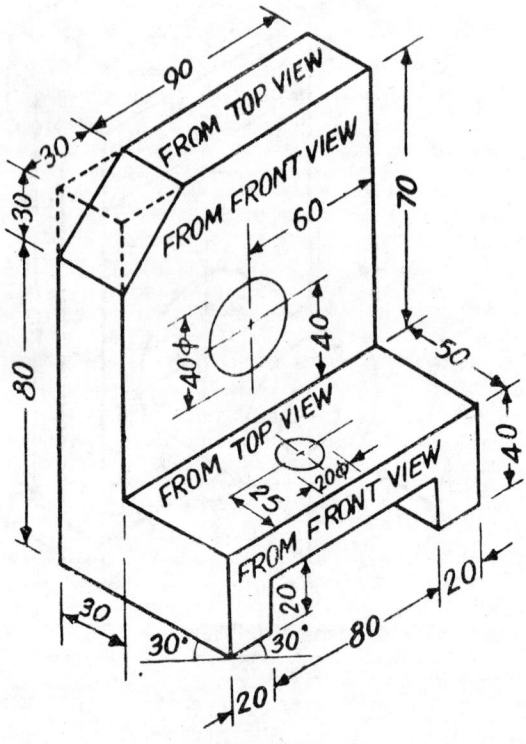

Fig. 3·29.

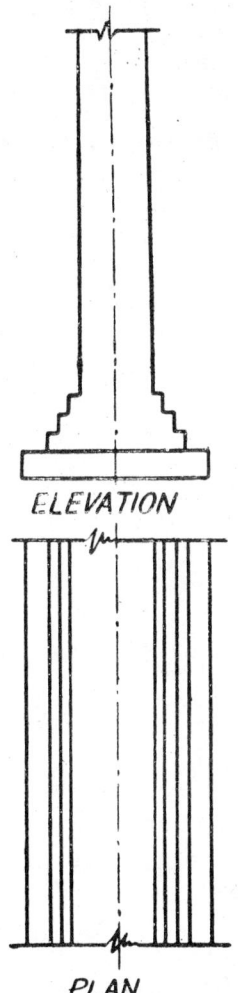

ELEVATION

PLAN

Spread Foundation of a wall
Fig. 3·30.

Problem 12. *Plan and elevation of a wall with its spread foundation are shown in Fig. 3·30. Draw its isometric view.*

Place the elevation on the left-hand isometric axis and project lines parallel to the right-hand isometric axis as shown in Fig. 3·31.

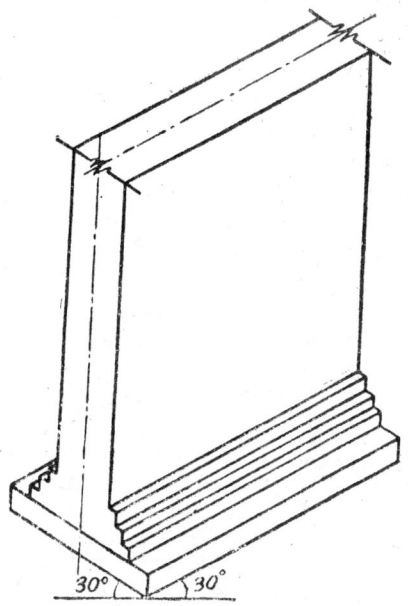

Isometric view of the Spread Foundation
Fig. 3·31.

Problem 13. *Plan and elevation of a stair are shown in Fig. 3·32. No dimension is given. It is required to draw its tsometric view by taking measurements from the orthographic views.*

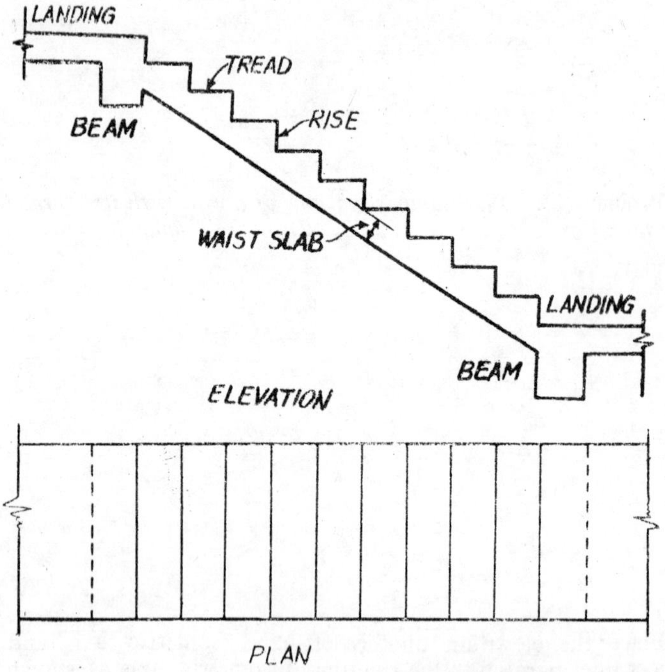

LANDING
TREAD
BEAM
RISE
WAIST SLAB
LANDING
BEAM
ELEVATION
PLAN

Plan and Elevation of a Stair
Fig. 3·32.

The isometric view of the stair is shown in Fig. 3·33.

LANDING
RISE
BEAM
TREAD
LANDING
BEAM
30°
30°

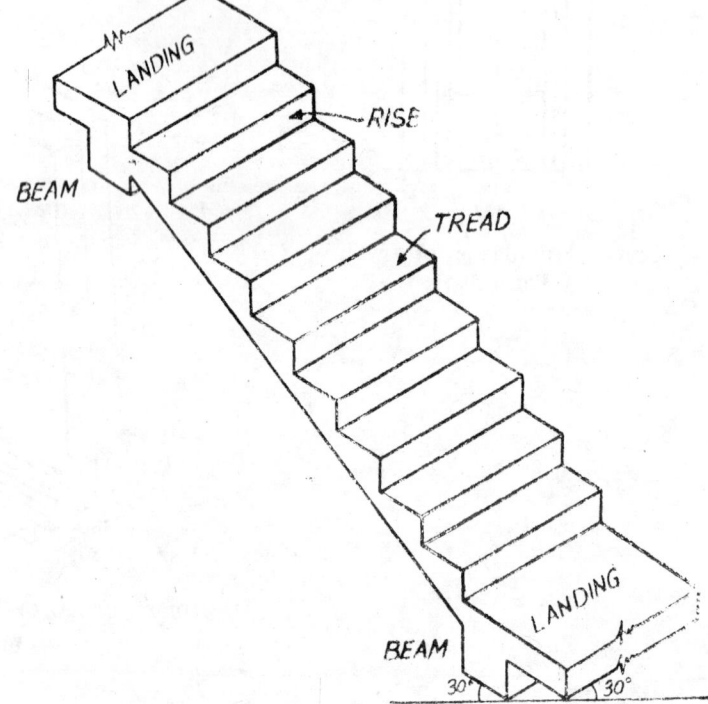

Isometric view of the Stair
Fig. 3·33.

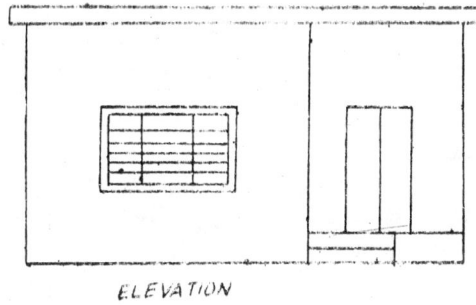

ELEVATION

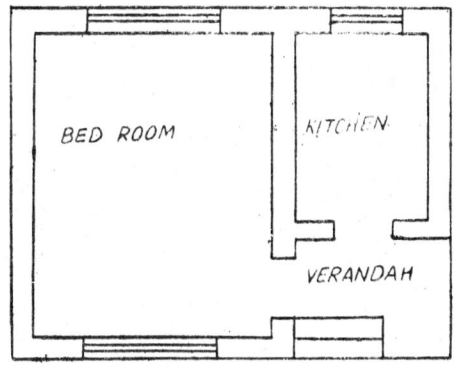

BED ROOM   KITCHEN

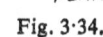

VERANDAH

PLAN

Fig. 3·34.

Poblem 14. *Plan and elevation of a watchman's quarter are shown in Fig. 3·34. Draw its isometric view by taking measurements from plan and elevation.*

First, draw the isometric view of the outer rectangle of the plan. Cut the Bed Room width (outer dimension) and construct the verandah in isometric view. Next, place the steps and window as shown. The doors of Bed Room and Kitchen will not come in isometric view (in this case). Put the slab over the top of the outer rectangle in isometric view, by projecting it on all sides as it appears in the elevation and thus complete the view as shown in Fig. 3·35. Remove all construction lines shown dotted in Fig. 3 35. and obtain the required isometric view of the quarter.

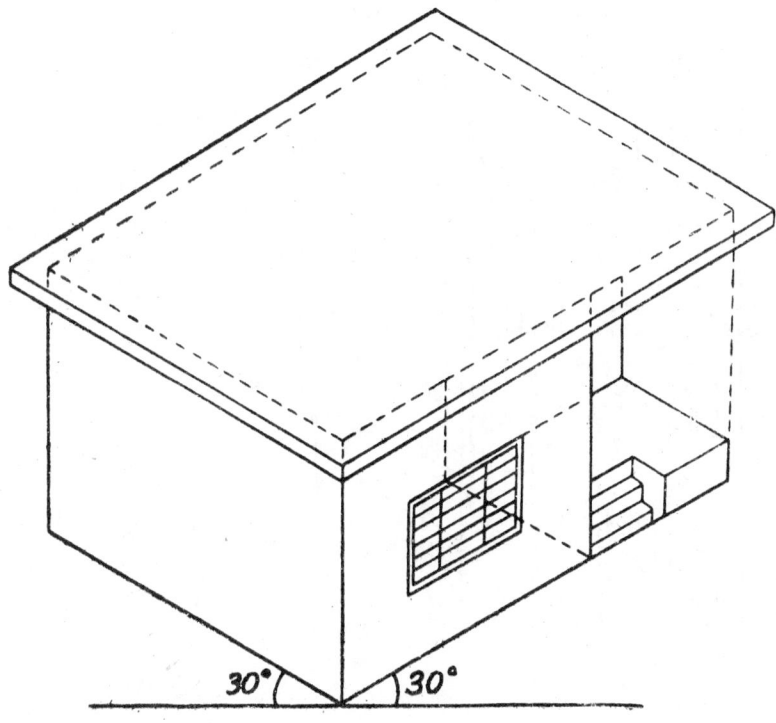

Construction lines shown dotted are to be removed.

30°   30°

Isometric view of the Quarter
Fig. 3·35.

Problem 15. *The ground floor plan and first floor plan of a building are identical. The plan is shown in Fig. 3·36. No dimension has been shown. You are to draw the isometric view of the building by taking measurements from plan and assuming normal room height, parapet wall, dimensions of attic, slab thickness, cornice projection, heights of doors and windows, dimensions of step, etc. etc.*

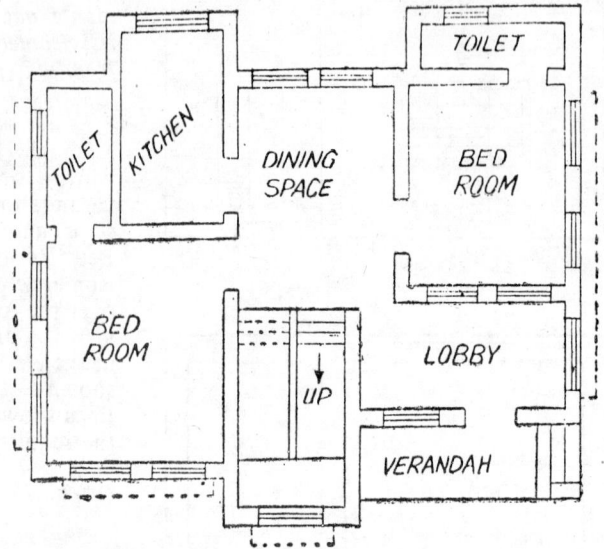

Plan of a Building
Fig. 3·36.

The required isometric view is shown in Fig. 3·37.

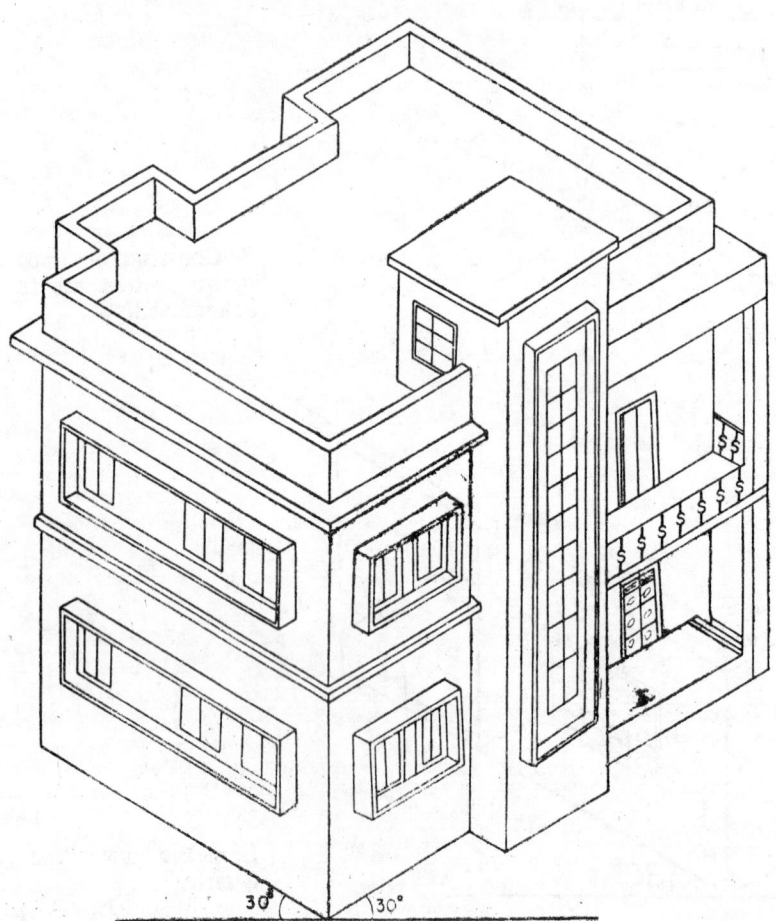

Fig. 3·37

Exercises on Chapter 3

1. Draw the isometric views of the following objects from their top view and front view as shown in figures 3/1 to 3/8. All dimensions are in mm. Put at least three dimensions on the isometric drawing. Construction lines should be shown in light pencil marks. Isometric scale may be used in drawing the isometric view.

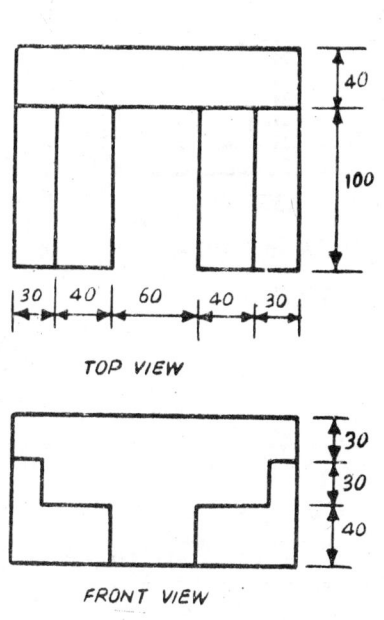

Fig. 3/1

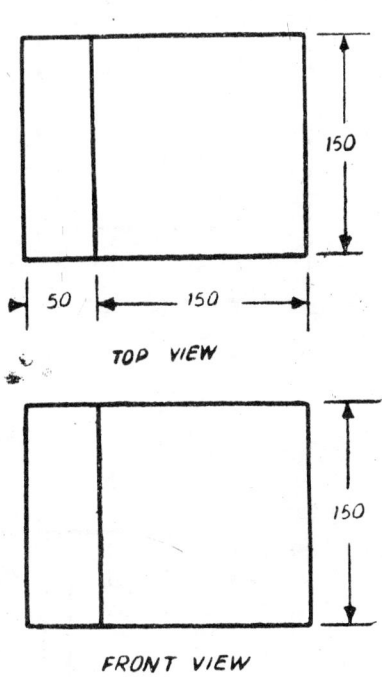

Fig. 3/2

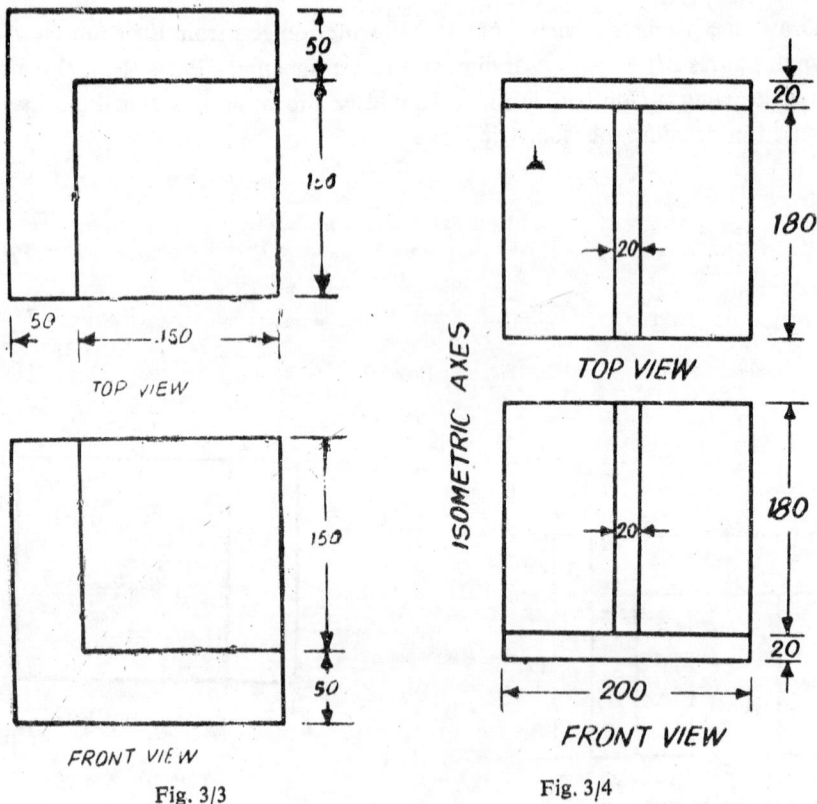

ISOMETRIC AXES

Fig. 3/3

Fig. 3/4

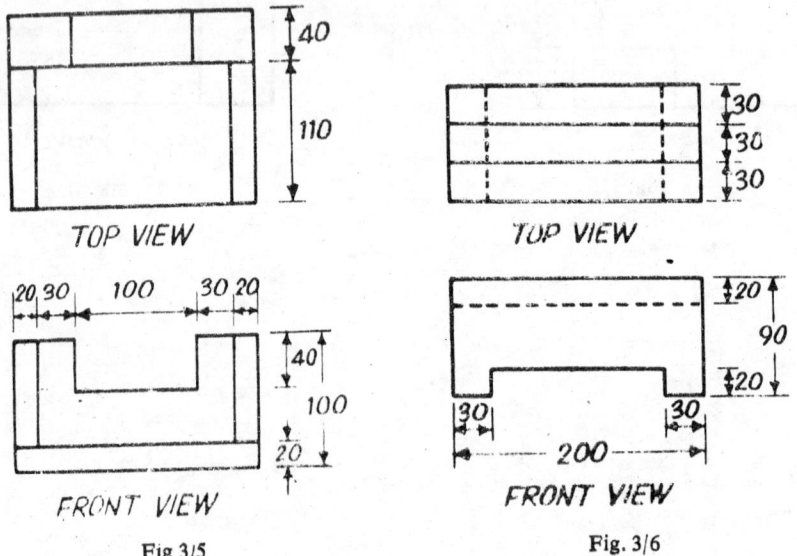

Fig 3/5

Fig. 3/6

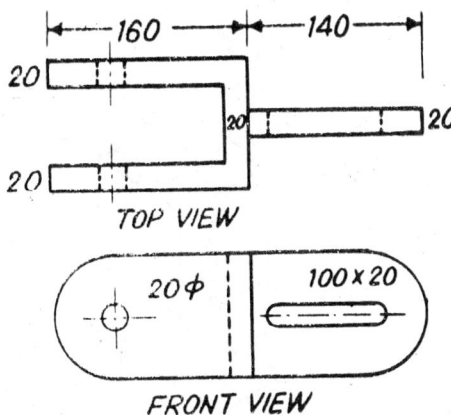

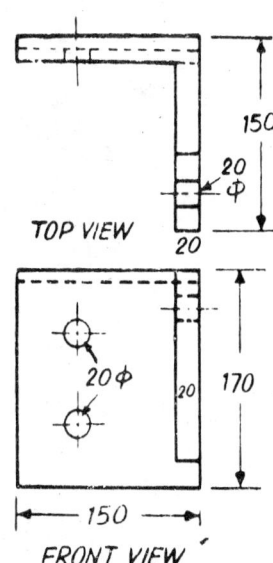

Fig. 3/7

2. Prepare freehand sketches showing any two orthographic views from the isometric views of the objects shown in Fig. 3/9 to 3/12. No dimension has been given. The sketches should be proportionate.

Fig. 3/8

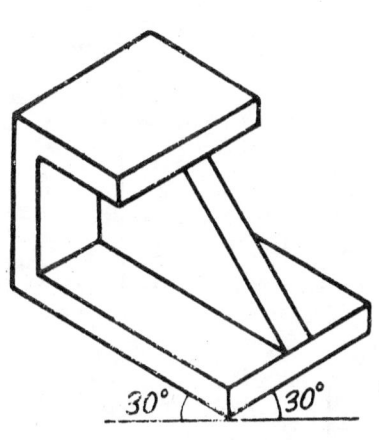

Fig. 3/9

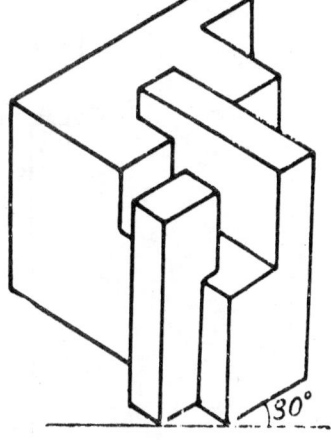

Fig. 3/10

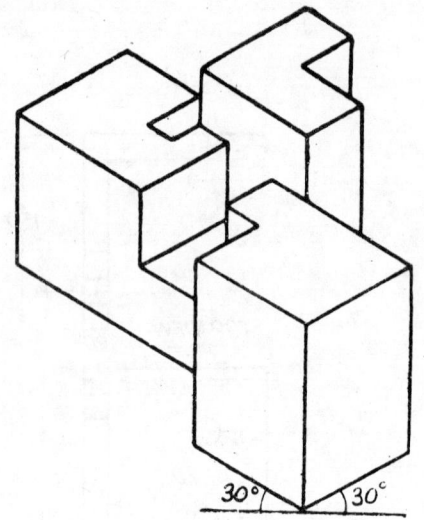

30°   30°   Fig. 3/11   30°   30°   Fig. 3/12

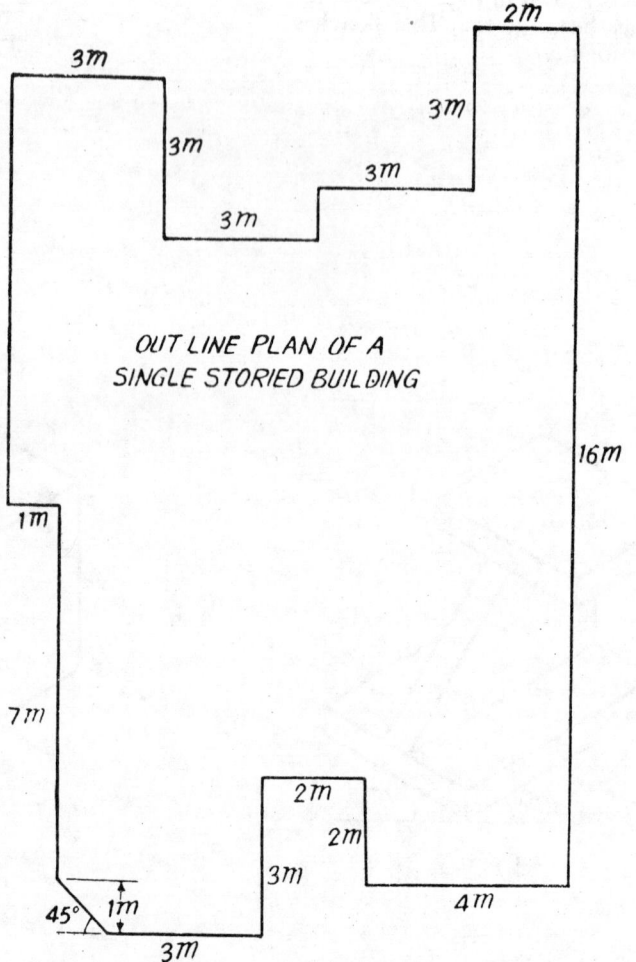

OUT LINE PLAN OF A
SINGLE STORIED BUILDING

Fig.3 /13

3. Draw isometric views of the following building from the plans shown in Fig. 3/13 to 3/15. Given : Height of building 4 m. including the plinth height ; Roof slab 100 mm. thick ; Doors : 1 m × 2 m ; Windows 1 m × 1·4 m ; R.C. chujja 0·5 m projected and 75 mm thick ; The tops of doors and windows are at same level ; There is no parapet wall ; cornice is 0·5 m projected, 100 mm thick and in continuation to the roof slab ; Assume reasonable dimensions that are not given. Each building is single storied.

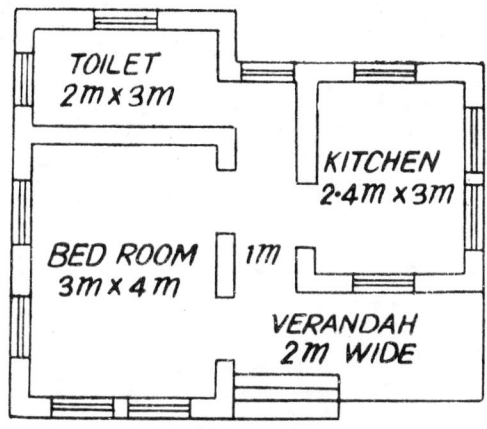

Plan of a single storied building for economically weaker section of people. All walls are 375 mm thick.

Fig. 3/14

All walls are 375 mm thick excepting partition walls which are 125 mm thick.

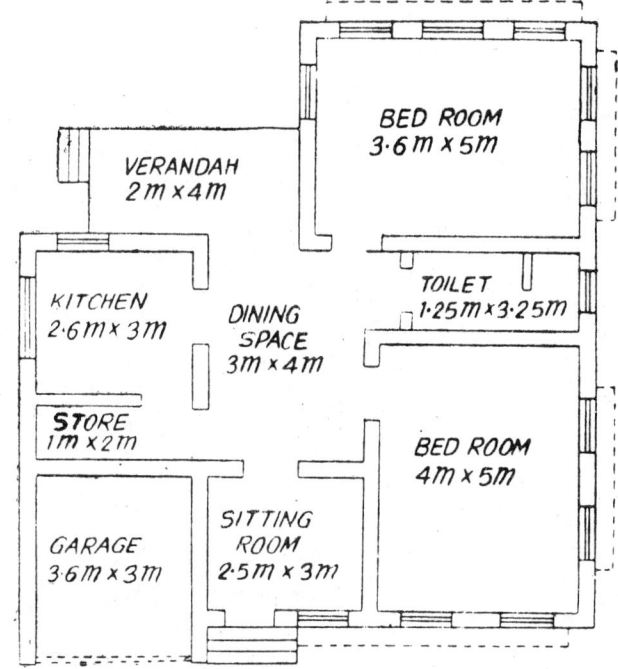

Plan of a Single Storied Building
Fig. 3/15

4

Perspective Projection

Perspective projection is one of the methods of pictorial projection which is easy to understand. In perspective projection the view of an object is obtained much as the human eye or camera would see it. A perspective drawing has three-dimensional effects and it is much used by artists, architects and draftsmen to draw the production illustrations, appearance of buildings and advertising pictures. It is rarely used by engineers in engineering drawing.

Actually, a perspective projection is a geometric method by which an object can be projected upon a picture plane in much the same way as in photography. True shape and size of an object are not obtained in a perspective drawing. It may be noticed that when an object goes away from one, it appears to get smaller. Railway lines which are parallel, appear to get closer and closer and converge at a point on the horizon. This is the *vanishing point* (V.P.).

The primary principle of perspective projection is that the lines converge to points on horizon or at the eye level of the observer. Thus, vanishing points are always at eye level or horizon. A perspective drawing is therefore drawn by looking an object with normal vision. The observer will imagine himself looking through a *picture plane*. It should be noted that the objects of same height intercept a greater distance on the picture plane when close to it than when further away. The point of observation is called the *station point*. The picture plane is usually placed between the object and the station point. Care should be taken in selecting the station point in order to have a good appearance of the perspective drawing. A poor location of the station point will result in a distorted perspective that will be very much displeasing to the eye. In general, the location of the station point should be slightly offset to one side and should be above or below the exact centre of the object. However, the centre of vision should be near the centre of interest in the object.

The station point should be thought of as a viewing point. The entire object can be viewed naturally without turning the head, if the distance between the station point and the picture plane is equal to twice the maximum dimension of the object or greater. With a view to having a good picture of the object, a wide angle should always be avoided. The best results are obtained when the visual rays from the station point to the object are kept within an angle of not more than 30° as shown in Fig. 4·1 (b). The object should be placed in relation to the picture plane such that both of the side faces do not make the same angle with the picture plane. As per common practice, the side angles are chosen to be 30° and 60° for rect-angular objects.

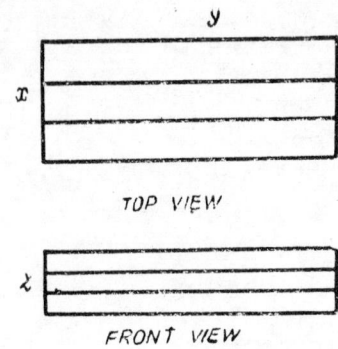

Top view and front view of a step
Fig. 4·1. (a)

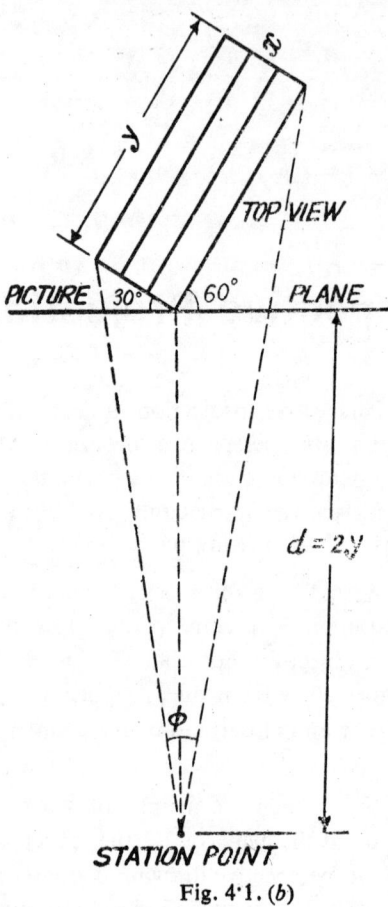

Fig. 4·1. (b)

Ref. to Fig. 4·1 (b).

Minimum distance, d, between the picture plane and the station point should be twice the maximum dimension of the object i.e., $2y$ in this case

The angle ϕ should not be greater than 30°.

Types of perspective

There are two types of perspective, in general. These are : *parallel perspective* and *angular perspective.*

In parallel perspective, one of the principal faces is parallel to the picture plane. All vertical lines are vertical in parallel perspective and the receding horizontal lines converge to a single vanishing point. The horizontals are truly horizontal. This type of perspective is also called *one vanishing point perspective* and it is used only when one plane of the object is of interest. This is the simplest type of perspective. See Fig. 4·2.

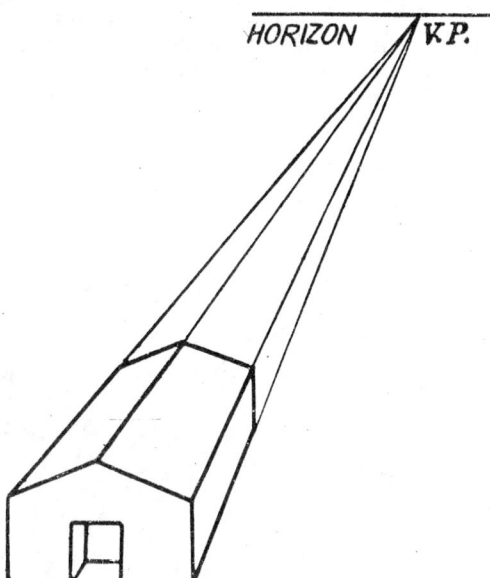

In angular perspective, the object is placed such that the principal faces are at angles with the picture plane. The horizontal lines converge at two vanishing points. The vanishing point lie on the same horizontal line. The vertical edges are truly vertical and parallel to each other, but the horizontal edges are not horizontal and they are not drawn parallel to eaee other. Fig. 4·3 shows two vanishing point perspectivᵒs when the object is above the horizon and below the horizon. About 80% of respectives are drawn by using two varnishing point.

One vanishing point perspectivo
(Parallel Perspective)
Fig. 4·2.

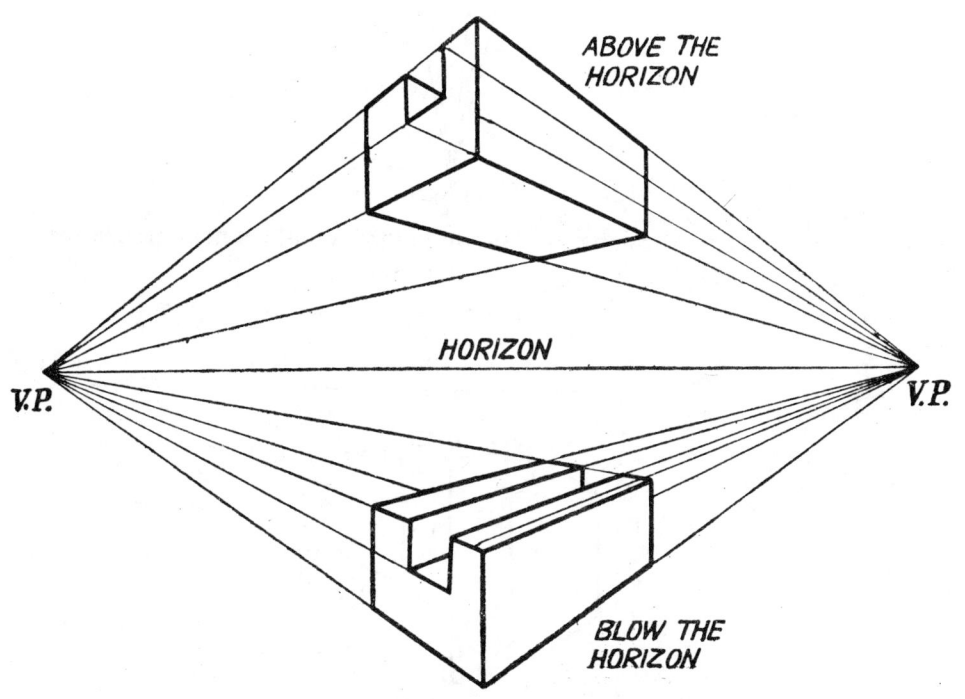

Two Vanishing Point Perspectives
Fig. 3·4.

Three vanishing point perspective is required to be drawn when the object is to be seen from a great height or a tall building is to be seen from street level. Here, the vertical edges of the object are not vertical and parallel to each other. The horizontal edges also are not horizontal and parallel to each other. See Fig. 4'4.

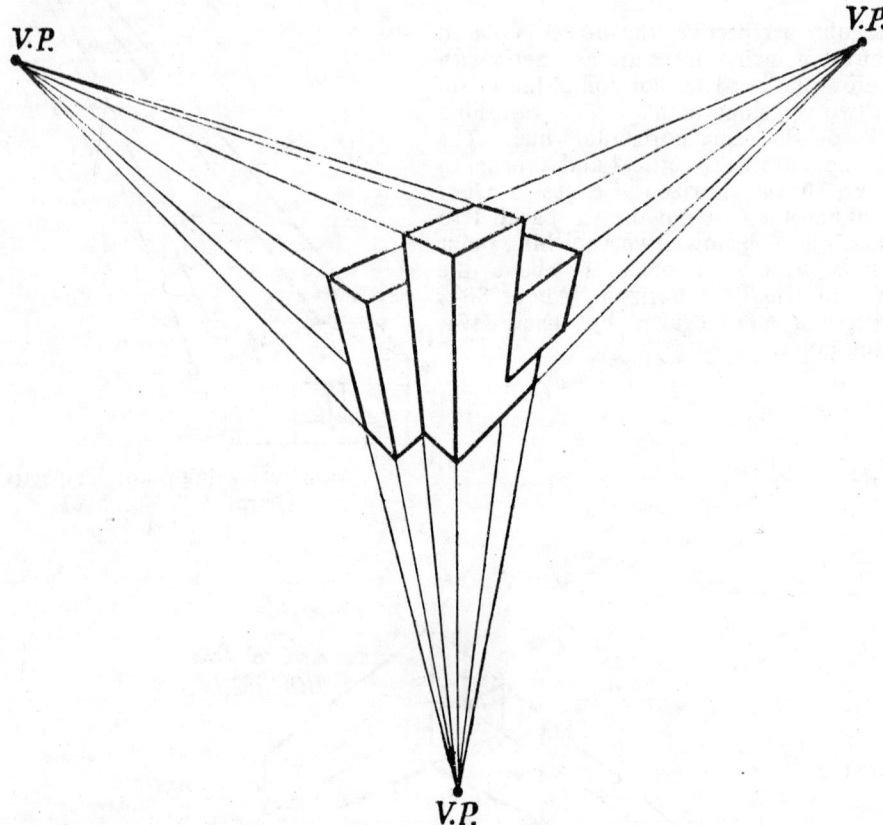

An Architectural composition shown in three vanishing point perspective
Fig. 4·4.

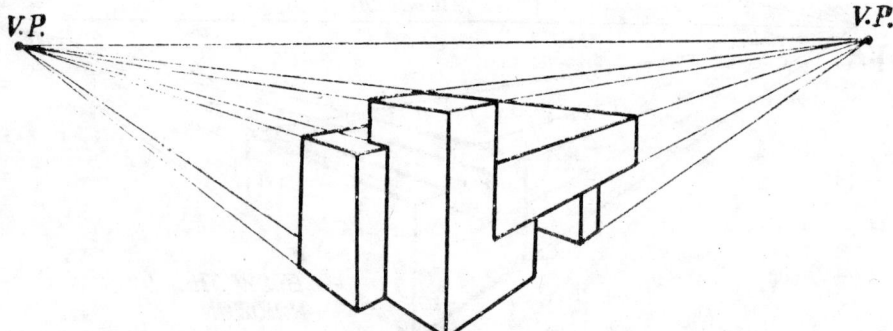

An Architectural composition shown in two vanishing point perspective
Fig. 4·5.

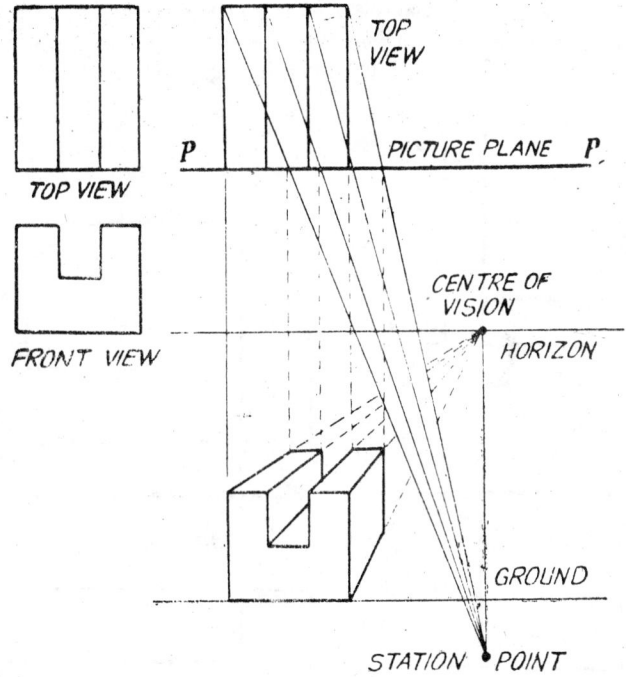

Construction of Parallel Perspective
Fig. 4·6.

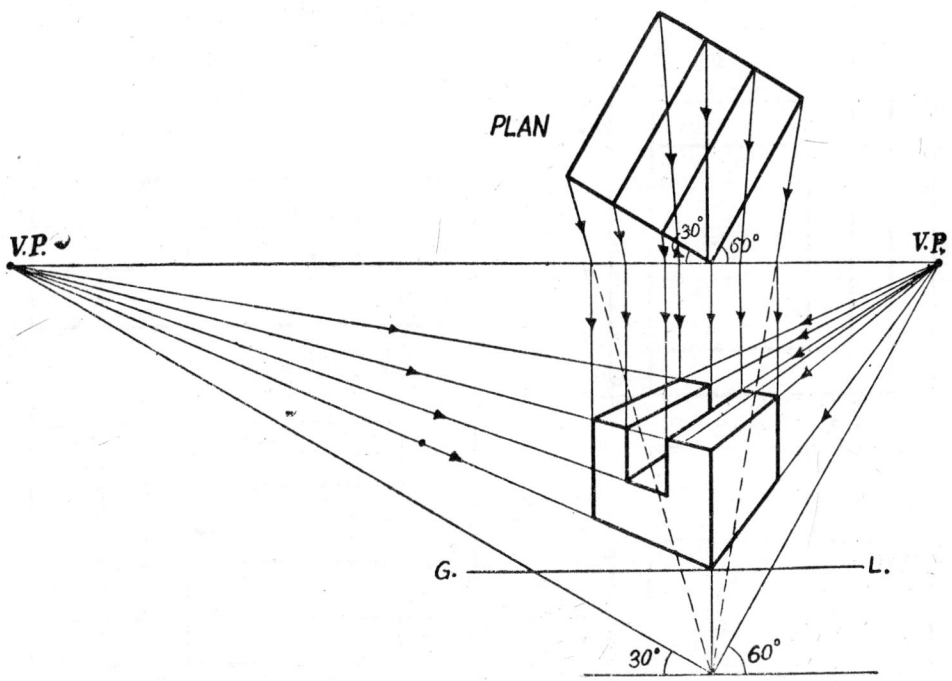

Illustration showing the method of preparation of a two-point perspective view
from the orthographic views of a channel section. Follow the arroheads.
Fig. 4·7.

Exercises on Chapter 4

1. Draw one vanisning and two vanishing point perspectives from the following following orthographic views of the objects shown in Fig. 4/1 to 4/4.

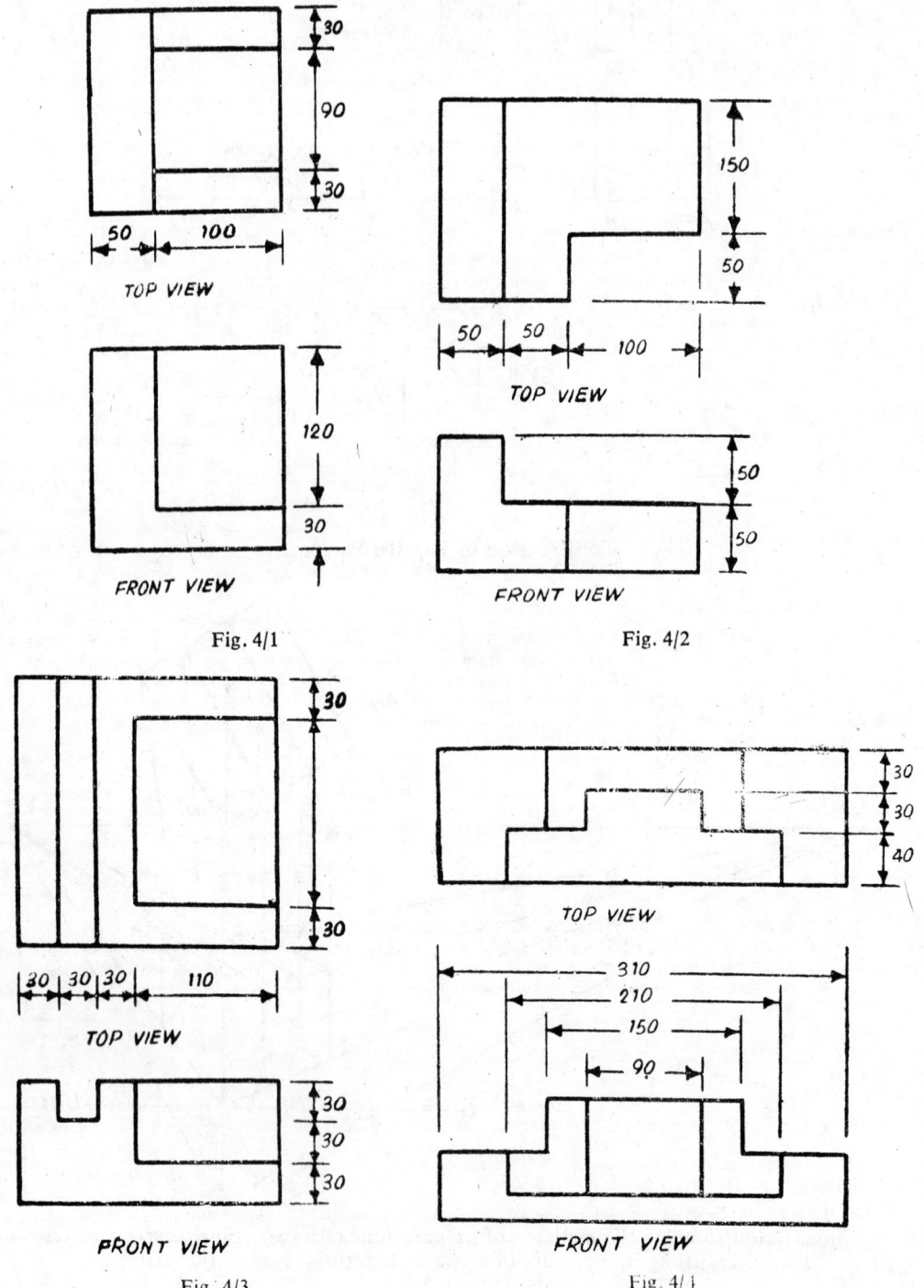

Fig. 4/1

Fig. 4/2

Fig. 4/3

Fig. 4/4

5

Construction of Arches

An arch is a curved structure spanning an opening to support the load of the structure constructed above the opening. Arches were extensively used in buildings of early ages. Arches were made of wedge shaped stones or bricks joined together with mortar. The bricks or stones in an arch are held in position by mutual pressure because of the wedge action.

The various terms applied to arch construction are given below :

'Springing line' is the horizontal line joining the two springing points *i.e.*, the starting points of the arch at the two ends.

'Span' of an arch is the horizontal distance of clear opening between the two end supports of the arch.

'Rise' is the height of the arch *i.e.*, the vertical distance from the springing line to the topmost point of the intrados.

'Intrados' is the inner curved surface of an arch. It is also known as 'soffit'.

'Extrados' is the external curve of the arch.

'Crown' is the highest point of the extrados.

'Voussoirs' are the wedge-shaped bricks or stones used in arch construction.

'Key brick' is the central voussoir *i.e.*, the last voussoir in arch construction.

'Spandril' is the triangular space enclosed by the extrados of two adjoining arches.

'Abutment' is the extreme end wall supporting one end of an arch.

'Pier' is an intermediate support between two spans *i.e.*, the common support of two adjoining arches.

'Skew Back' is the splayed surface of the abutment or pier from which the arch starts.

'Skew Brick' is the special brick used in skew back.

A semi-circular arch showing arch nomenclatures is illustrated in Fig. 5.1.

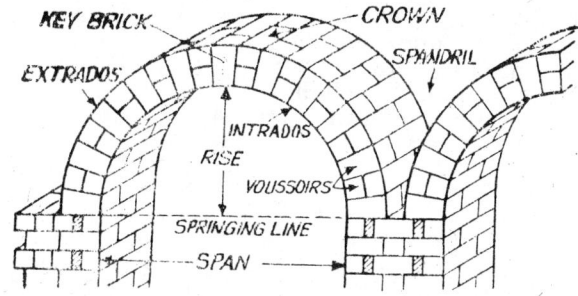

A Semi-Circular Arch Showing Arch Nomenclatures
Fig. 5.1.

There are various forms of arches used in architectural composition for buildings. The construction of these arches is shown in Fig. 5·2. The sketches shown in Fig. 5·2 are self-explaining.

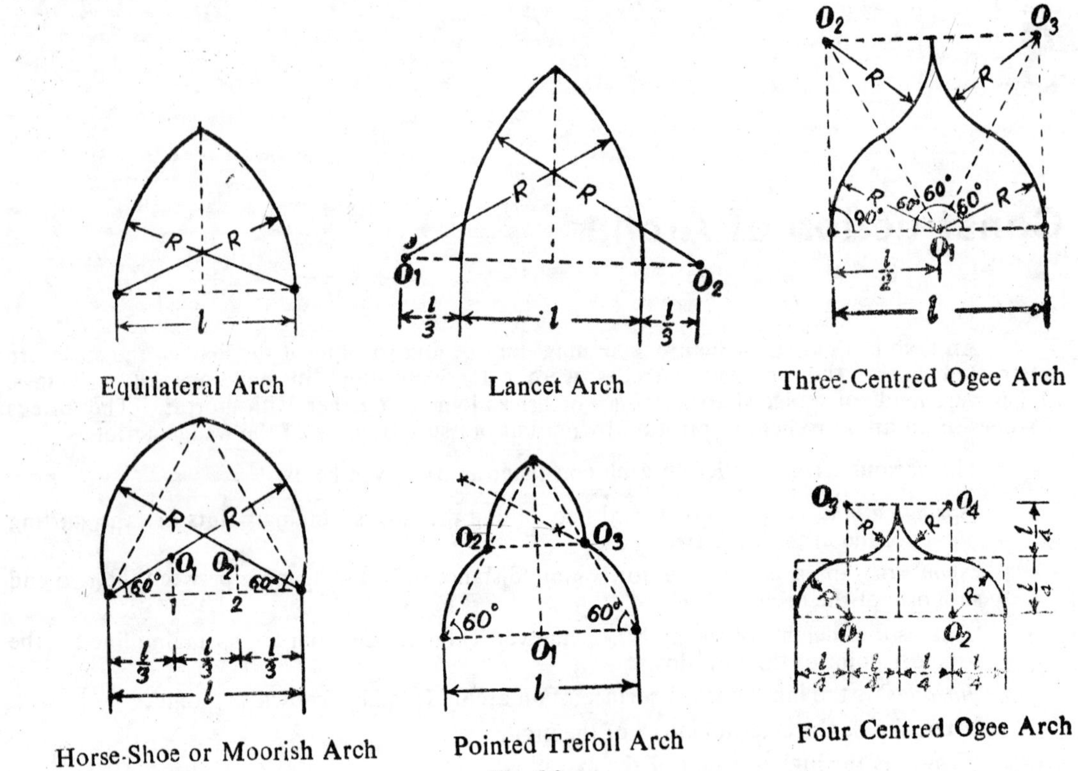

Equilateral Arch

Lancet Arch

Three-Centred Ogee Arch

Horse-Shoe or Moorish Arch

Pointed Trefoil Arch

Four Centred Ogee Arch

Fig. 5·2.

Equilateral Arch :

An equilateral arch is a two-centred arch. The two centres lie at the two ends of the springing line. With these two centres and radius equal to the span length, draw two arcs which will meet at the point C forming an equilateral arch. The bricks are laid radially from the corresponding centre as shown in Fig. 5·3.

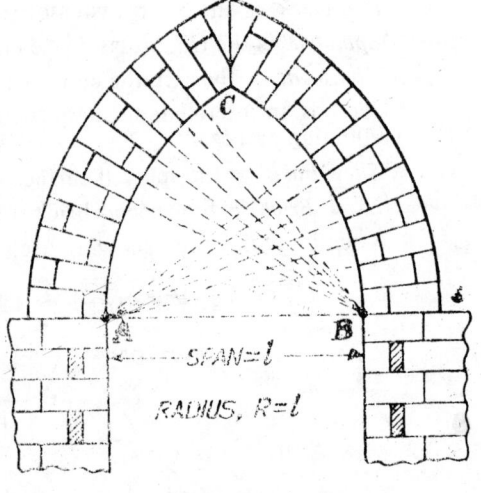

Equilateral Arch

Fig. 5·3.

Flat Arch :

A flat arch is constructed by subtending an angle of 60° at the centre, as shown in figure. Bricks are laid radially from the centre. See Fig. 5·4. The arch remains in position by wedge action.

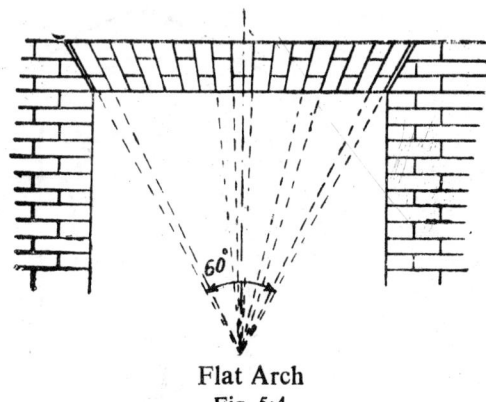

Flat Arch
Fig. 5·4.

Semi-Elliptical Arch :

Semi-elliptical arches may be three-centred, four-centred, five-centred or seven-centred. A four-centered semi-elliptical arch is shown in Fig. 5·5.

Divide the span into four equal parts and construct the curves as shown in Fig. 5·5 with centres 1, 2, 3 and 4. Then, place the voussoirs radially.

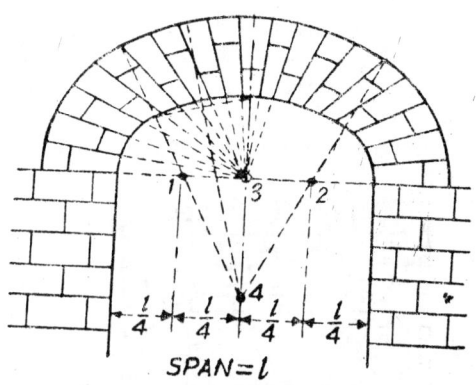

Semi-Elliptical Arch
Fig. 5·5.

Tudor Arch :

Divide the span into four equal parts. With centres 1 and 2 and radius equal to $l/4$ draw two circles as shown. Obtain points 3 and 4 by intersecting lines at 45°. With centres 3 and 4 and radius equal to $3x$ or $4y$, ($\because 3x = 4y$) complete the arch. Next, provide arch thickness as shown in Fig. 5·6.

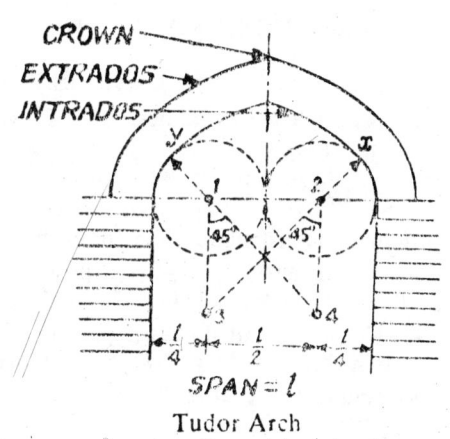

Tudor Arch
Fig. 5·6.

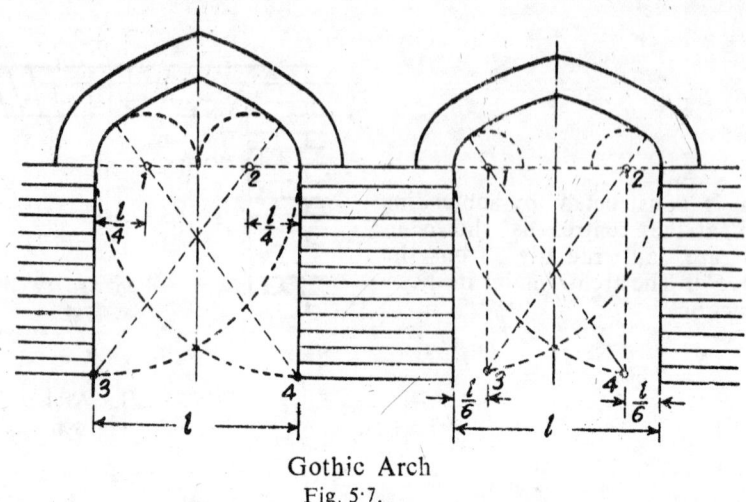

Gothic Arch
Fig. 5·7.

Gothic Arch :

Two forms of four-centred gothic arch are illustrated in Fig. 5·7. The construction technique for both the forms is also shown.

Relieving Arch : It is usually a segmental arch constructed over a flat arch to give relief to the later. See Fig. 5·8.

Circular Arch : The construction of a circular arch is shown in Fig. 5·9.

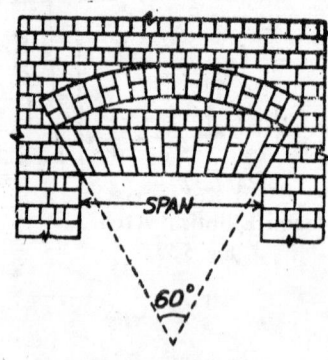

A Flat Arch with a Releiving Arch at Top
Fig. 5·8.

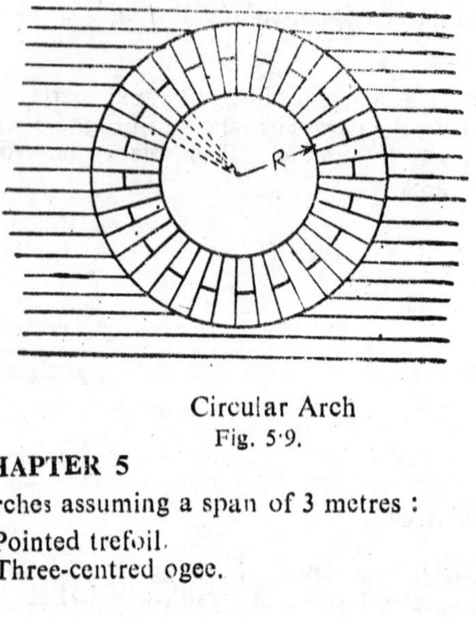

Circular Arch
Fig. 5·9.

EXERCISES ON CHAPTER 5

1.   Construct the outlines of the following arches assuming a span of 3 metres :

(a) Lancet.   (b) Pointed trefoil.
(c) Moorish.   (d) Three-centred ogee.
(e) Four-centred ogee.

Show the construction lines of each.

2.   Construct a pointed trefoil arch and a three-centred ogee arch for a clear span of 2 metres, the arch thickness being 375 mm   Show the placement of bricks.

3.   Construct and show the placement of bricks in the following arches for a span of 1·8 m, the arch thickness being 375 mm (1½ brick).

(a) Equilateral arch.   (b) Segmental arch.
(c) Semi-elliptical arch.   (d) Tudor arch.
(e) Flat arch.   (f) Gothic arch.

6

Bonds in Brickwork

Bonds in brickwork are essential to tie the bricks with each other and to avoid the continuous vertical joints. This also helps in distributing the load coming over the brickwall. A good bonding depends upon good workmanship and uniform size of bricks. Commonly used terms in brick bonding are given below.

'Stretcher' is the longitudinal face of the brick.

'*Header*' is the cross face of the brick.

'*Stretching Bond*' is a bonding of bricks with continuous layers of stretchers in elevation.

'*Heading Bond*' is a brick bond with continuous layers of headers in elevation.

Various forms of bricks that are required to make different brick bonds are illustrated in Fig. 6·1. These forms are made to shape and size by cutting the bricks.

'*Bat*' is a portion of brick cut to size from a full brick. Three-fourth bat, half bat and bevelled bat are shown in Fig. 6·1.

'*Closer*' is a portion of brick by cutting a full brick longitudinally. King-closer, queen-closer, bevelled closer and queen-closer quarter are shown in Fig. 6·1.

'*Quoin*' is the brick used at the corner of a wall.

'*Squint Brick*' is the brick moulded specially or cut to required size from a full brick. These bricks are used at the corner of two walls meeting at an angle other than 90°.

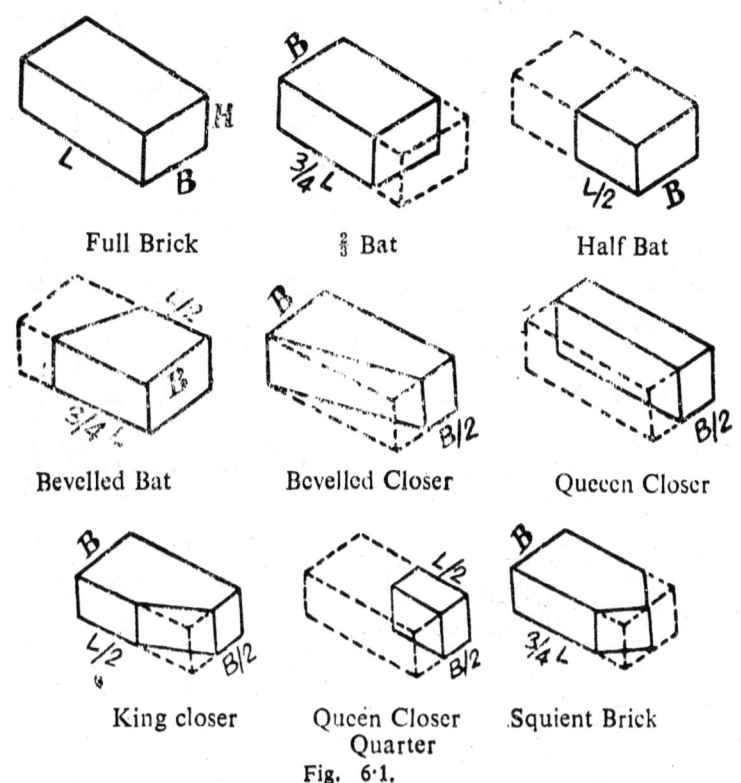

Full Brick	⅜ Bat	Half Bat
Bevelled Bat	Bevelled Closer	Queeen Closer
King closer	Queen Closer Quarter	Squient Brick

Fig. 6·1.

VARIOUS BONDS

Different bonds used in brickwork are stretching bond, Heading bond, English bond, Flemish bond, English-Garden wall bond, Flemish-Garden wall bond, Raking bond, Herring-Bone bond, Dutch bond. etc.

Stretching Bond

In this bond, bricks in all layers are arranged as stretchers. This is used in one-brick partition only.

Heading Bond

In this bond, bricks in all layers are arranged as headers. This is used in curved walls and very thick walls.

English Bond

The stretchers and headers are placed in alternate courses. Queen closers are used to break the continuous vertical joints.

Flemish Bond

Each course in this bond consists of stretchers and headers alternately. Queen closers and bats are used to break the continuous vertical joints. Every header is centrally placed over a stretcher. Flemish bonds may be single or double. In single Flemish bond the face elevation is of Flemish bond and the filling is of English bond. In double Flemish bond headers and stretchers are placed alternately in front and rear elevation of the wall.

English-Garden wall bond :

There is one heading course to three or five stretching courses.

BRICKS WALLS IN ENGLISH BOND :

There is one header to 3 or 5 stretchers in each course. In every alternate course 3/4 bat is used just after the header at end.

Bricks Walls in English Bond :

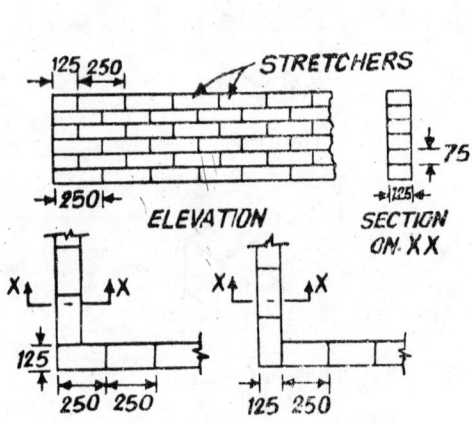

Courses 1,3,5.... Courses 2,4,6,...
90° JUNCTION OF 125 MM
WALLS IN ENGLISH BOND.
Fig. 6·2,

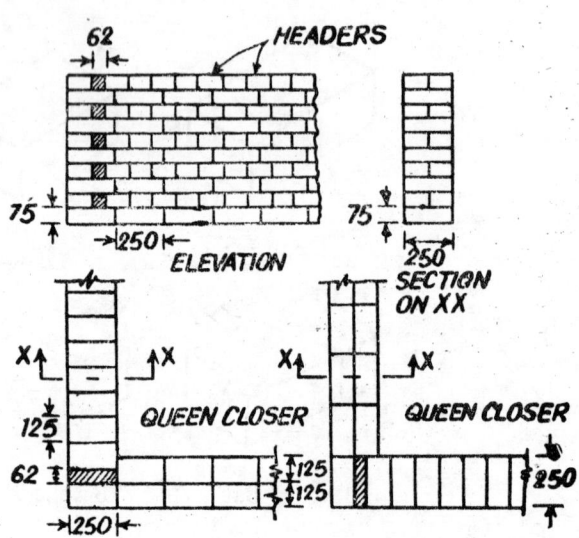

Courses 1,3,5,... Courses 2,4,6,...
90° JUNCTION OF 250 MM WALLS
IN ENGLISH BOND.
Fig. 6·3.

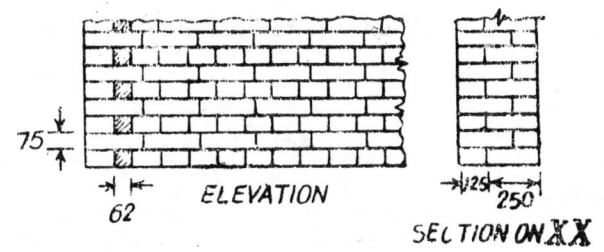

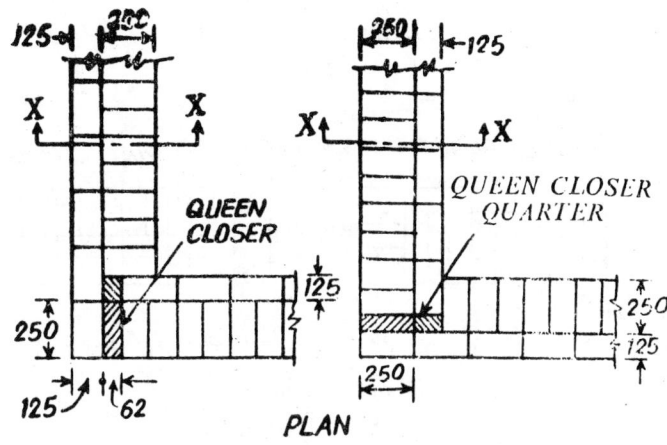

Courses 1,3,5,...   Courses 2,4,6,...

90° JUNCTION OF 375 MM WALLS IN ENGLISH BOND.

Fig. 6·4.

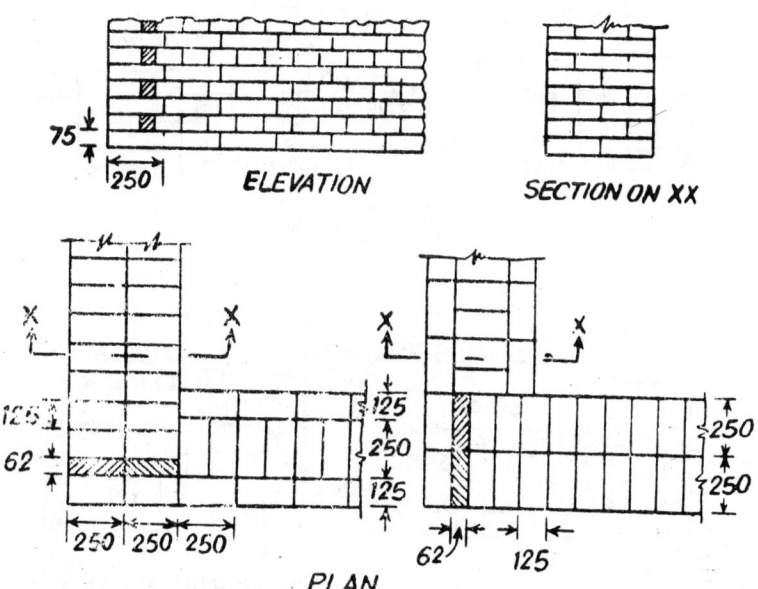

Courses 1, 3, 5, ..   Courses 2, 4, 6, ..

90° JUNCTION OF 500 MM WALLS IN ENGLISH BOND.

Fig. 6·5.

SINGLE FLEMISH BOND

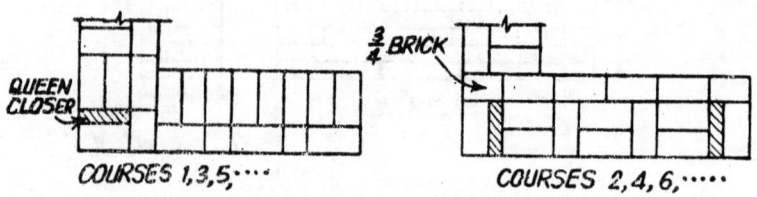

375 mm. Thick wall

Fig. 6·6.

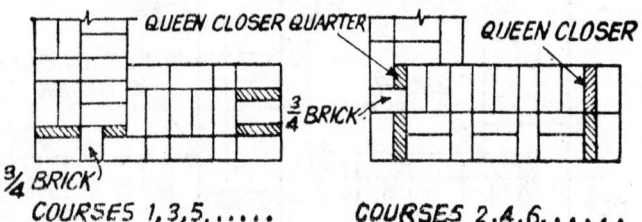

500 mm. Thick wall

Fig. 6·7.

DOUBLE FLEMISH BOND

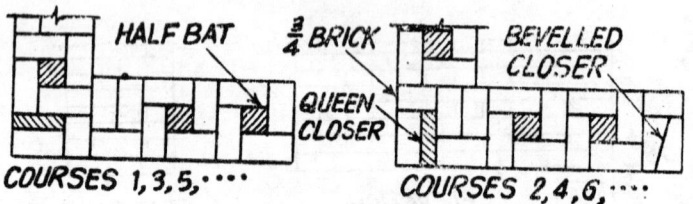

375 mm. Thick wall

Fig. 6·8.

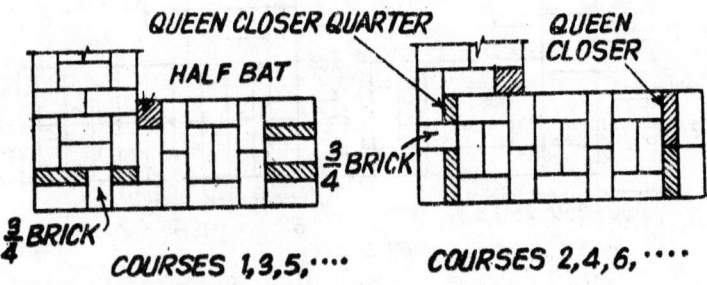

500 mm. Thick wall

Fig. 6·9.

BRICK WALL IN FLEMISH BOND

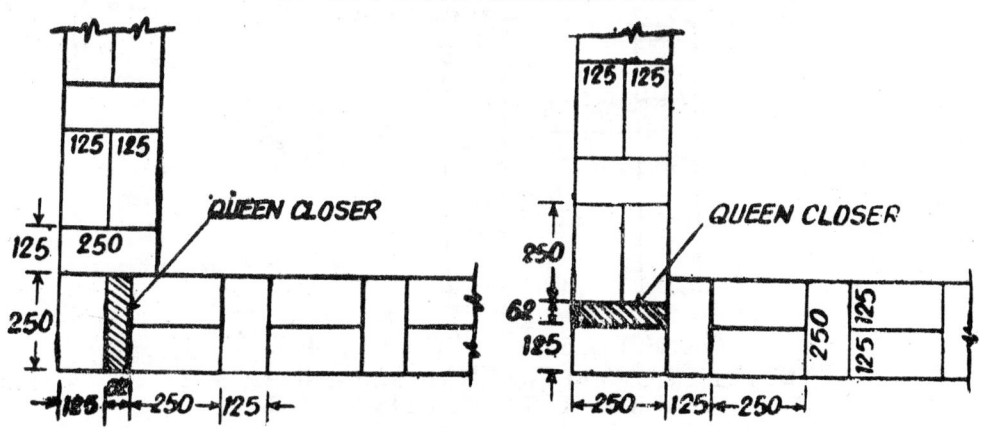

Courses 1, 3, 5, ..   Courses 2, 4, 6,...

90° JUNCTION OF 250 MM WALLS IN FLEMISH BOND
Fig. 6·10.

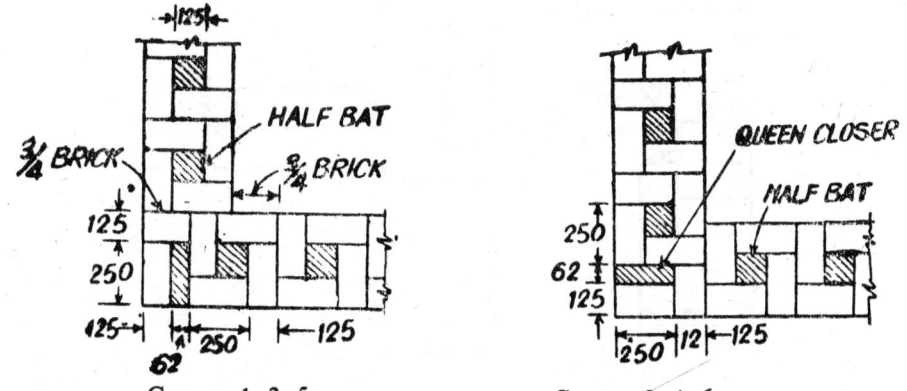

Courses 1, 3, 5, ...   Courses 2, 4, 6, ...

90° JUNCTION OF 375 MM WALLS IN FLEMISH BOND
Fig. 6·11.

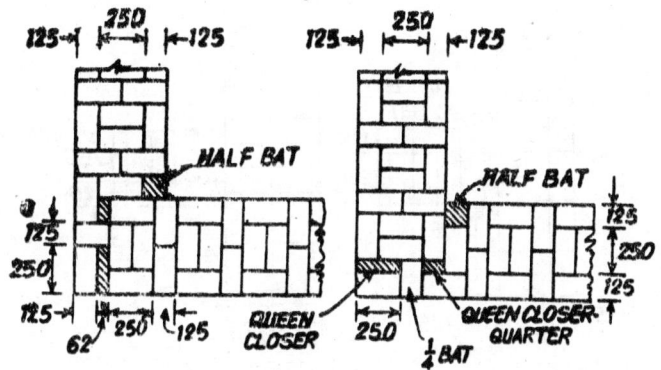

Courses 1, 3, 5, ...   Courses 2, 4, 6, ...

90° JUNCTION OF 500 MM WALLS IN FLEMISH BOND
Fig. 6·12.

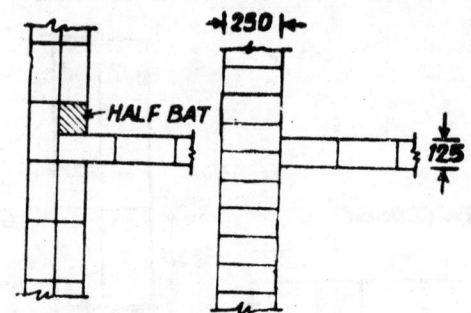

Courses 1, 3, 5, ...   Courses 2, 4, 6, ...

TEE JUNCTIONS BETWEEN 250 MM AND 125 MM
THICK WALLS IN ENGLISH BOND

Fig. 6·13.

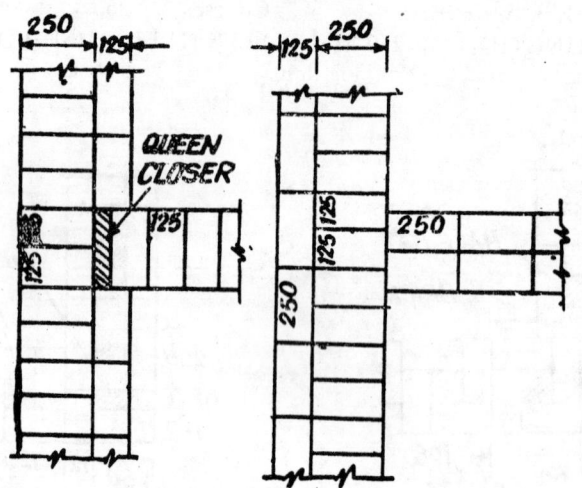

Courses 1, 3, 5, ...   Courses 2, 4, 6, ...

TEE JUNCTION BETWEEN 250 MM 375 MM WALLS IN ENGLISH BOND

Fig. 6· 14.

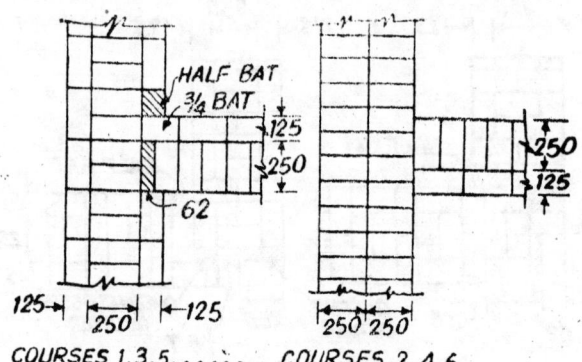

COURSES 1,3,5,.....   COURSES 2,4,6,.....

TEE JUNCTION BETWEEN 375 MM AND 500 MM
THICK WALLS IN ENGLISH BOND

Fig. 6·15.

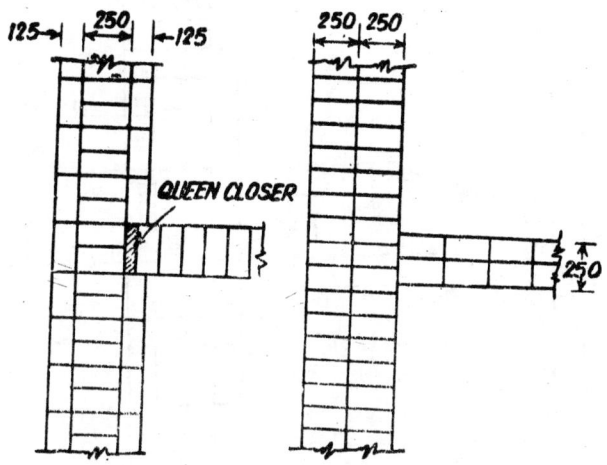

Courses 1, 3, 5, ...   Courses 2, 4, 6, ...
TEE JUNCTION BETWEEN 250 MM AND 500 MM
WALLS IN ENGLISH BOND

Fig. 6·16.

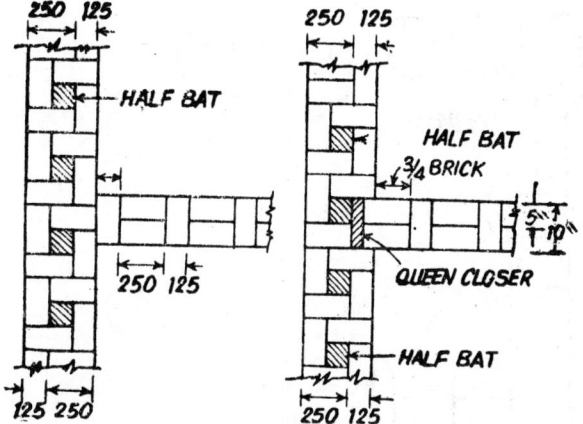

Courses 1, 3, 5, ...   Courses 2, 4, 6, ...
TEE JUNCTION BETWEEN 250 MM AND 375 MM
WALLS IN FLEMISH BOND

Fig. 6·17.

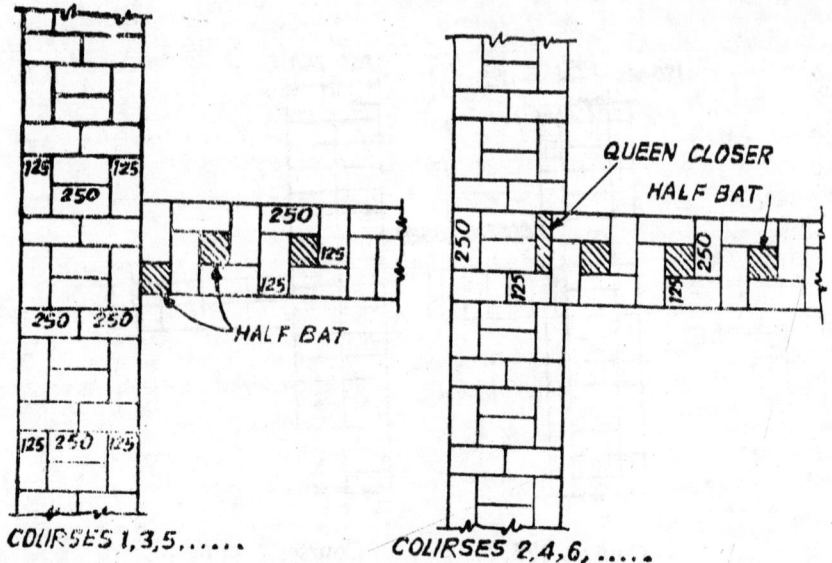

TEE JUNCTION BETWEEN 375 MM AND 500 MM WALLS IN FLEMISH BOND
Fig. 6·18.

CROSS WALLS IN ENGLISH BOND

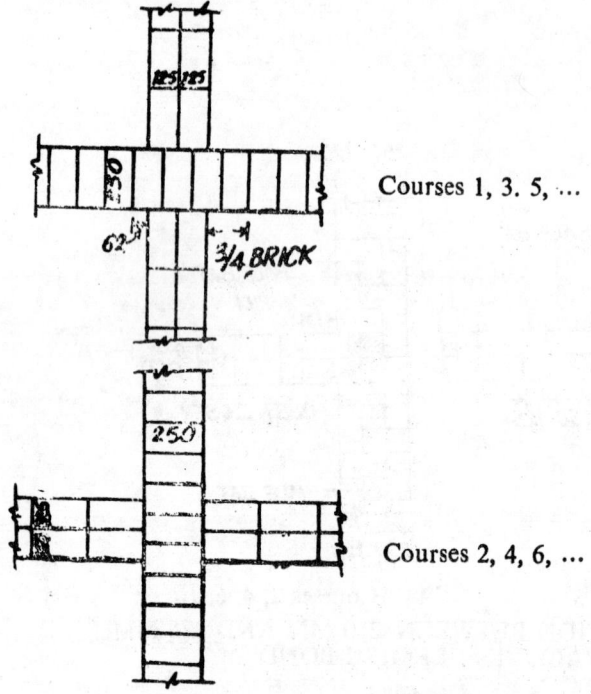

Courses 1, 3. 5, ...

Courses 2, 4, 6, ...

**ALTERNATE LAYERS OF 250 MM CROSS WALLS
IN ENGLISH BOND**
Fig. 6·19.

CROSS WALLS IN ENGLISH BOND

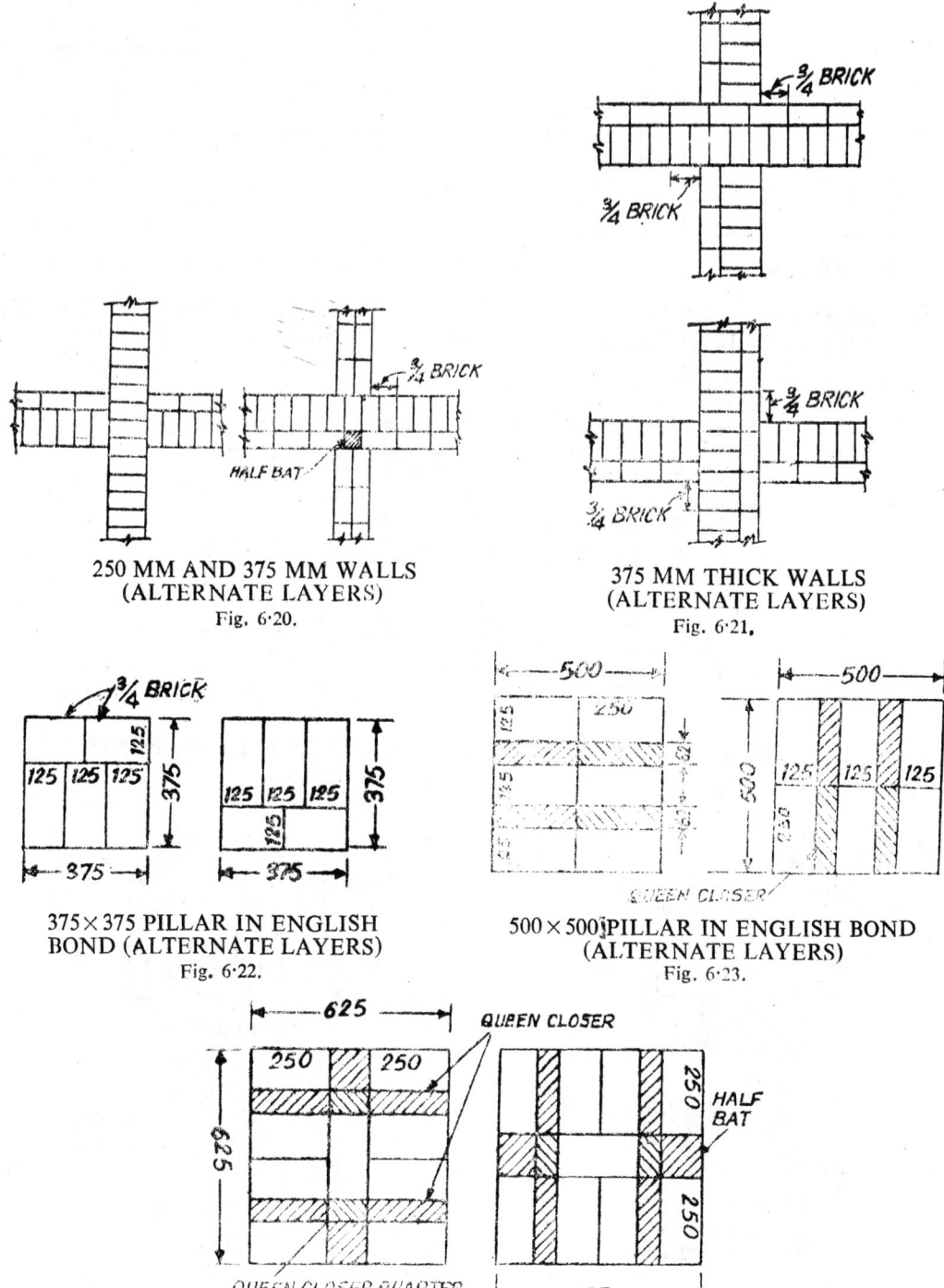

250 MM AND 375 MM WALLS
(ALTERNATE LAYERS)
Fig. 6·20.

375 MM THICK WALLS
(ALTERNATE LAYERS)
Fig. 6·21.

375×375 PILLAR IN ENGLISH
BOND (ALTERNATE LAYERS)
Fig. 6·22.

500×500 PILLAR IN ENGLISH BOND
(ALTERNATE LAYERS)
Fig. 6·23.

ALTERNATE LAYERS OF 625×625 PILLAR IN ENGLISH BOND
Fig. 6·24.

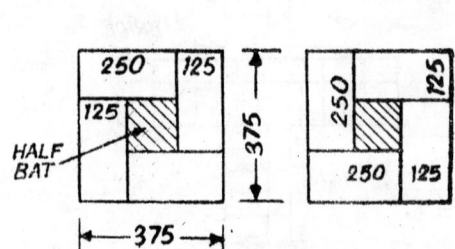

ALTERNATE LAYERS OF 375×375
PILLAR IN FLEMISH BOND
Fig. 6·25.

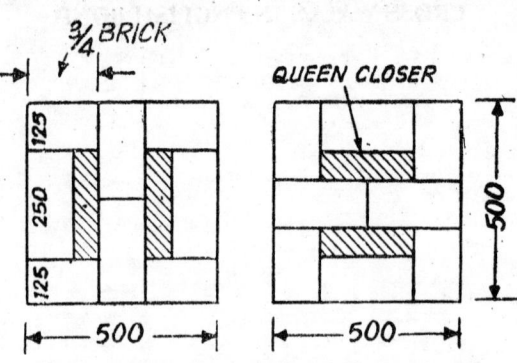

ALTERNATE LAYERS OF 500×500 IN
PILLAR IN FLEMISH BOND
Fig. 6·26.

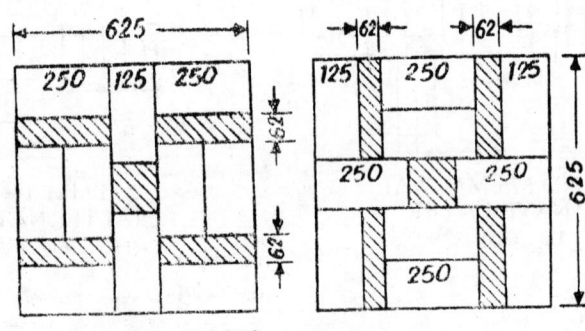

ALTERNATE LAYERS OF 625×625 PILLAR IN FLEMISH BOND
Fig. 6·27.

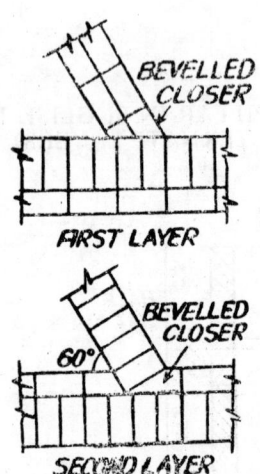

SQUINT JUNCTION BETWEEN
250 MM AND 375 MM WALLS
Fig. 6·28.

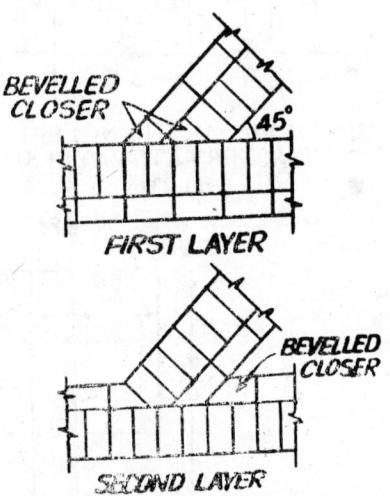

SQUINT JUNCTION BETWEEN
375 AND 375 WALLS
Fig. Fig. 6·29.

Raking Bond

In this bond bricks are laid inclined to the wall face. Raking bonds are Diagonal bond, Herring-bone bond and zig-zag bond. These bonds and chiefly used in construction of thick walls and in brick pavements.

In diagonal bond all bricks are laid inclined in one direction only. In Herring bone bond bricks are laid at 45° angle from centre in both directions. In zig-zag bond bricks are laid in zig-zag pattern.

Dutch Bond

This is a modified form of English bond. This bond is used to strengthen the corner of the wall. Alternate courses are of headers and stretchers. In every alternate stretcher course, a header is introduced next to 3/4 bat.

English Garden wall bond

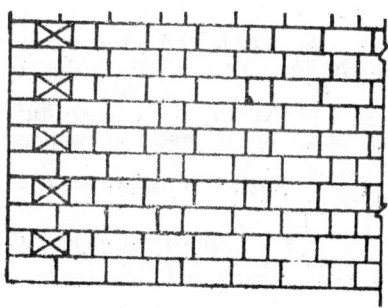

Flemish Garden wall bond

Dutch wall bond

Zig-zag bond

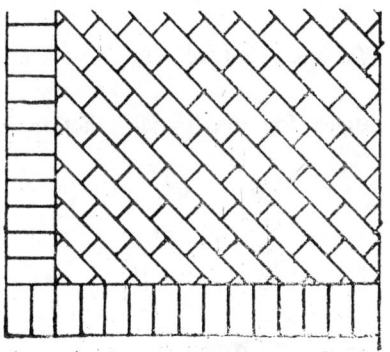

Diagonal bond

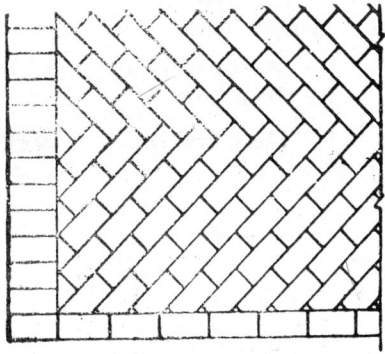

Herring Bone bond

Fig. 6·30.

Facing Bond

This bond comprises one heading course to several stretching courses. This is used for facework of walls and where the bricks are of assorted thickness.

English Cross Bond :

This is very close to English bond. In this bond alternate course of stretchers and headers are used which produces a good elevation.

EXERCISES ON CHAPTER 6

1. Make neat sketches of the following brick pieces and show their dimensions :

Bevelled bat, King closer, Queen closer, Queen closer quarter, Bevelled closer and squint brick.

2. Sketch the following bonds used in brick work : Stretching bond. Heading bond, English Garden wall bond, Flemish Garden wall bond and Dutch wall bond.

3. Sketch any two types of brick bond used in flooring and pavement. What do you mean by brick-on-edge and brick on-end ?

4. Draw two consecutive layers of 625 × 625 pillers in English bond and in Flemish bond.

5. Fig 6·1 shows one layer of bricks for a 500 × 500 pillar in English bond and in Flemish bond. Draw the alternate layers.

Fig. 6·1.

6. Draw two consecutive layers of 375 × 375 piller in English bond and in Flemish bond.

7. Draw two consecutive layers for making Tee junctions between 250 mm and 375 mm and between 375 mm and 500 mm. walls in English bond and Flemish bond respectively.

8. Draw alternate layers for cross walls of 250 mm and 375 mm and 375 mm and 375 mm walls in English bond.

9. Draw two consecutive courses for making 90° junction of 375 mm walls and 500 mm walls in Flemish bond.

10. Sketch alternate courses of 375 mm and 500 mm thick walls in single Flemish bond and double Flemish bond.

11. Make neat sketches showing alternate courses for making 90° junction in English bond.

7

Joinery in Woodwork

INTRODUCTION

Joinery in woodwork is a technology practised by the carpenters. This is an art of joining wooden pieces or members to form a structure subjected to load. For joinery works in

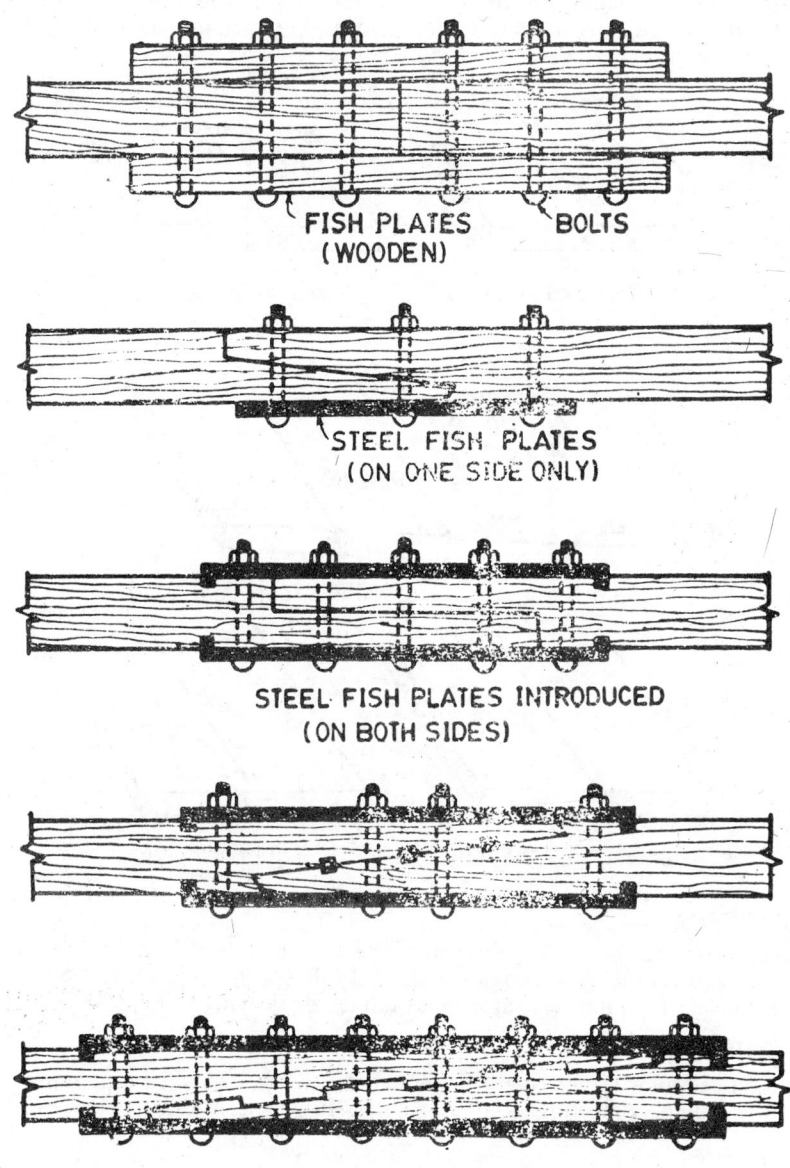

FISH PLATES (WOODEN)   BOLTS

STEEL FISH PLATES (ON ONE SIDE ONLY)

STEEL FISH PLATES INTRODUCED (ON BOTH SIDES)

Longthening Joints
Fig. 7·1.

woodwork, skilled labour is required. The carpenters therefore should have knowledge of cutting, shaping, joining and finishing the work as per the requirement. The formation of grooves, projections, beads, mitres, etc. to the exact shape and size and making the pieces assembled to form one member speak of skill and experience of the craftsmen. Various methods are adopted for joining pieces of timber depending upon the nature of frame and load conditions. A good craftsman know the technique to be adopted for a specific job. Various types of joint employed in timber work are explained here with their illustrations.

LENGTHENING JOINTS

The joints are used when it is required to lengthen a wooden member. These joints may be lapped, fished or scarfed. The pieces are secured with the help of bolts and nuts.

Fig. 7.1 illustrates various ways of forming lengthening joints.

WIDENING JOINTS

These joints are employed for widening the wooden board or planks required in flooring and roof ceilings. Various ways of forming such joints are illustrated in Fig. 7.2.

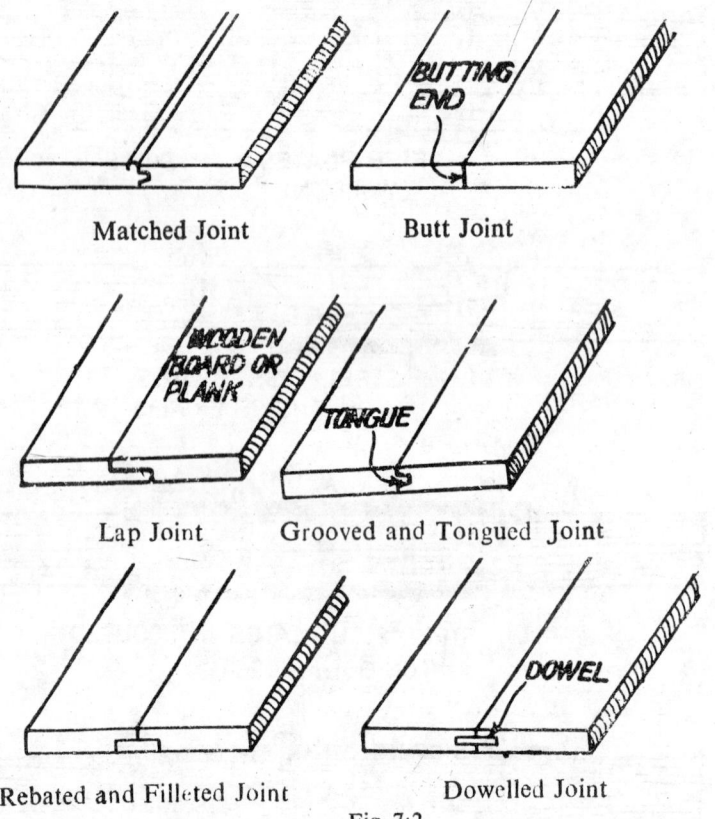

Matched Joint   Butt Joint

Lap Joint   Grooved and Tongued Joint

Rebated and Filleted Joint   Dowelled Joint

Fig. 7.2.

BEARING JOINTS

These are the joints required to impart adequate strength at the junction when two members usually meet at right angles. Halved joint, cogged joint, housed joint, Mortice and Tenon joint, Dovetailed joint, etc. are the examples of Bearing joints.

HALVED JOINT

In a halved joint the pieces to be joined together are halved and lapped. This type of joint is used in members meeting at an angle, crossing each other or when they are in the same alignment. Each member is cut to half of its thickness and lapped as shown in Fig. 7.3 to 7.5. Lapped Mitred joint as shown in 7.7 is also a form of halved joint. In a Tenoned Mitred joint fixity developed is more. See Fig. 7.6.

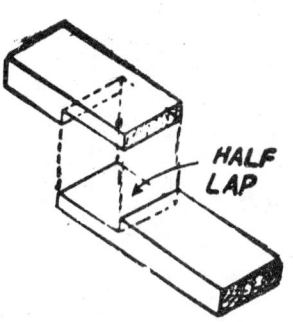

Half Lap Joint

Fig. 7.3.

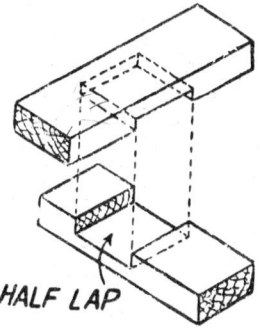

Half Lap Cross or
Cross Halving Joint
Fig. 7.4.

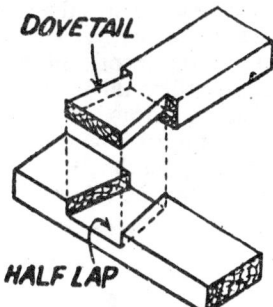

Half Lap Dovetail joint or
Dovetail Halving Joint
Fig. 7.5.

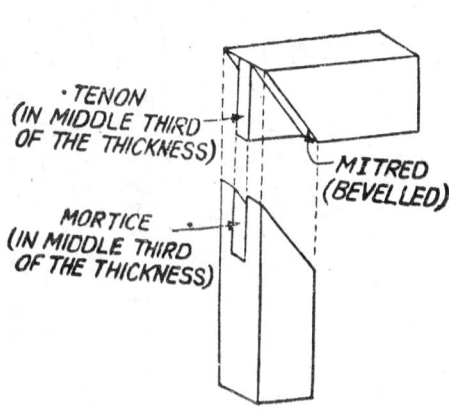

Tenoned and Mitred Joint
Fig. 7.6.

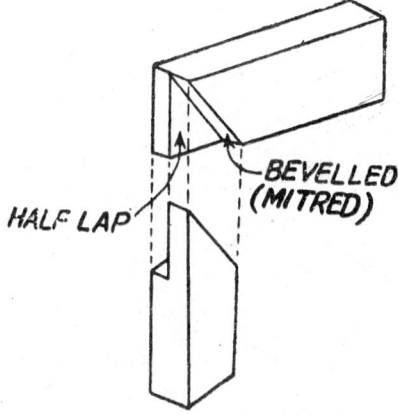

Lapped Mitred Joint
Fig. 7.7.

BRIDE JOINT

This is actually a form of Mortice and Tenon joint. The bridle or projection made in one piece fits into the Mortice of the other. See Fig. 7.8.

HOUSING JOINT

In this joint, the projections made in one member fits into the notch or groove of the other. See Fig 7.9 which illustrates the formation of housing joints.

MORTICE AND TENON JOINT

This type of joint is made by forming a projection called 'tenon' in one piece and introducing it into the groove called 'mortice' formed in another piece.

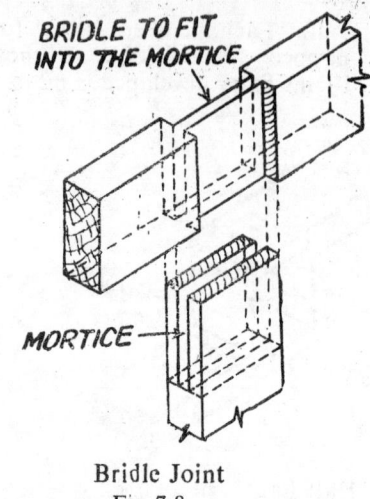

The exploded view of an open mortice and Tenon joint which is the simplest form of mortice and tenon joint is shown in Fig. 7·10. Usually, the depth of mortice and tenon is made one-third thickness of each part Fig. 7·11 is also an exploded view to illustrate how a haunched mortice and tenon joint is formed. For developing the fixity in mortice and tenon joints, wedges and dowel pins are used.

Bridle Joint
Fig. 7·8.

TUSK-TENON JOINTS

Fig. 7 12 presents a pictorial view of a tusk-tenon joint. The use of dowel pin is shown here This joint is commonly used for joining cross timber beams with main beams required to form a timber floor or ceiling.

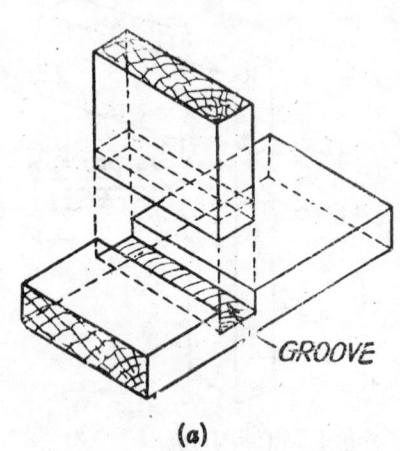

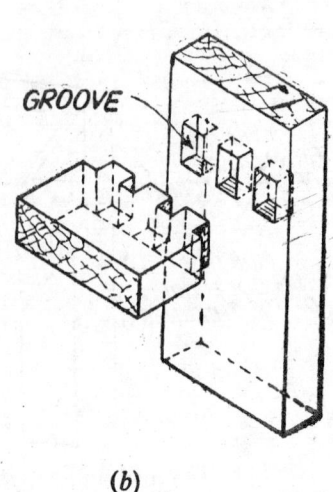

(a)   Housing Joints   (b)
Fig .7.9.

DOVETAILED JOINTS

To form a dovetailed joint wedge shaped pieces looking like dovetail are cut out of each member and fitted together by introducing the projected dovetails into the dovetailed grooves.

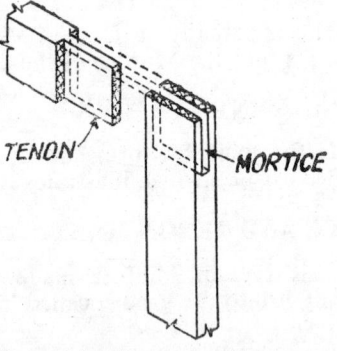

Open Mortice and Tenon Joint
Fig. 7·10.

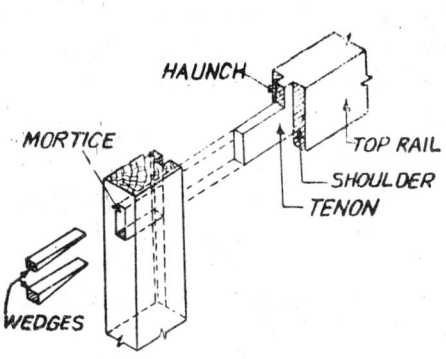

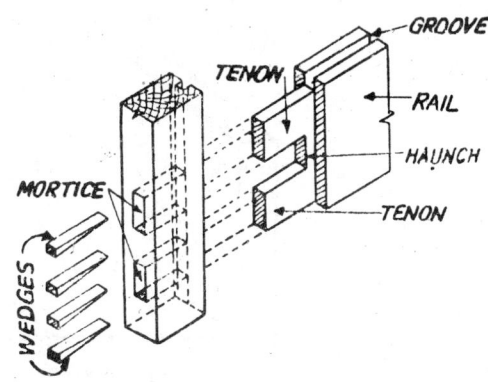

Haunched Mortice and Tenon Joint
Fig. 7·11.

Fig. 7 13 shows a common dovetail joint. A secret dovetail or double lap dovetail joint is to be made by fitting the two pieces as shown in Fig. 7 14. The formation of a mitred dovetail joint is presented in Fig. 7.15. This type of joint needs great skill. The interlocking arrangement in this type of joint is very interesting.

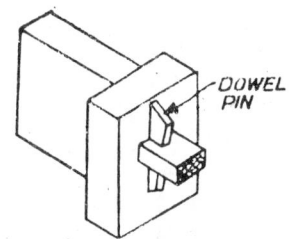

Tusk Tenon joint
Fig. 7·12.

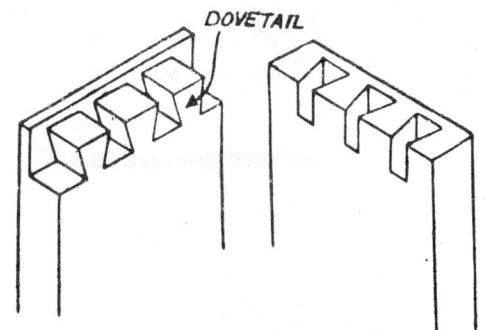

Secret Dovetail or Dovetail Bat Dovetail Secript
Fig. 7·14.

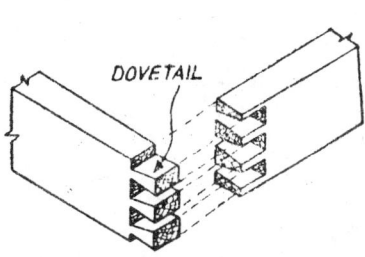

Common Dovetail Joint
Fig. 7·13.

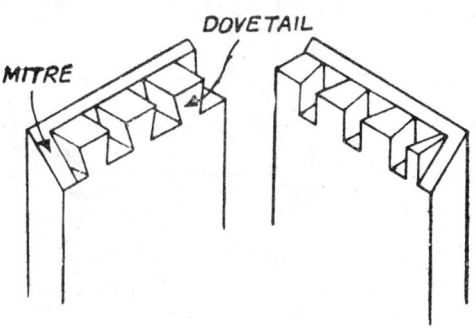

Mitred Dovetail joint
Fig. 7·15.

Board Joints at Right Angles

Various ways of joining two planks or boards meeting at right angles are illustrated in Fig. 7 16.

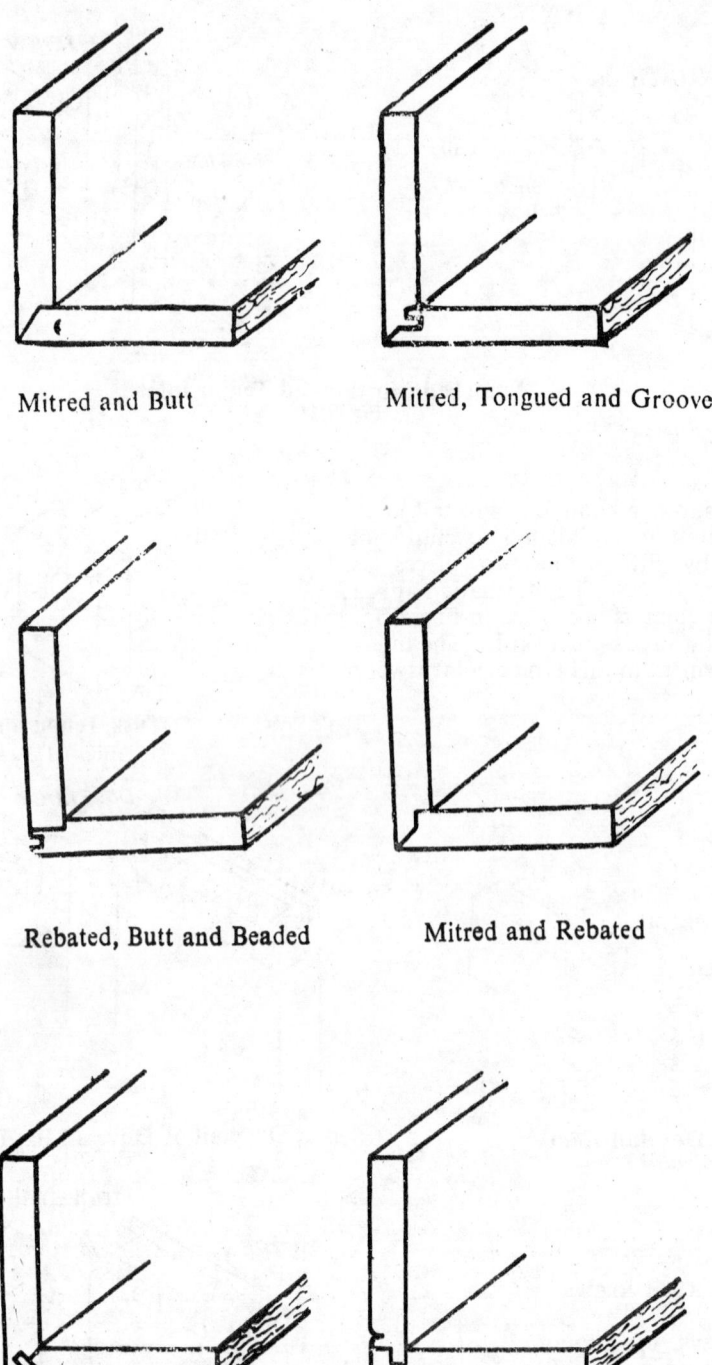

Mitred and Butt   Mitred, Tongued and Grooved

Rebated, Butt and Beaded   Mitred and Rebated

Mitred and Feathered   Grooved, Tongued and Beaded
Board Joints at right angles.
Fig. 7·16.

EXERCISES ON CHAPTER 7

1. Make neat sketches of the following :

    (a) Grooved and tongued joint for floor boards.

    (b) Scarfed or fished joint for lengthening two pieces of timber.

    (c) Rebated, butt and beaded joint for boards meeting at right angles.

    (d) Grooved, tongued and beaded joint for planks meeting at 90°.

2. Sketch assembled views of the following :

    (a) Lapped mitred joint.

    (b) Tenoned and mitred joint.

    (c) Bridle joint.

    (d) Mortice and Tenon joint.

    (e) Housing joint.

    (f) Dovetailed joint.

    (g) Tust Tenon joint.

8

Wooden Floors and Partitions

Introduction

In areas of cold climate, hilly areas and in places where building materials other than wood are not readily available, timber is chiefly employed for building construction. Building walls, floors, roofs and partitions are constructed with various types of timbers that are available in plenty in the locality. The construction of timber floors, partitions, roofs and ceilings need some special technique of which framing and joinery works are important and interesting as well.

Wooden floors are also in use especially for dance and drama stages, auditoriums, concert halls and specially built halls for other purposes.

Wooden floors consist of wooden boards or planks supported by wooden or steel joists. In general, there are three classes of wooden floors :

1. Single floors.
2. Double floors.
3. Framed floors.

In all these floors, boards rest directly upon bridging joists. The timbers of upper storey floors are to carry a ceiling also. In ground floors where there is a space below and no ceiling, intermediate walls called 'dwarf' or 'sleeper walls' or piers are constructed to support the bridging beam or joists at intervals.

Single floors

In single flooring, bridging joists span the distance from wall to wall and rest upon the wall plates. Fig. 8·1 shows a sectional part of a single floor with wooden boards in ground floor.

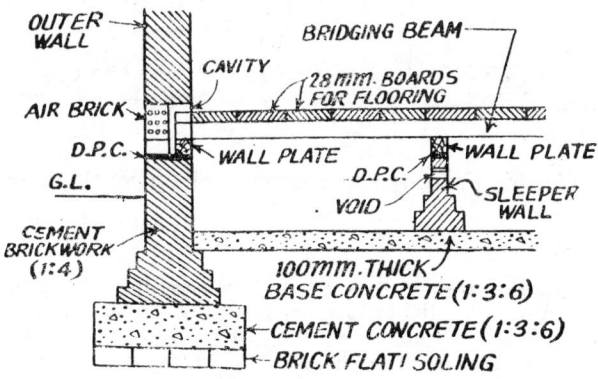

Single Floor (Ground floor) wooden Boards.
Fig. 8·1.

The minimum thickness of boards should be 25 mm. The bridging beam should rest on wall plates as shown. The base concrete under the wooden floor is provided to prevent unwholesome exhalations from being drawn up from subsoil into the room. The damp proof courses (D.P.C.) are to prevent damp from rising into the walls from subsoil. Air bricks and cavity are provided at side walls for good ventilation. Sleeper walls or dwarf walls should preferably be placed at a centre to centre distance of about 2 m. The bridging joists are placed across the sleeper walls usually at a centre to cantre distance of 30 to 45 cm and secured to the wall plates.

Single floors are always cheapest and simplest. They distribute load very equally over the walls upon which they rest and hold the sides of the building together.

Fig. 8·2 shows the plan of a single floor above ground floor level. In single flooring no ceiling joist is used. The laths are nailed to the underside of the bridging beams which are all of same depth. The beams bear equally on all parts of the walls.

Bridging beams are usually spaced at 30 to 40 cm c/c distance, In order to avoid failure of a beam by sagging and cracking, the span of a bridging beam for single floors should not usually exceed 2·5 metres. The depth of a beam should be 2 to 4 times its width. When the span exceeds 2·5 m., the beams are strengthened to avoid sidewise buckling. Strenngthening is done either by herring bone strutting or by solid strutting at the middle of the span. Herring bone strutting is stronger than solid strutting. In herring bone strutting, timber struts are placed diagonally between the bridging beams in such a manner that they cross each other and are secured to the beam ends by nails. For additional strength, wedges should be inserted between

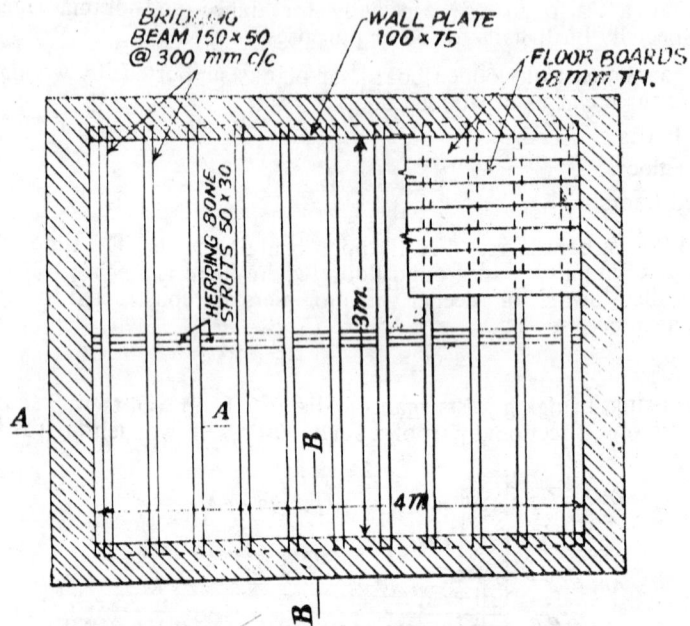

Wooden Flooring above Ground Floor (Single Floor)
Fig. 8·2.

the wall and the end joist as shown in Fig. 8·3. Solid strutting is shown in Fig. 8·5 and 8·6. Solid struts are placed in between the beams and these are strengthened further by means of tie rods (usually 12 mm to 20 mm diameter) secured at the ends. For making single floors sound proof, sometime sound absorbing materials are filled in spaces between the ceiling and the floor boards.

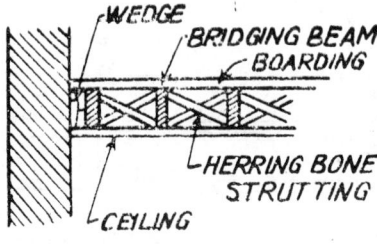

Section AA
Fig. 8·3.

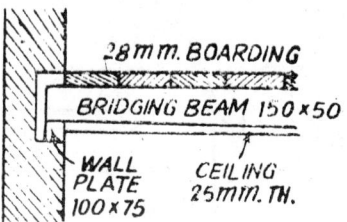

Section BB
Fig. 8·4.

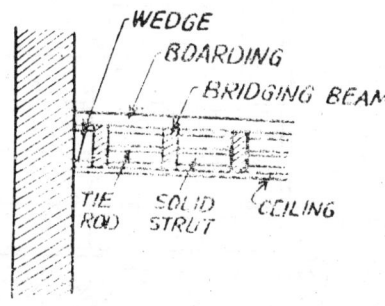

Solid Strutting in place of
Herring Bone Strutting.
Fig. 8·6.

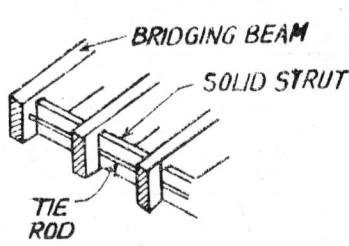

Isometric view of solid
Strutting.
Fig. 8·6.

Double floors

In double floors, bridging joists instead of spanning the distance from wall to wall, are supported by intermediate beams called 'binders' or 'binding' joists. The space between two binders is called 'Case Bay' and that between the binder and wall "Tail Bay'. The binders should be at 10 to 15 cm apart and should bear either upon wall plates running the whole lengths of the wall or upon templates. The massive binders are of great assistance to the walls of the building in tying them together. But, the binding joists bring the total weight of the floor (self load+live load) to bear upon a few points. Moreover, double flooring cause the floor to be very deep which adds to the height of the walls and cost of building. Double floors are heavy, complicated in construction and not economical. On the other hand, double floors are strong and rigid and have good sound insulation property. Plan of a double floor with its components are illustrated in Fig. 8·7. The detailed views by taking sections on XX and on YY are shown in Fig. 8·8 and 8·9 respectively.

Framed floors

In this type of floors bridging joists rest on binders, but the binders in their turn, are supported by girders. If the girders are made of balk of timber, the binders are framed into them by double tusk tenons. The distance between the girders should not exceed 3 m. The girders rest upon templates. Two typical sections of framed floors are shown in Fig. 8.10.

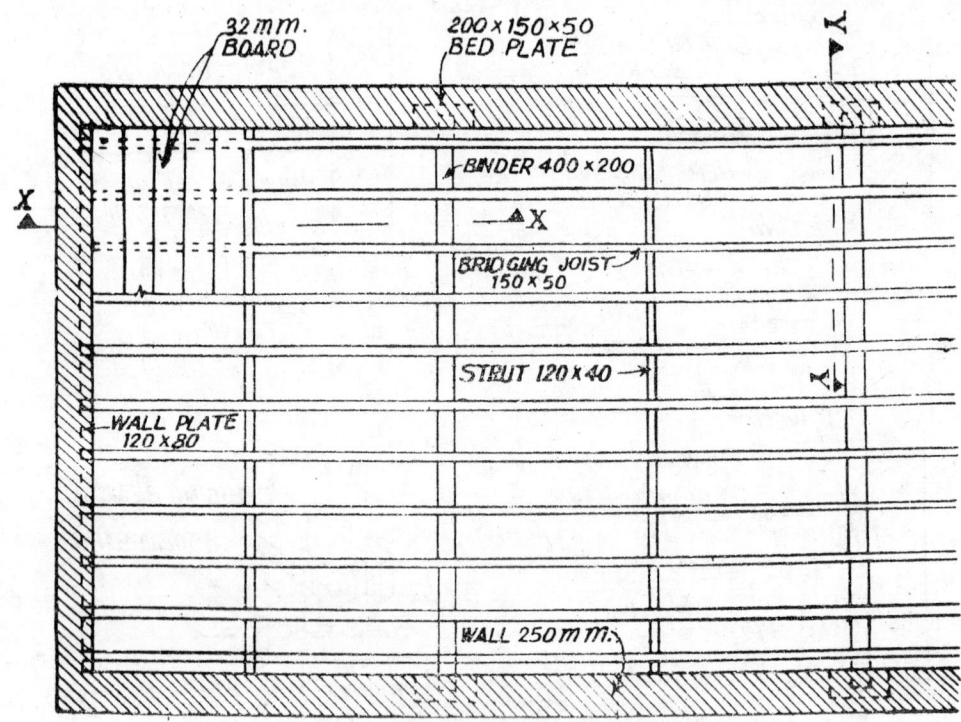

PLAN SHOWING DOUBLE FLOOR

Fig] 8·7.

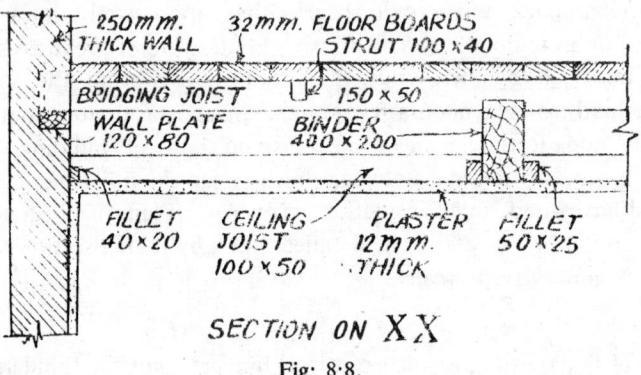

SECTION ON XX

Fig· 8·8.

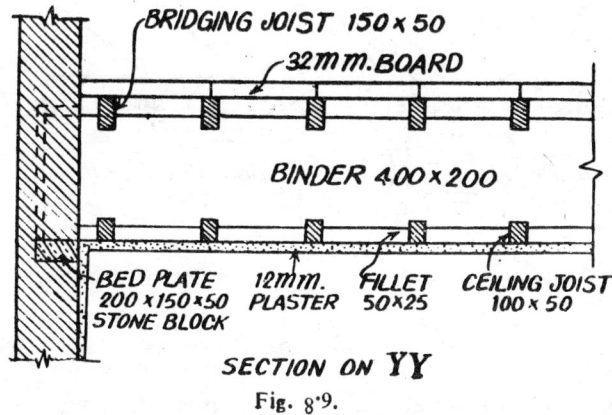

SECTION ON YY

Fig. 8·9.

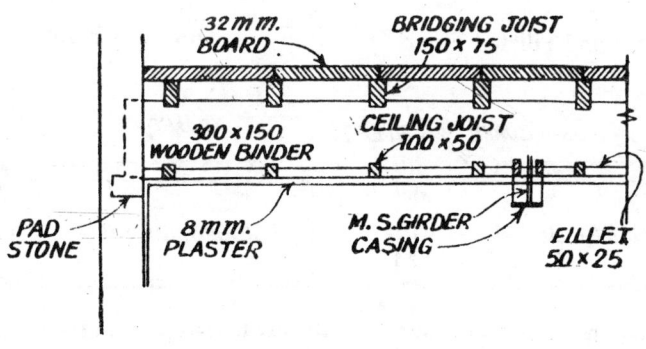

(a)

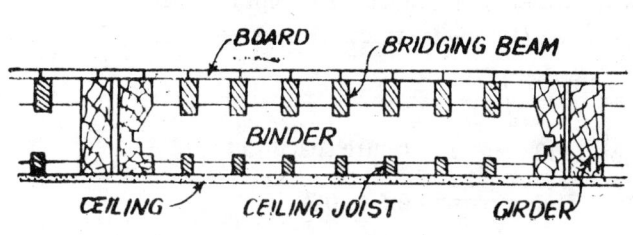

(b)

Sections of Framed Floors

Fig. 8·10.

Laying and fixing of boards to the bridging beams with various types of joints are shown in Fig. 8·11.

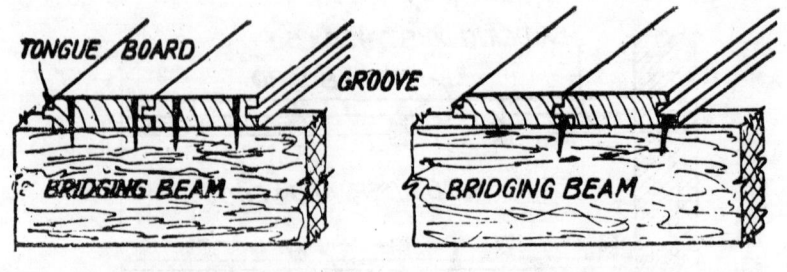

Plain Jointed Floor   Rebated Floor

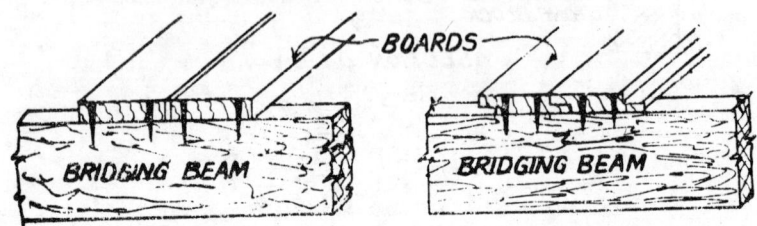

Rebated and Filleted Floor   Dowelled Floor

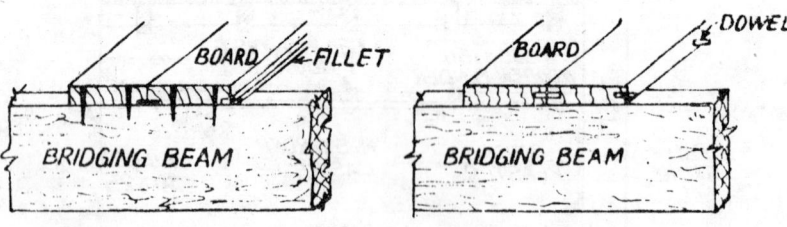

Grooved and Tongued Floor   Rebated, Grooved and Tongued Floor

Fig. 8·11.

Rules for scantling of Floor Timbers

L=length in ft ; B=breadth in inch ; D=depth in inch.

For bridging joists (12″ centre to centre spacing) :

$$D=\sqrt[3]{\frac{L^2}{B}}\times 2 \text{ to } 2\cdot 5$$

For ceiling joists (12″ centre to centre spacing) ;

$$D=\frac{L}{\sqrt[3]{B}}\times 0\cdot 6 \text{ to } 0\cdot 7$$

For ceiling joists 2″ wide, half the length of bearing in feet will give the depth in inches ; e.g. for 12 ft. span, depth is 6″.

For binders (6′ 0″ c/c distance) :

$$D=\sqrt{\frac{L^2}{B}}\times 3\cdot 4 \text{ to } 3\cdot 5 \qquad \text{and} \quad B=\frac{L^2}{D^3}\times 40 \text{ to } 45$$

For Girders : $D=\sqrt[3]{\frac{L^2}{B}}\times 4 \text{ to } 4\cdot 4$; $B=\frac{L^2}{D^3}\times 70 \text{ to } 80$

Note : Convert ft. dimension into m. dimension by multiplying ft. dimension by 0·3 and convert inch dimension ino mm. dimension by multiplying inch dimension by 25.

Wooden Partitions

Wooden Partitions in buildings are required to divide rooms from one another, instead of making walls, to save space and expense and to make the construction light and removable as and when required.

Quartered partitions consist of frames filled in with light scantlings or quarterings of inferior wood upon the sides of which laths are nailed. A quartered partition sometimes rests on the Cross and party wall of the ground floor. The weight of the partition wall should not be allowed to come directly on the floor. The above arrangements are not good, A better arrangement is to suspend the partition wall from the floor or the roof above. By far the best plan, is to make the partition wall self-supporting, depending on the main walls carrying its ends and forming, in fact, a deep truss. Thus, wooden partitions may be classified into two categories as follows :

    1. Common partition   2. Trussed partition.

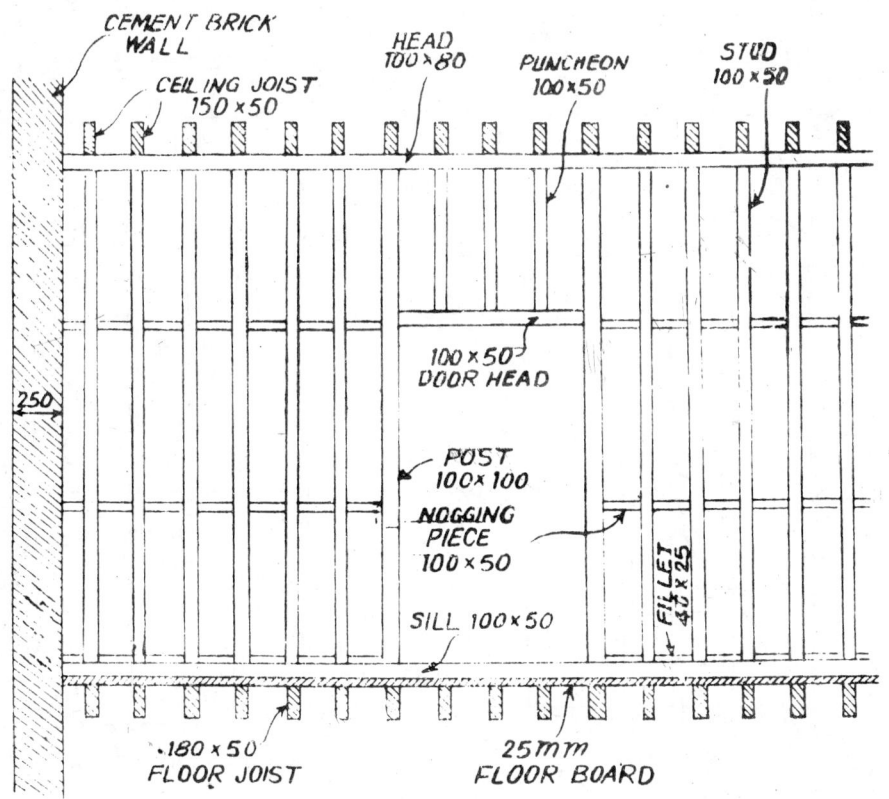

Common Wooden Partition

Fig. 8·12.

Common Partitions

Common partitions rest upon a dwarf wall at ground floor level and in upper storeys on walls, floors and rolled steel joists or beams immediately below them. Fig. 8·12 shows the framework of a common partition. It consists of a framework of head, sill and upright members called studs or quarters. The studs are usually 100 mm × 50 mm spaced at 30 cm to 50 cm apart. The short studs above the door head are called puncheons. The studs are fixed to the two horizontals, the head and the sill. The studs are either housed or stump-tenoned into the head and sill as shown in Fig. 8·12. The nogging should be fixed between studs as shown and these should not come up flush with the face of the stud on either side. The usual length of stud is 1 to 1'2 m. Although the spacing of the studs is mentioned above, the centre to centre distances of the studs should be such that the ends of the laths (covering sheath) should fall upon every third or fourth stud. The principal stud, door studs (posts) and quarterings should be removed through the head and sill. When the partition rests on the floor below, the sill would project inconveniently above the floor in the doorway. This portion of the sill should therefore be cut after the partition is fixed. When extra strong and sound-proof partitions are required, the studs should not be filled in between the framing, but nailed on the outside as battens.

Trussed Partition

A trussed partition supporting two floors with the framework is illustrated in Fig. 8'13. Trussed partitions are constructed where partitions are restricted to impart load directly to the

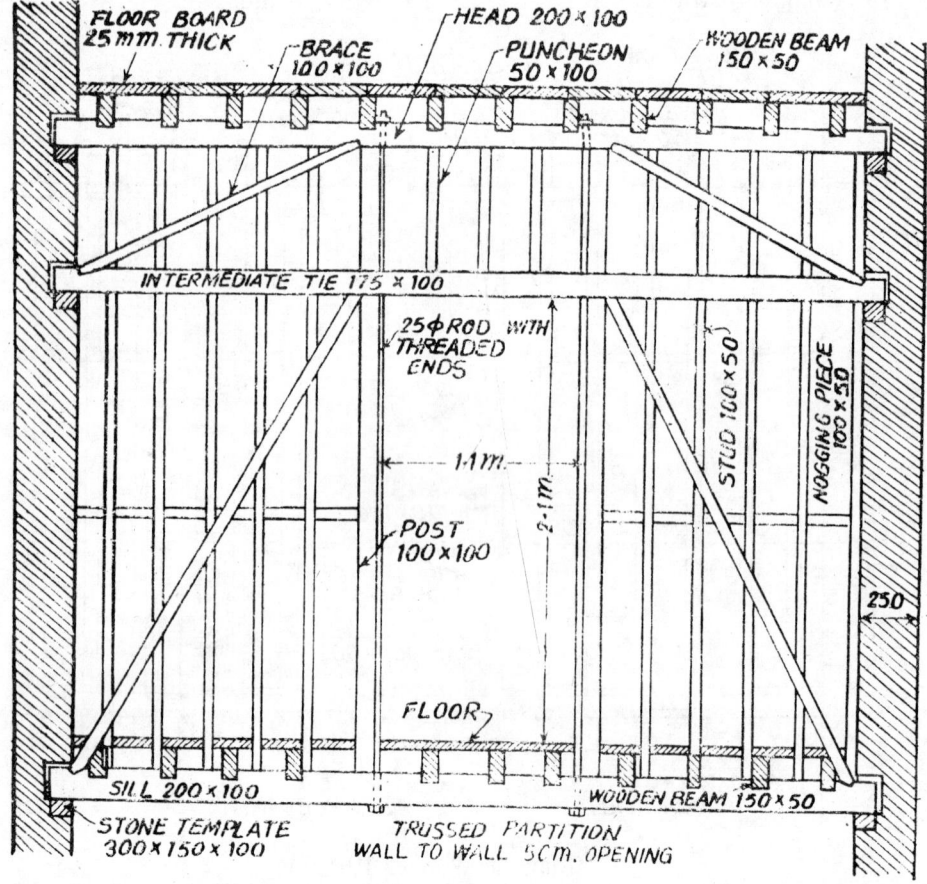

Fig. 8·13.

floor. These partitions are supported at their ends and these are capable of carrying some floor loads in addition to their self-weights.

A trussed partition is usually composed of a head, a sill, door posts, studs, braces (inclined members), nogging pieces, etc. An intermediate horizontal member, called intertie is generally provided when the partition height exceeds 3 m. The frame is further strengthened by long bolts on each side of the door post as shown in Fig. 8'13. Framing may also be done by following the principle of a king post or queen post truss. Iron straps and bolts may be used as practised in making roof trusses.

9

Doors and Windows
(*Timber and Steel*)

Doors

As a rule, doors should open inwards from a person entering a room. The doors should be so placed as to conceal as much as possible of the room, when they are partly open. 'Vitruvius' gives as a rule for internal doors that their height, to give the best architectural effect should be $\frac{4}{7}$ th that of the room. The width of door opening starts from 600 mm. When a door is more than 1.0 m wide, it should as a rule be hung in two halves. By this arrangement shutters become light and less space is required in opening the shutters. Average size of doors in residential buildings is 1·1 m × 2·1 m and height of doors is given by thumb rule as : Height=width+1 metre. Doors are usually placed near the corner of a room or close to one end or side of a room, at a distance of about 30 to 60 cm from the corner.

Types of Doors

Doors receive their distinctive names according to the nature of their construction. Various types of doors that are most commonly used are :
1. Ledged door
2. Ledged and braced door
3. Ledged and framed door
4. Ledged, framed and braced door
5. Flush door
6. Panelled door
7. Panelled and glazed door
8. Panelled and venetian door
9. Wire gauged door.

Ledged door

A Ledged door is the simplest kind of door used for temporary constructions and low-cost housing in developing and under-developed countries. It consists of vertical boards butted against one another and connected by two or three horizontal pieces called 'ledges' nailed across the back. In superior quality of ledged doors, the boards are grooved or ploughed and tongued together and sometimes united by rebated joints. The boards are usually 12 to 15 mm thick and 100 to 150 mm wide. The ledges or battens are usually 15 mm thick and 150 mm wide. A ledged shutter is shown in Fig. 9·1.

A ledged door (frame with shutter) with required fittings and fixtures is shown in Fig. 9·3. The ledges shown here are cut off at the ends so as not to fit into the frame. In some cases the ledges are made of length equal to the width of the door and recesses are cut out in the frame beyond the rebate to receive them where they occur. The rebate in the frame is to be cut of a depth equal to the thickness of the boards *i.e.*, the thickness of the door shutter.

Ledged and Braced door

A ledged and braced door is similar to a ledged door with braces diagonally across the back of the shutter. The ledges and braces are bevelled or beaded and the boards are ploughed, tongued and beaded on both sides. The braces are usually 30 mm thick and 100 to 50 mm wide. The braces give additional strength to the door shutter. They are housed into the ledges. The braces should be fixed in such a way that they incline downwards toward the side on which the door is to be hung. Fig. 9·2 shows a ledged and braced shutter. A complete ledged and braced door (frame and shutter) with various fittings is shown in Fig. 9·4.

Ledged and Framed door

In this type of door, a framework is provided for shutters with stiles and lock rail. The stiles are usually 100×40. Ledges are secured to the battens as usual.

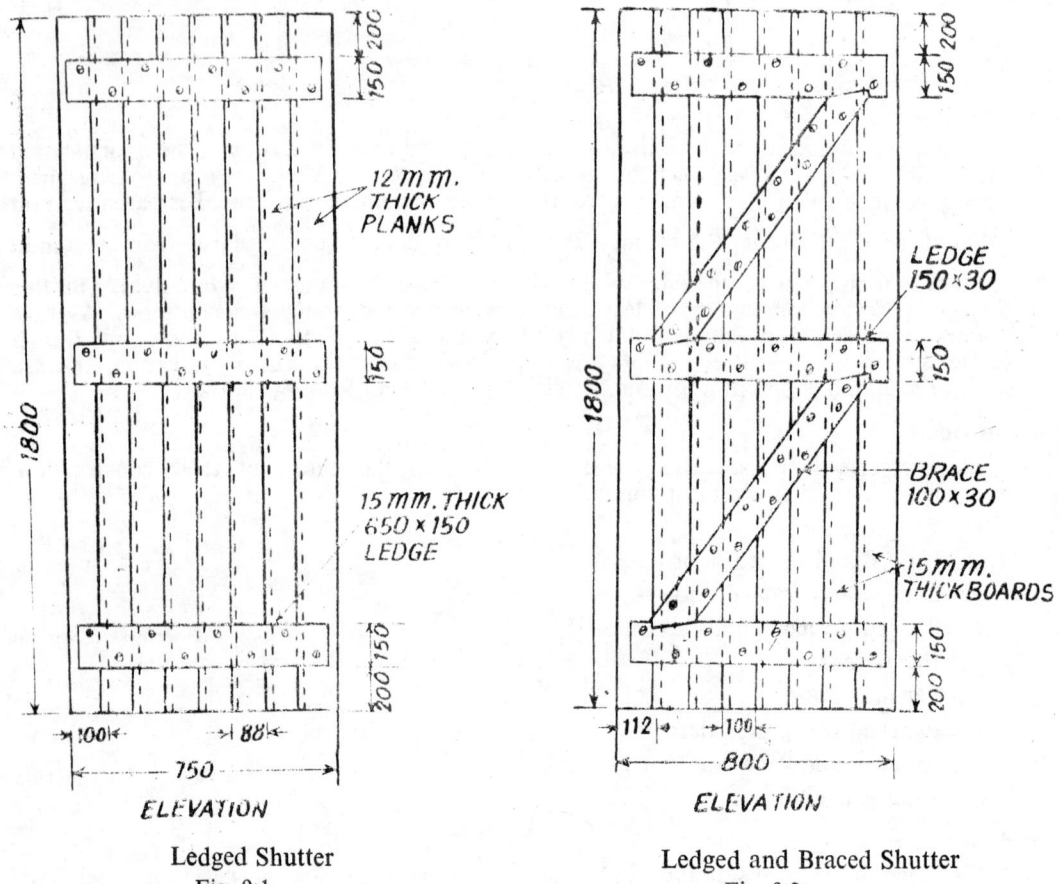

Ledged Shutter
Fig. 9·1.

Ledged and Braced Shutter
Fig. 9·2.

Braced and Framed door

The door consists of a frame strengthened by a lock rail and diagonal braces. The frame consists of stiles, a top rail and a bottom rail. The end of the braces are tenoned into the stiles and rails. The braces should incline upward from the hanging side. They should be connected at the upper end with the rails only and the lower ends may abut partially upon the hanging stile. The braces and the lock rail are thinner than the other components of the frame.

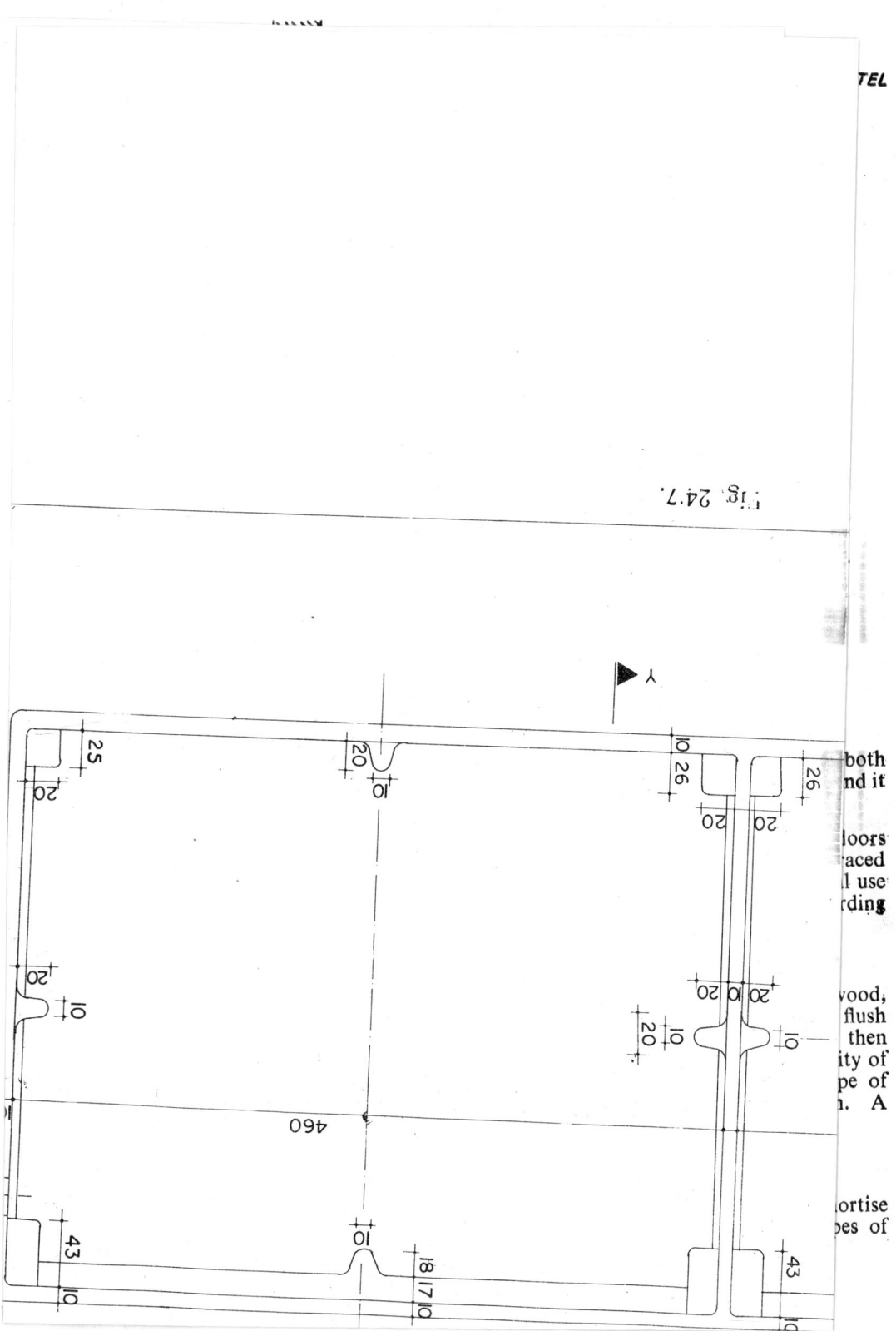

Fig. 24.7.

both
nd it

loors
aced
l use
rding

vood,
flush
then
ity of
pe of
n.   A

ortise
es of

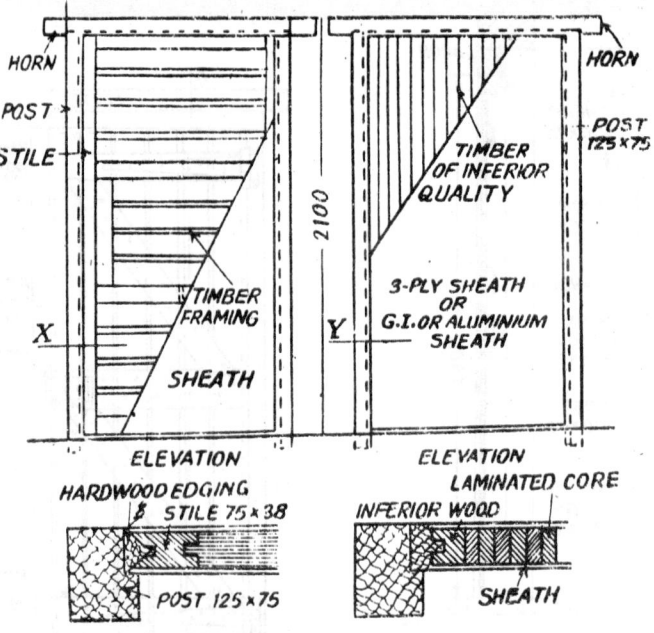

Details at X
Solid or Laminated flush door
Fig. 9·5.

Details at Y
Framed Flush door
Fig. 9·6.

In a frame, the vertical pieces are called 'stiles' and the horizontal pieces 'rails'. The inner edges of the stiles and rails are grooved to a depth of about 12 mm to receive the panels, which should fit so tightly as not to rattle.

There are various forms of panel. *Square and flat panels* are boards of same thickness throughout, thinner than the frame and sunk square below its surface. In *moulded and flat* and *moulded and square panels*, the edge of the panel, close to the framing is ornamented by a moulding either 'planted' or 'stuck' on to the inner edge of the frame. *Flush panels* have their surface 'flush' with the frame surface. The flushing may be on one side or on both sides. In almost all forms of flush panelling, the edges of either the frame or the panel are ornamented by a bead, groove, moulding or chamfer. *Solid pannels* are formed in one piece of same thickness as the frame and they flush on both sides with the surface of the frame. *Bead flush panels* have a bead all round close to the inner edge of the framing. All such panelling are shown in Fig. 9·8.

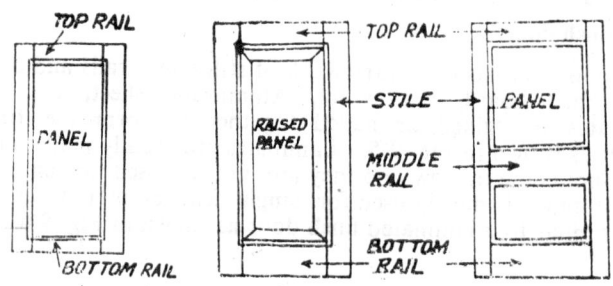

Single Panelled   Double Panelled
Various types of panelled door
Fig. 9·7.

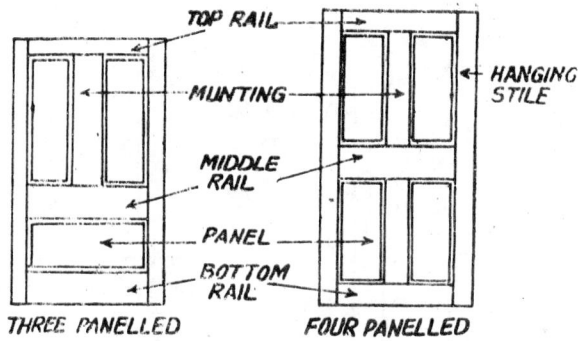

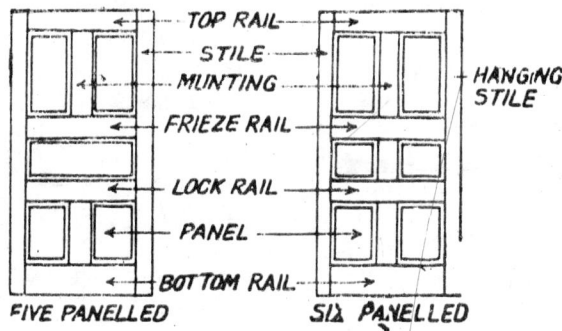

Various types of panelled door.
Fig. 9·7. (contd.)

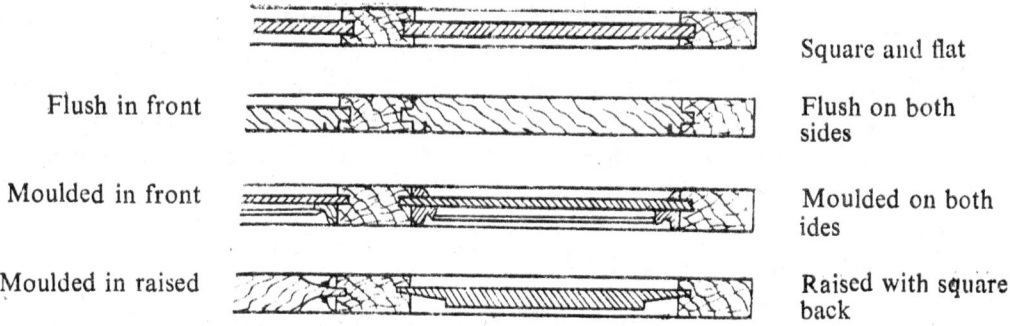

Horizontal sections showing different forms of Panelling.
Fig. 9·8.

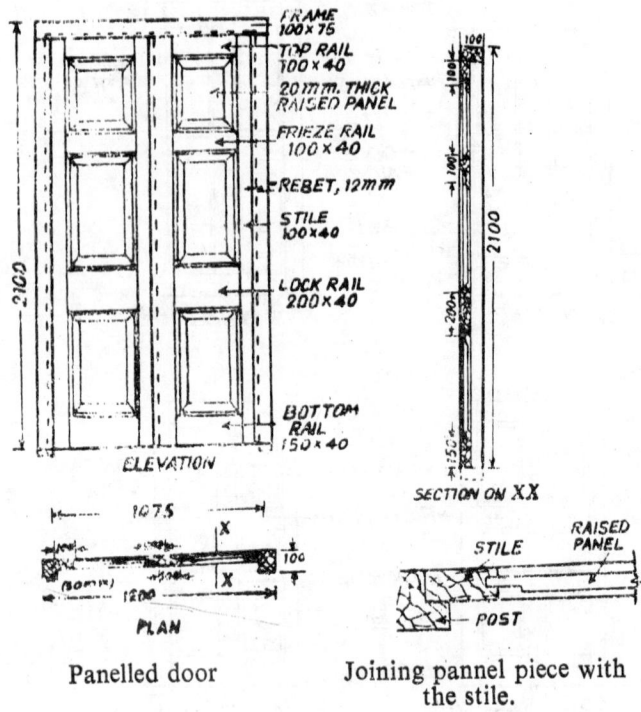

Panelled door   Joining pannel piece with
                                          the stile.

Fig. 9·9.

Plan Elevation and a Sectional view of a panelled door are shown in Fig. 9·9.

Glazed and Panelled Door

Glazed doors are chiefly emploped in hospitals, libraries, show rooms, office buildings, banks, shopping centres, etc. for visibility and entry of light. Sometimes, partly panelled and partly glazed doors are used. The ratio of height of panelled portion to glazed portion is generally kept 1 : 1 or 2 : 1. In such doors, glazing is done in the upper portion of the doors. The glass pane is received in the rebates provided in the sush bar and secured either by nails and pully or by wooden beads fixed to the frame. Fig. 9·10 shows ⅓rd glazed and ⅔rd panelled door.

Panelled and Venetion Door

Fig. 9·11 shows a door with a fanlight at top. The door is ⅓rd panelled and ⅔rd venetian (movable louvres). Louvres are provided at such an inclination that horizontal vision is obstructed, but air and light can pass through the louvres by adjusting the louvres with the help of the monkey bar. The louvres are hinged to the monkey bar. The fanlight provided at the top of the door ensures entry of light and it helps in ventilation of the room.

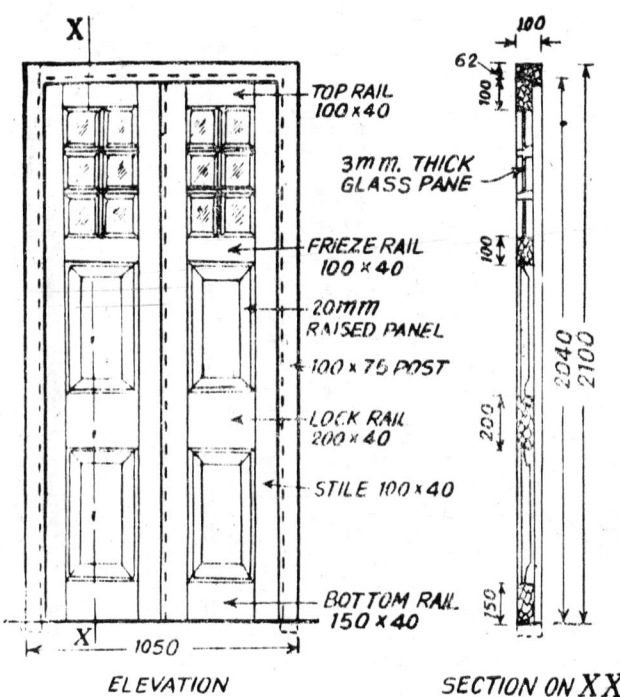

TOP RAIL 100 × 40

3 m.m. THICK GLASS PANE

FRIEZE RAIL 100 × 40

20 mm RAISED PANEL

100 × 75 POST

LOCK RAIL 200 × 40

STILE 100 × 40

BOTTOM RAIL 150 × 40

1050

ELEVATION

SECTION ON XX

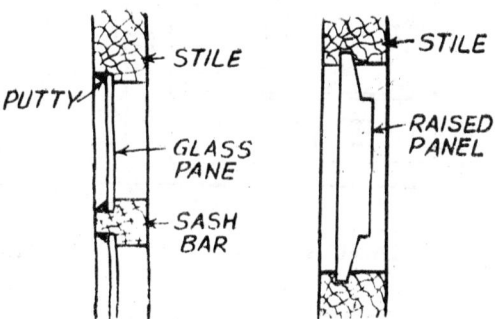

STILE

PUTTY

GLASS PANE

SASH BAR

STILE

RAISED PANEL

Details showing Glazed Panelling and Wooden Panelling.
⅓rd Glazed and ⅔rd Panelled door.

Fig. 9·10.

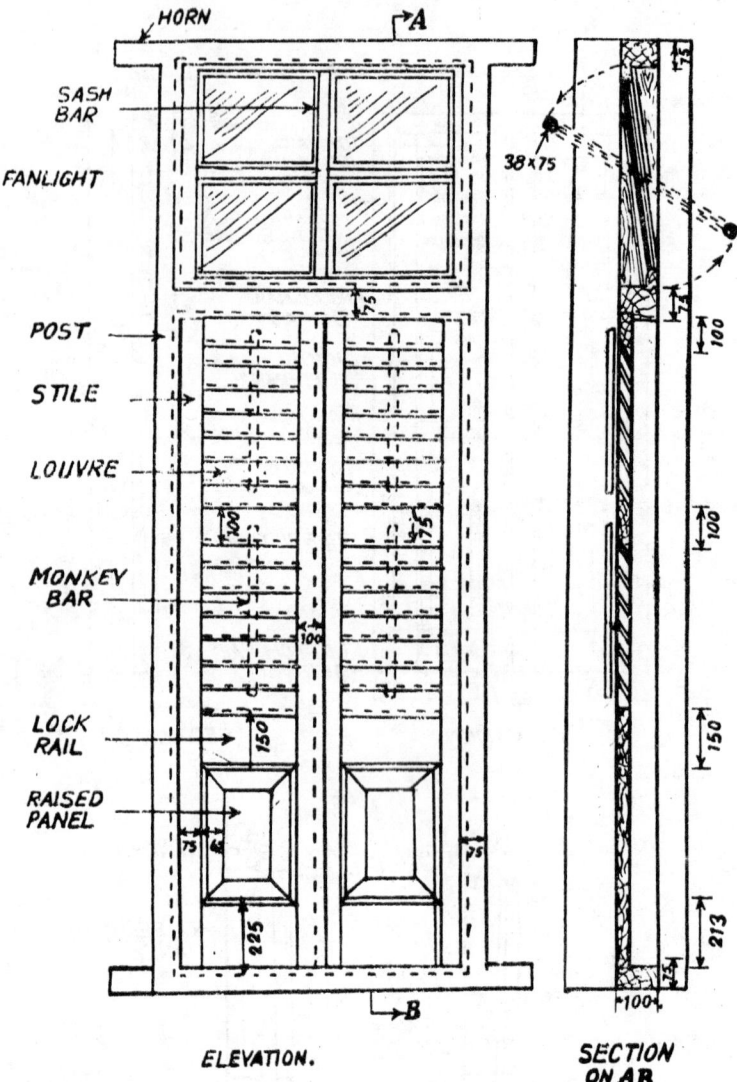

ELEVATION.   SECTION
                                        ON AB

One-third panel, two-third venetian door (1050 × 2100)
with a fanlight.
Fig. 9·11.

Wire Gauge Door

Sometimes, wire gauge is used instead of glass or wooden panel as shown in Fig. 9·12. This type of door facilitates in free circulation of air with no entry of flies and mosquitos in addition to the entry of light.

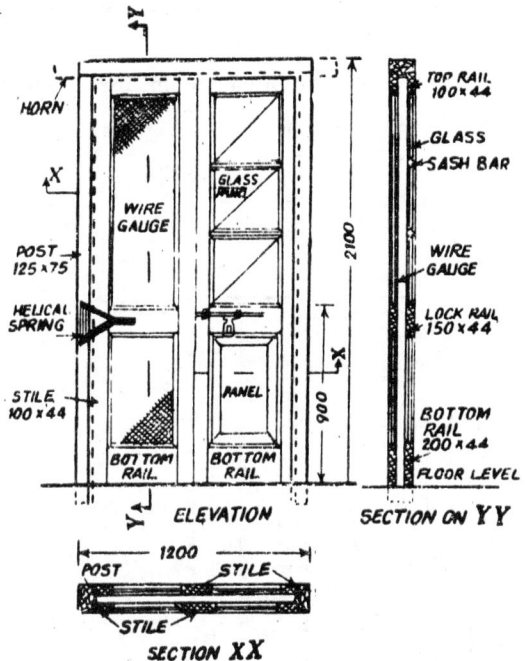

ELEVATION   SECTION ON YY

SECTION XX

Helical springs are provided
on the upright post of frame
to make it self closing.

Panelled and Glazed door

wire gauge shutter is provided on the opposite side of the same door
Wire gauge is fixed in between the stiles and rails.

Fig. 9·12.

Windows

Construction of windows are similar to the construction of doors. The sill (bottom) of windows is usually kept at bed level and the height of windows is such that the tops of windows and doors are at the same level. The sizes of windows are regulated both by their external appearance and by the arrangements required for light and ventilation in rooms. There are several rules in determining the area and size of windows for rooms, of which the useful ones, to regulate the size of windows as regards internal arrangement, are mentioned below.

According to Morris ;

$$\text{Area of light} = \sqrt{\text{cubic content of room.}}$$

Chambers mentioned :

$$\text{Breadth of window} = \frac{1}{8} \text{ (width of room + height of room).}$$

$$\text{Height of window} = 2 \text{ to } 2\cdot5 \text{ times the breadth of window.}$$

Galton opined :

There should be 1 foot superficial of window space to every 100 or 125 cubic feet contents of the room in dwelling houses and 1 foot superficial to 50 or 55 cubic feet in hospitals.

The window sill should be about 2'−6" above the floor level.

There are various types of windows commonly in use. These are : Panelled, Ledged, Ledged and Braced, Glazed, wire gauged, etc.

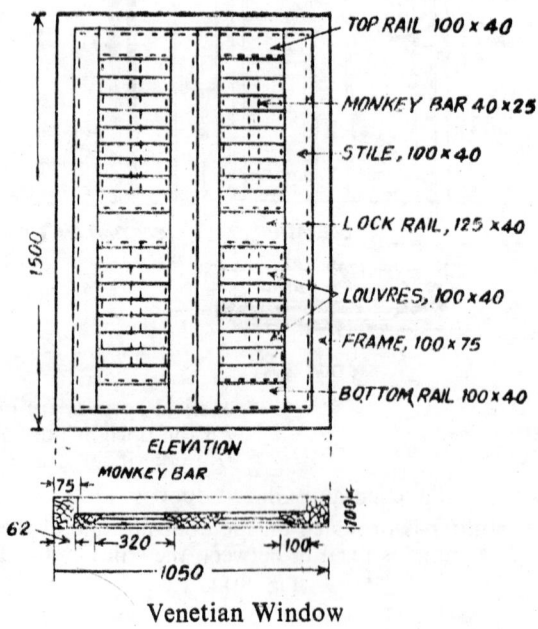

Venetian Window

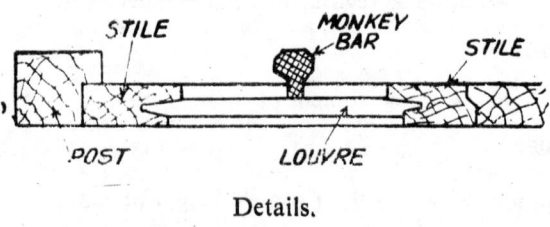

Details.

Fig. 9·13.

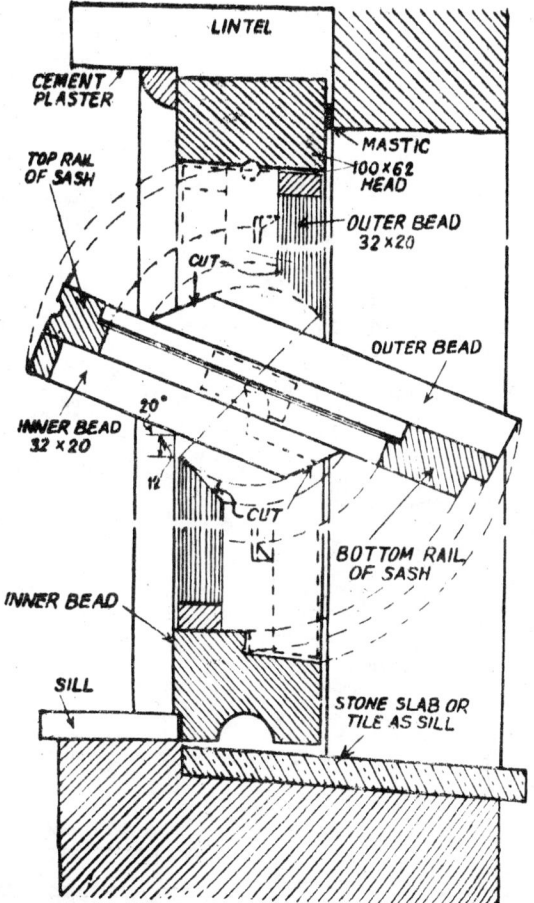

Sectional view of a clerestory window.
Fig. 9·14.

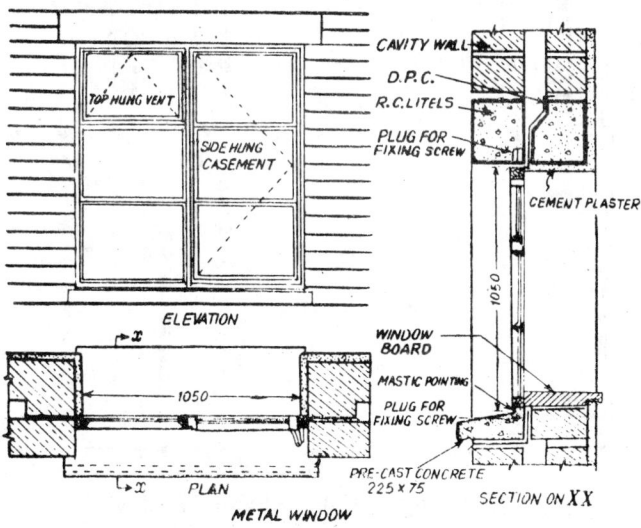

Timber Casement Window 1500 × 1350
Fig. 9·15.

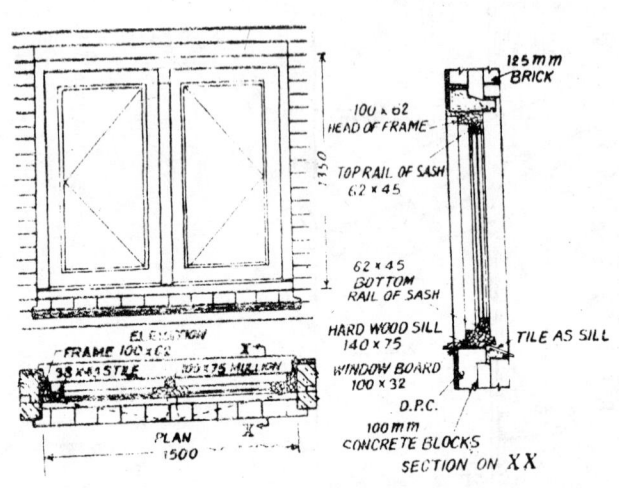

125 m m
BRICK

100 x 62
HEAD OF FRAME

TOP RAIL OF SASH
62 x 45

62 x 45
BOTTOM
RAIL OF SASH

HARD WOOD SILL
140 x 75

TILE AS SILL

WINDOW BOARD
100 x 32

D.P.C.

100 m m
CONCRETE BLOCKS

SECTION ON XX

ELEVATION

FRAME 100 x 62

33 x 45 STILE   100 x 75 MULLION

PLAN
1500

130.0

Metal Window.
Fig. 9.16.

HARDWOOD
FRAME MEMBER

HEAD

BRACKET FIXING
FOR FRAME

PRESSED ALUMINIUM
SUB FRAME AND
TRANSOME CASING

UPPER
TRANSOM

125

MAIN FRAME OF
R. S. SECTIONS

LOWER
TRANSOM

250

SILL

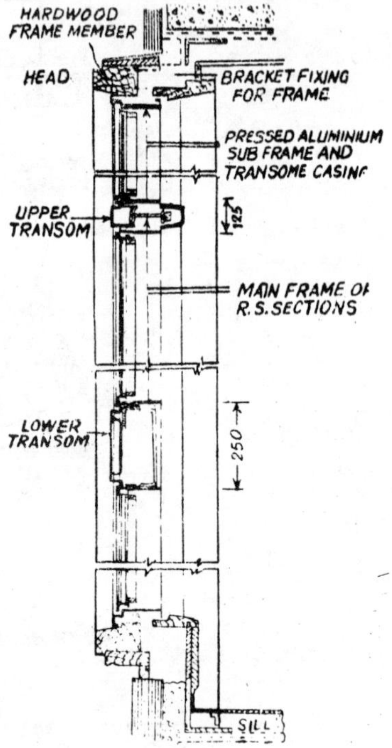

Large Metal Window.
Fig. 9.17.

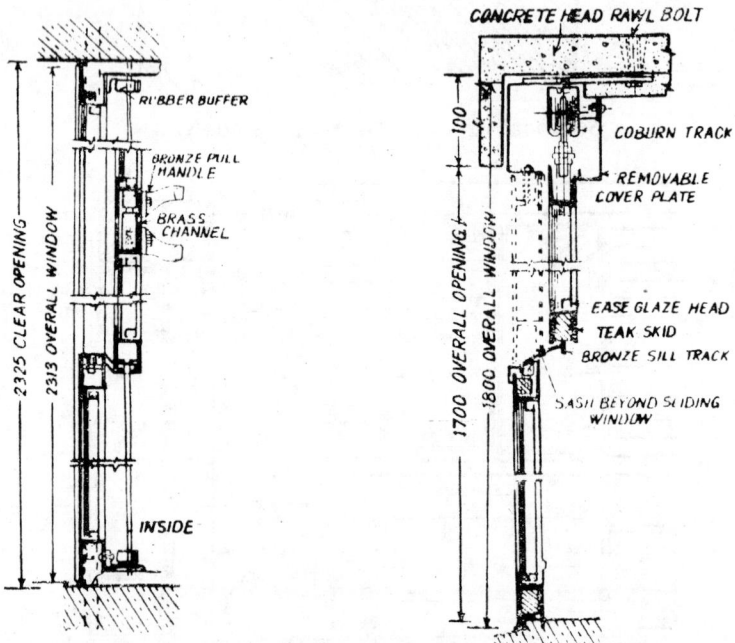

RUBBER BUFFER

BRONZE PULL
HANDLE

BRASS
CHANNEL

2325 CLEAR OPENING

2313 OVERALL WINDOW

INSIDE

CONCRETE HEAD RAWL BOLT

COBURN TRACK

REMOVABLE
COVER PLATE

100

1700 OVERALL OPENING

1800 OVERALL WINDOW

EASE GLAZE HEAD

TEAK SKID

BRONZE SILL TRACK

SASH BEYOND SLIDING
WINDOW

Metal Signal Cabin Sliding Window.

Fig. 9.18.   Fig. 9.19.

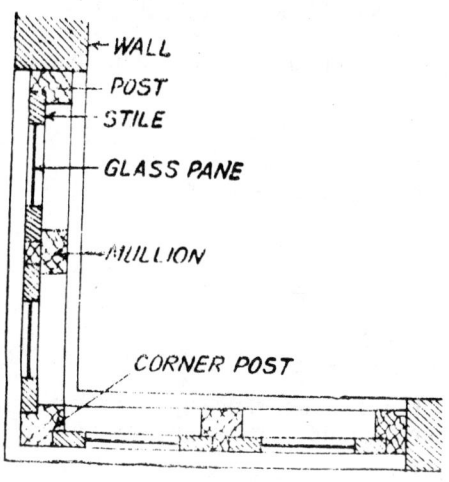

Corner Window
Fig. 9·20.

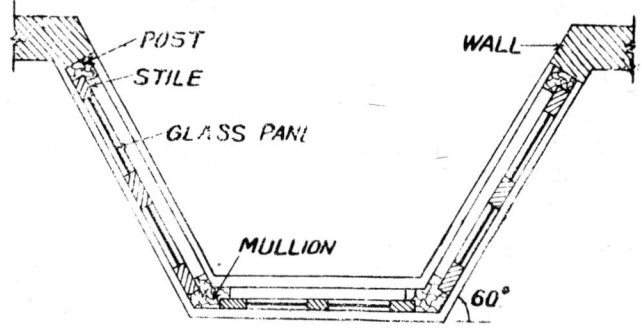

Bay Window
Fig. 9·21.

EXERCISE ON CHAPTER 9

1. Design and draw a half-panelled and half glazed door with the following dimensions :

   door size : 1200×2100

   door frame : 75×100

   top rail and bottom rail : 150×40 each.

   Lock rail : 200×40

   Stiles : 100×40

   Panelling : 25 mm raised panel

   Glass panes : 5 mm thick

   Show plan, elevation and a vertical section.

2. Draw a ledged and braced door showing all fittings and fixtures with the following dimensions : Assume suitable dimensions which are missing.

   door frame : 75×75

   ledge : 150×30 at top and 175×30 at middle and bottom

   brace : 100×30

   boards : 100×20.

3. Design and draw a six-panelled, double leaf door for a residential building. Show at least three views.

4. Design and draw a panelled door with a fanlight for a hospital. Show at least two views with necessary dimensions.

5. Design and draw a venetian window with the following dimensions :

   Window size 1050×1500

   Frame   75×100

   Top rail : 105×90

   Stiles : 100×40

   Bottom rail : 100×40

   Middle rail : 100×40

   Louvre   100×40

   Monkey bar : 40×25

   Show at least two views of the window.

6. Design and draw a fully-glazed window showing necessary dimensions for a room in a residential house.

7. Sketch the following :

   Bay window, corner window, metal window and laminated flush door.

8. Design and draw a half-wire gauged and half-glazed window 1050×1500 for a nursing home.

10

Graphical Solution of Trusses

Introduction

A roof truss, a bridge girder or a framed structure is made up of several members joined together by welding, by rivets or by hinges. If a frame consists of the least number of members without having any distortion when loaded, it is called a 'perfect frame'. In a frame, each joint is called a 'node'. In a perfect frame if n be the number of nodes, then the number of members in the frame will be $2n-3$. If the number of members be less than $(2n-3)$ then the frame is said to be 'deficient'. If this number be more than $(2n-3)$, the frame is called 'redundant'. No force comes to play in a redundant member and consequently no stress is developed in it. In a frame, the member under tension is called a 'tie' and the member under compression is called a 'strut'. Analysis of trusses is made with a view to find out the reactions at supports and forces in the members of a truss or frame. This may be done by two methods :

(a) Analytical method and
(b) Grapical method.

The analytical method of analysis itself has three different approaches.

(i) Method of joints or method of resolution ;
(ii) Method of section or method of moments ;
(iii) Method of substitution ; either by substituting the form of bracing or by substituting an imaginary member.

Representation of a Force or Stress

Usually, a force or stress is represented by a straight line with an arrowhead. The arrowhead speaks of the sense of the force or stress. Fig. 10·1 shows representation of nature

TENSION   COMPRESSION

Fig. 10·1

of two forces, tension and compression. The member subjected to tension as shown in the left figure is called a tie or a tensile member and the member subjected to compression as indicated in the right figure is called a strut or a compressive member. In graphical analysis i.e., in graphic statics, the length of the line represents the magnitude of force or stress.

Method of Joints (Analytical method)

To explain this method, let us take a simple truss as shown in Fig. 10·2. First of all, let us name the joints 1, 2 and 3. Next, let us name the spaces between pairs of adjacent forces or between pairs of adjacent members or between a force and a member, like A, B, C and D. By symmetry, $R_1=R_2=2t$. R_1 is denoted by the spaces A and B. Similarly, R_2 is denoted by the spaces C and A. Thus, the horizontal member is (A, D) and the inclined members are (B, D) and (D, C).

Let us now consider joint 1 in equilibrium under the reaction R_1 and the internal forces in the members BD and DA. So, ΣV and ΣH must be equal to zero for equilibrium.

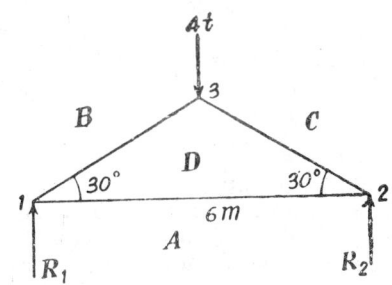

Fig. 10·2

Refer to Fig. 10.2, the senses of the internal forces *i.e.*, the directions of the internal forces in members BD and DA are given arbitrarily. The vertical component of the force in BD = P $\sin 30° = \frac{1}{2}$P where P is the force in BD.

$$\therefore \quad \Sigma V = 0, \qquad \frac{P}{2} + 2 = 0 \qquad \text{or,} \qquad P = -4t$$

The −ve sign indicates that the direction of force is incorrect and it should be downwards.

$$\therefore \quad P = 4t$$

By correcting the sense of internal force in BD, the horizontal component of the force P is P cos 30° which must be equal to the force in the horizontal member DA, since $\Sigma H = 0$.

Thus, force in DA = P cos 30°

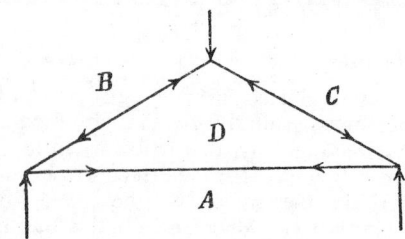

$$= 4 \times \frac{\sqrt{3}}{2} = 3.464t$$

By symmetry, force in DC = force in BD = 4t

The members BD and DC are therefore struts as they are in compression. The member DA is a tie *i.e.*, under tension.

Fig. 10·3

In applying method of joints for analysis of trusses it should be remembered that a joint having more than two unknown forces should not be considered first. A joint having three or four unknown forces may be reduced to two unknown forces by taking first the other joints having two unknowns only.

Method of Sections (Analytical method)

In this method, the frame or truss is cut off into two portions by an imaginary section line passing through the members, the stresses or forces in which are to be determined. The section line should pass through the frame in such a way that it will not cut more than three members with unknown forces at a time. Next, the conditions of equilibrium ($\Sigma H = 0$, $\Sigma V = 0$ and $\Sigma M = 0$) are to be applied to one of the sectioned portion, which is assumed to be in equilibrium under the action of the external load applied and the forces exerted by the members. Here, the other sectioned part is taken to be vanished. It is also convenient to determine the forces in members in the sectioned part by taking moments about a point, for which this method is also known as method of moments.

Method of Substitution (Analytical method)

The use of this method is restricted to certain cases where the form of bracing is substituted in a frame or an imaginary member is called for and substituted in the frame in order to facilitate easy solutions. This method is rarely adopted as it is complicated and cumbersome.

Graphical Method

This is a method of solving statical problems with the help of geometrical figures. This method is also known as Graphic Statics. In this, no calculation is involved. The analysis of a truss or frame is done by drawing figures graphically and the magnitudes of forces or stressess in the members of the frame or truss are obtained by measuring lengths of the lines in the force diagram or stress diagram. The sense of a force or stress in a member come out from this diagram.

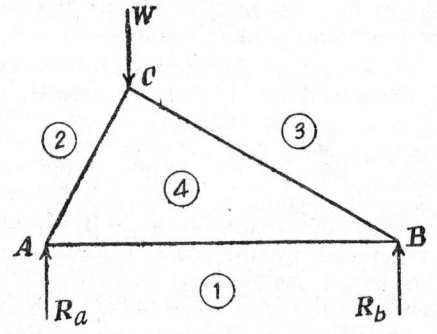

Bow's Notation

Bow's notation is used to designate a force or a member. The spaces between the pairs of adjacent forces or pairs of adjacent members or between a member and a force are named by

Fig. 10.4

Bow's notation. Thus, R_A is designated by the spaces 1 and 2, member AC is desigated by 2 and 4 and so on.

The members and the forces shown in Fig. 10·4 may be designated by Bow's notation as tabulated below :

Member or force	Bow's Notation	Member or force	Bow's Notation
AB	1, 4 or 4, 1	W	2, 3
AC	2, 4 or 4, 2	R_a	1, 2
BC	4, 3 or 3, 4	R_b	3, 1

In Graphical method, first of all, the truss or frame with the loading is drawn to a suitable scale. This diagram is called 'Space diagram'. Reactions are determined by taking moments and the spaces are named according to Bow's notation. Next, force diagram or stress diagram is to be drawn. To do this, when the reactions and load applied are vertical, plot them to a convenient scale on a vertical line accordingly as they are vertically upward or downward.

Now, select any joint having two unknown forces only. Move round the joint in clockwise direction, take members one by one and draw lines parallel to the members in the force diagram. Thus, for each joint, a geometrical figure will be obtained. Put arrowheads on the members in the space diagram close to the join considered. by moving round the joint in clockwise direction Arrowheads are to be put accordingly as they are upward or downward in the force diagram. The magnitude of force is determined simply by measuring the corresponding length in the force diagram. The sense of force *i e.*, nature of force is indicated by the arrowheads.

To explain it more clearly let us take a problem and solve it step by step.

Problem 1. *Determine by graphical method the forces in the members of the N-girder shown in Fig. 10·5 and tabulate the results by indicating nature of the forces.*

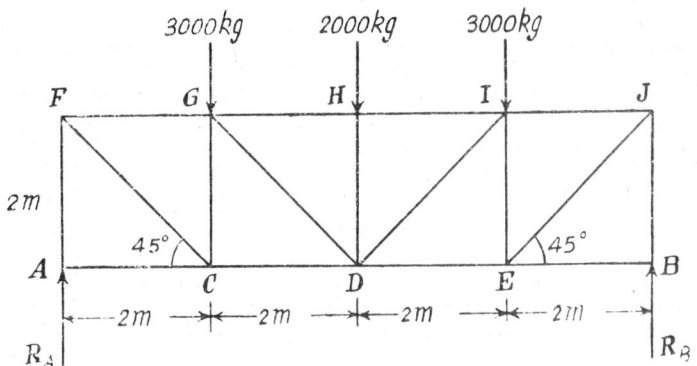

3000kg   2000kg   3000kg

N-Grider with symmetrical Loading
Fig. 10·5

Solution. Total load applied = 3000 + 2000 + 3000 = 8000 kg.

By symmetry, $R_A = R_B = 4000$ kg. Draw the girder frame to a suitable scale and name the spaces between a force and a force, a member and a force as shown in Fig. 10·6. The spaces are named by numbers within circles. Thus, each member has a notation and it is noted

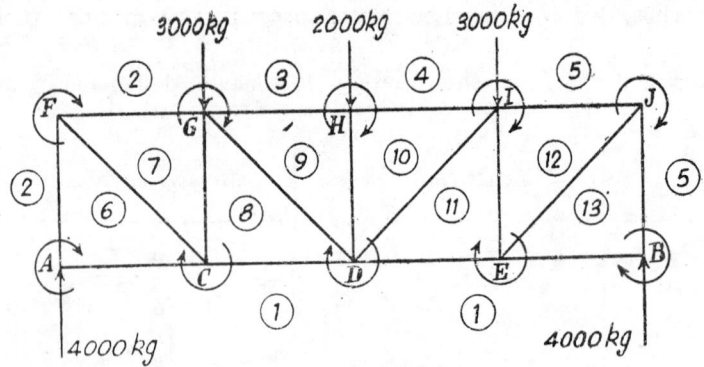

Space diagram
Fig. 10·6

in the table. Next, force diagram or stress diagram is to be drawn. Mark that, there are three
loads vertically downwards and two reactions
vertically upwards. Draw a vertical line
(parallel to the veritcal members of the frame)
and plot the loads to a convenient load scale
(say, 1 cm.=100 kg) as shown in Fig. 10·7.
In space diagram, loads are : (2, 3), (3, 4)
and (4, 5) acting vertically downwards.
Accordingly, in stress diagram these are drawn
vertically downwards from 2 to 3, 3 to 4
and 4 to 5. Plot the reactions vertically
upwards from 5 to 1 and 1 to 2. Thus, 1
becomes the mid-point of the line 5 to 2.

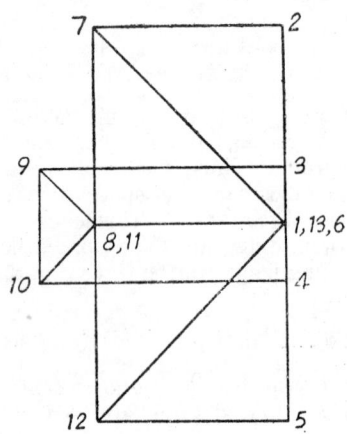

Force diagram or stress diagram representing
the force in members
Fig. 10 7

Next, joints are to be considered one
by one. At any time, a joint having more
than two unknowns cannot be considered.
So, first of all, select a joint with two unknowns only. Here, in this problem, either
joint A or joint B may be considered first.
Remember that you will have to move your
pencil round each joint considered, in clock-
wise direction. Thus, by taking joint A, forces and members designated are (1, 2), (2, 6) and
(6, 1). Already we know the point 1 in the force diagram. Now, (2. 6) is a vertical line
starting from the point 2 in the force diagram and (6, 1) is a horizontal line starting from point
1. Thus, 1 and 6 lie on the same point as the point 6 lies both in horizontal as well as in
vertical line. Hence, (6, 1) is a redundant member *i.e.*, there is no force in this member.
Similarly, considering joint B, it can be seen that 6 and 13 lie on the same point. Next, we can
not take joints C and E, as each of them has three unknowns. So, we will have to consider either
joint F or joint I. By considering joint F, member (2, 7) is horizontal and member (7, 6) is
inclined. In stress diagram, points 2 and 6 are already located. So, draw a horizontal line
passing through point 2 and an inclined line passing through point 6 and parallel to the member
FC.

The meeting point of these two lines is point 7. Thus, the lengths of lines (2, 7) and
(6, 7) in stress diagram give the magnitude of forces in members FG and FC respectively. In
similar way, you consider joints C, G, H, D, I E and J one after another and complete the
force diagram. Mark that each and every line in the force diagram will have to be made parallel
to the corresponding line in the space diagram.

Now, we shall discuss the way of putting arrowheads in the space diagram to represent
the nature of force in each member of the frame. Let us first consider joint A. The forces are
(1, 2), (2, 6) and (6, 1). Out of these (1, 2) is known and (6, 1) is zero. The line 2 to 6 in

force diagram is vertically downward. So, put the arrowhead vertically downward on FA and close to the joint A which is considered, as shown in Fig. 10.8. In joint F, forces are (6, 2), (2,7)

Member in the truss	Bows Notation for the member	Force in the member (kg.)	Nature of force
AC	1, 6	0	Redundant member
BE	1, 13	0	,,
CD	1, 8	4000	Tension
DE	1, 11	4000	,,
FG	2, 7	4000	Compression
JI	5,12	4000	,,
GH	3, 9	5000	,,
HI	4, 10	5000	,,
AF	2, 6	4000	,,
BJ	5, 13	4000	,,
CG	7, 8	4000	,,
EI	11, 12	4000	,,
DH	9, 10	2000	,,
CF	6, 7	5700	Tension
EJ	12, 13	5700	,,
DG	8, 9	1400	,,
DI	10, 11	1400	,,

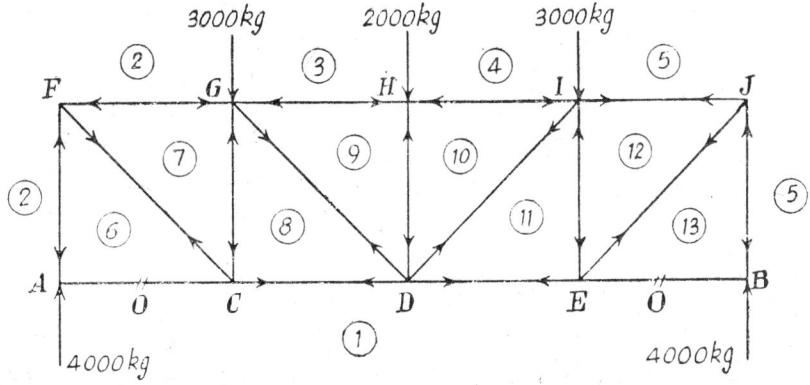

Space diagram showing nature of forces in the members of the truss.
Fig. 10·8.

and (7, 6). As obtained io the force diagram, the line 2 to 7 is towards left, 7 to 6 downwards and 6 to 2 upwards. So, put the arrowheads on the corresponding members near the joint F, and accordingly as they are towards left or downwards or upwards in the force diagram. In this way, put arrowheads in all the members of the frame and measure all the lines in the force diagram. The lengths of lines will indicate magnitudes of forces and the arrowheads will give senses or natures of forces. Complete table as shown

Problem 2. *The diamensions and loading of a simply supported roof tress are shown in figure. Find the magnitude and nature of forces in all the members. Indicate the results in a tabular form.*

Solution. First, let us find out the reactions R_A and R_B.

By taking moments about B,

$$R_A \times 12 = 2 \times 9 + 2 \times 6$$

or, $R_A = 1\cdot5 + 1 = 2\cdot5$ tonnes ;

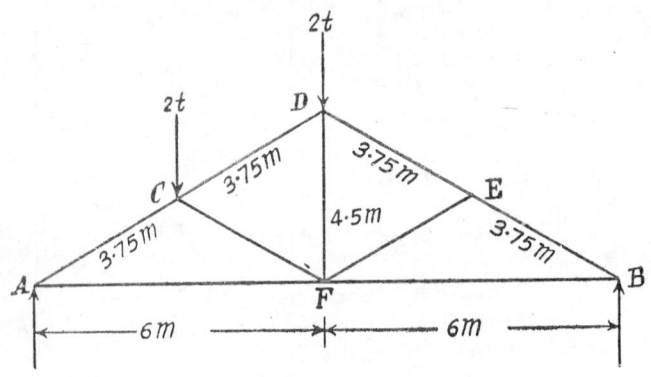

Fig. 10 9.

Total load $= 2 + 2 = 4$ tonnes

∴ $R_B = $ Total load $- R_A = 4 - 2\cdot5 = 1\cdot5$ tonnes.

Next, the space diagram is to be drawn to a convenient scale. Plot all the loads to a suitable scale. 2 to 3 is $2t$. downwards ; 3 to 4 is $2t$. downwards ; 4 to 1 is 1 5 t upwards and 1 to 2 is $2\cdot5$ t. upwards.

First, let us take the joint A. Draw lines parallel to AC and AF through points 2 and 1 respectively. The meeting point of these two lines is point 5.

Similarly, by taking joints C, D, F, E and B one after another, complete the diagram. As points 7 and 8 are coinciding, the member 7, 8 *i.e.*, EF is carrying no load and thus it is a redundant member. The results *i.e.*, nature and magnitude of forces are shown in table. The space diagram and stress diagram are shown in Fig. 10˙10 and Fig. 10˙11 respectively.

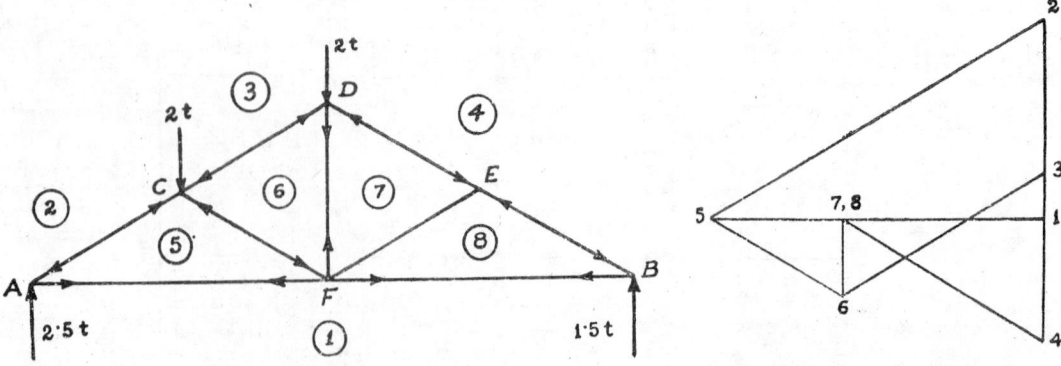

Fig. 10·10. Space diagram.   Fig. 10·11. Stress diagram.

Result :

Member	Notation	Magnitude of force	Nature of force
AF	5, 1	3·33 tonnes	Tension
BF	8, 1	2.00 ·,	,,
DF	6, 7	1·00 ,,	,,
AC	2, 5	4·16 ,,	Compression
CD	3, 6	2·50 ,,	,,
CF	5, 6	1·66 ,,	,,
BE	4, 8	2·05 ,,	,,
ED	4, 7	2·05 ,,	,,
EF	7, 8	0	Rudundant Member

Problem 3. *The dimensions and loading of a simply supported roof truss are shown in Fig. 10·12. Find the magnitude and nature of the force developed in the members.*

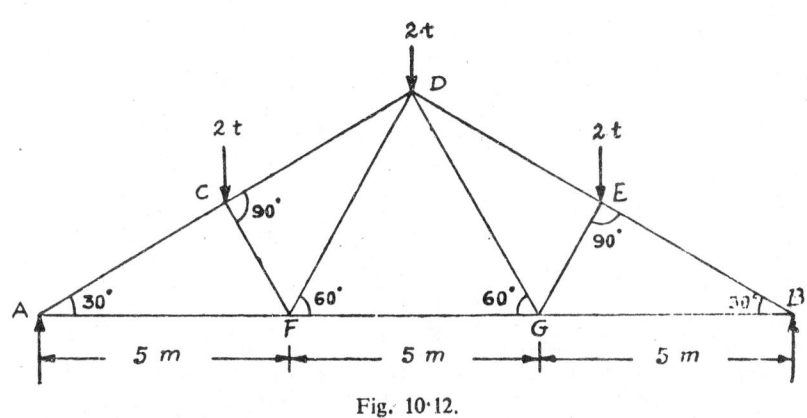

Fig. 10·12.

Solution. Draw the space diagram to a suitable scale as shown in Fig. 10·13.

Next draw the stress diagram (Fig. 3·14), considering the joints A, C, F, D, G and E serially one after another.

From the stress diagram, put the arrowheads on members in space diagram to indicate the nature of force. Measure the magnitudes of the forces from the stress diagram. Prepare the table by studying both space diagram and stress diagram simultaneously.

Result :

Member	Notation	Magnitude of force	Nature of Force
AC	2, 6	6 tonnes	Compression
CD	3, 7	5 ,,	,,
BE	5, 10	6 ,,	.,
DE	4, 9	5 ,,	,,
CF	6, 7	1·75 ,,	,,
EG	9, 10	1·75 ,,	,,
AF	1, 6	5·25 ,,	Tension
BG	1, 10	5·25 ,,	,,
FG	1, 8	3·50 ,,	,,
DF	7, 8	1·75 ,,	,,
DG	9, 8	1·75 ,,	,,

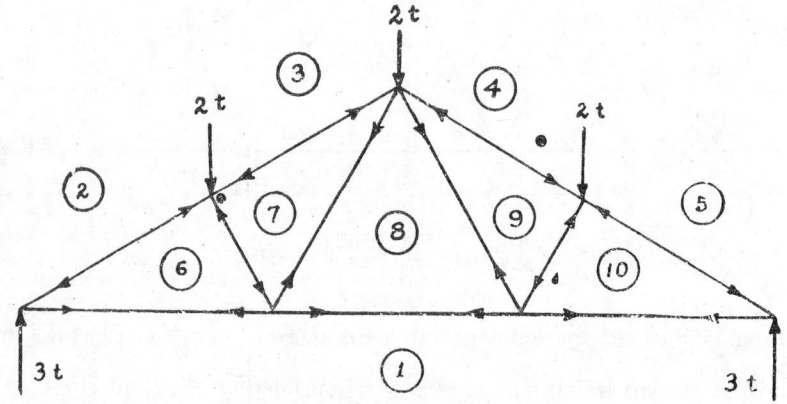

Space diagram
Fig. 10·13.

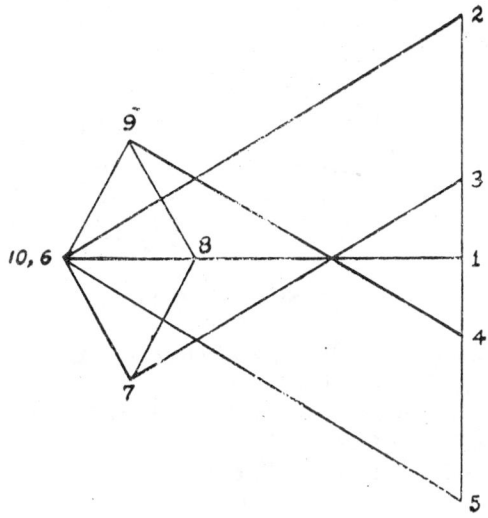

Stress Diagram
Fig. 10·14.

Problem 4. *Find graphically the value and the nature of stress in the different members of the following roof truss with the loads as shown in Fig. 10·15.*

Solution. From symmetry, $R_A = R_B = 3.75$ tonnes

Stresses developed in the members found out from the stress diagram are as follows :

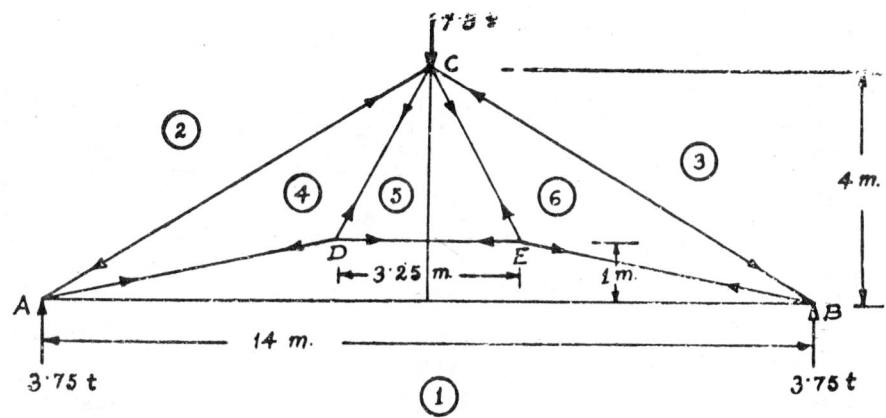

Space Diagram
Fig. 10·15.

$BC = AC = 12.25$ tonnes (compressive) ;

$CD = CE = 2.90$ tonnes (tensile) ;

$AD = BE = 9.00$ tonnes (tensile) ;

$DE = 7.75$ tonnes (tensile) ;

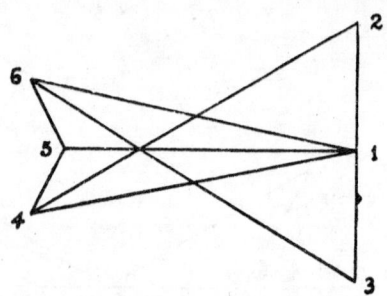

Stress Diagram
Fig. 10·16.

Problem. 5. *Evaluate forces in all the members of the truss shown in Fig. 10·17. Present the results in table form, indicating nature of forces.*

Solution. Fig. 10·18 shows the space diagram drawn to a convenient scale. The stress diagram for the truss is shown in Fig. 10·19.

To draw the stress diagram, joints A, F, C, G, D, H, E and I are considered one after another. In each case, the joint having only two unkeown forces is to be chesen first. Move your pencil around each joint in clockwise direction and put the arrowhead in the space diagram accordingly as they are in the stress diagram. Lastly, prepare the table as shown.

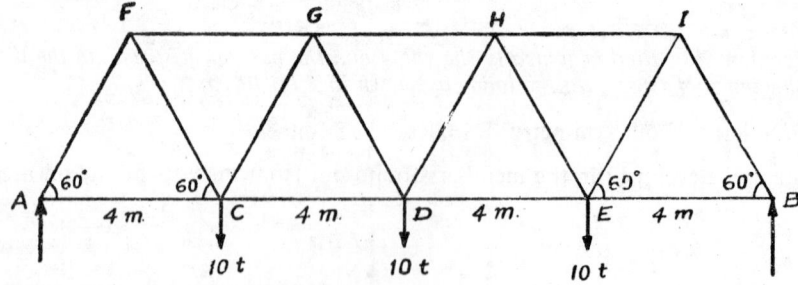

Fig. 10·17.

Member by notation	Magnitute of Force (tonnes)	Nature of Force
1,6	17·25	Compression
1,12	17·25	,,
6,7	17·25	Tension
11,12	17·25	,,
7,8	6·00	Compression
10,11	6·00	,,
8,9	6·00	Tension
9,10	6·00	,,
2,6	8·50	,,
5,12	8·50	,,
3,8	19·75	,,
4,10	19·75	,,
1,7	17·00	Compression
1,11	17·00	,,
1,9	22·00	,,

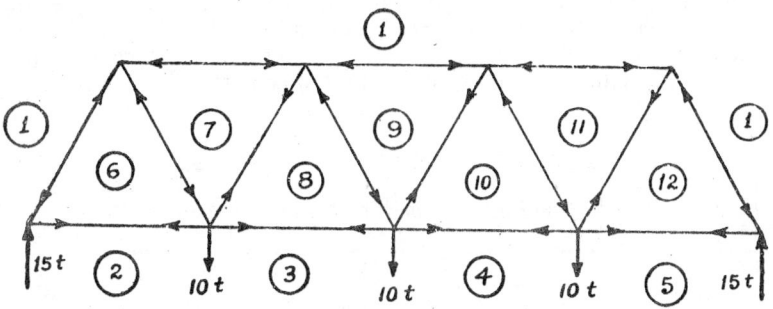

Space Diagram
Fig. 10·18.

Problem 6. *Fig. 10·20 shows a girder simply supported at the ends and loaded as shown. Determine graphically magnitude and nature of force in each member of the girder.*
Solution. Taking moment about B,

$$R_A \times 8 = 1 \times 6 + 2 \times 4 + 3 \times 2$$

or, $\quad R_A = \dfrac{20}{8} = 2\cdot5$ tonnes

Total load $= 6$ tonnes

$\therefore \quad R_B = 6 - 2\cdot5 = 3\cdot5$ tonnes.

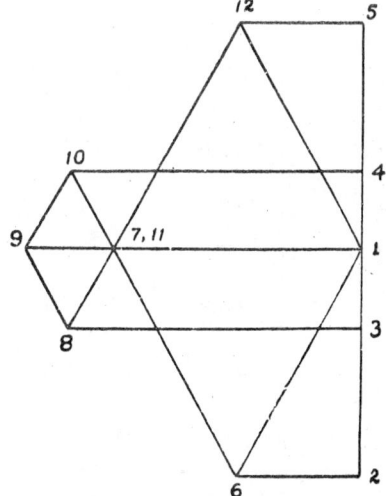

Stress Diagram
Fig. 10·19.

Draw the space diagram to a convenient scale and name the spaces. As the loads are vertical, plot these on a vertical line choosing a suitable scale. Considering the joint A first,

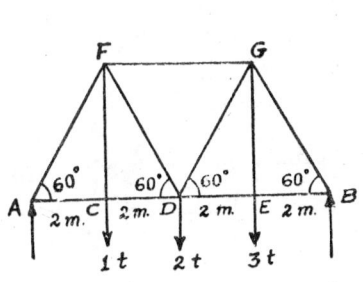

Fig. 10·20.

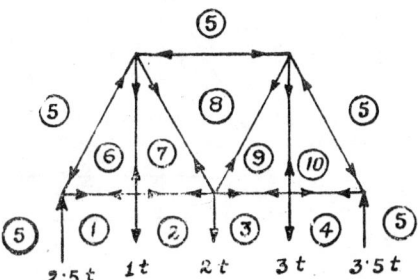

Space diagram
Fig. 10·21.

determine magnitude and nature of force in members AF and AC as in previous cases. When joint A is solved, proceed with joints C, F, D, G and E one after another. Thus, the stress diagram will be completed. With the help of stress diagram, mark the arrowheads in the space diagram to indicate the nature of forces in different members of the girder. Prepare the table as shown.

Member	Notation	Magnitude of Force	Nature of Force
AC	6,1	1.7 tonnes	Tension
CD	7,2	1.7 ,,	,,
DE	9,3	2 ,,	,,
EB	10,4	2 ,,	,,
BG	5,10	4 ,,	Compression
GF	5,8	2.3 ,,	,,
FA	5,6	3 ,,	,,
FC	6,7	1 ,,	Tension
FD	7,8	1.8 ,,	,,
DG	8,9	0.6 ,,	,,
GE	9,10	3 ,,	,,

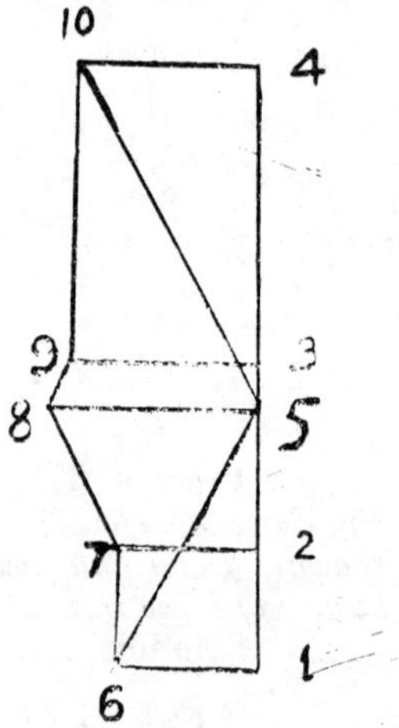

Stress Diagram
Fig. 10·22.

Problem 7. *Find graphically the magnitude and nature of the forces in the truss shown below. Indicate the results in the tabular form.* (*A.M.I.E., May 1965*)

Solution. Taking moment about D,

$$R_A \times 15 = 3 \times 10 + 6 \times 2\text{·}5$$

or, $R_A = 3$ tonnes ∴ $R_D = 6$ tonnes.

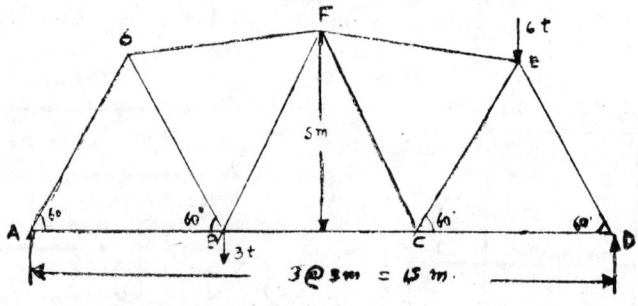

Fig. 10·23.

Draw the space diagram as shown in Fig. 10·24 to a convenient scale. Plot the vertical loads to a suitable scale on a vertical line and complete the stress diagram by considering joints one after another as it is explained in previous problems. The stress diagram is shown in Fig. 10·25. With the help of stress diagram and considering the joints one by one, put the arrowheads in the space diagram which will indicate the nature of force in each number. The stress in each member is found out by measuring the distance in stress diagram and these are presented in a table.

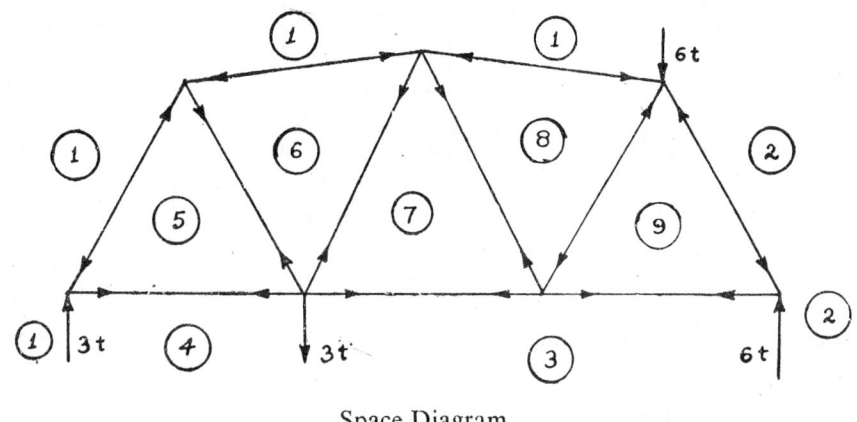

Space Diagram
Fig. 10·24.

Results

Member	Magnitude of Force	Nature of Force
4,5	1·75 tonnes	Tension
3,7	3·00 ,,	,,
3,9	3·50 ,,	,,
1,6	3·25 ,,	Compression
1,8	3·25 ,,	,,
1,5	3·50 ,,	,,
5,6	3·00 ,,	Tension
6,7	0·50 ,,	,,
7,8	0·50 ,,	,,
8,9	0·50 ,,	Compression
9,2	7.00 ,,	,,

Stress Diagram
Fig. 10·25.

Problem 8. *A railway bridge girder is shown in Fig. 10·26. Find out the magnitude and nature of force in each of the members of the girder and show the results in tabular form.*

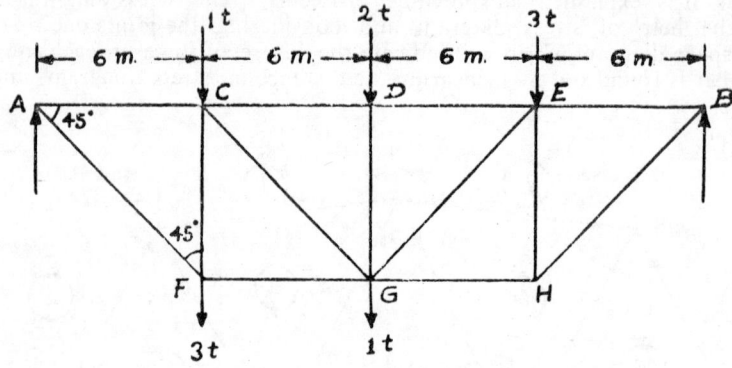

Fig. 10·26.

Solution. First, the reactions at the end supports are found out. Taking moment about B,

$$R_A \times 24 = 4 \times 18 + 3 \times 12 + 3 \times 6$$

or, $4R_A = 12 + 6 + 3 = 21$

∴ $R_A = \dfrac{21}{4} = 5·25\ t$

Total load $= 1 + 2 + 3 + 3 + 1 = 10\ t$

∴ $R_B = 10 - 5·25 = 4·75\ t.$

Draw the space diagram to a suitable scale as shown in Fig. 10·27 and name the spaces. Plot all the vertical loads to a scale, on a vertical line. Next, start from either of the support

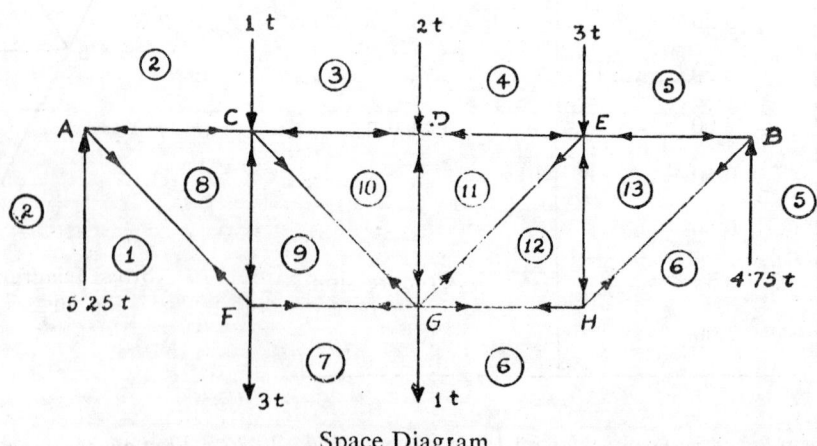

Space Diagram
Fig. 10·27.

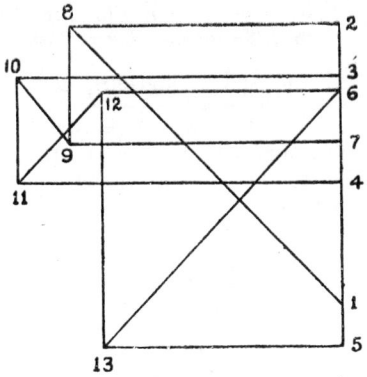

Stress Diagram
Fig. 10·28.

joints or any other joint, where there are two unknowns only. Go round each joint in clock wise direction and draw lines parallel to the members shown in the space diagram. Never take a joint, where number of unknown forces is more than two.

To indicate nature of force in each member, put arrowheads accordingly as they are in the stress diagram. To put arrowheads go round each joint in clockwise direction and mark the directions shown by the lines in the stress diagram. Put the arrowheads near the joint considered. Lastly, prepare the table showing results.

Results :

Member	Notation	Force (t)	Nature of Force
AC	2,8	5·20	Compression
CD	3,10	6·45	,,
DE	4,11	6·45	,,
EB	5,13	4·75	,,
FG	9,7	5·20	Tension
GH	12,6	4·75	,,
CF	8,9	2·25	Compression
DG	10,11	2·00	,,
EH	12,13	4·75	,,
AF	1,8	7·30	Tension
CG	9,10	1·75	,,
EG	11,12	2·30	,,
BH	6,13	6·55	,,

Problem 9. *A specially-built roof truss is shown in Fig. 10·29. It is required to find out the forces (both magnitude and nature) in different members of the truss.*

Solution. Fig. 10·30 shows the space-diagram. The stress-diagram for the truss is shown in Fig. 10·31.

To draw the stress diagram, proceed to the joint A, E, C, F, G and D one after another. Move around each joint in clockwise direction and never consider a joint where there are more than two unknowns.

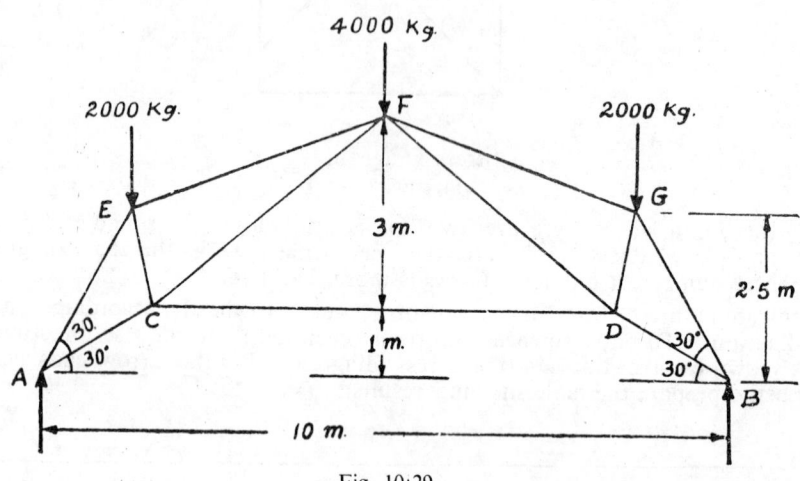

Fig. 10·29.

Results :

Member	Force (kg)	Nature of Force
7,3	6,700	Compression
6,10	6,700	,,
3,1	3,800	Tension
1,6	3,800	,,
8,4	4,150	Compression
9,5	4,150	,,
,4	2,300	Tension
5,6	2,300	,,
4,2	600	Compression
5,2	600	,,
1,2	4,250	Tension

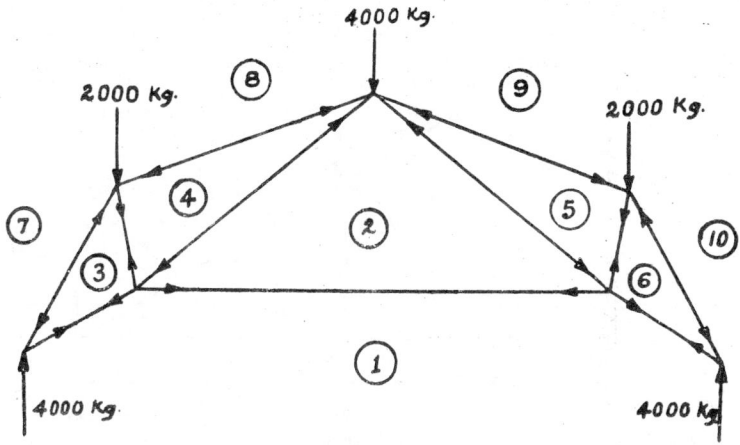

Space Diagram
Fig. 10·30.

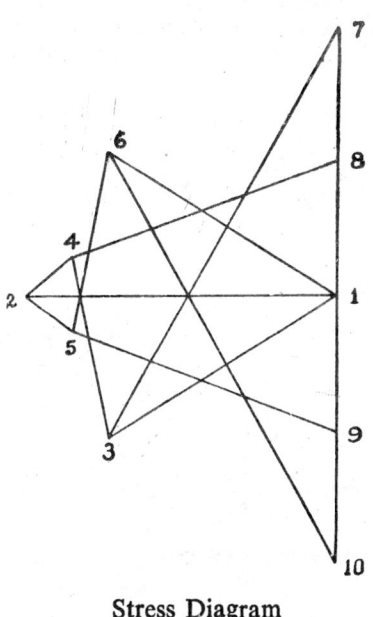

Stress Diagram
Fig. 10.31.

Problem 10. *Determine the forces in all the members of the pin-jointed truss shown in Fig. 10·32.*

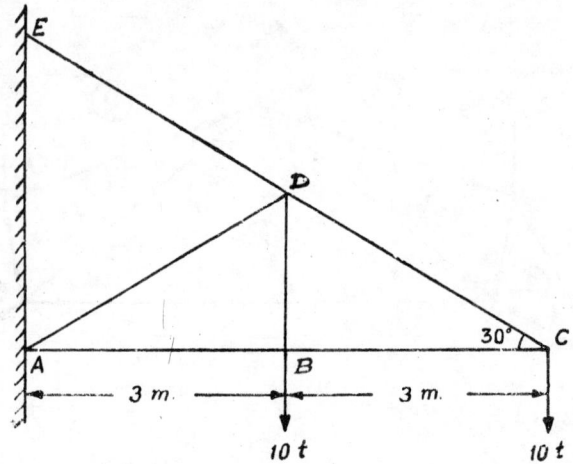

Fig. 10·32.

Solution. The space diagram shown in Fig. 10·33 is drawn first to a convenient scale and the spaces are named as per Bow's notation. Next, the stress diagram is drawn to a suitable scale as shown in Fig. 10·34 by following the procedure as explained in previous problems. From the stress diagram one can easily find out the magnitude of force or stress in each member of truss. During construction of the stress diagram, arrowheads are given in the space diagram accordingly, which indicate the nature of force in each member. Last of all magnitude and nature of force in each member are tabulated.

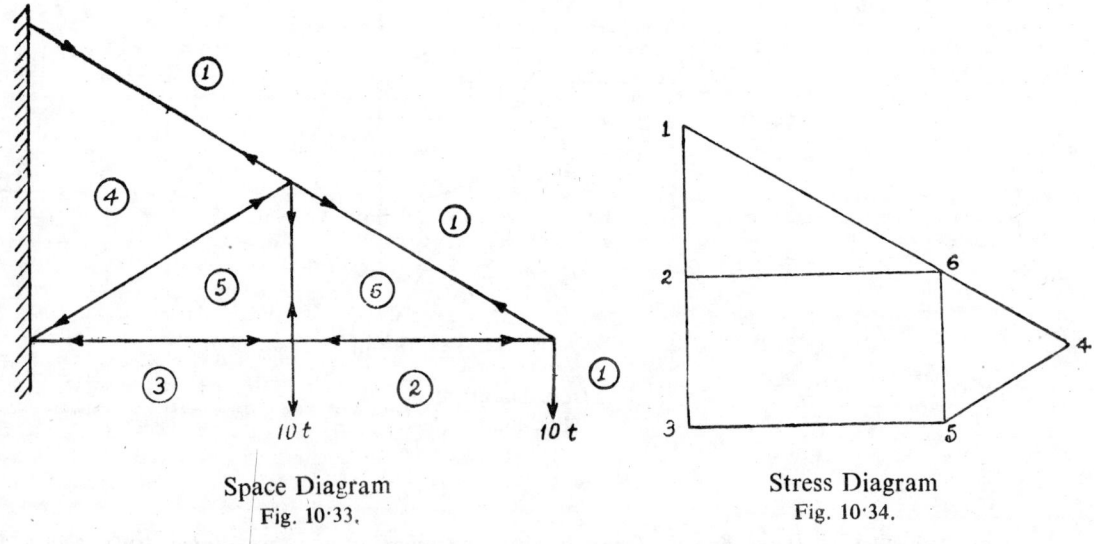

Space Diagram
Fig. 10·33.

Stress Diagram
Fig. 10·34.

Table

Member	Notation	Nature of Force	Magnitude of Force, tonnes
AB	5,3	Compressive	17·3
BC	6,2	,,	17·3
CD	6,1	Tensile	20·0
DE	4,1	,,	30·0
BD	5,6	,,	10·0
AD	4,5	Compressive	20·0

Problem 11. *Find out the forces in all the members of the truss shown in Fig. 10·35.*

Solution. The space diagram is shown in Fig. 10·35 and Fig. 10·36 shows the stress diagram. Both the figures are drawn to scales. The forces with their nature are given in the table.

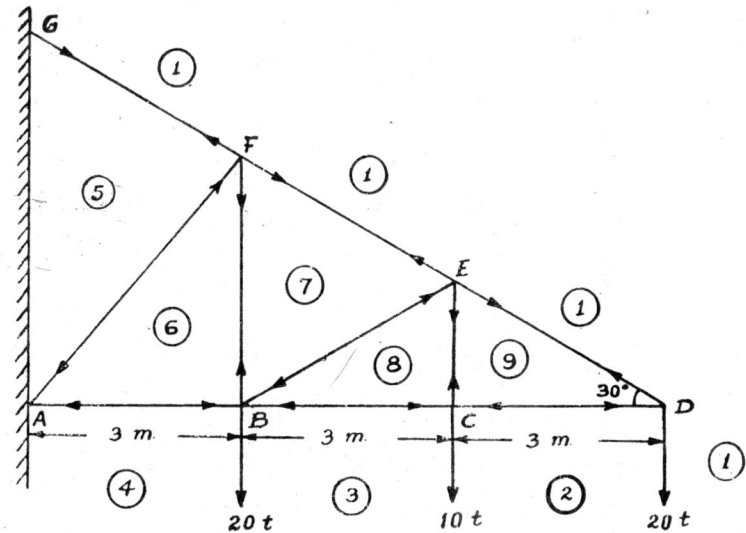

Fig. 10·35. Space Diagram.

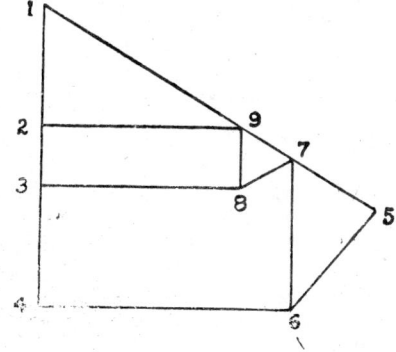

Fig. 10·36. Stress Diagram.

Table :

Membor	Force (tonne)	Nature of force	Member	Force (tonne)	Nature of force
2,9	33·75	Compressive	1,9	38·75	Tensile
3,8	33·75	,,	6,7	25·00	,,
4,6	42·50	,,	8,9	10·00	,,
1,5	45·00	Tensile	5,6	20·00	Compressive
1,7	48·75	,,	7,8	10·00	,,

Problem 12. *Find the forces in all the members of the truss shown in Fig. 10·37. Indicate results in a tabular form.*   (*A.M.I.E., Nov. 1963*)

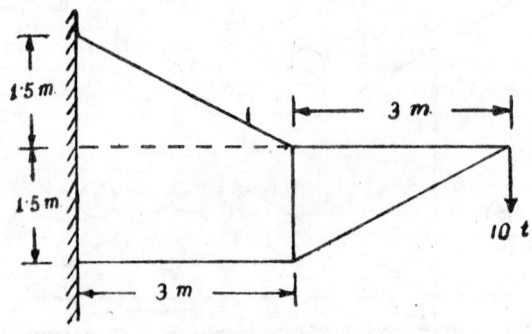

Fig. 10·37.

Solution. The procedure is same as for previous problems.

Table :

Member	Force	Nature of force
1,3	22·5 tonnes	Tension
1,4	20·0 ,,	,,
2,3	20·0 ,,	Compression
2,4	22·5 ,,	,,
3,4	10·0 ,,	Tension

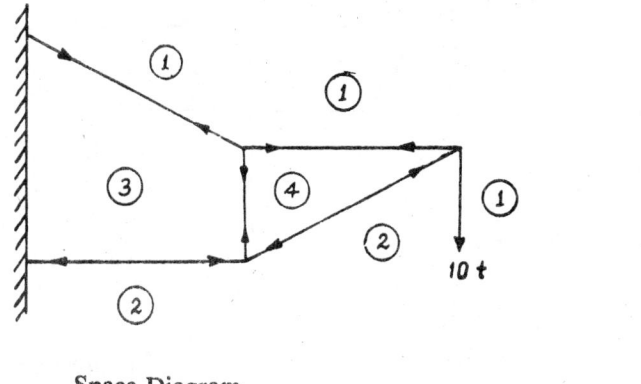
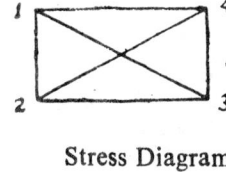

Space Diagram	Stress Diagram
Fig. 10·38.	Fig. 10·39.

Problem 13. *Determine the forces in various members of a pin-jointed framework shown in Fig. 10·40. Designating the members by Bow's notation, tabulate the forces stating whether they are in tension or in compression.* (*A.M.I.E.*)

Solution. The space diagram designating the members by Bow's notation and indicating the nature of forces by arrowheads is shown in Fig. 10·40. Fig. 10·41 shows the stress diagram drawn to a suitable scale.

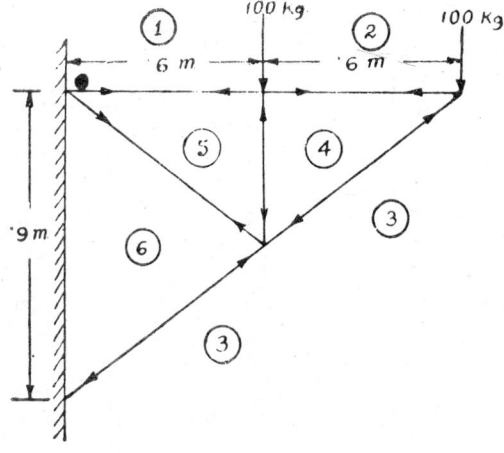
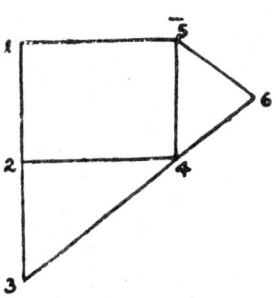

Space Diagram.	Stress Diagram.
Fig. 10·40.	Fig. 10·41.

Table

Member	Force (kg)	Nature of force
3, 4	164·70	Compression
3, 6	248·03	,,
4, 2	131·30	Tension
5, 1	131·30	,,
4, 5	100·00	Compression
5, 6	88·30	Tension

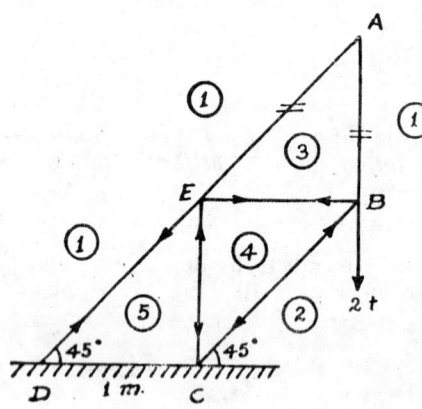

Space Diagram.
Fig. 10·42.

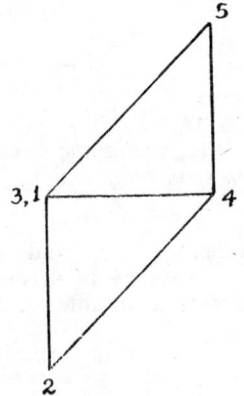

Stress Diagram.
Fig. 10·43.

Solution. Fig. 10·43. shows the stress diagram drawn to a suitable scale following the previous procedure.

Table

Members	Force	Nature of force
1, 3	0	—
2, 4	2·83 t	Compression
5, 1	2·83 t	Tension
3, 1	0	—
4, 3	2 t	Tension
4, 5	2 t	Compression

Problem 15. *A pin-jointed frame is shown in Fig. 10·44. It is hinged at A and loaded at D. A horizontal chain is attached at C and pulled so that AD is horizontal.*

Determine the pull on the chain and also the force in each member, stating whether it is in tension or in compression. (*A.M.I.E.*)

Solution. Let P be the pull in the chain.

Taking moment about A, $P \times 0.9 = 2 \cos 45° \times 1.2$

$$\text{or} \quad P = \frac{2 \times 0.707 \times 1.2}{0.9} = 1.885, \text{ say } 1.90 \text{ tonnes.}$$

Plot to a convenient scale the forces 5, 1 and 1, 2 both in magnitude and direction Complete the stress diagram as shown in Fig. 10·46 choosing a suitable scale and present the results in tabular form.

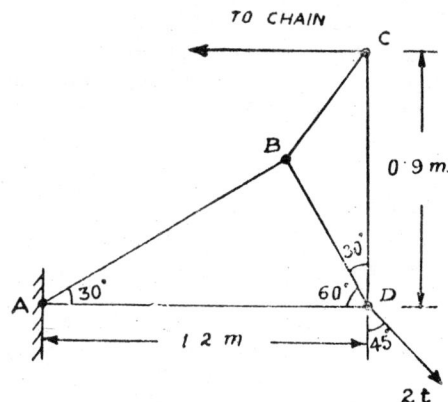

Fig. 10·44.

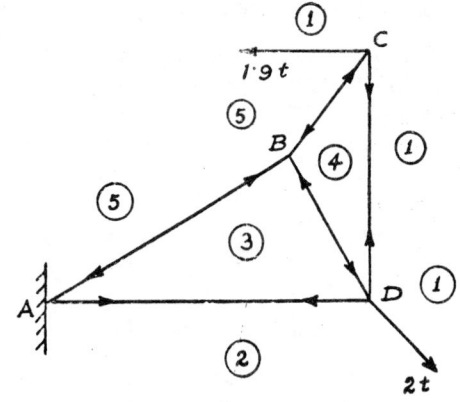

Fig. 10·45. Space Diagram.

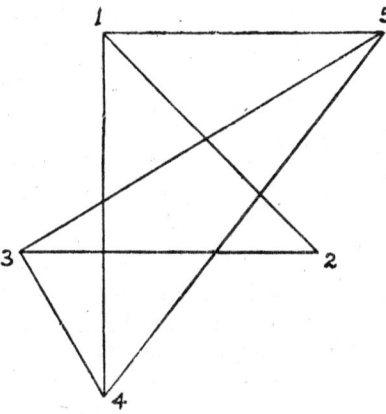

Fig. 10·46. Stress Diagram.

Table

Member	Notation	Forces (tonnes)	Nature of force
AB	3, 5	2 90	Compression
BC	4, 5	3·00	,,
CD	1, 4	2·40	Tension
BD	4, 3	1·15	Compression
DA	2, 3	2·10	Tension

Problem 16. *Fig. 10·47 shows a double cantilever roof truss. Determine and present in a tabular form the magnitude and nature of force in each of the members of the truss.*

Solution   The space diagram with Bow's notation drawn to a convenient scale is shown in Fig. 10·48. The procedure for drawing the stress diagram is same as before. Only, here it should be started from either of the free ends of the truss, where there are two unknown forces.

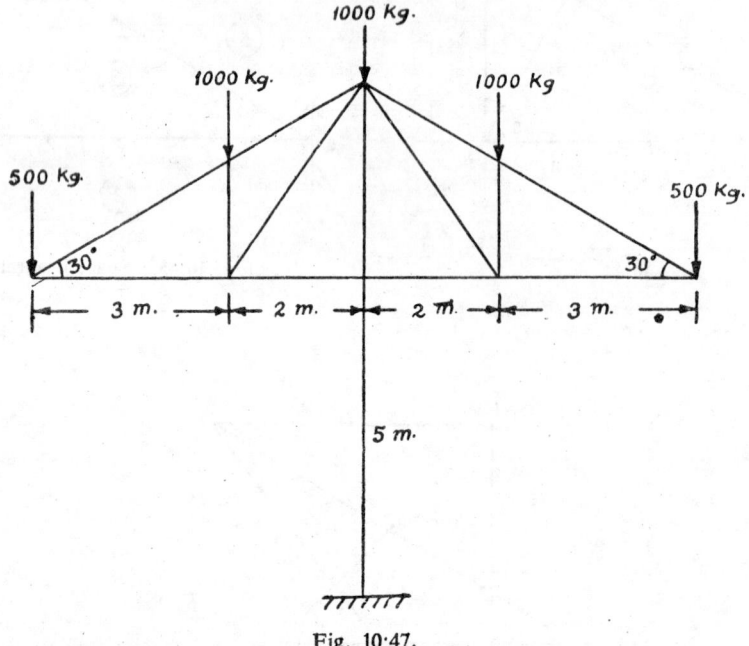

Fig. 10·47.

During construction of the stress diagram as shown in Fig. 10·49, arrowheads are marked in the space diagram accordingly which indicate the nature of force in each member of the truss. Care should be taken to draw lines for the stress diagram making parallel to the members in the space diagram.

Measure the lengths in the stress diagram for each member, note the arrowheads given in the space diagram and complete the table as shown.

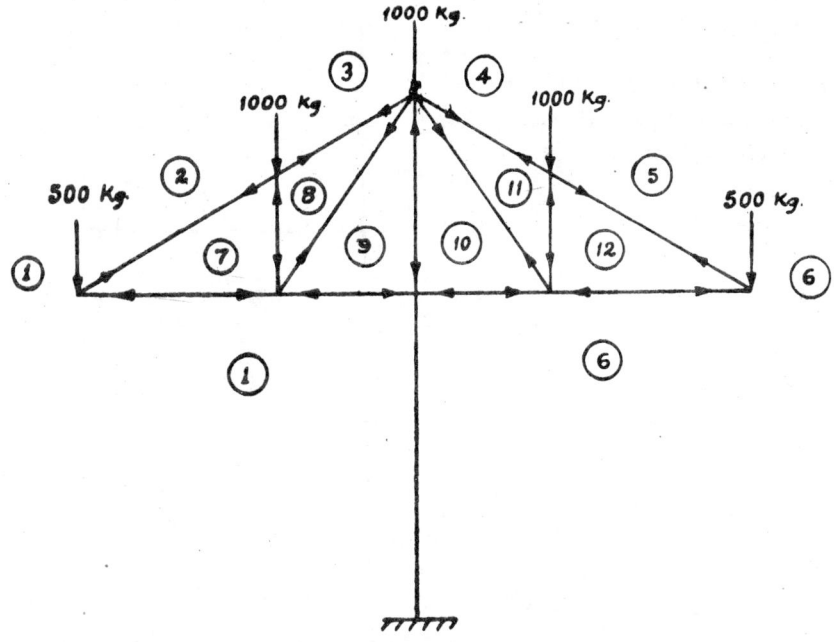

Fig. 10·48. Space Diagram.

Table

Member	Force (kg.)	Nature of force
2, 7	1000	Tension
5, 12	1000	,,
3, 8	1000	,,
4, 11	1000	,,
7, 8	1000	Compression
9, 10	4000	,
11, 12	1000	,,
8, 9	1250	Tension
10, 11	1250	,,
1, 7	875	Compression
6, 12	875	,,
1, 9	1625	,,
6, 10	1625	,,

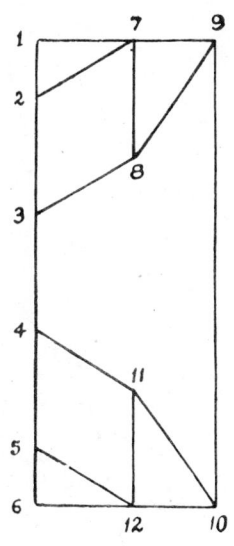

Stress Diagram.
Fig. 10·49.

Problem 17. *A triangular space frame is shown in Fig. 10·50. The frame is subjected to a horizontal load of 1000 kg at its topmost point. Draw the stress diagram and find out the magnitude and nature of force in each of the members of the frame.*

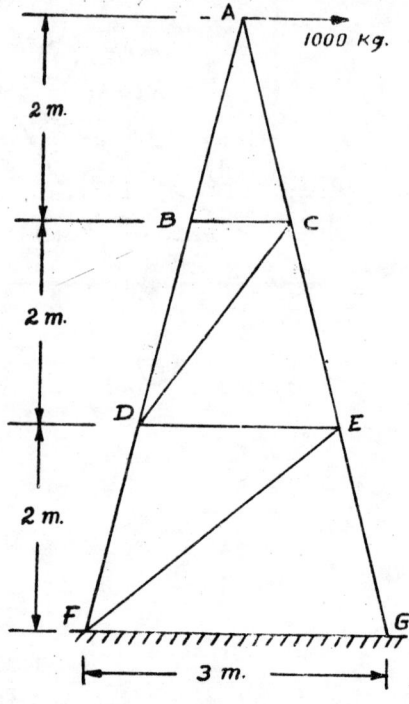

Fig. 10·50.

Solution. Fig. 10·51 shows the space diagram (drawn to a convenient scale) with Bow's notation. By plotting the horizontal force 1, 2 and then drawing 1, 3 and 2, 3 parallel to the top members and taking the joints one after another, complete the stress diagram as shown in Fig. 10·52. The stress diagram is a simple one, indicating that no bracing is required for this frame, for the given loading.

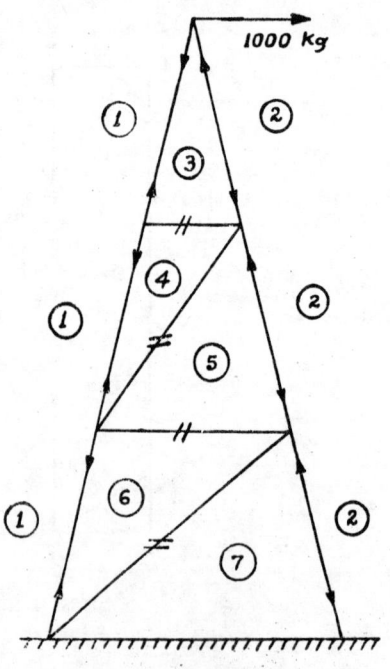

Fig. 10·51. Space Diagram.

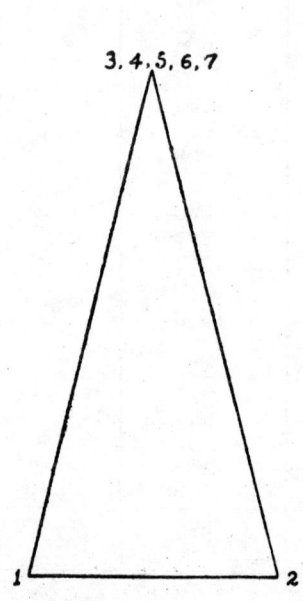

Fig. 10·52. Stress Diagram.

Table

Member	Notation	Force (kg)	Nature of force
AB	1, 3	2000	Tension
BD	1, 4	2000	,,
DF	1, 6	2000	,.
AC	2, 3	2000	Compression
CE	2, 5	2000	,,
EG	2, 7	2000	,,
BC	3, 4	0	Redundant member
DE	5, 6	0	,,
CD	4, 5	0	,,
EF	6, 7	0	,,

Problem 18.   *A frame as shown in Fig. 10·53 carries a vertical load of 1000 kg. at point A and an equivalent horizontal thrust due to wind of 500 kg. at point B.   Find graphically the stresses in the inclined members of the frame.*

Solution.   The space diagram is shown in Fig. 10·54. Fig. 10·55 shows the stress diagram for the frame.   To draw the stress diagram, plot the loads 500 kg (5, 1) and 1000 kg (1, 2) both in magnitude and direction.   Next, start from joint A and proceed to joints C, B, D, E and F one after another.   You need not find out reactions at G and H.

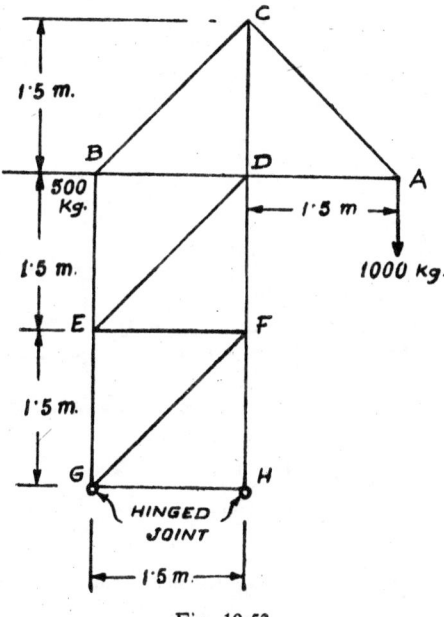

Fig. 10·53

Table

Member	Notation	Force	Nature of force
AC	1, 3	1400 kg	Tension
BC	1, 4	1400 kg	,,
DE	6, 7	833 kg	,,
FG	8, 9	833 kg	

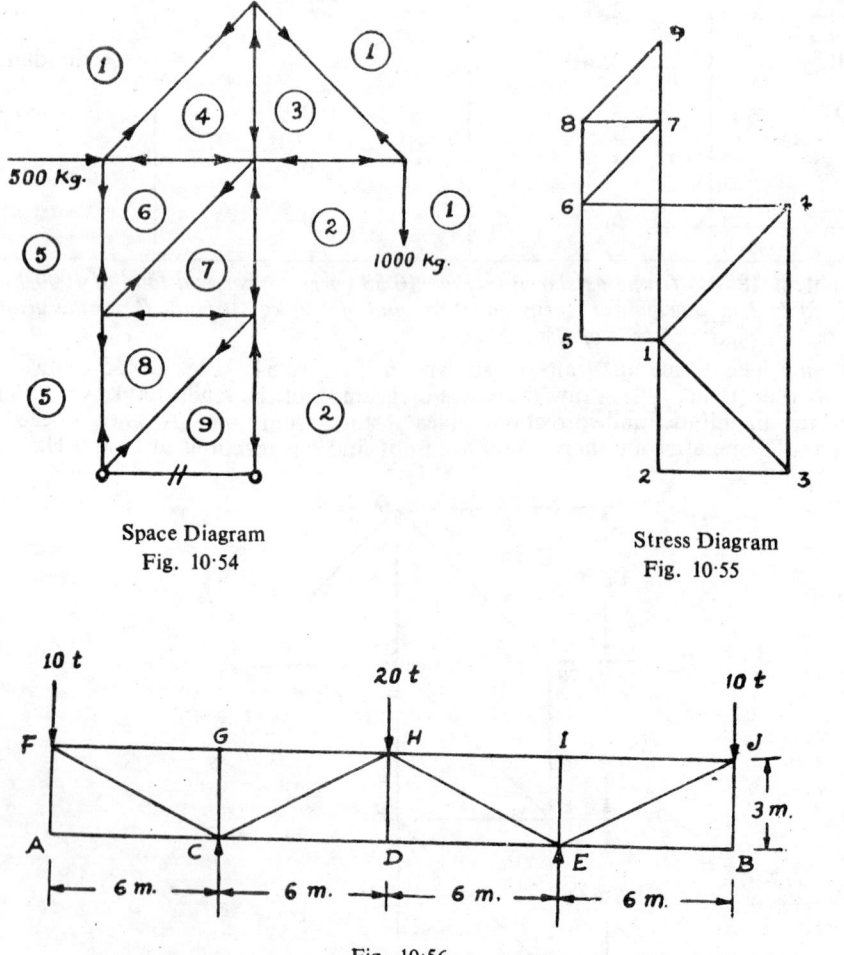

Space Diagram
Fig. 10·54

Stress Diagram
Fig. 10·55

Fig. 10·56

Problem 19. *Evaluate forces in all the members of the truss shown in Fig. 10·56 and present the results in a tabular form.*

Solution. From symmetry, $R_C = R_E = 20$ tonnes. Draw the space diagram and name the spaces as shown in Fig. 10·57.

To draw the stress diagram, start either from A or from B. Considering joint A, we have two members AF and AC. AF is 5, 6 by Bow's notation and it is vertical. AC is 6, 5 and it is horizontal. Therefore, 5 and 6 are the same point and thus AF and AC are carrying no load *i.e.*, these two members are redundant. Next, proceed to joints F, G, C, D, H, I, E and J one after another and complete the stress diagram as shown in Fig. 10·58.

Fig. 10·59 shows the truss by excluding the redundant members. The results are given in a table.

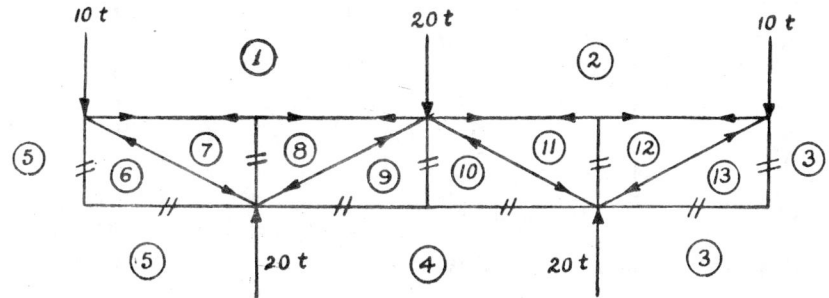

Space Diagram
Fig. 10·57

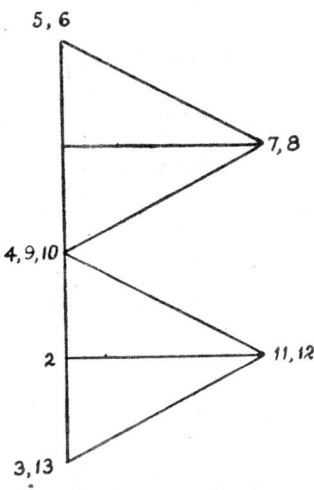

Fig. 10·58. Stress Diagram

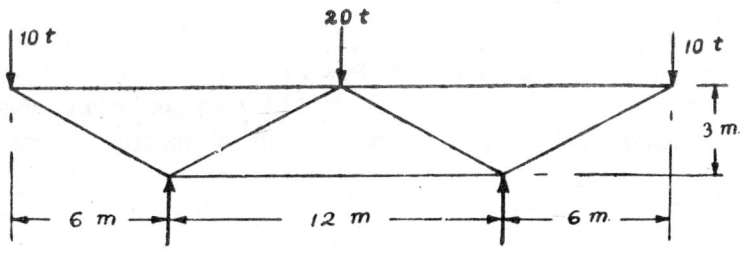

Fig. 10·59

Table

Member	Notation	Force (tonnes)	Nature of force
AC	5,6	0	Redundant member
CD	4,9	0	,,
DE	4,10	0	,,
EB	3,13	0	,,
AF	5,6	0	,,
CG	7,8	0	,,
DH	9,10	0	,,
EI	11,12	0	,,
BJ	13,3	0	,,
FG	1,7	19·5	Tensile
GH	1,8	19·5	,,
HI	2,11	19·5	,,
IJ	2,12	19·5	,,
CF	6,7	22	Compressive
CH	8,9	22	,,
EH	10,11	22	,,
EJ	12,13	22	,,

Problem 20. *Fig. 10·60 Shows a truss having equal overhangs. Ii is required to draw the stress diagram for the truss and to present the magnitude and nature of force in different members, in a tabular form.*

Solution. The space diagram with Bow's notation is shown in Fig. 10·60, The stress diagram (Fig. 10·61) is developed as usual following the procedure of graphicstatics. The magnitude and nature of force in different members of the truss are presented in a tabular form.

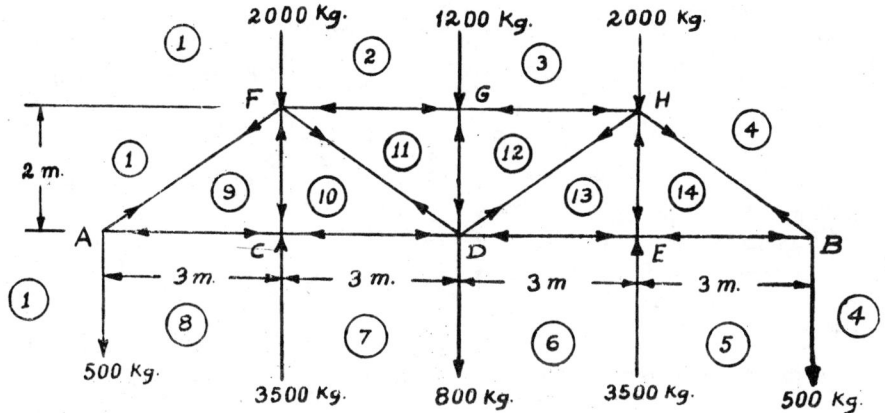

Space Diagram
Fig. 10·60

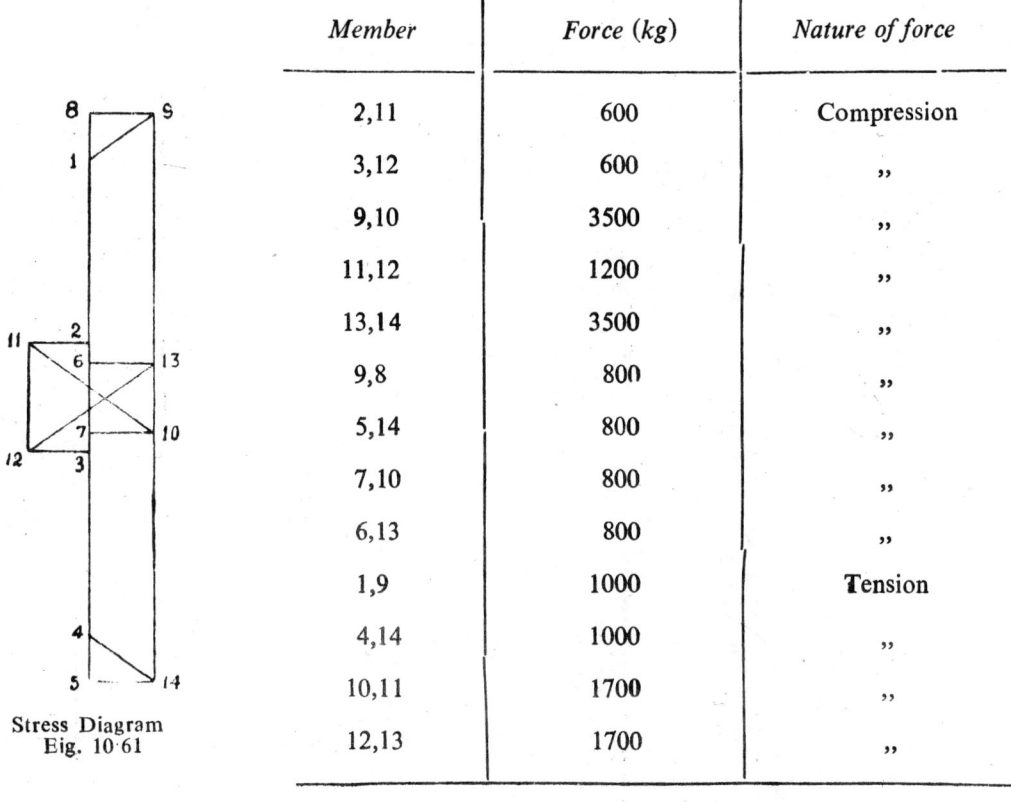

Stress Diagram
Eig. 10·61

Member	Force (kg)	Nature of force
2,11	600	Compression
3,12	600	,,
9,10	3500	,,
11,12	1200	,,
13,14	3500	,,
9,8	800	,,
5,14	800	,,
7,10	800	,,
6,13	800	,,
1,9	1000	Tension
4,14	1000	,,
10,11	1700	,,
12,13	1700	,,

Problem 21. *Determine the stresses in all the members of the water tower truss shown in Fig. 10·62.*

(A.M.I.E., May 1952)

Solution. The truss consists of six joints and it is subjected to two vertical and two horizontal loads.

Draw the space diagram and name the spaces as per Bow's notation, as shown in Fig. 10·63. You need not find out reactions at E and F.

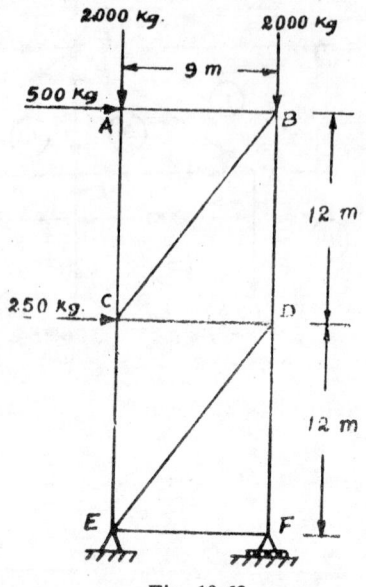

Fig. 10·62

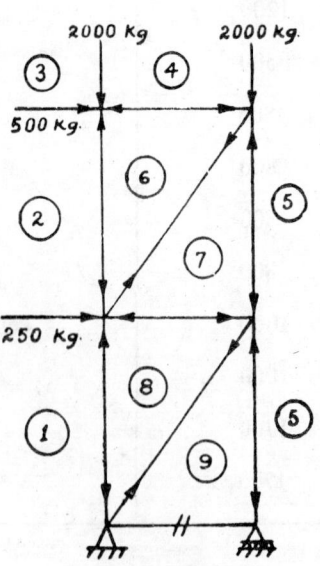

Fig. 10·63.   Space Diagram

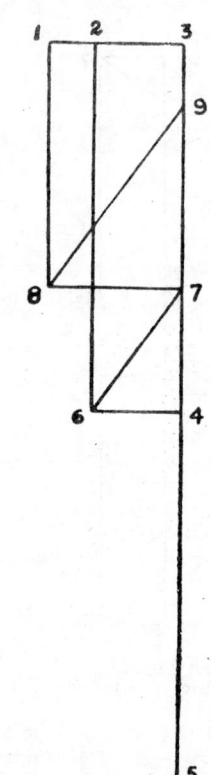

Fig. 10·64.   Stress Diagram

To draw the stress diagram, start from joint A and then proceed to joints B, C and D. It is not necessary to solve joints E and F. Complete the stress diagram as shown in Fig. 10·64 and put arrowheads in the space diagram to indicate the nature of force. Get the magnitude of force in each member of the truss by direct measurement from the stress diagram and prepare the Table :

Member	Notation	Force (kg.)	Nature of force
AB	4,6	500	Compression
CD	7,8	750	,,
BC	6,7	825	Tension
DE	8,9	1275	,,
AC	2,6	2000	Compression
CE	1,8	1350	,,
BD	7,5	2650	,,
DF	9,5	3850	,,

Problem 22. *A North-light roof truss is shown in Fig. 10·65. Tabulate the magnitude and nature of forces in different members of the truss.*

Solution. To find reactions, taking moments about B,

$$R_A \times 8 = 500 \times 6 + 500 \times 4 + 500 \times 2 - 500 \times 2 = 5000$$

$$\therefore \quad R_A = 625 \text{ kg. and } R_B = 2500 - 625 = 1875 \text{ kg.}$$

Draw the space diagram as shown in Fig. 10·66 and name the spaces according to Bow's notation.

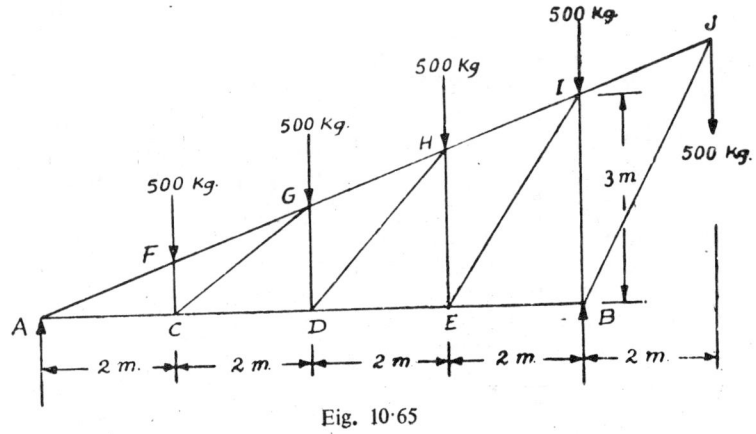

Fig. 10·65

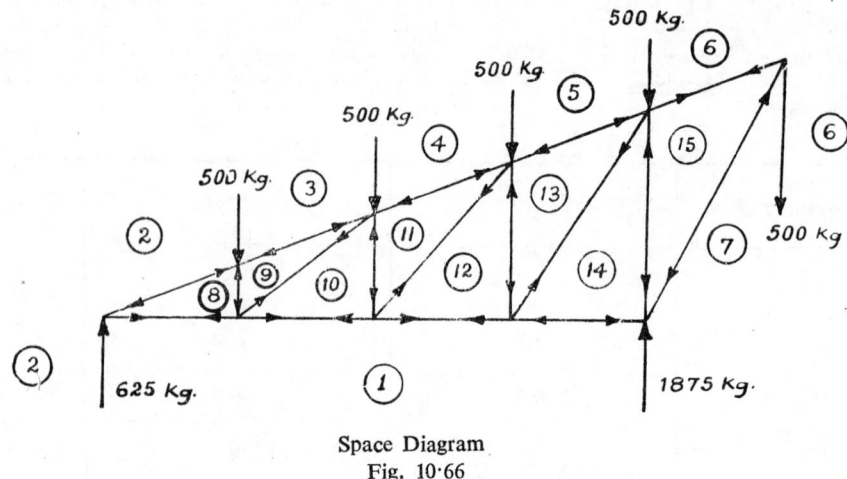

Space Diagram
Fig. 10·66

To draw the stress diagram, start either from A or from J and complete the diagram   as shown in Fig. 10·67.

The results (magnitude and nature of force in each member are shown in the Table).

Table

Member	Force (kg.)	Nature of force	Member	Force (kg)	Nature of force
AC	1650	Tension	CF	500	Compression
CD	990	,,	DG	750	,,
DE	300	,,	EH	990	,,
EB	330	Compression	BI	1230	,,
AF	1770	,,	CG	840	Tension
FG	1770	,,	DH	1020	,,
GH	1650	,,	EI	1200	,,
HI	340	,,	BJ	700	Compression
IJ	340	Tension			

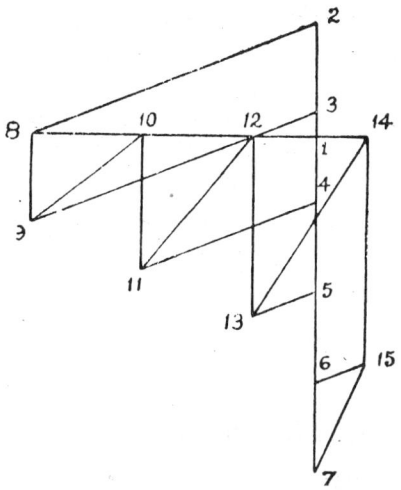

Stress Diagram.
Fig. 10·67.

Problem 23. *The horizontal truss shown in Fig. 10·68 is made of bars 2 m long. Find the loads carried by members AB and CD. Would the truss be statically determinate if there were a bar connecting F and G ? Why ? There is hinge at F and roller support at G.*

(A.M.I.E. May,. 1969)

Solution. The perpendicular distance of line of action of 25 kgf. from F

$$= 2 \sin 60° = 2 \times \frac{\sqrt{3}}{2} = \sqrt{3} \text{ m.}$$

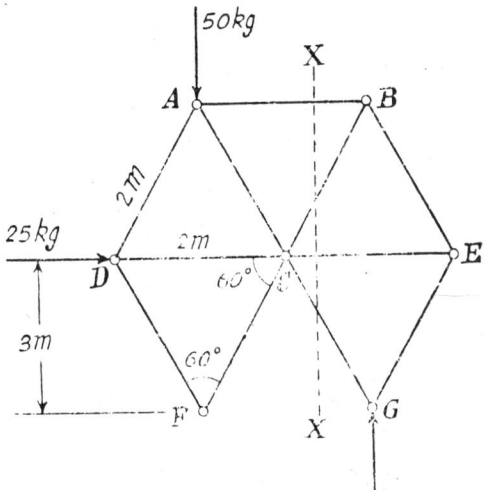

Fig. 10·68.

Let R be the reaction at the roller support at G. Then, by taking moment about **F**,
$$25 \times \sqrt{3} = R \times 2 \qquad \text{or,} \qquad R = 12·50\sqrt{3} = 21·65 \text{ kg.}$$

Let F_{ab} be the force in the member AB. Taking moments about C, $F_{ab} \times \sqrt{3} = 12·5 . 3$,

$$\therefore \qquad F_{ab} = 12·5 \text{ kg (Compression)}$$

Let the members AD and AC be under compression. Considering joint A, vertical components of F_{ad} and F_{ac} must balance 50 kg. f, since $\Sigma V = 0$

So, $F_{ad} \sin 60° + F_{ac} \sin 60° = 50$

or, $(F_{ad} + F_{ac}) = \dfrac{100}{\sqrt{3}}$ ...(1)

The horizontal components of F_{ad} and F_{ac} are $F_{ad} \cos 60°$ and $F_{ac} \cos 60°$ respectively.

$\therefore$ $\Sigma H = 0$, $F_{ad} \cos 60° = F_{ac} \cos 60° + 12.5$

or, $(F_{ad} - F_{ac}) = 25$ ...(2)

Solving equations (1) and (2) we have $F_{ad} = \frac{1}{2}\left(\dfrac{100}{\sqrt{3}} + 25 \right)$

or, $F_{ad} = 41.4$ kg. (Compression).

Let us now consider joint D. Equating the vertical components to zero,

$F_{df} \sin 60° = F_{ad} \sin 60°$.

or, $F_{df} = F_{ad} = 41.4$ kg. (compression).

Equating the horizontal components to zero, we have

$25 + F_{ad} = F_{ad} \cos 60° + F_{df} \cos 60°$.

$= (F_{ad} + F_{df}) \cos 60° = 2 \times 41.4 \times \frac{1}{2} = 41.4$ kg.

$\therefore$ $F_{cd} = 41.4 - 25 = 16.4$ kg. (tension).

Problem 24. *A plane truss is loaded and supported as shown in Fig. 10·69. The ball E rests on the joint D, and its weight Q=1000 kgf. Determine the nature and magnitude of force in the members 1, 2 and 3.*   (*A.M.I.E. Nov., 1969*)

Solution.   Length $AC = \sqrt{AB^2 + AC^2} = \sqrt{16a^2 + 9a^2} = 5a$

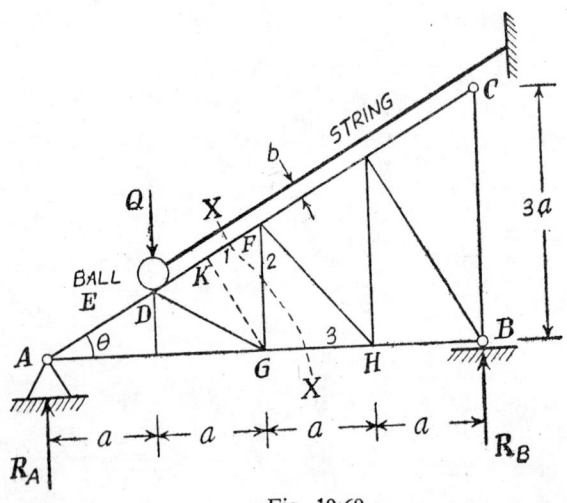

Fig. 10·69.

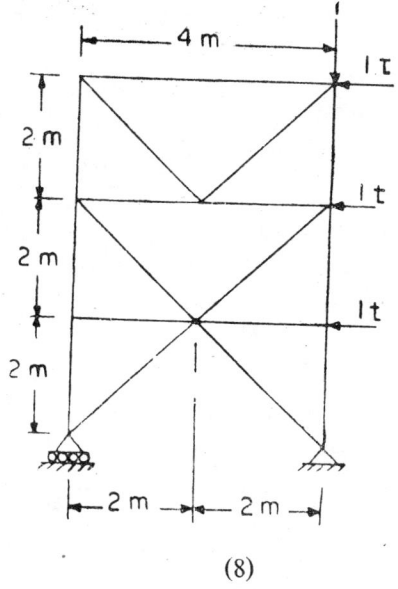

(8)

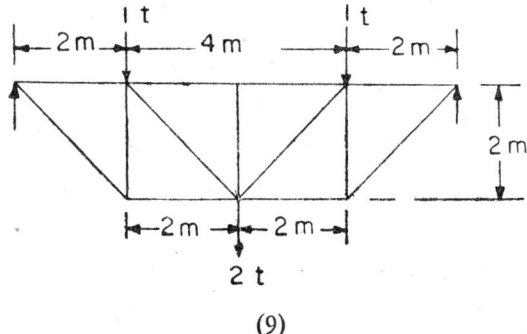

(9)

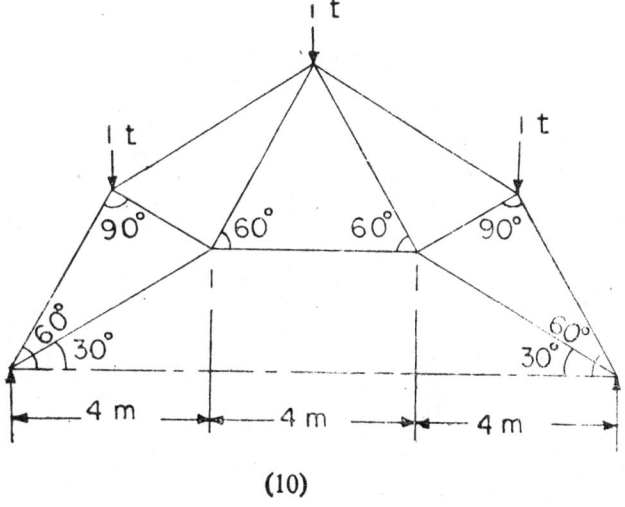

(10)

11

ROOF TRUSSES
(*Timber and Steel*)

The roof is intended to cover the top of a building to keep out the weather. The roofs may be made flat or pitched (sloped). In places subjected to heavy rainfall or snowfall pitched roofs are preferred. Such roofs throw off rain and snow easily and for this, the sides of the roof are tilted up to give them a slope. The inclination of the sides of a roof to the horizontal plane is called the 'pitch' and it is expressed either in degrees or in the ratio of horizontal to vertical. A roof whose sides slope at $26\frac{1}{2}°$ to the horizon, has a rise equal to $\frac{1}{4}$th of the span, and is called a roof of $\frac{1}{4}$ or $26\frac{1}{2}°$ pitch. However, the slope of a roof depends upon the climate, span of the roof and the type of covering to be used. The framework for building such a roof is called 'roof truss'. There are various types of roof trusses depending upon the span to be covered.

Technical Terms used in Roof Trusses

1. *Span* : The horizontal distance between the end supports is called span.

2. *Wall plates* : These are timber blocks provided over the top of a wall to support the end of a roof truss.

3. *Rise* : It is the vertical distance (height) between the wall plate and the ridge top.

4. *Pitch* : It is the slope or angle of inclination of the common rafer with the horizontal tie.

5. *Ridge* : It is the apex of a roof truss.

6. *Ridge Piece or Ridge Board* : It is the piece of timber provided at the ridge of a truss, to which common rafter is fixed.

7. *Valley* : The acute angle or the 'V' formed by the intersection of two slopes in pitched roofing is called valley.

8. *Eaves* : It is the lower edge of the sloped roof surface.

9. *Gable* : It is the triangular portion of a wall formed at the end of a sloped roof.

10. *Eaves Board* : It is a wooden board or plank fitted to the eave in order to throw down the rain water or snow.

11. *Fascia* : It is similar to the Eaves Board. This board is fixed to the ends of the rafters.

12. *Eaves Gutter* : It is the gutter or channel fitted to the eaves board to collect rain water and to pass on to the drainage conduit.

13. *Parapet Gutter* : It is the gutter fitted to the parapet wall built at the side of a sloped roof.

14. *Gutter Bearer* : It holds the gutter.

15. *Riser* : It is the wooden piece required to support the gutter bearer.

16. *Hip* : It is the angle formed at the intersection of two roof slopes.

17. *Collar* : It is a beam used in a truss about halfway up the rafters.

18. *Tie Beam* : It is a beam subjected to tensile stress only and is used in a truss to hold the feet of the rafters to prevent them from spreading out.

19. *King post* : It is a vertical post which acts as a Tie in a roof truss to hold up the centre of the tie beam and prevent it from sagging.

20. *Queen post* : It is similar to a king post.

21. *Principal Rafter* : The principal rafter carries the purlins which are notched to fit them. Each principal rafter is supported near its centre, close below the purlin by a strut tenoned into it.

22. *Strut* : A strut is a compression member in a roof truss which prevents sagging of the principal rafters. The head of a strut is tenoned into the principal rafter and its foot into the foot of the king post. The strut should be almost immediately under the purlin.

23. *Purlin* : A purlin is a beam running longitudinally from principal to principal as support for the common rafter. Purlins are sometimes framed in between the principal rafters. The purlins are generally notched where they rest upon the principal rafters, so as to keep the latter rigidly apart.

24. *Cleat* : It is a wooden block housed into the back of rafters to support the purlins.

25. *Common Rafters.* : These are wooden members laid from the ridge to the eaves and parallel to the roof pitch. They are supported by purlin at their centre. Common rafters are bevelled at the upper end to abut against the ridge piece and nailed to it.

26. *Boarding* : Roof boarding is nailed upon the common rafters to receive the roof covering. Boarding generally runs parallel to the ridge.

DIFFERENT TYPES OF TIMBER ROOF TRUSSES

Lean to Roof or Pent Roof

This type of sloped roofing is intended to cover an open verandah. Fig. 11.1 shows the sectional view of such a roof. The rafters are placed between two walls, one higher than the other and are nailed to the wall plates, the higher end being secured to the wall by bolts.

Couple Roof

This is the simplest form of sloped roofs with two sides sloping and it is formed by the meeting of two rafters as shown in Fig. 11.2. At the apex, the rafters are secured to the ridge board and their feet are nailed and notched upon a wall plate embedded on the wall top. In such a roof, the rafters have the tendency of spreading out and there is nothing to prevent it. For this, the span of such roof should be restricted to 5.4 m.

Tentative Scantlings for couple Roofs (Rise=$\frac{1}{4}$th span)

Span from centre to centre of wall plates, m.	Rafters (mm × mm)	Ridge Board (mm × mm)
2·4	75 × 50	175 × 40
3·0	90 × 50	175 × 40
3·6	100 × 50	175 × 40
4·2	115 × 50	175 × 40
4·8	125 × 50	200 × 40
5·4	150 × 50	200 × 40

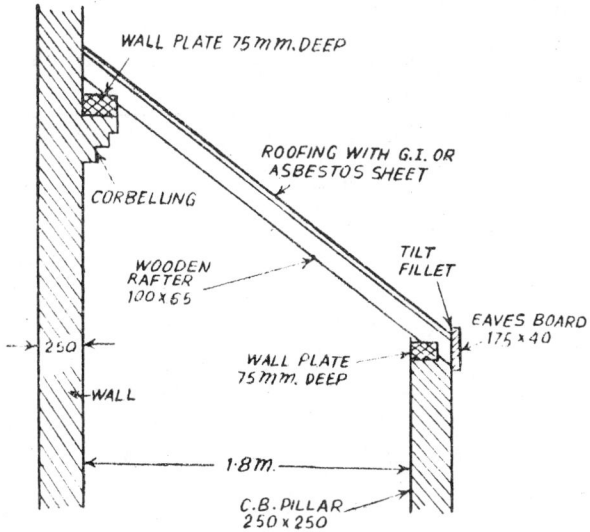

LEAN TO OR PENT ROOF
Fig. 11·1.

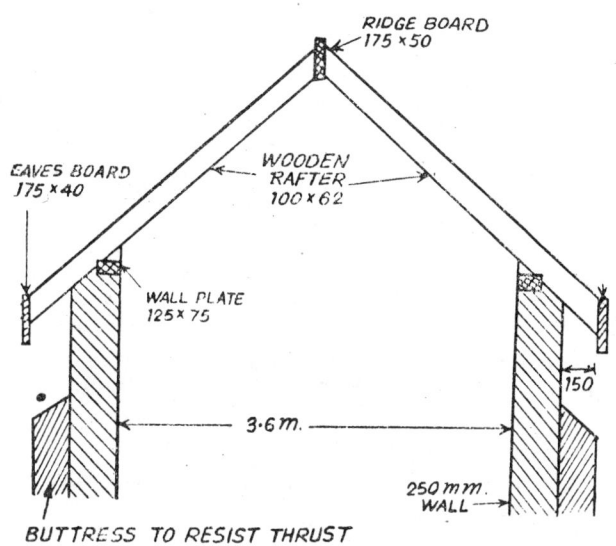

SIMPLE COUPLE ROOF
Fig. 11·2,

Couple close Roof

In couple close roof the spreading out of the rafters is prevented by providing a **Tie** beam (horizontal member in this case) to hold the rafters at their feet. The tie beam also acts a ceiling joist. The sectional view of a couple close roof is shown in Fig. 11˙3.

Collar Tie Roof

Where considerable head room is required, a collar tie beam is placed higher up as shown in Fig. 11.4.

Tentative Scantlings for Collar Tie Roof

Span in m.	Rafters (mm × mm) 300 mm apart	Collars (mm × mm) 1/4 to 1/2 way up
2·4	55 × 80	45 × 60
3·0	55 × 100	50 × 60
3.6	55 × 115	50 × 70
4·2	60 × 120	50 × 75
4·8	65 × 130	50 × 85
5·4	65 × 150	50 × 100

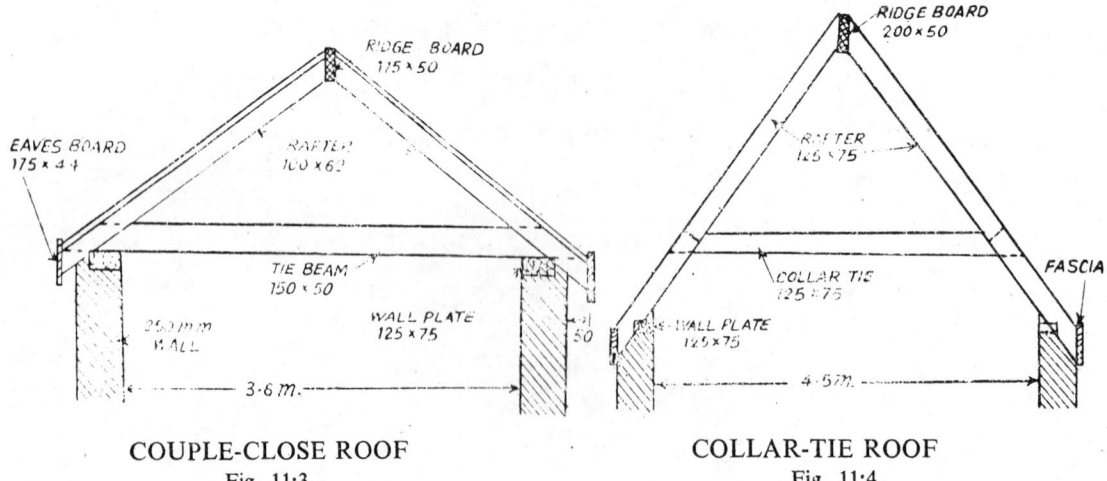

COUPLE-CLOSE ROOF
Fig. 11·3.

COLLAR-TIE ROOF
Fig. 11·4.

Framed House Roof

This type of Timber roofing frame has been thought of for wall to wall span of about 10 m. with a central pillar as shown in Fig. 11·5. The frame consists of the following members :

Tie beam—Resting upon wall plates ;

Rafters—2 Nos, held in position by the tie beam and the ridge piece ;

Ridge piece—Holding the rafters at the ridge ;

Hangers—2 Nos, acting as tensile members which prevents the central deflection of the tie beam ;

Struts—2 Nos, acting as compression members to support the rafters ;

Purlin and Cleats—used to support the rafters ;

Collar Bolt and Tie Truss

This type of truss as shown in Fig. 11.6 is a combination of wooden and iron truss and is suitable for a clear span of about 12 m. The principal rafters are held firmly by metal shoes at the two ends. The tie rods are inclined and are tied with the king bolt. The collar beam is provided at half way up and is tied with the tie rods and the king bolt. The common rafters are supported by purlins as shown in figure.

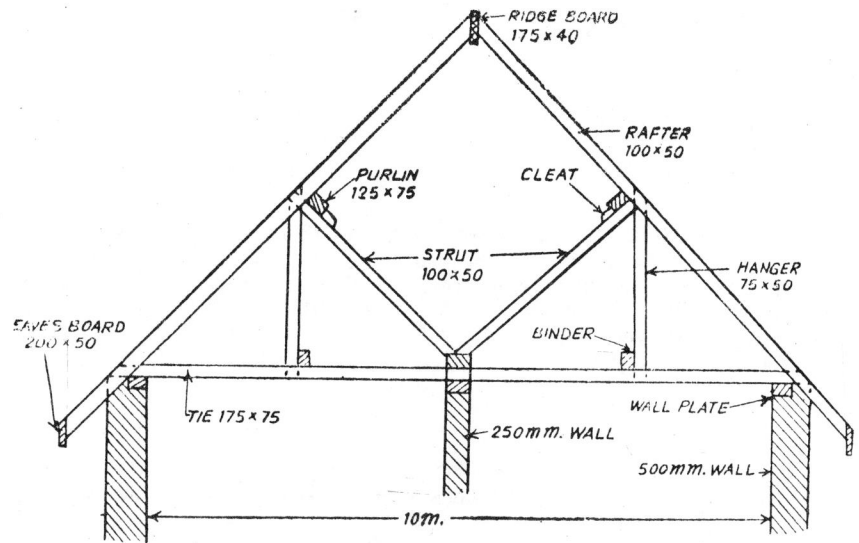

FRAMED HOUSE ROOF
Fig. 11·5.

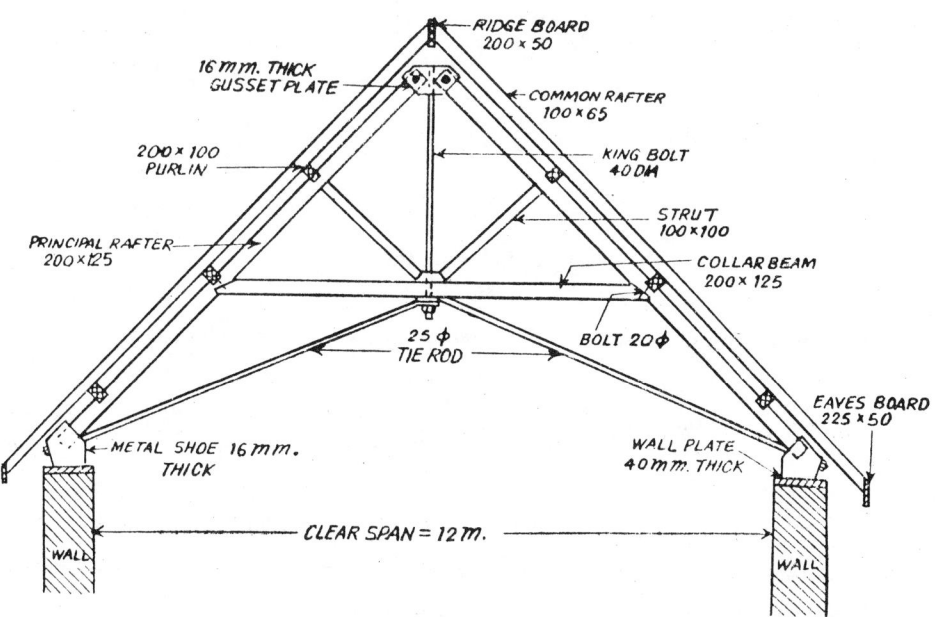

COLLAR BOLT AND TIE TRUSS
Fig. 11·6.

King Post Roof Truss

A king post roof truss is suitable for spans from 5 m to 9 m. This truss consists of the following members :

1. Tie beam
2. King post

3. Principal rafters

4. Struts

5. Common rafters

6. Purlins and cleat

7. Ridge Board

8. Struts.

A king-post roof truss for a clear span of 6 m with various members with their sizes are illustrased in Fig. 11·7. The tentative dimensions of different members of a king post truss for various spans are given in a tabular form. Details of iron fittings for a king post truss of various spans are also given in tabular form. The junction of the king post at the top with the ridge board, common rafters and principal rafters is detailed in Fig. 11·8. The junction of the king post at the bottom with the Tie, and the struts is shown in details in Fig. 11·9.

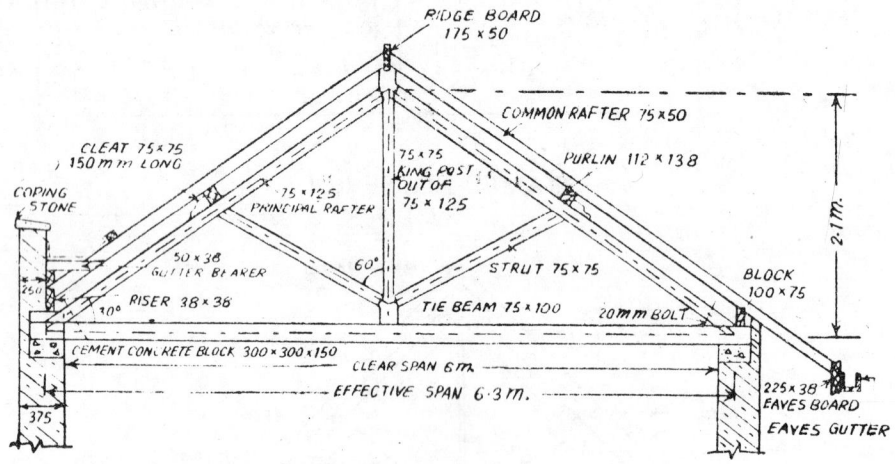

KING POST ROOF TRUSS
Fig. 11·7.

Tentative dimensions of different members of a king post-truss.

Clear Span	Tie beam	Principal rafter	Common rafter	King post	Strut	Purlin	Ridge board
4·5 m.	75×75	75×100	75×50	75×125	75×50	100×112	50×175
6 m.	75×100	75×125	75×50	75×150	100×75	112×138	62×175
7·5 m.	100×125	100×125	100×60	100×150	125×75	125×150	75×150
9 m.	125×150	100×150	100×75	125×175	150×100	150×175	100×200

Dimensions are in mm.

Details of iron fittings for a king-post truss.

Clear span	Junction of the King post and the tie beam				Junction of the tie beam and the principal rafter			
	Dia. of bolt in mm.	Iron strap		Distance between end of strap and bolt mm.	Dia. of bolt mm.	Iron strap		Distance between end of strap and bolt, mm.
		Thick-ness mm.	Width mm.			Thick-ness mm.	Width mm.	
4·5 m	16	6	25	20	20	6	32	25
6 m	20	10	38	20	25	10	44	32
7·5 m	22	10	50	22	25	10	50	44
9 m	25	12	60	25	25	12	50	50

Dimensions are in mm.

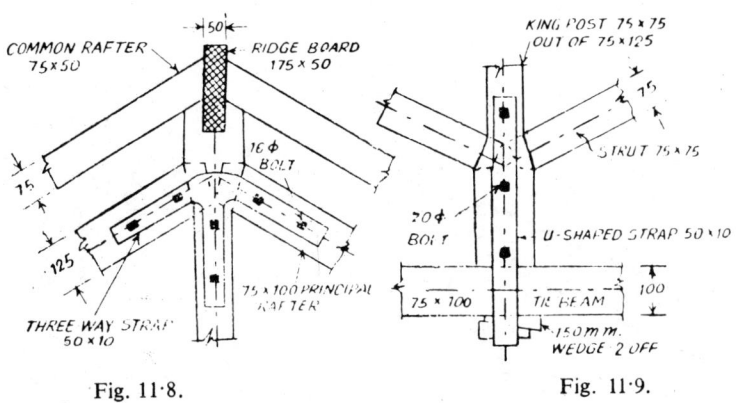

Fig. 11·8.   Fig. 11·9.

DETAILED JOINTS OF A KING POST TRUSS

Fig. 11·10 shows in details the joining of different members at the end support of a king post truss. The figure also shows the placement of parapet gutter with its support.

The joining of strut with the principal rafter and supporting of the common rafter by the purlin just above the joint are presented in Fig. 11·11. Fig. 11·12 shows how a strut is housed into the principal rafter.

Fig. 11·13 illustrates the way of joining the struts and the tie beam with the king post.

The two ways of joining the principal rafter with the tie beam are shown in Fig. 11·14 and Fig. 11·15.

The joining of the principal rafers with the king post at top is shown in Fig. 11·16.

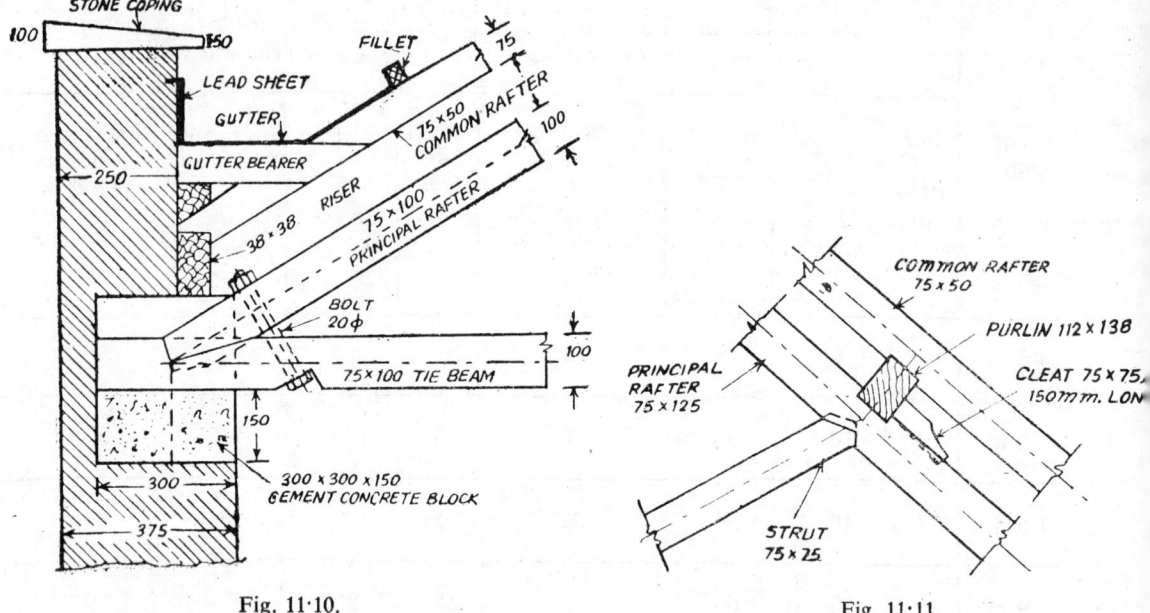

Fig. 11·10.

Fig. 11·11.

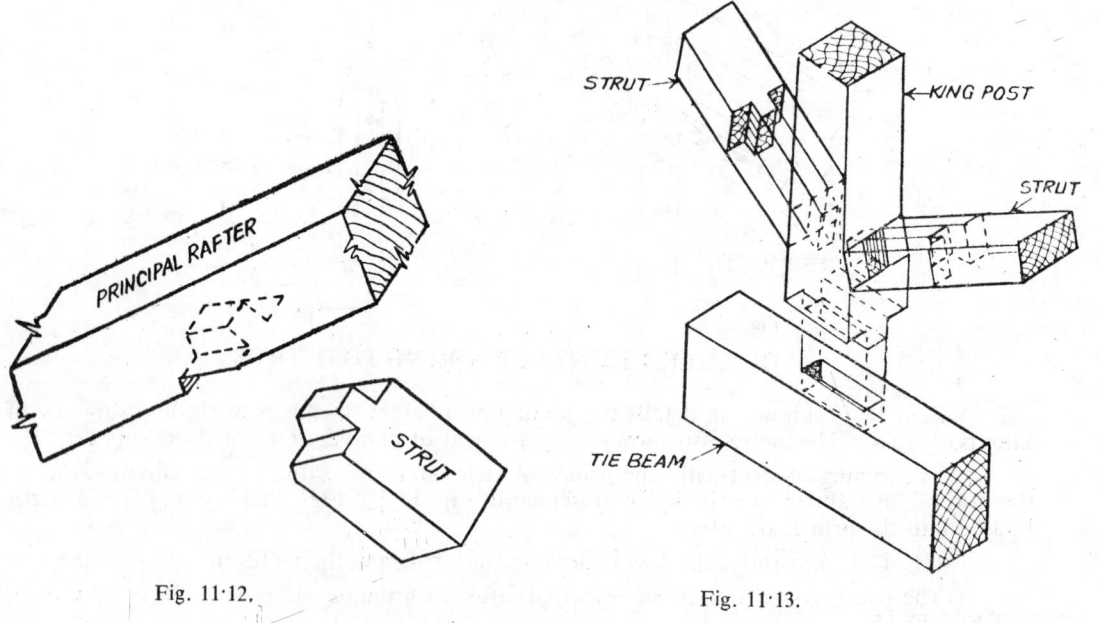

Fig. 11·12.

Fig. 11·13.

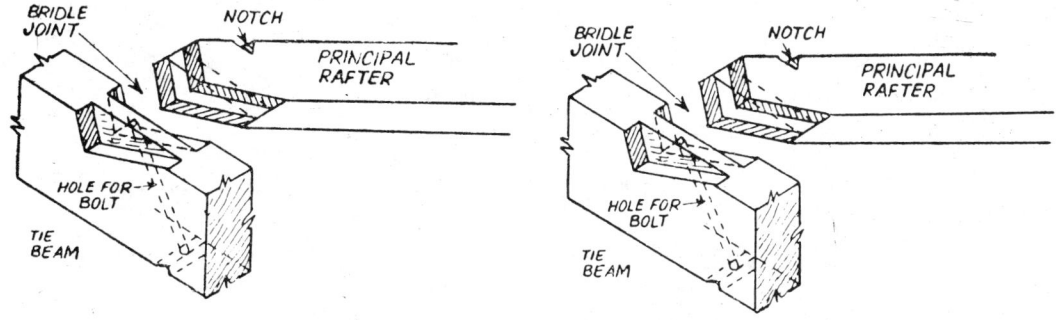

Fig. 11·14. Fig. 11·15.

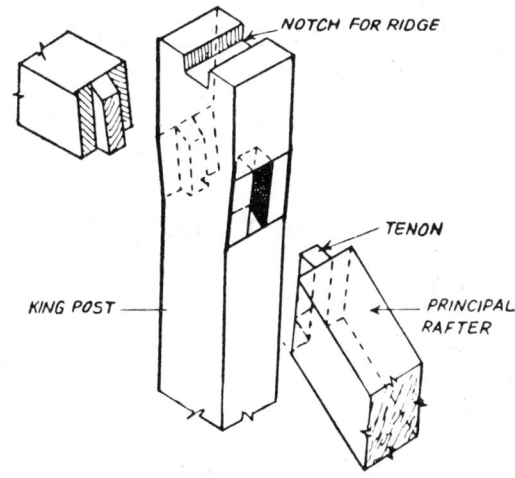

Fig. 11·16.

Queen Post Roof Truss :

A Queen-post roof truss comprises of the following members :

1. A tie beam, the bottommost horizontal member.

2. Queen post—2 Nos. of the vertical member.

3. A straining beam between two queen posts at their top ; This is a horizontal member.

4. Principal rafter—2 Nos. of inclined members. Each rafter is joined with the queen post at top and with the beam at bottom.

5. Strut—2 Nos. supporting the principal rafters at their centres.

6. Common rafter-2 Nos. of inclined members running from ridge to eave. This is supported by pole plates and purlins at intermediate positions.

Fig. 11·17 illustrates a portion of a queen post truss for a clear span of 12·6 m. This type of truss is most suitable for a clear span ranging from 9 m. to 12 m. The span should not exceed 14 m. in any case. Tentative dimensions of different members of a queen-post roof truss for various spans are given in tabular form. The dimensions of bolts and straps required for a queen post truss for different spans are also presented in tabular form.

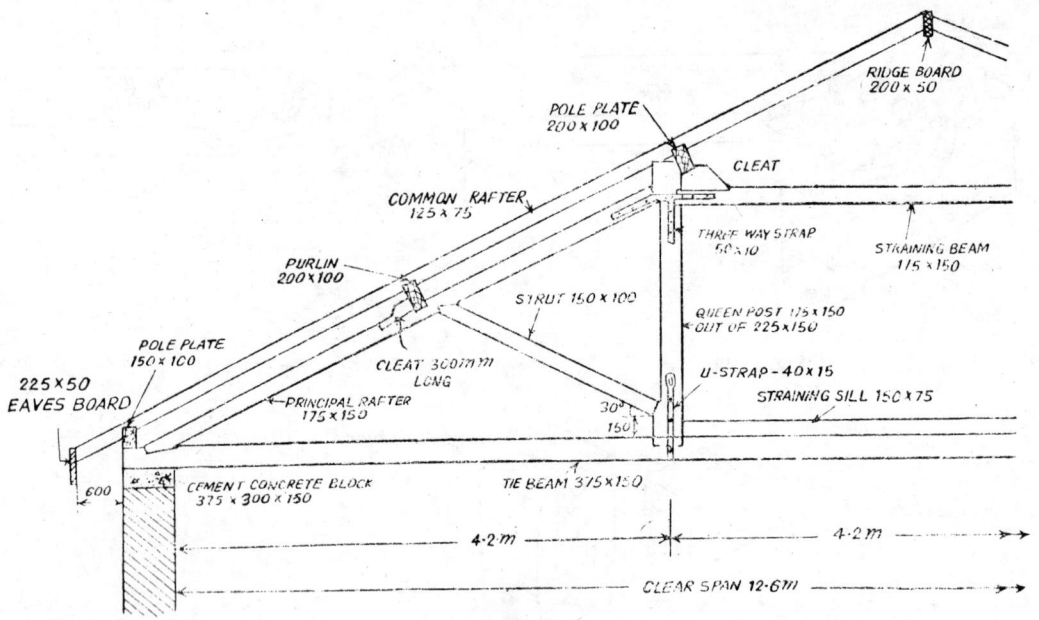

QUEEN POST ROOF TRUSS
Fig. 11·17.

Tentative Dimensions of Different Members of a Queen-post Truss

Clear Span	Queen Post	Straining beam	Straining Sill	Strut	Tie beam	Principal rafter	Common rafter	Purlin	Ridge piece
9 m.	100×125	100×125	100×50	100×75	100×100	100×150	50×75	125×138	62×175
10·5 m.	100×125	100×125	100×50	100×75	100×125	100×175	50×75	125×138	62×175
12 m.	125×150	125×150	125×50	125×75	125×125	125×175	50×87	125×150	62×175

Dimensions are in mm.

Dimensions of Bolts and Iron straps for a Queen-post Truss

| Clear Span | Bolts at the junction | | | | Iron strap at the junction | |
| | Principal rafter and tie beam | | Tie beam and Queen post | | Principal rafter and tie beam | Tie beam and Queen post |
	Diameter	Distance between end of strap and bolt	Diameter	Distance between end of strap and tie bolt		
9 m.	25	44	20	20	56×12	25×10
10·5 m.	28	50	22	25	62×15	32×12
12 m.	32	56	25	25	70×20	38×15

Dimensions are in mm.

Different forms of eaves are presented in Fig. 11·18.

Fig. 11·19 illustrates in details a truss end. It shows fixing of boards on to the purlins supported by the common rafter. The fixing of eaves gutter with the fascia, gutter bearer, etc. is also shown.

A mansard roof truss which in essence is a combination of king-post and queen-post trusses is shown in Fig. 11·20.

A bracketted hammer-beam roof truss is shown in Fig. 11·21. This is an ornamental roof truss, the upper part of which is a king-post truss. This type of truss in built in churches.

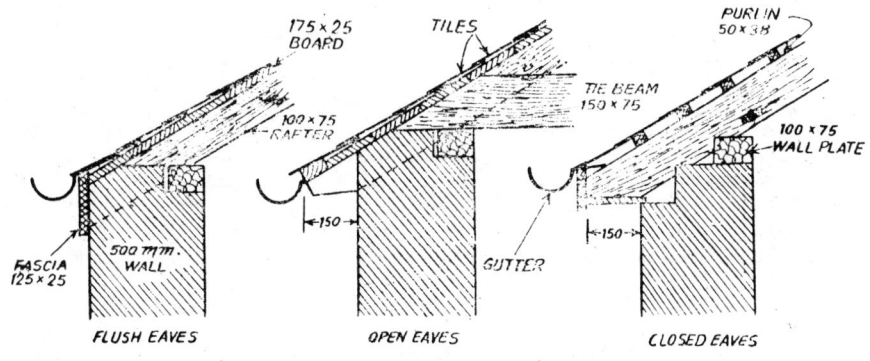

Different forms of Eaves
Fig. 11 18.

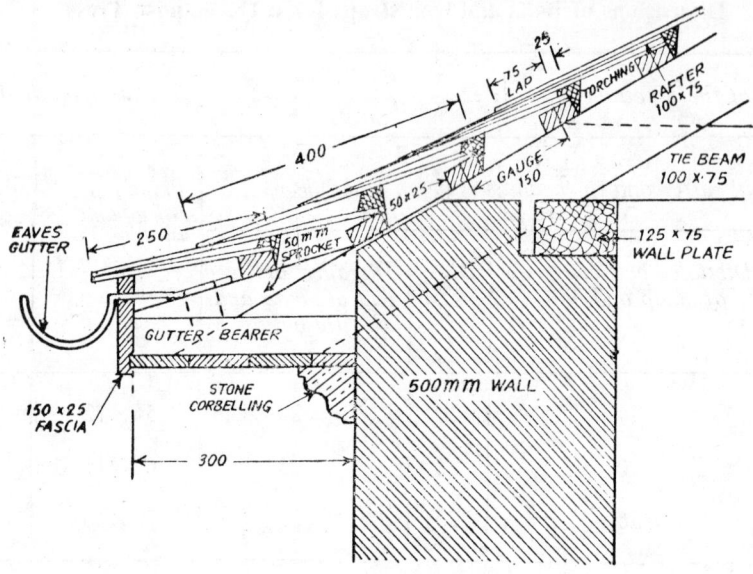

Details at Truss End
Fig. 11·19.

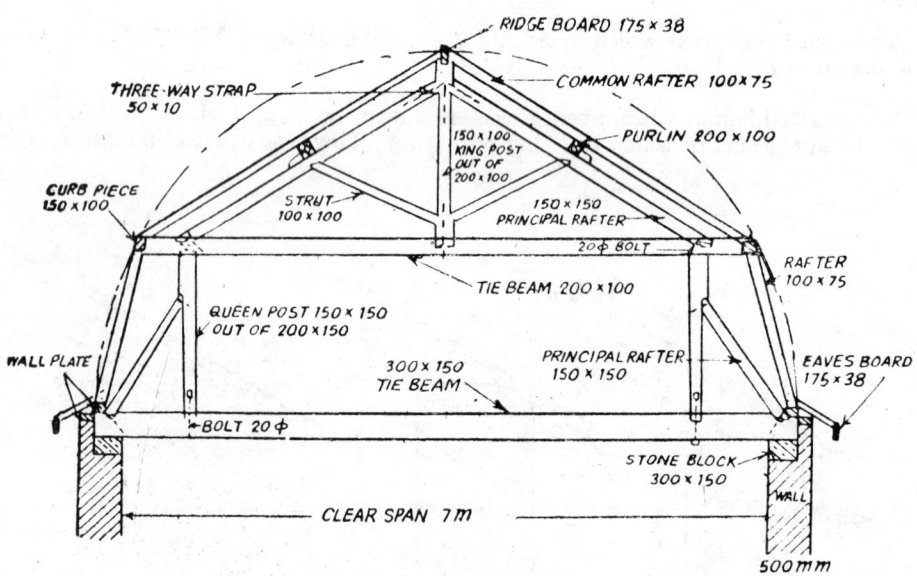

MANSARD ROOF TRUSS
Fig. 11.20.

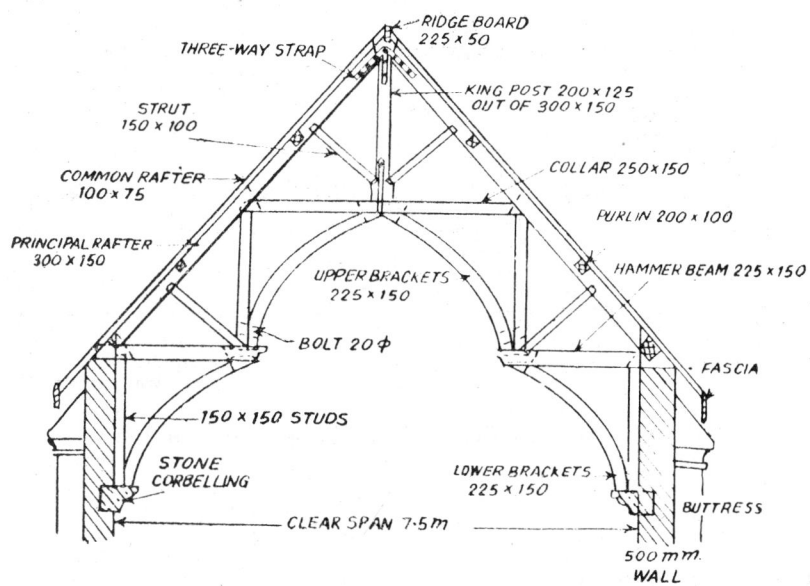

BRACKETED HAMMER-BEAM ROOF TRUSS
Fig. 11·21.

STEEL ROOF STRUSSES

Steel roof trusses have superseded to a great extent the timber roof trusses, especially in trusses of large span. For large spans beyond 9 m., timber trusses are not used because of heavy sections of timber, difficulty in erection, uncertainty in stress distribution, question of durability and involvement of high cost (capital and maintenance costs).

There are various forms of steel roof trusses designed to suit various spans. Mainly, they may be classified as :

(*a*) Truss with arched rafters ;   (*b*) Truss with straight rafters ;   (*c*) Special trusses ;

Truss with Arched Rafters

The simplest form of arched roof consists of G.I. sheets riveted or screwed together into the form of an arch as shown in Fig. 11·22. The edges of the arched sheet are secured at the springing to wall plates, angle iron, etc. This type of roofing is recommended for a span of about 3 m., beyond which tie rods, king bolts and struts are used. The ends of

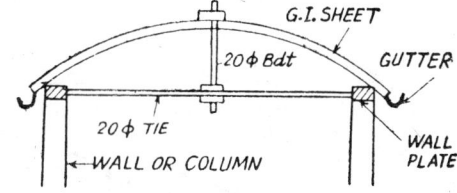

Fig. 11·22.

the tie rod are secured to plates on the heads of the columns or walls supporting the roof as shown in Fig. 11.23. Such roofs are suitable for spans of 6 m. to 9 m. But, for spans beyond 9 m, curved principals must be used with purlins to carry the roof covering. Tie rods are usually spaced 2 to 2·4 m. apart.

Roofs with Straight Rafters

In trussed rafter roofs, the principal rafters are supported by one or more than one struts at right angles or nearly at right angles to them which together with the tension rods form the principal rafters into a pair of trusses joined at the ridgr of the roof. In some trusses, the struts supporting the principal rafters are placed vertically. Various forms of trussed rafter steels roofs of spans varying from 4·5 m. to 13·5 m. are illustrated in Fig. 11·24 and steel trusses for spans 12 m to 18 m. are shown in Fig. 11·25.

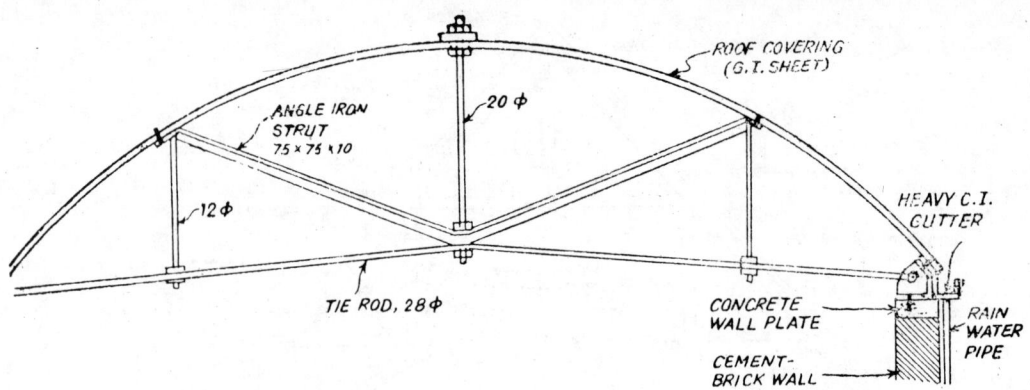

TRUSS WITH ARCHED RAFTERS (Span 6 m. to 9 m.)
Fig. 11·23.

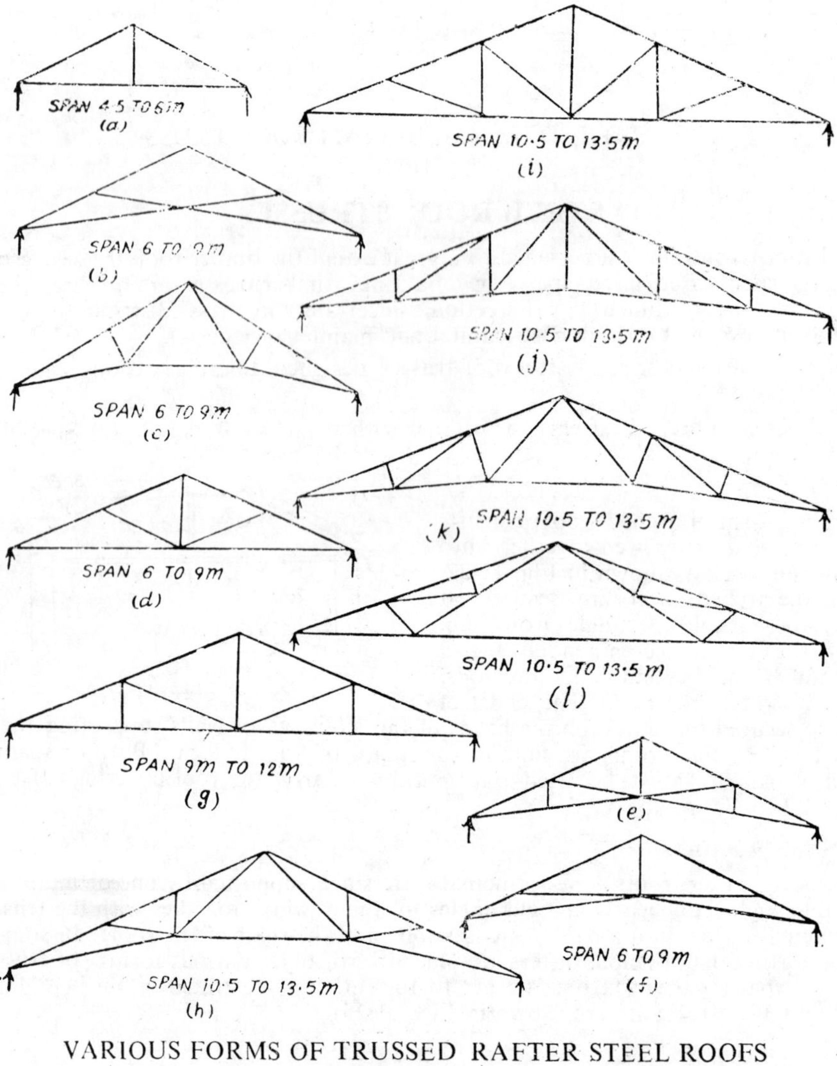

SPAN 4 5 TO 6 m
(a)

SPAN 6 TO 9 m
(b)

SPAN 6 TO 9 m
(c)

SPAN 6 TO 9 m
(d)

SPAN 9 m TO 12 m
(g)

SPAN 10·5 TO 13·5 m
(h)

SPAN 10·5 TO 13·5 m
(i)

SPAN 10·5 TO 13·5 m
(j)

SPAN 10·5 TO 13·5 m
(k)

SPAN 10·5 TO 13·5 m
(l)

SPAN 6 TO 9 m
(e)

SPAN 6 TO 9 m
(f)

VARIOUS FORMS OF TRUSSED RAFTER STEEL ROOFS
(Spans of 4·5 m to 13·5 m)
Fig. 11·24.

King rod or king bolt roof truss and its modifications are shown in Fig. 11'24 (d), (e) and (g). Here, the struts abut against one another and rest upon a nut at the lower end of the king bolt, which can be screwed up to tighten the roof when needed. The rafters are of angle iron or T-iron, united at the apex by a pair of overlapping plates riveted to both and from which is suspended the king bolt, the head of which is forked. The tie rod is bolted to the lower end of the rafter and is supported at the centre by a double nut at the foot of the king bolt. The lower end of the rafter is secured to the head of the column supporting the truss.

A queen-rod or queen-bolt truss and its modified forms are presented in Fig. 11'24 (i), (j) and in Fig. 11'25 (a), (b).

Forms of special trusses for spans 12 m. to 18 m. are illustrated in Fig. 11'25 (c), (d) and (e).

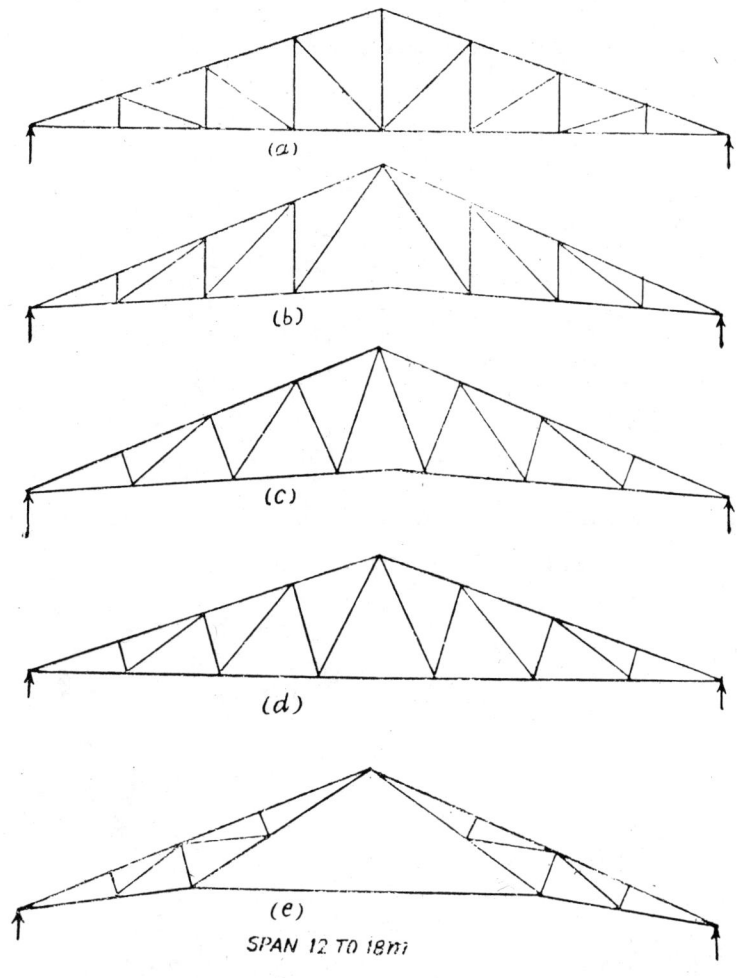

(a)

(b)

(c)

(d)

(e)

SPAN 12 TO 18m

VARIOUS FORMS OF STEEL TRUSSES
Fig. 11·25.

Fig. 11'26 shows two forms of steel trusses for spans 15 m. to 22'5 m.

A tubular roof truss and its junction details are presented in Fig. No. 11'27, 11'28 and 11'29.

A saw tooth (north light) roof truss with its junction details is shown in Fig. 11'30.

Various other forms of steel roof trusses and their junction details are illustrated in Fig. 11'31 through 11'35.

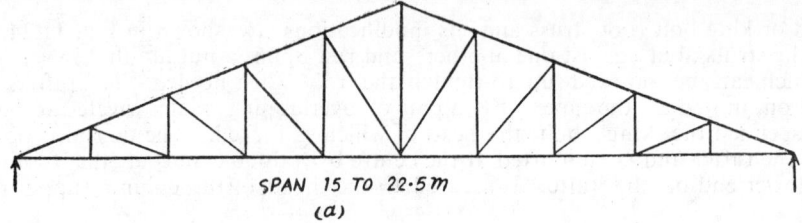

SPAN 15 TO 22·5m
(a)

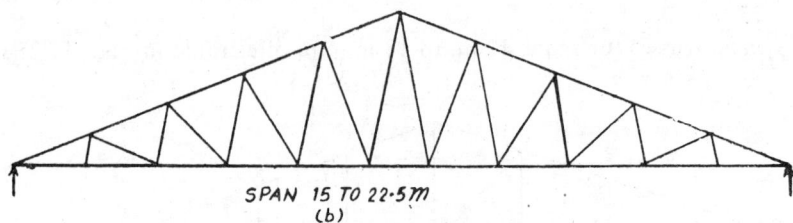

SPAN 15 TO 22·5m
(b)

STEELS TRUSSES FOR LARGE SPANS
Fig. 11·26.

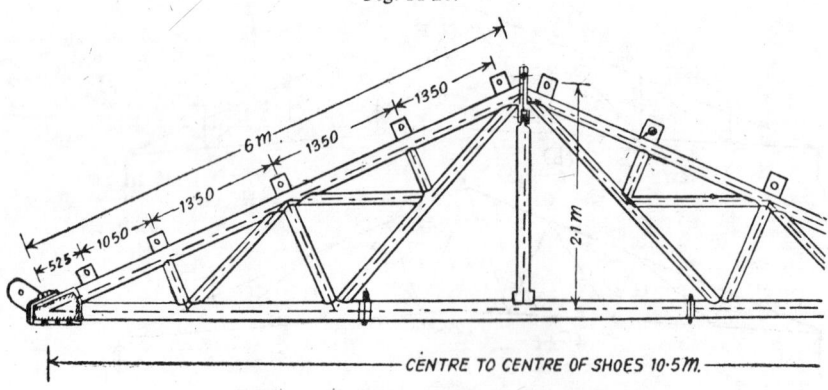

TUBULAR STEEL ROOF TRUSS
Fig. 11·27.

NUT WELDED TO PLATE AND
END OF PURLIN

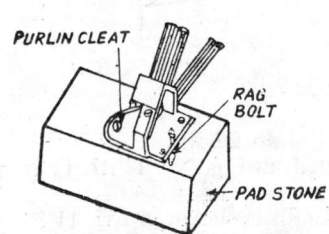

PURLIN CLEAT

RAG BOLT

PAD STONE

DETAIL OF TRUSS FIXING TO PADSTONE
Fig. 11·28.

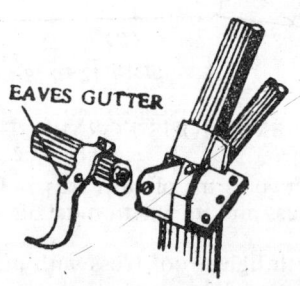

EAVES GUTTER

DETAIL OF CONNECTIONS OF TRUSS TO COLUMN CAP & PURLIN TO TRUSS
Fig. 11·29.

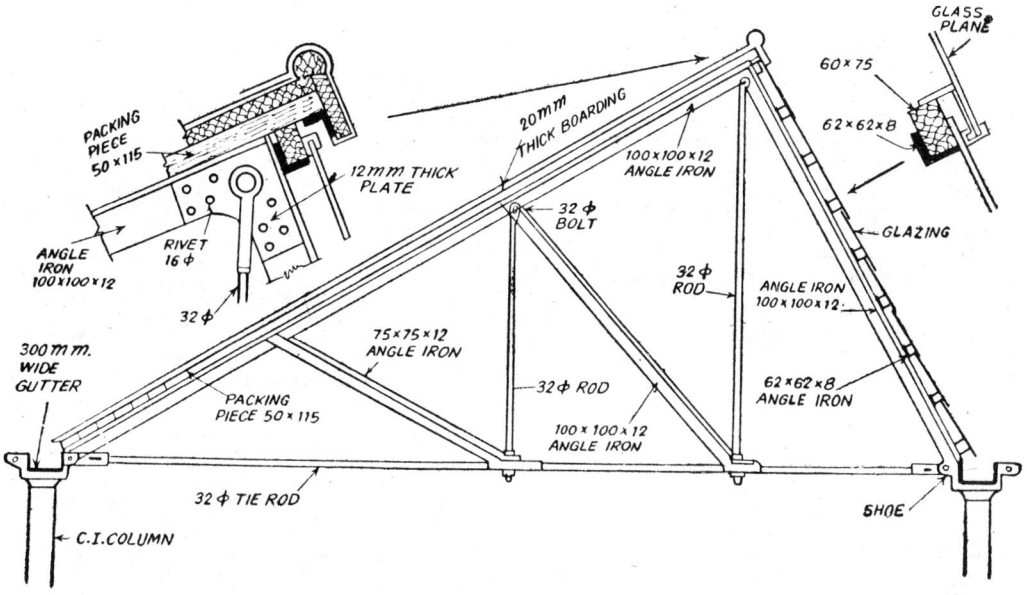

NORTH LIGHT ROOF TRUSS
Fig. 11·30.

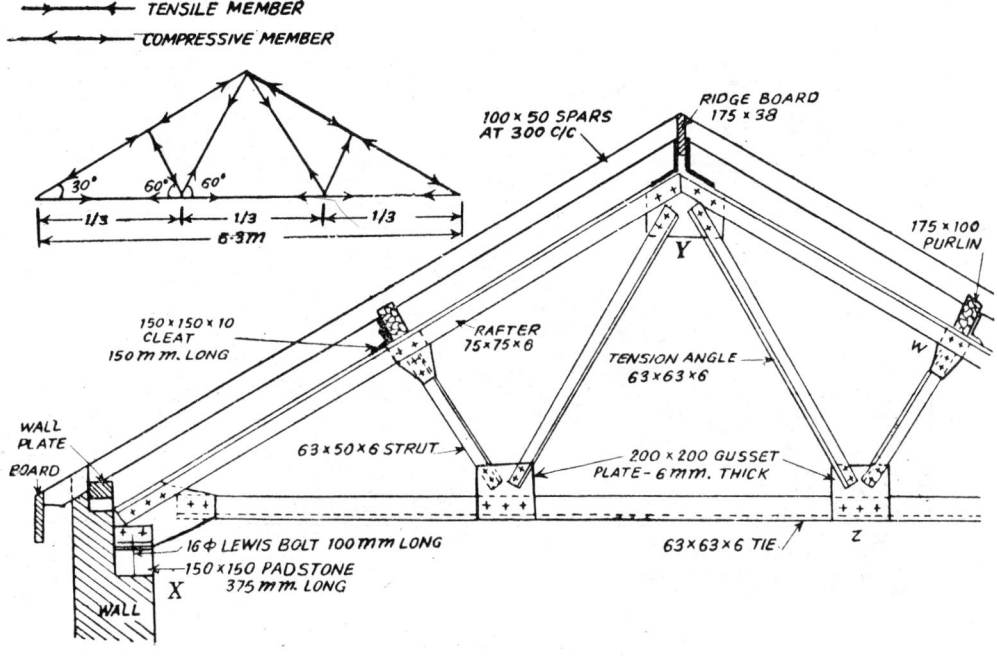

Fig. 11·31.

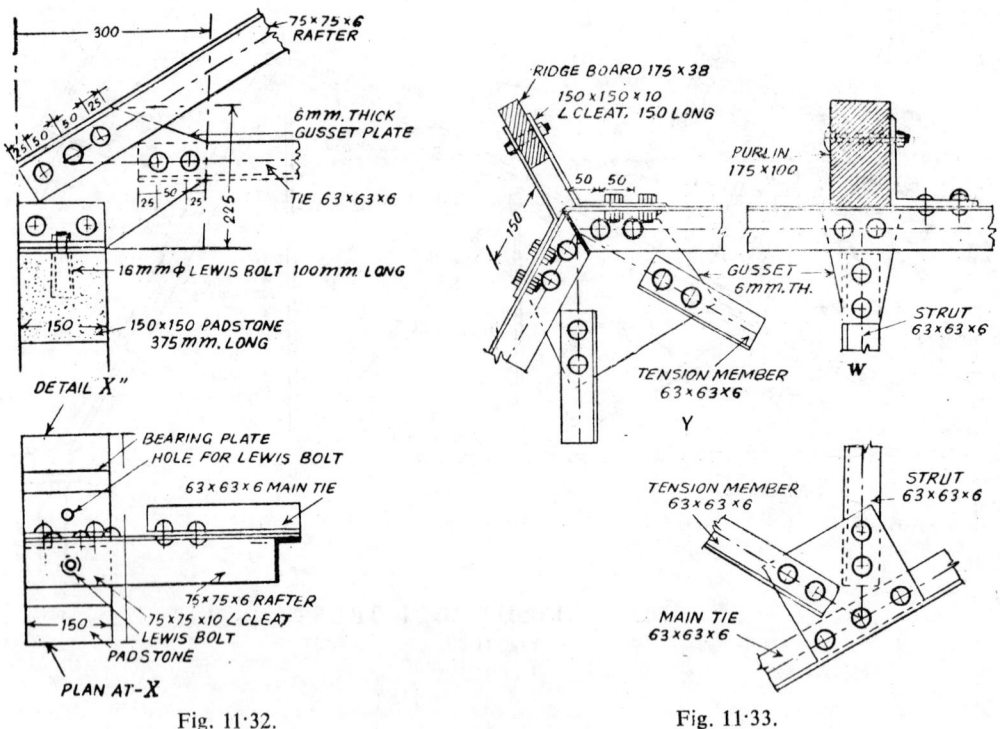

Fig. 11·32.

Fig. 11·33.

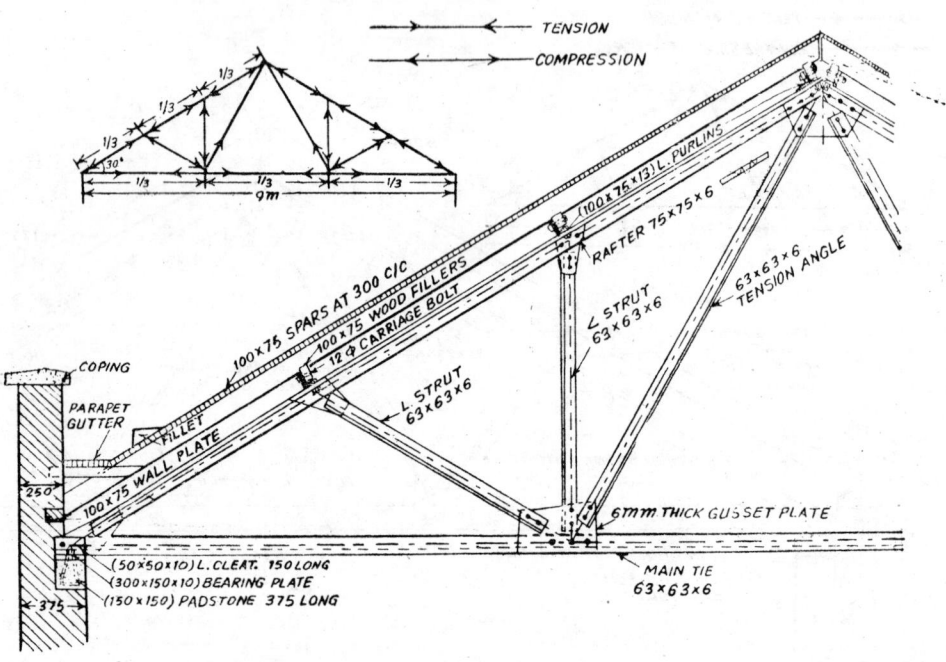

Fig. 11·34.

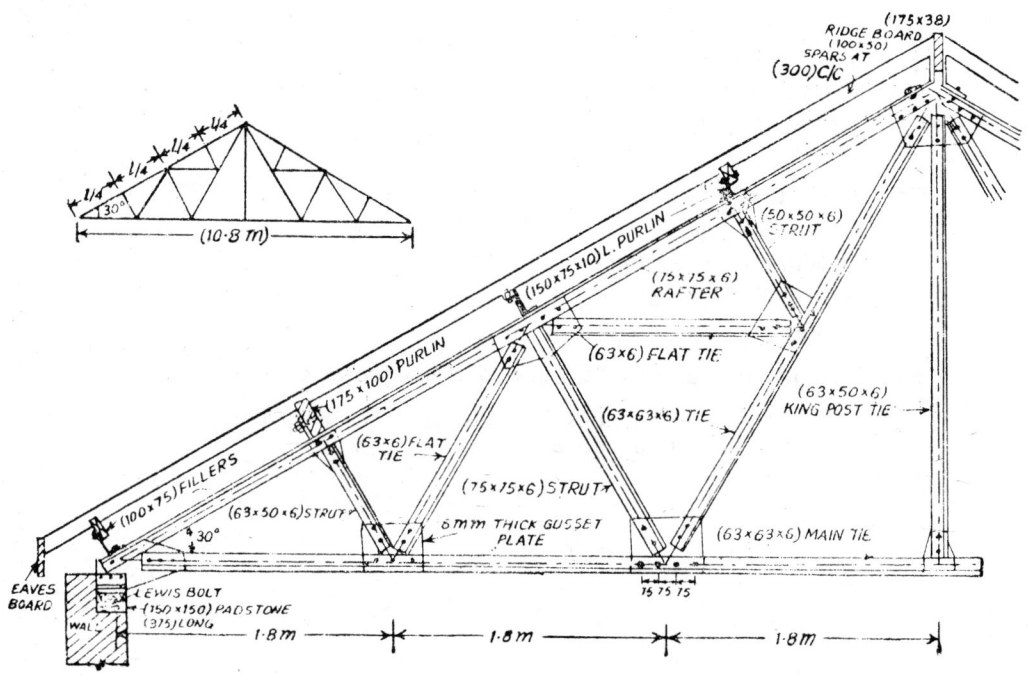

Fig. 11·35.

EXERCISES ON CHAPTER 11

1. Draw a couple-close roof for a clear span of 4 m., with the following dimensions :
   Principal rafter 125×60
   Tie beam 175×60
   Ridge board 175×50
   Eaves board 175×44
   Assume rise, wall thickness, wall plate, boarding etc.

2. Show the junction details of a king post at its top and bottom, when used in a timber roof truss. The following dimensions are given :
   King post 80×80 out of 80×130
   Tie beam 80×125
   Strut 75×75
   Principal rafter 80×125
   Common rafter 80×50
   Ridge board 175×50
   U-Strap 50×12
   Three-way strap 50×12
   Bolts 20ϕ
   Assume all other dimensions required to make the sketch.

3. Make neat sketches of three important junction of a queen post timber truss. The clear span of the truss is 12 m.

4. Draw the end portion of a king post timber roof truss showing the junction details of the following :

Tie beam 125×150
Principal rafter 100×150
Common rafter 100×75
Gutter bearer 60×40
Riser 40×40
Parapet gutter 400 mm. wide
Parapet wall 250 thick
Superstructure wall 375 thick
C.C. block 300×300×150
Bolt 25 mm ϕ.

5. Sketch the junction details of the apex of a saw tooth steel truss.

6. Sketch the details at junctions A, B and C of the given skeleton frame of a steel roof truss. Assume suitable dimensions.

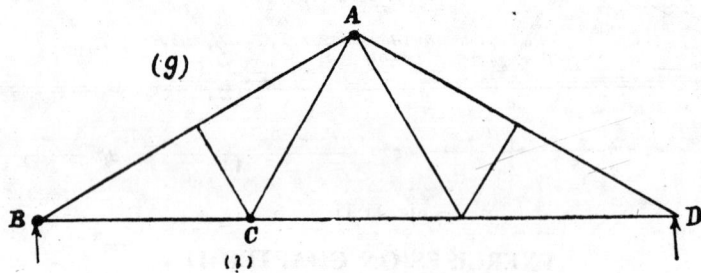

12

Building Foundations

A structure erected on ground must have a foundation for its stability and for the distribution of loads coming over it. Thus, a building requires a foundation. The width of foundation needed for a building depends upon the following two factors :

(1) Load to be transmitted, and

(2) Bearing power of soil.

There are other various factors related to the above two factors which are to be considered prior to selection of a particular type of foundation for a specified class of building to be erected at a predetermined site. This will involve site investigation and soil testing. Therefore, soil characteristics, sub-soil water condition, soil stability, etc. are to be determined first.

The foundations for most structures are constructed below the ground surface. A foundation should be built on a firm soil which will not cause unequal settlement of the structure.

Various types of building foundations are :

1. Spread footing foundation
2. Raft foundation
3. Inverted Arch foundation
4. Well foundation
5· Grlllage foundation

Spread Footing Foundation

This is the simplest form of foundation commonly used in building construction. A footing is an enlargement of the base of a wall or column to spread the load over greater area so that the building load is well distributed to the soil without causing any marked settiement. The base of the wall or the column is gradually made wider in a series of steps.

A footing that supports a single column is called an 'isolated footing' or 'spread footing' or 'individual footing'. The footing beneath a wall is known as 'wall footing' or 'continuous footing'. When a footing supports more than one column, it is called a 'combined footing''. An isolated footing for a column is shown in Fig. 12·1. A spread footing foundation for a wall (continuous footing) is illustrated in Fig. 12·2. Where the spreading of the wall base is restricted in one side, offset foundation is provided as shown in Fig. 12·3.

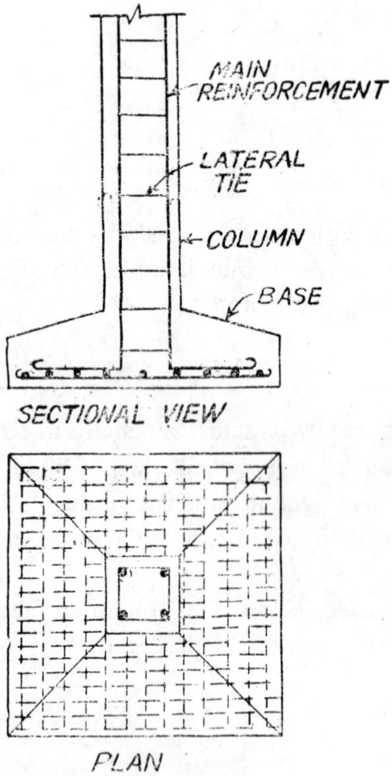

MAIN
REINFORCEMENT

LATERAL
TIE

← COLUMN

BASE

SECTIONAL VIEW

PLAN

R.C.C. Column Foundation
(Isolated Footing)
Fig. 12.1.

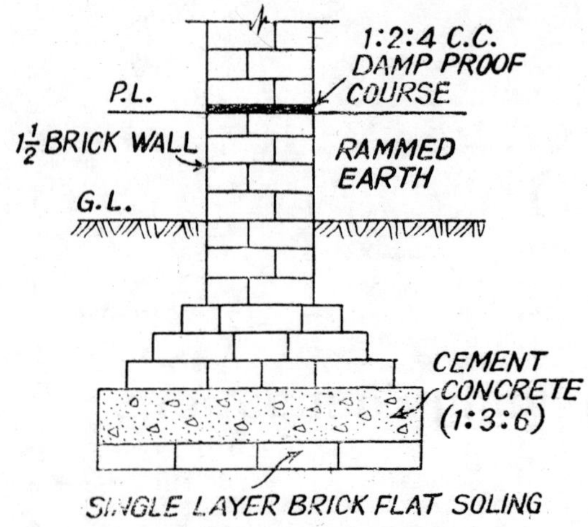

1:2:4 C.C.
DAMP PROOF
COURSE
P.L.
1½ BRICK WALL
RAMMED
EARTH
G.L.
CEMENT
CONCRETE
(1:3:6)
SINGLE LAYER BRICK FLAT SOLING

Spread Footing Foundation
(Continuous Footing)
Fig. 12·2.

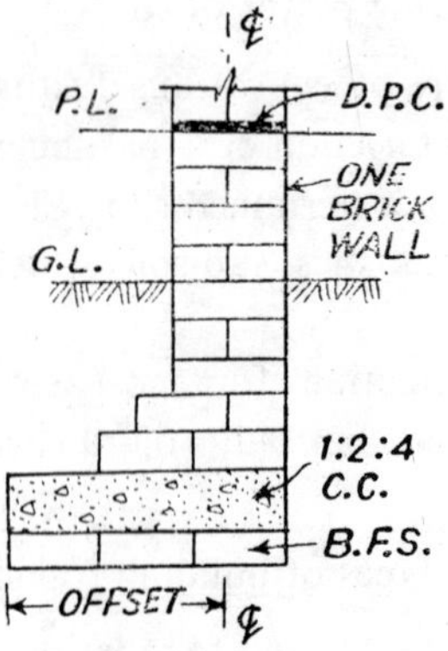

¢
P.L.   D.P.C.
ONE
BRICK
WALL
G.L.
1:2:4
C.C.
B.F.S.
├── OFFSET ──┤ ¢

Offset Foundation

Fig. 12·3.

Combined R.C. footing and combined steel footing are shown in Fig. 12·4 and 12·5 respectively.

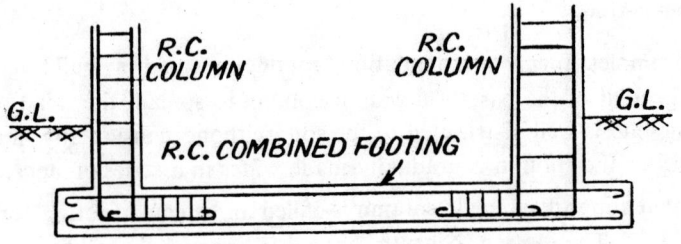

R.C.
COLUMN
R.C.
COLUMN
G.L.
G.L.
R.C. COMBINED FOOTING

Combined R.C. Footing (Sectional view)
Fig. 12·4.

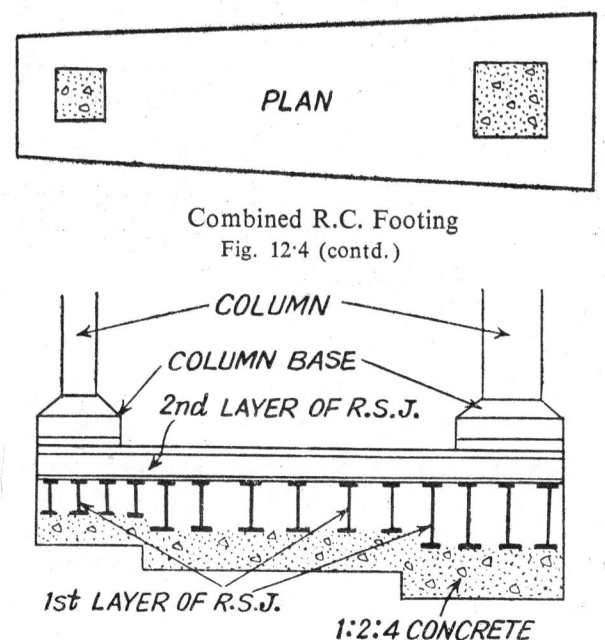

Combined R.C. Footing
Fig. 12·4 (contd.)

Combined Steel Footing
Fig. 12·5.

In combined footings, loads from several colmns are transmitted to the same footing. Therefore, the footing should be proportioned so that its centroid coincides with the centre of gravity of the column loads and the maximum pressure beneath the foosing does not exceed the allowable bearing power of soil.

A cantilever footing a typical form of combined footing in which one of the columns supports an exterior wall as shown in Fig. 12·6.

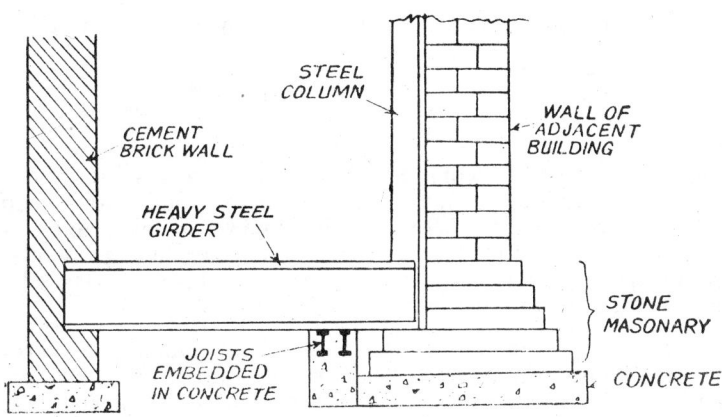

Cantilever Footing supporting exterior column of a Building.
Fig. 12·6.

Width of Foundation :

To determine the base area of foundation, the total load acting at the base of a column or wall including its self load is divided by the allowable bearing power of soil. The base area is then divided by the unit length of wall and the width of foundation is determined. The size of a square footing is found out by taking the square root of the base area.

Usually, the following widths of foundation are considered safe by taking 1 ton/$\square'$ as the bearing power of soil.

Single-storied building   : 0'6 to 0'75 m.

Two-storied building   : 0'9 to 1'05 m.

Three-storied building   : 1'2 to 1'35 m.

Also, the following thumb rule may be used :

Width of foundation $= 2 \times$ thickness of wall above plinth $+ 30$ cm.

Depth of Foundation

The depth of foundation of a wall or a column is found out by using Rankine's formula :

$$\text{Maximum depth of foundation, } D = \frac{W}{A.\omega} \left\{ \frac{1 - \sin \phi}{1 + \sin \phi} \right\}^2$$

$$\text{or,} \quad D = \frac{p}{W} \left\{ \frac{1 - \sin \phi}{1 + \sin \phi} \right\}^2 ;$$

where $W =$ total load to be transmitted ;

$A =$ base area of foundation ;

$\omega =$ density or unit weight of earth ;

$p =$ bearing power of soil *i.e.*, intensity of upward pressure of soil ;

$\phi =$ angle of repose of soil.

Usually, the angle of repose of soil is 30°. For 30° angle of repose, minimum depth of foundation is given by $D = \dfrac{p}{9\omega}$.

Raft Foundation

A raft or mat foundation is a combined footing that covers the entire area beneath a building and supports all the walls and columns over it. A raft foundation is more economical than individual or continuous footings, where the bearing power of soil is too low compared to the building loads. This type of foundation is also called a floating foundation, as if the concrete mat floats over the subsoil and the structure rests on this mat. The self weight of the raft is not considered in the structural design.

If the centre of gravity of the loads from the individual columns or walls coincides with the centroid of the raft, the upward load is regarded as a uniform pressure equal to the sum total of the downward loads divided by the area of the raft. It is customary to reinforce the raft more heavily than required, since the moments and shears due to differential settlement are not taken into account in the design. Raft foundations are also used to reduce the settlement of a structure. A raft foundation for a building is shown in Fig. 12'7.

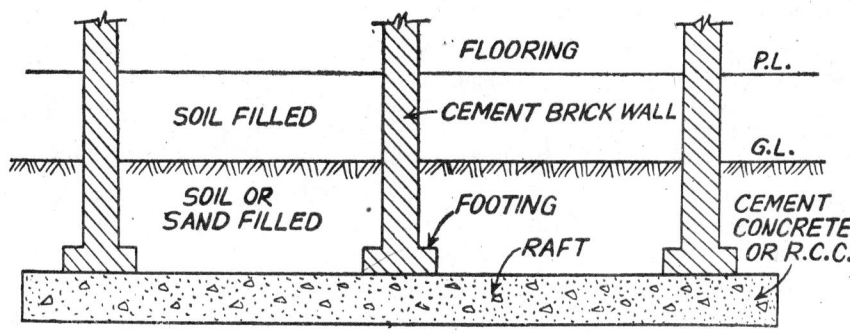

Raft or Mat or Floating Foundation
Fig. 12·7.

Sometimes, the depth at which a raft is established is so great that the weight of the building plus that of the raft is compensated by the soil excavated. The settlement of the building is then insignificant. In situations where complete compensation is impracticable, a shallower raft is made acceptable.

Well Foundation

This type of foundation is adopted in a very soft soil and in marshy lands. A caisson as shown in Fig. 12·8 is driven into the ground either by its own weight or by loading. A well curb with the cutting edge is laid on the ground and masonry work is built over it. The soil within the hollow inside and underneath the edges is excavated. Thus, the well gradually sinks down and when the well rests on a sufficiently stable and hard stratum, it is plugged at the bottom by pouring concrete and the hollow portion is filled with sand. At the end, the top is plugged. Two or three such wells are built and connected together by R.C.C. well cap as shown in Fig. 12·8.

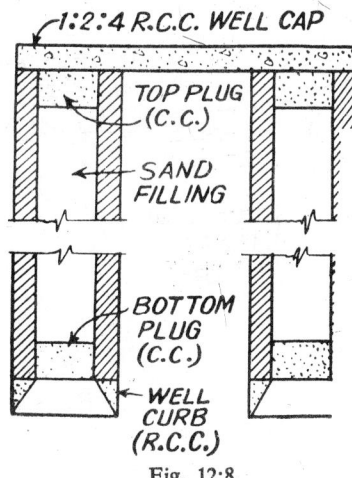

Fig. 12·8.

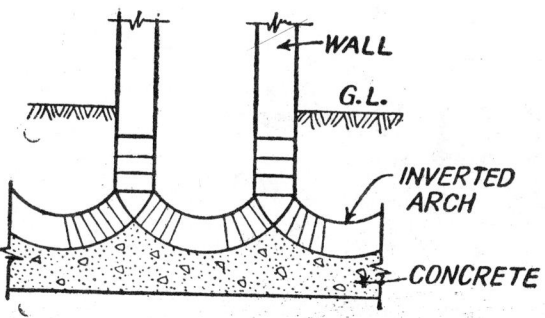

Inverted Arch Foundation
Fig. 12·9.

Inverted Arch foundation

An inverted arch foundation as shown in Fig. 12·9 are found in buildings of earlier days. Now-a-days, this type of foundation is hardly adopted in building construction. With inverted arch foundation in soft soils, the depth of foundation is reduced.

Grillage Foundation

One of the earliest attempts to enlarge the areas of footings without corresponding increase in weight, was grillage foundation made of oak timbers beneath the permanent ground water table and construction of conventional masonary footing on it. One of the earliest construction on wooden grillage foundation is still standing and it is the pullman buildiug at chicago constructed in 1884. Fig. 12·10 shows a grillage foundation made of timber.

In 1891, a grillage foundation consisting of old railroad steel rails embedded in concrete was devised. This was an improvement over the timber grillage and an important forward step in grillage foundation. Fig. 12·11 shows a grillage foundation made of old railroad rails.

During the end of nineteenth century between 1895 and 1900 A.D., railroad rails were superseded by rolled steel joists and the grillage became more economical. A grillage foundation made of R.S.J. is shown in Fig. 12·12.

With advent of reinforced cement concrete shortly after 1900 A.D., grillage foundation was superseded by reinforced concrete footings which are still dominant. A reinforced concrete footing has been illustrated in Fig. 12·1.

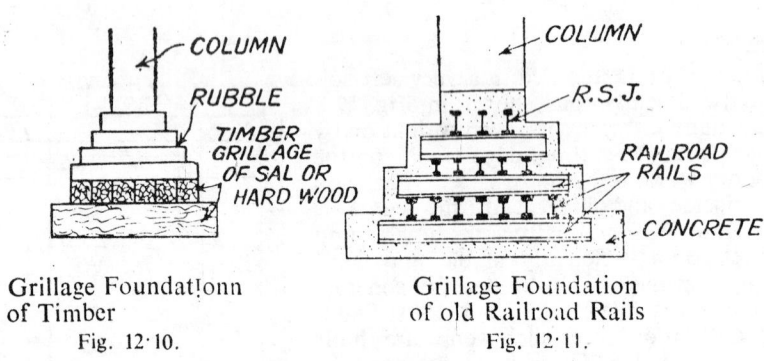

Grillage Foundatlonn
of Timber
Fig. 12·10.

Grillage Foundation
of old Railroad Rails
Fig. 12·11.

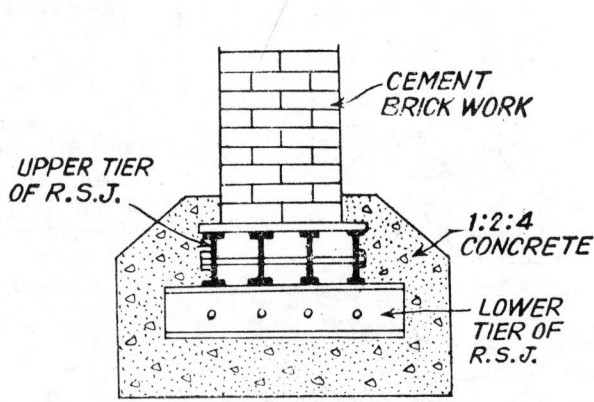

A Grillage Foundation made of R.S.J.
Fig. 12·12.

Pile Foundation

Pile foundations are adopted for heavy loaded structures which are to be erected in soils having low bearing capacity. The loads exerted by a structure are transferred to groups of piles through pile cap. There are various forms of pile foundations, of which only two types are shown in Fig. 12'13.

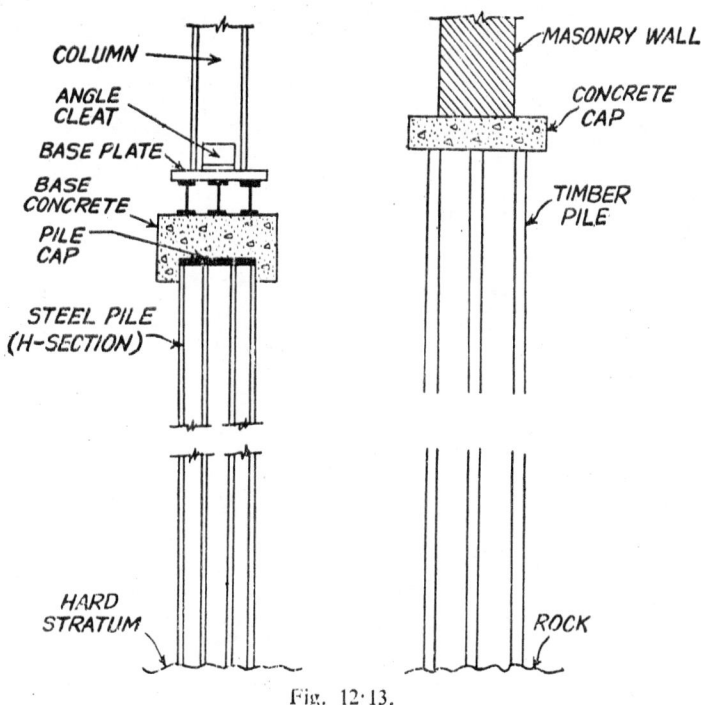

Fig. 12·13.

13

Flooring and Roofing

Flooring :

Floors divide a building horizontally at different levels, usually of same height, with the idea of providing accomodation for various activities. The bottom most floor just above the ground surface and at the plinth level is called ground floor and the subsequent top floors are called 1st floor, 2nd floor and so on until reaches the roof. The floor below ground level is termed as basement floor. The ground floor and the basement floor should have to be made damp proof and the top floors should be designed to take the load coming over it. Fig. 13·1 shows the usual flooring at ground floor level in most buildings. The usual flooring at top floor level in most buildings is presented in Fig. 13·2.

Brick flooring is adopted where well burnt good bricks of uniform shape, size and colour are readily available at a low price. This type of flooring may be used at groud floor level for low cost housing. Fig. 13·3 shows brick flooring. Usually, brick-on-edge flooring either in herring-bone bond, diagonal bond or zig-zag bond over a lime concrete base is made. The gaps in bricks should be filled with cement slurry.

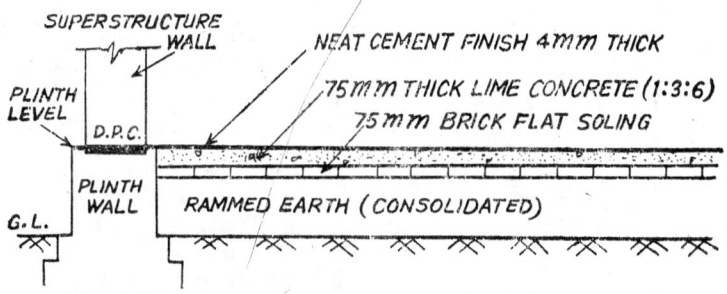

Usual Flooring at Ground Floor level in most Buildings
Fig. 13·1.

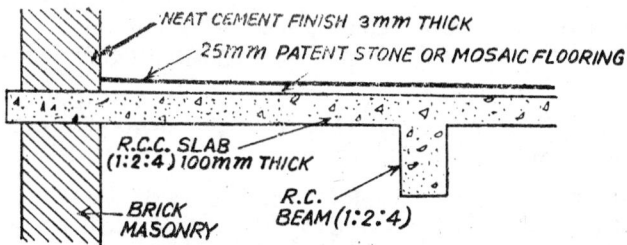

Usual Flooring at Top Floor level in most Buildings
Fig. 13·2.

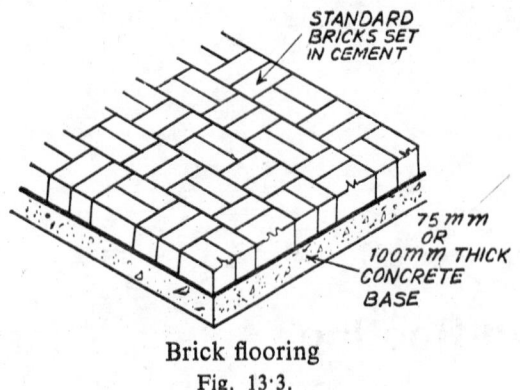

Brick flooring
Fig. 13·3.

Wooden Flooring has already been discussed in details in chapter 8·0. In strip flooring narrow and thin strips of good quality timber are joined to each other by tongued and grooved joints. The widths of strip are 60 mm. to 100 mm. Strip flooring is used in thicknesses of 20 to 25 mm.

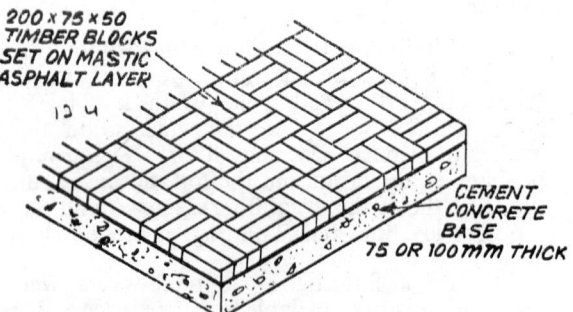

Flooring with Timber blocks
Fig. 13·4.

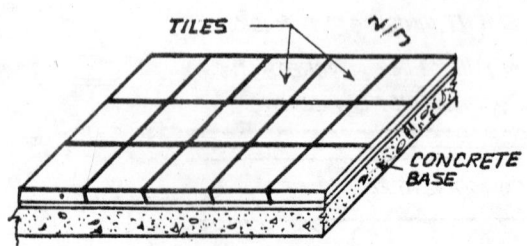

Flooring with Stone slabs or Mosaic Tiles
Fig. 13·5.

Wooden flooring is also made with fabricated heavy wood blocks which are set on mastic asphalt layer over a concrete base as shown in Fig. 13·4. Flooring with stone slabs or mosaic tiles is shown in Fig. 13·5. The slabs or tiles are set in cement over a concrete base. This type of flooring may be used at all levels. The concrete base should be of cement concrete or lime concrete. The joints are flush pointed with cement mortar (1 : 3) the floor should have a slope of 1 in 40. The tiles may be set in various pantters to form a design in flooring.

In concrete flooring, a wearing course, usually 25 mm thick patent stone or mosaic flooring, rests over a concrete base. This type of flooring is most commonly used in buildings as shown in Fig. 13·2.

R.B.C. flooring as shown in Fig. 13·6 is hardly used now-a-days. In this type of flooring, bricks are reinforced with concrete.

Hollow tile flooring is similar to R.B.C. flooring as shown in Fig. 13·7. The hollow files give the effect of sound and heat insulations. A hollow tile flooring may also be made without any reinforcement, if used at ground floor level.

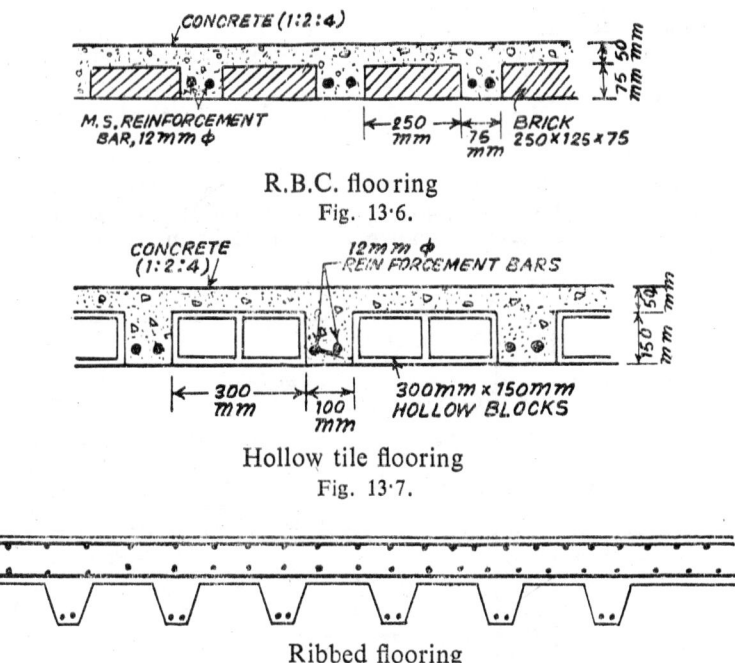

R.B.C. flooring
Fig. 13·6.

Hollow tile flooring
Fig. 13·7.

Ribbed flooring
Fig. 13·8.

Roofing :

The function of a roof is to protect a building from sun, rain, snow and other weather actions. A roof is the topmost covering of a building. It may be pitched (sloped), flat or a dome. Pitched or sloped roofs by forming timber or steel roof trusses have already been discussed and illustrated in chapter 11'0. Domes are beyond the scope of this book. This chapter shall deal with various forms of flat roofs and slant roof coverings showing the arrangements of their fixing.

Fig. 13'9 shows the construction of a first class mud roof (flat roof) which consists of two layers of tiles 30 cm × 15 cm. × 3 cm in size resting on 75 mm × 50 mm wood burgahs or battens supported by 375 mm × 200 mm. wooden beams for a specific room size.

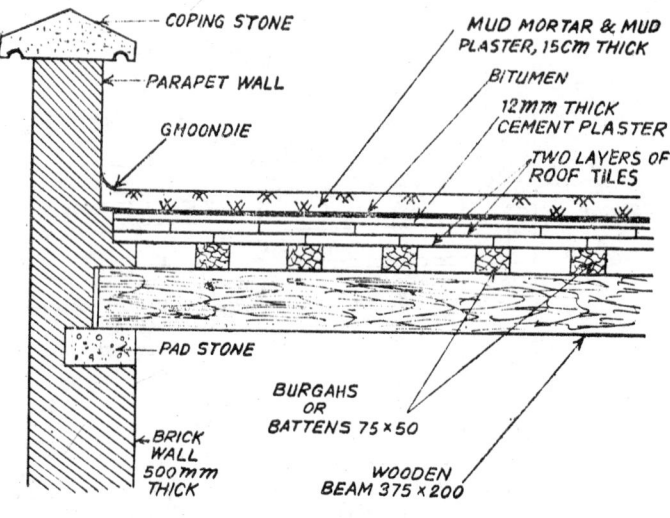

Mud Roof
Fig. 13·9.

The tile joints are sealed with cement plaster and the top surface is plastered with 12 mm thick cement plaster. After laying down bitumen over the plastered surface 15 cm. thick mud and mud plaster shall be laid to ensure the water tightness. The centre to centre distance of the battens should be 30 cm to hold the tiles. The space of the end batten should not be less than 7·5 cm and not greater than 22·5. cm.

Fig. 13·10 shows details of jack arch roofing. Brick arches are built between R.S. joists embedded in concrete. To hold the joists in position and to take the thrust by the R.S.J.S., tie rods of 20 to 25 mm ϕ are used as shown. About 125 mm. thick average lime terrracing is provided at the top of the arches.

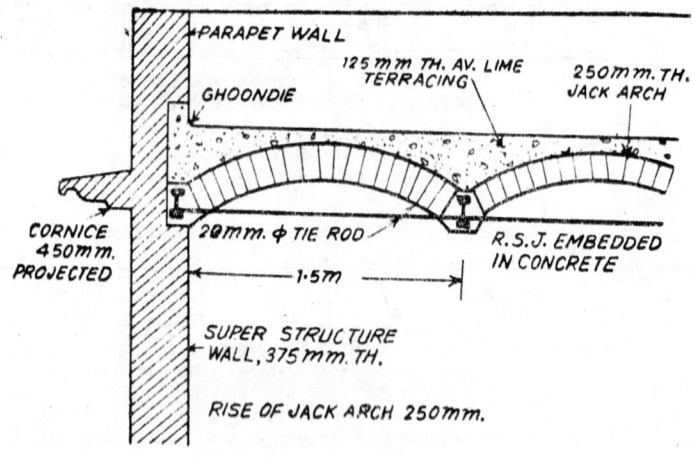

Details of Jack-Arch roof.
Fig. 13·10.

Fig. 13·11 illustrates a flat roof constraction with R.S.J.S, wooden burghs (battens) and roof tiles. This is similar to a mud roof. But, here lime terracing is used at the top surface of the tiles instead of the use of mud mortar. Sometimes the R.S.J.S are replaced by wooden beams of heavy section. This type of roofing is quite old and now-a-days it is not practised.

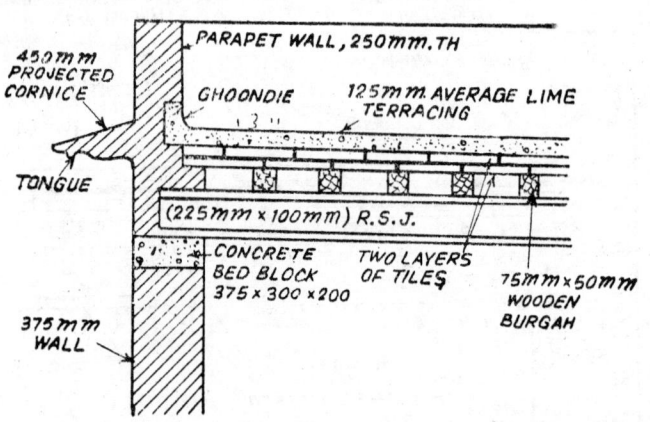

Details of roofiing with R.S.J sections, wooden burgahs, tiles and lime terracing.
Fig. 13·11.

The three types of flat roofs described above are replaced by concrete flat roofs which are simple, cheap and durable. R.C.C. flat roofing with R.C.C. beams in place of R.S.J.S is shown in Fig. 12·12.

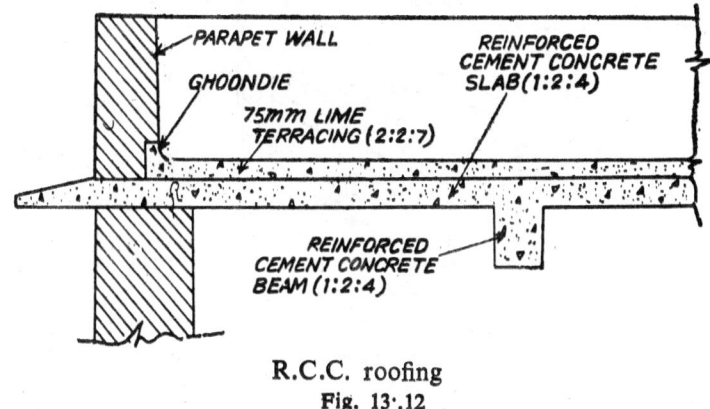

R.C.C. roofing
Fig. 13·.12

Fig. 13·13 shows a sectional view by taking section through parapet wall with pillasters, R.C. slab with cornice, lime terracing and superstructure wall.

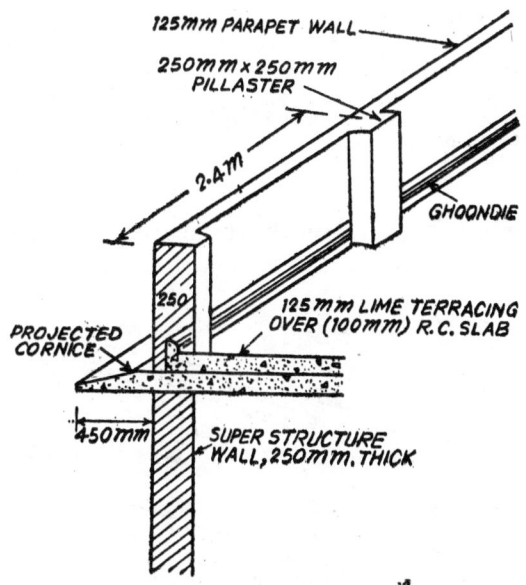

Section through parapet wall with pillasters, R.C. slab with cornice, lime terracing and superstructure wall.
Fig. 13·13.

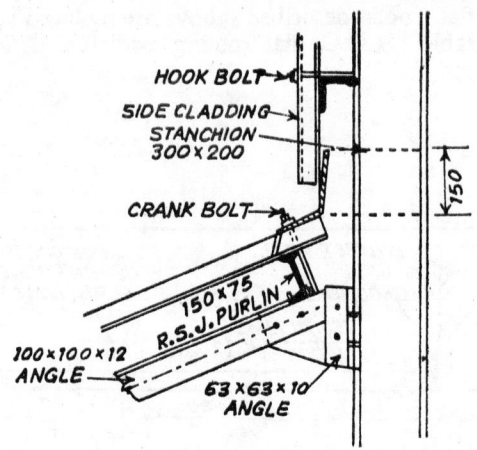

Cement-asbestos roofing over Lean-to-roof.
Fig. 13·14.

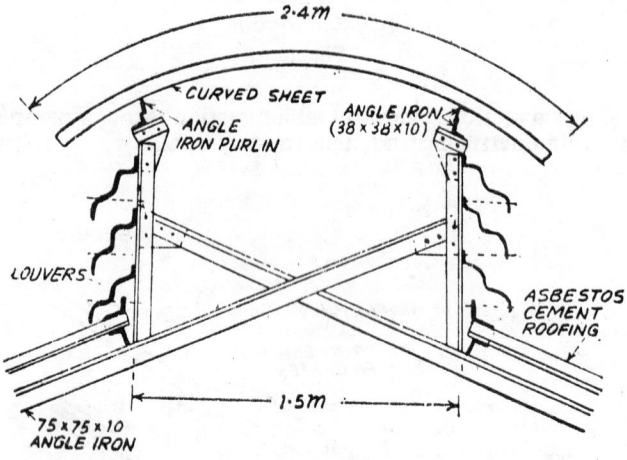

Ventilator type roof with cement asbestors sheets.
Fig. 13·15.

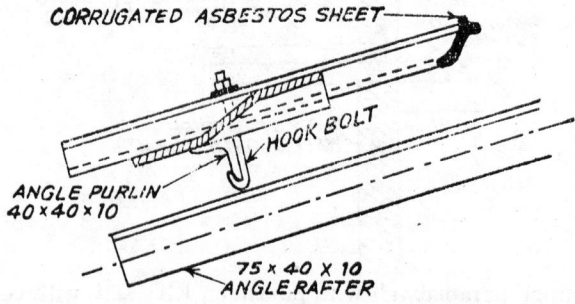

Fixing corrvgated cement-asbestos sheets over angle-iron rafters.
Fig. 13·16.

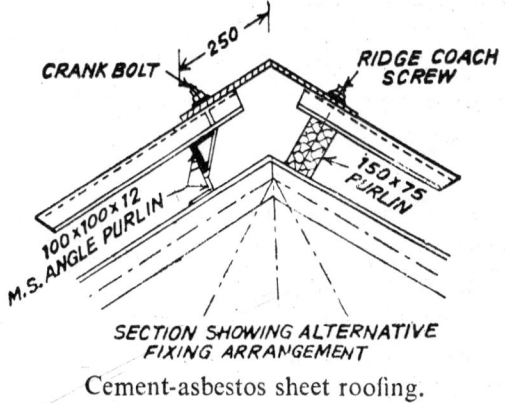

SECTION SHOWING ALTERNATIVE
FIXING ARRANGEMENT

Cement-asbestos sheet roofing.
Fig. 13·17.

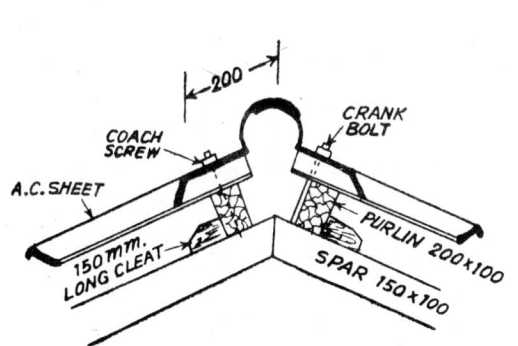

A.C. Sheet Roofing with details at Ridge
showing alternative fixing accessories.
Fig. 13·18.

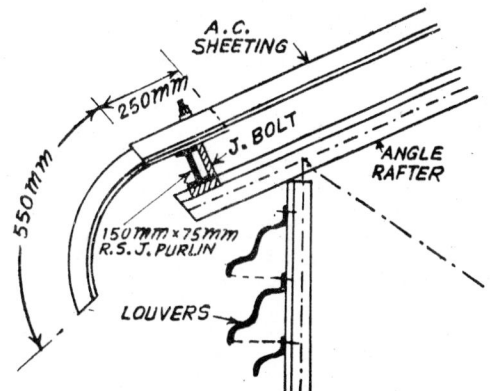

Bent-asbestos with louvres and
ventilator curves.
Fig. 13·19.

14

Stairs

A stair may be defined as a set of steps leading from one floor to the other to afford a means of communication between the various floors of a building. The space in a building where the stair is located, is called staircase. A staircase should be placed at such a location in a building that it fascilitates all the the occupants or persons from outside in easy ascending and descending from all parts of building without having too much of criss-cross travel from one end to the staircase. Therefore, for multi-storied flat buildings the staircase should be located either centrally or two staircases at two ends of the building.

A stair consists of treads, risers, stringers, newel posts, handrails, landing, and balusters. The materials used in stairs are concrete, bricks, timber, C.I, W.I., M.S., stone slabs, Aluminium bars, etc.

Some important technical terms used in connection with stairs are given below. Fig. 14·1 illustrases stair nomenclatures.

Flight—A continuous set of steps from floor to floor, floor to landing or landing to landing is called a flight.

Landing—A platform at the end of a series of steps is called landing. Depending upon the arrangement of steps it may be half-landing or quarter-landing.

Tread—It is the horizontal part of a step on which the foot rests.

Rise—It is the vertical distance between two consecutive steps or treads.

Riser—The vertical member between two conscutive treads is known as riser.

Nosing—It is the projected edge of a tread.

Stringer—It is the sloping member in a stair which supports the steps.

Newel—This is a post of heavy section set at the two ends of a handrail.

Balusters—These are intermediate vertical members supporting the handrail.

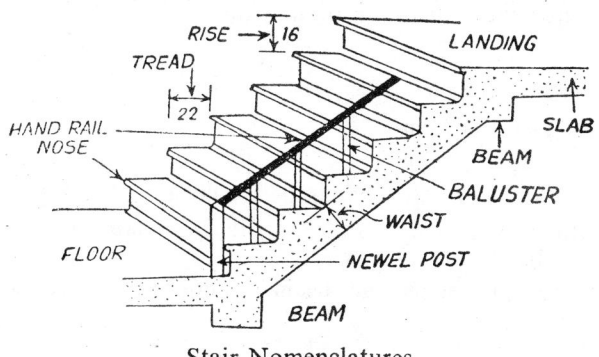

Stair Nomenclatures
Fig. 14·1.

Handrail—It is a rail of metal or wood provided at the side of a stair for safety and is fixed at about waist height parallel to the line of nosing.

Line of Nosing—This is an imaginary line joining the nosing points and is parallel to the slope of the stair.

Essential Requirements :

Apart from structural design there are certain essential requirements for a stair.

These are :

(1) The width of a stair should be 0·9 m for residential building and 1·35 m for public buildings.

(2) The slope of the stair should not be greater then 45° and not less then 25° with the horizontal.

(3) The stair should be well-lighted and ventilated.

(4) All the risers and treads should be made uniform.

(5) The number of steps in a flight should not be more than 12.

(6) The headroom in a stair should be at least 2 m.

(7) The width of tread should be between 225 m m and 275 mm. and the rise should not be more than 175 mm. and not less than 150 mm.

(8) The width of landing should be equal to the width of the stair.

(9) The nosing should not project beyond 15 mm.

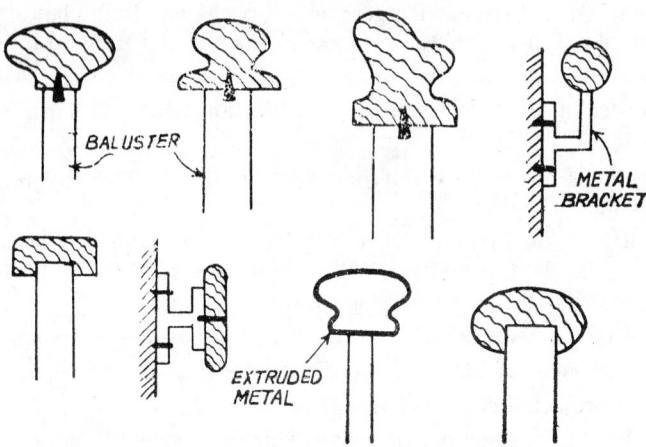

Various forms of Balusters and Stair Handrails.

Fig. 14·2.

There are some thumb rules for proportioning steps.

These are :

(a) Rise+Tread=42·5 to 45 cm.

(b) 2 Rise+Tread=58·0 to 62 cm-

(c) Rise×Tread=420 to 460 sq. cm.

Various forms of balusters and stair handrails are illustrated in Fig. 14·2.

Wooden stair, stone stair, metal stair and R.C.C. stair are shown in Figs. 14·3, 14·4, 14·5 and 14·6 respectively.

Typical joints between the steps and beam or floor in R.C.C. stair are presented in Fig. 14·7.

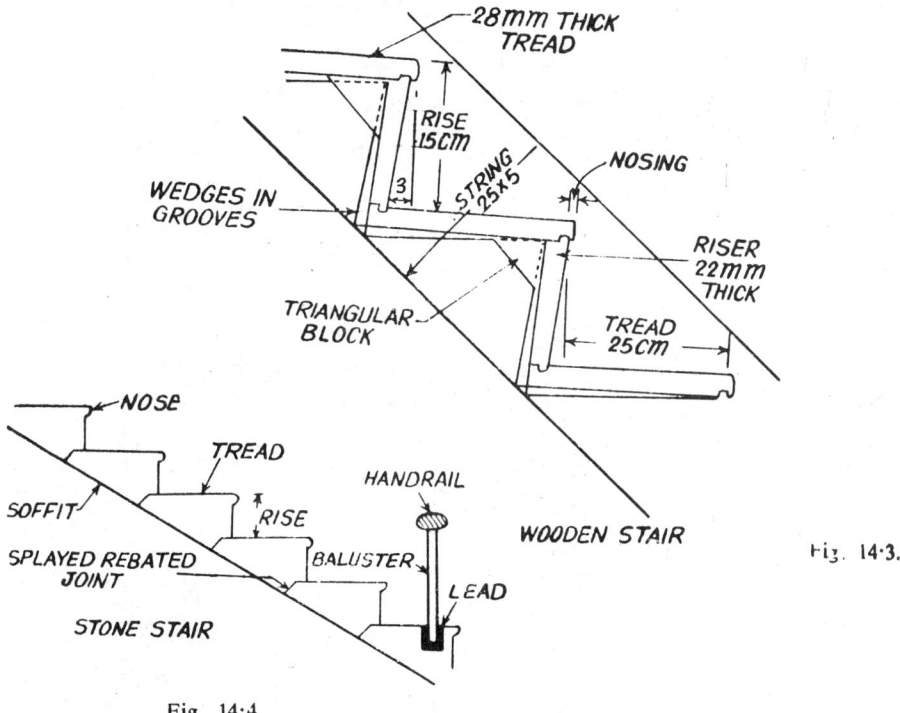

Fig. 14·3.

Fig. 14·4.

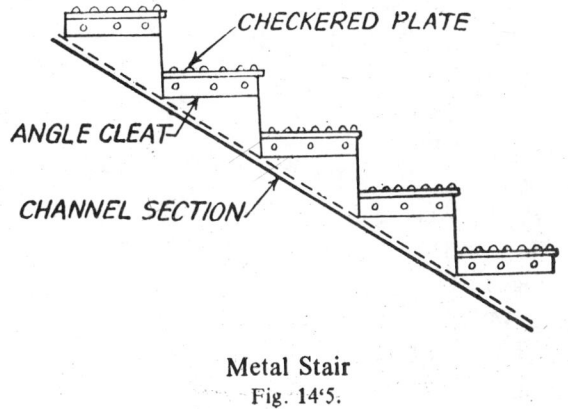

Metal Stair
Fig. 14·5.

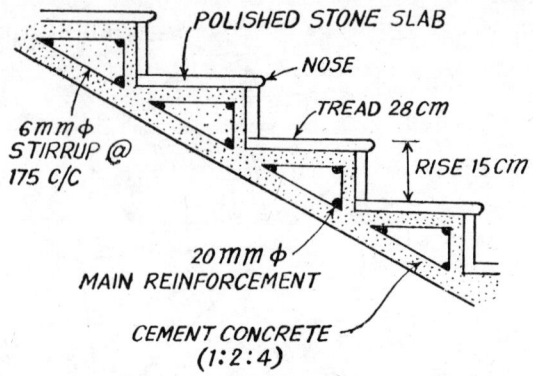

R.C.C. Stair with Stone Treads
Fig. 14·6.

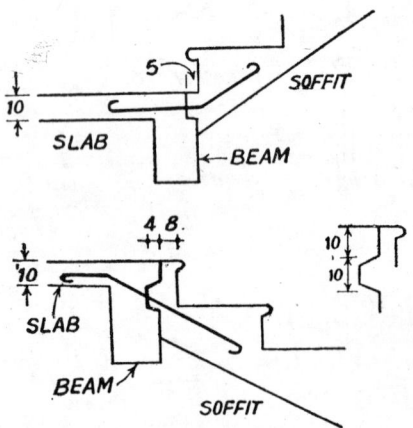

Typical joints between the steps and Beam or Floor showing additional Bond bar used in R.C.C. stair.
Fig. 14·7.

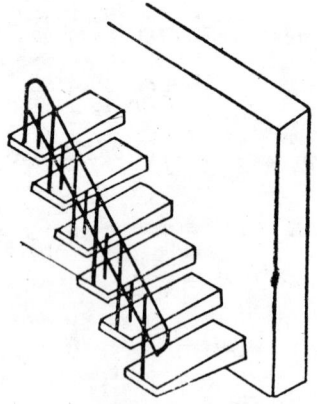

Cantilever Steps
Fig. 14·8.

The different types of stairs are :

(1) Straight flight or single flight (Fig. 14·9)

(2) Open-well (Fig. 14·10)

(3) Quarter-Turn (Fig. 14·11)

(4) Dog-Legged (Fig 14·12)

(5) Half-Turn open well (Fig. 14·13)

(6) Bifurcated (Fig 14·15)

(7) Four-flight (Fig. 14·16)

(8) Segmental (Fig. 14·17)

(9) Circular (Fig. 14·18)

(10) Geometrical (Fig. 14·19)

(11) Octagonal (Fig. 14·20)

(12) Hexagonal (Fig. 14·21)

(13) Spiral (Fig 14·23)

The type of stair to be provided in a building depends upon the nature of building and the availabilty of space for the staircase. In public buildings like schools, colleges, hospitals, offices, auditoriums etc, bifurcated stair, stair with lift well and half turn open well stair are used. In domestic buildings, dog-legged stair, Quarter-turn stair and open well stair are adopted. Circular stair and Spiral stairs are installed as an additional stair at the rearside of a building. Octagonal and hexagonal stairs also spiral brick stairs are used for minarets Straight flight stair is provided only when a narrow long strip of space is available for the stair. A four-flight stair is considered where a square space is available for the stair. Geometrial and segmental stairs are hardly used.

To-day, stairs are mostly made of concrete with reinforcement steel. In hilly areas and in forests where timber is readily available in plenty, wooden stairs are built in residential quarters and forest bungalows. For low-cost housing, precast light-weight concrete steps are used. Stone steps are used very seldom. It is especially built in construction of temples.

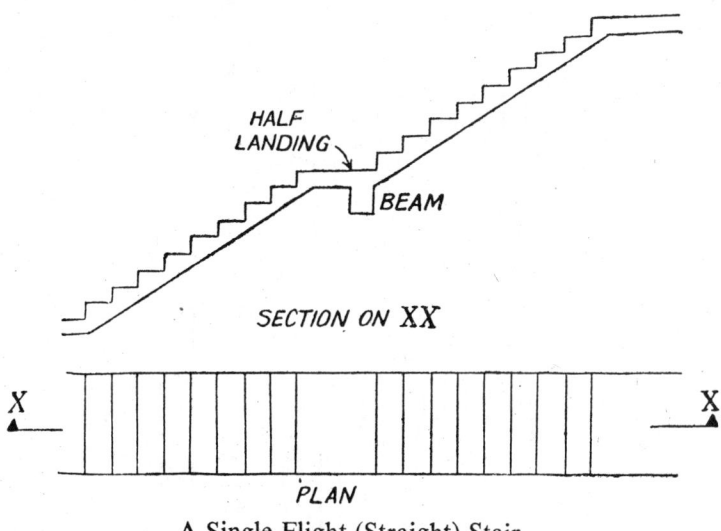

A Single Flight (Straight) Stair
Fig. 14·9.

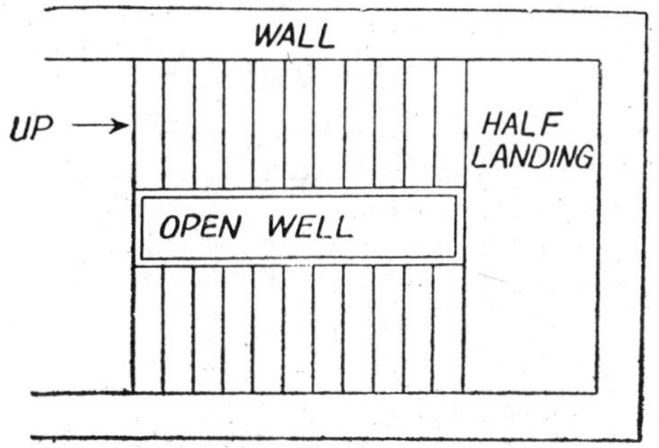

Plan of an-open well Stair
Fig. 14·10.

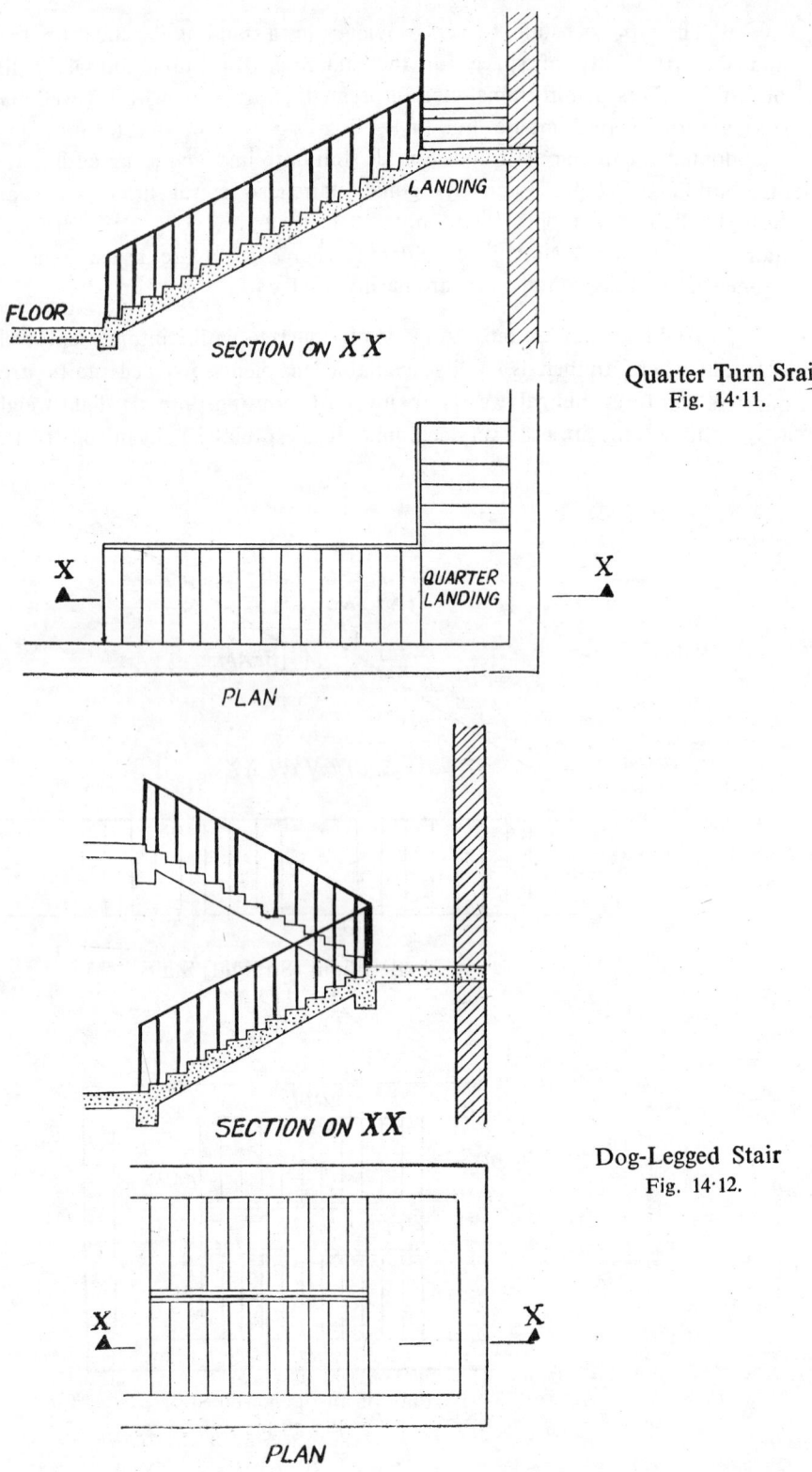

SECTION ON XX

FLOOR

LANDING

Quarter Turn Srair
Fig. 14·11.

X
QUARTER
LANDING

X

PLAN

SECTION ON XX

Dog-Legged Stair
Fig. 14·12.

X   X

PLAN

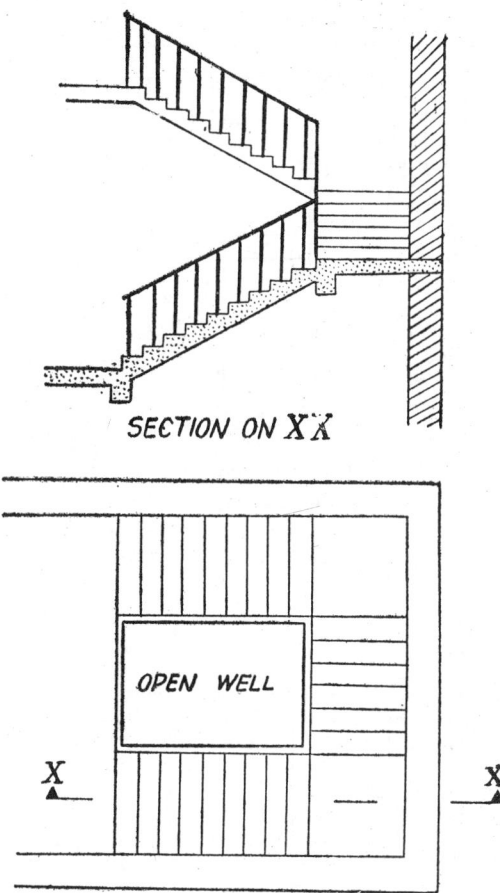

SECTION ON XX

PLAN
Half-Turn open-well Stair
Fig. 14.13.

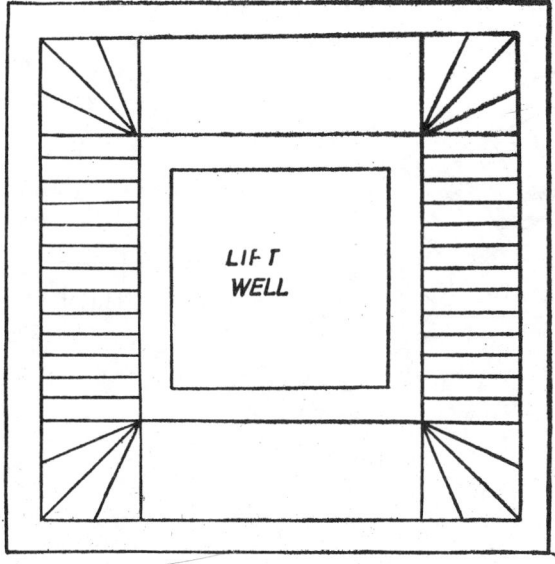

A Lift well encased in a Staircase
Fig. 14·14.

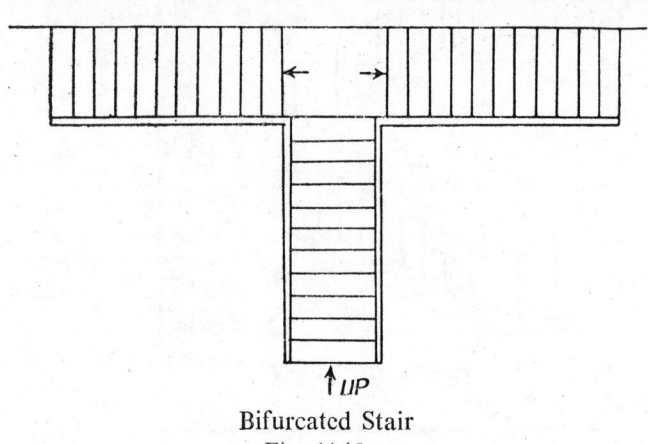

Bifurcated Stair
Fig. 14·15.

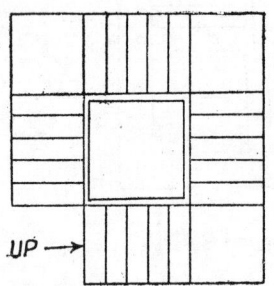

Plan of a Four-Flight Stair
Fig. 14·16.

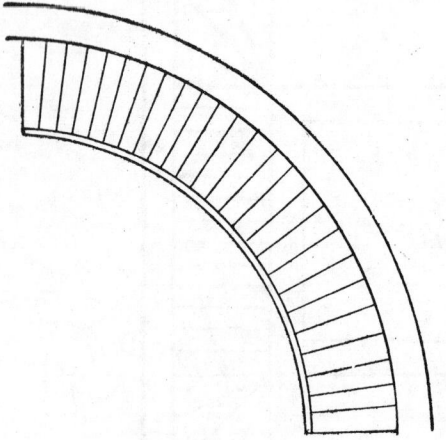

Plan of a Segmental Stair
Fig. 14·17.

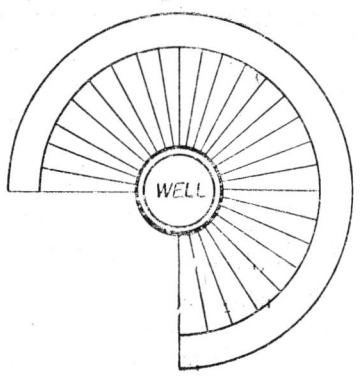

Plan of a Circular Stair
Fig. 14·18.

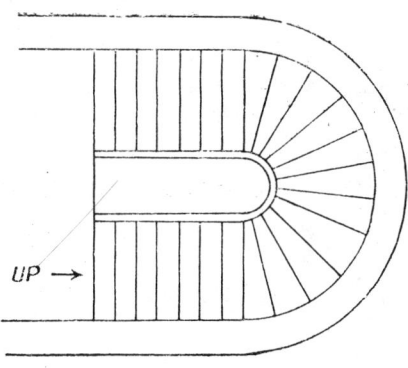

Plan of a Geometrical Stair
Fig. 14·19.

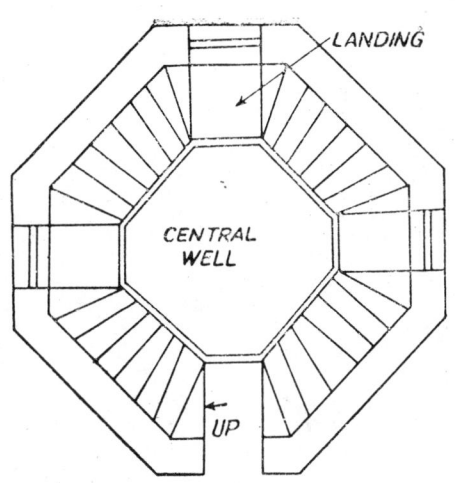

Octagonal Stair in Plan
Fig. 14·20.

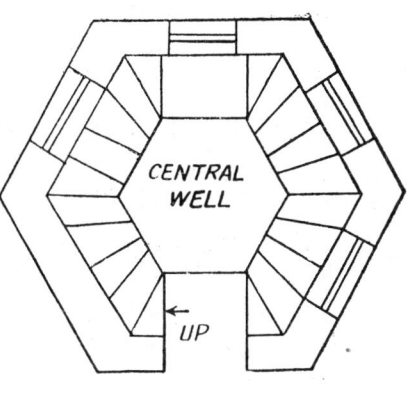

Hexagonal Stair in Plan
Fig. 14·21.

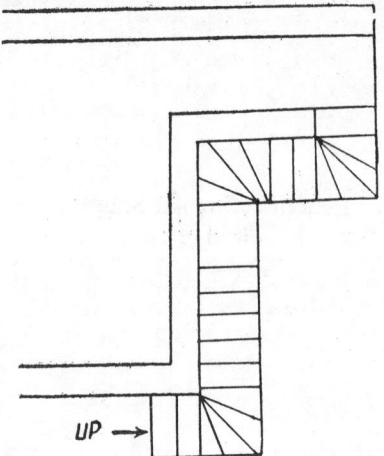

Provison of Stair by following the outside
wall where space is not available for
staircase.
Fig. 14·22.

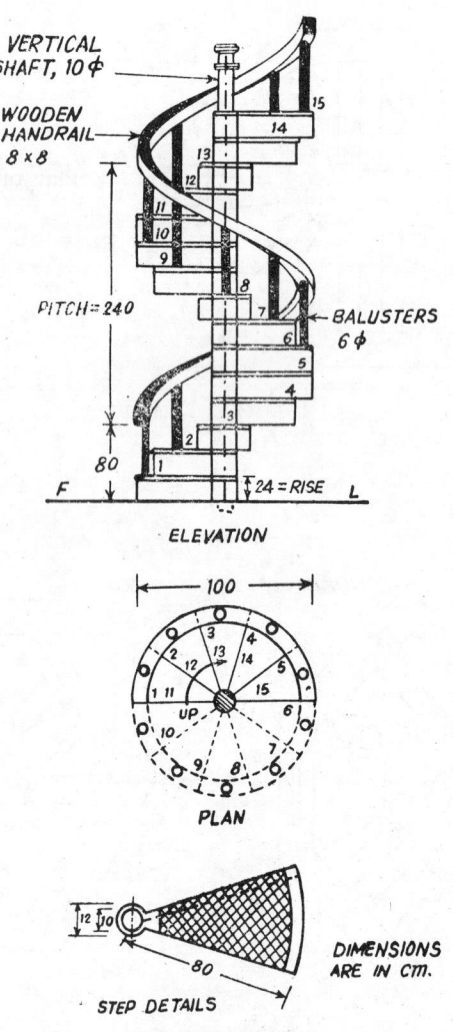

Steel Spiral Stair
Fig. 14·23.

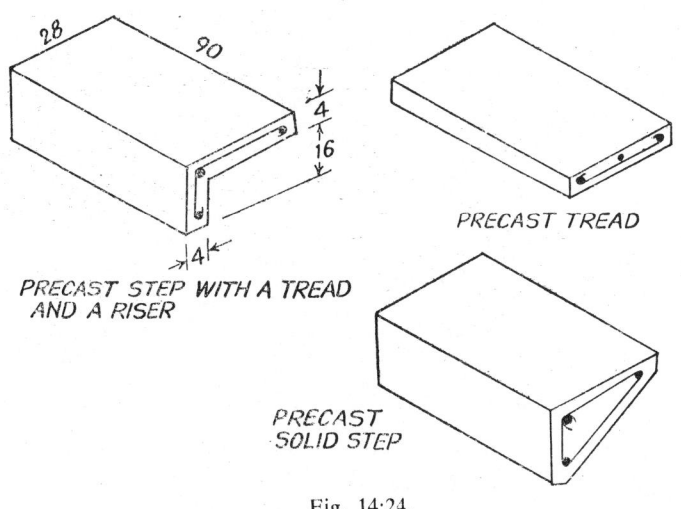

Fig. 14·24.

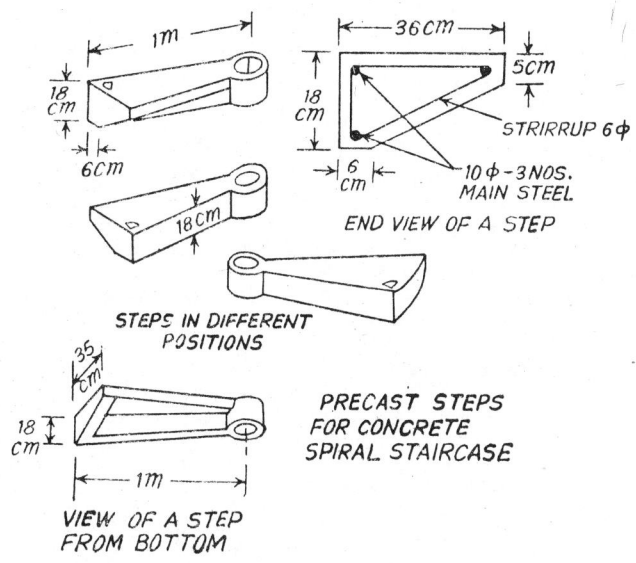

Fig. 14·25.

15

Chujjas, Grills and Gates

Chujjas, Grills and gates impart beauty to a building in addition to safety and security. Chujjas are made of reinforced cement concrete in conjunction with the R.C. lintel. The function of a chujja is to protect the windows from sun and rain and not to allow rain inside the room when the windows are kept open during a stormy weather. A chujja is also known as sunshade. In some buildings a continuous chujja is provided projecting it from the band lintel. But, the continuity of chujja around all the outside walls does not bring the beauty. Moreover, a continuous chujja is expensive, and it does not serve any purpose beyond the windows.

In modern day practice, chujias with some vertical and horizontal finges thin mouldings with R.C.C.) act as appendage in buildings which produce varieties of architectural effects. Thus, even a very simple and most ordinary old building can be beautified with the addition of proper appendages.

In recent days sun breakers have come in use instead of chujjas alone. In modern day architecture, sun breakers with chujjas are applied to front view of a building facing west with the object of breaking the scorching rays of sun. The sunrays will not get entry into the building due to these sunbreakers, but there will be natural light in the building coming from outside.

Fig. 15·1 shows the proper use of projected R.C. chujja, R.C. chujja with vertical moulding and R.C.C. sunbreakers (vertical finges) in plan. The sunbreakers may also be made normal to the outside walls.

Fig. 15·2 shows the pictorial views of various forms of R.C.C. chujja with different mouldings. The most ordinary type of R.C.C. chujja commonly used are shown in Fig. Nos. 15·2 (a) and (d). Box mouldings as shown in Fig. Nos. 15·2. (e) and (f) are also found in domestic buildings.

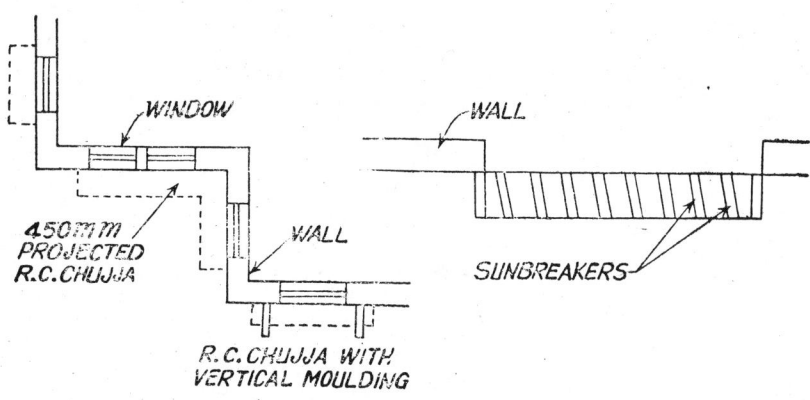

R.C. Chujja with vertical moulding
Fig. 15·1.

Grills are made of mild steel flats, squares and rods. The grills are used in window openings. The function of a grill is to provide safety and security of a room when windows are kept open and also to impart beauty to the building. Grills are also used in verandah railing, staircase railing, doors and gates. A grilled door permits the entry of wind and light in addition to sefety and security. Grills used in stairs and verandah give an additional beauty in addition to safety, security and visibility. Various forms of grills that are commonly used in buildings are shown in Fig. 15·3 through 15·12.

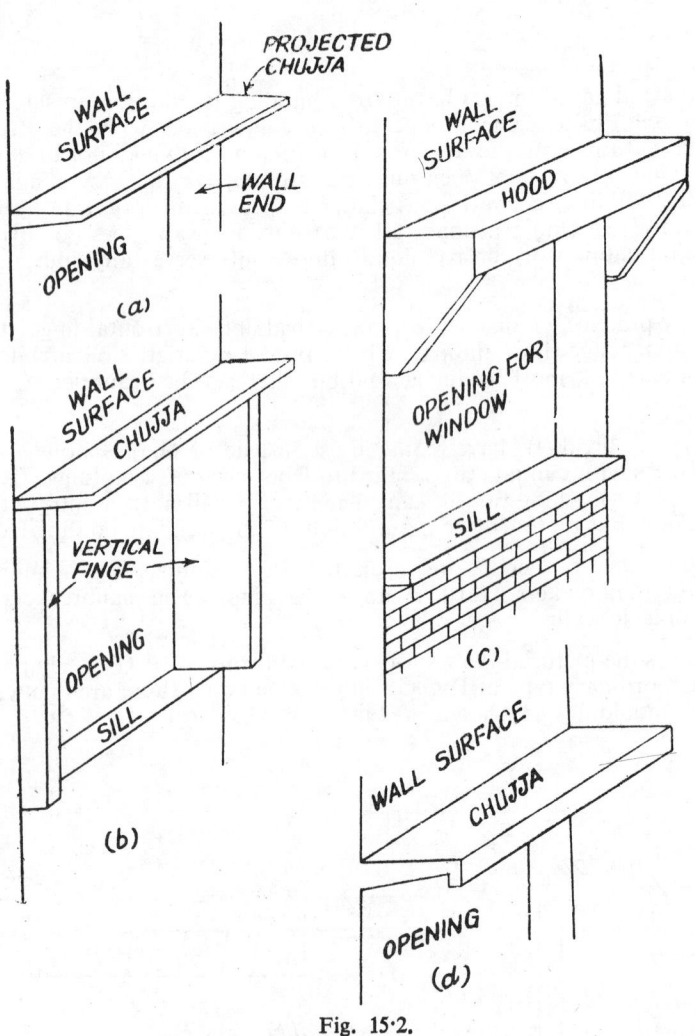

Fig. 15·2.

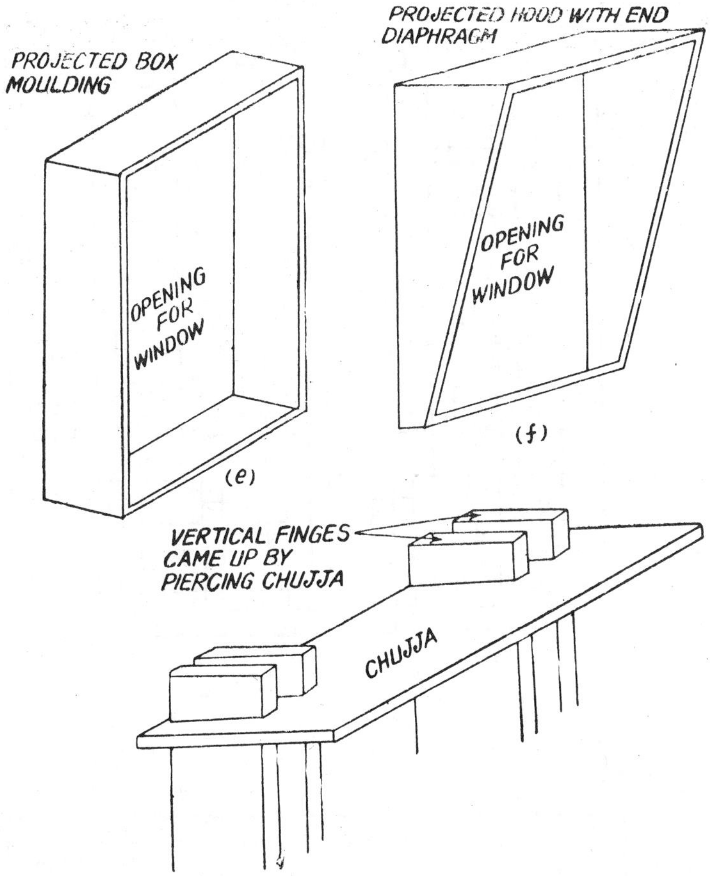

PROJECTED BOX
MOULDING

PROJECTED HOOD WITH END
DIAPHRAGM

OPENING
FOR
WINDOW

OPENING
FOR
WINDOW

(e)

(f)

VERTICAL FINGES
CAME UP BY
PIERCING CHUJJA

CHUJJA

Fig. 15·2. (Contd.)

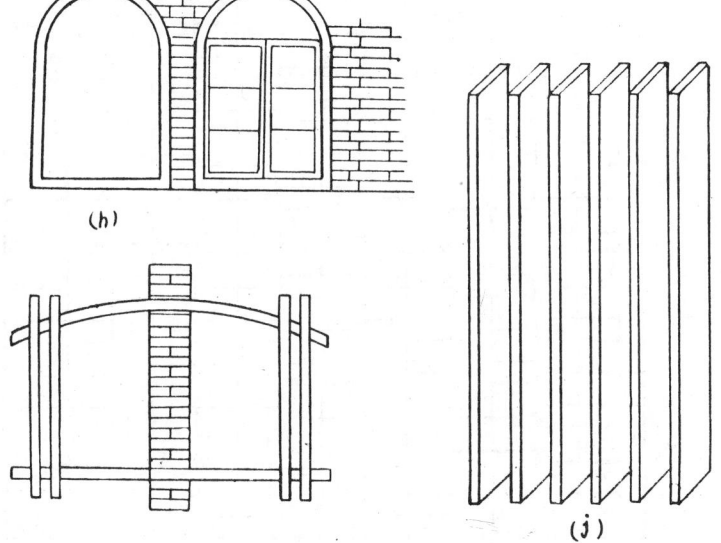

(h)

(j)

Fig. 15·2. (Contd.)

Fig. 15·3.

Fig. 15·4.

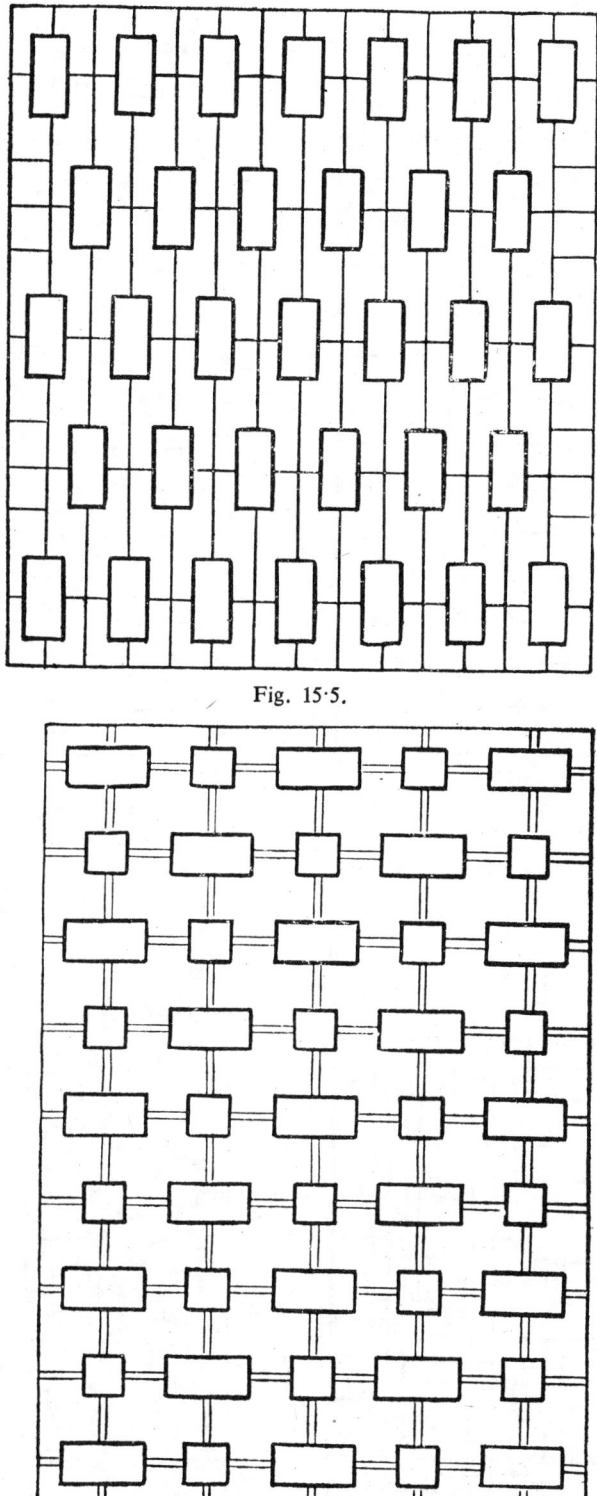

Fig. 15·5.

Fig. 15·6.

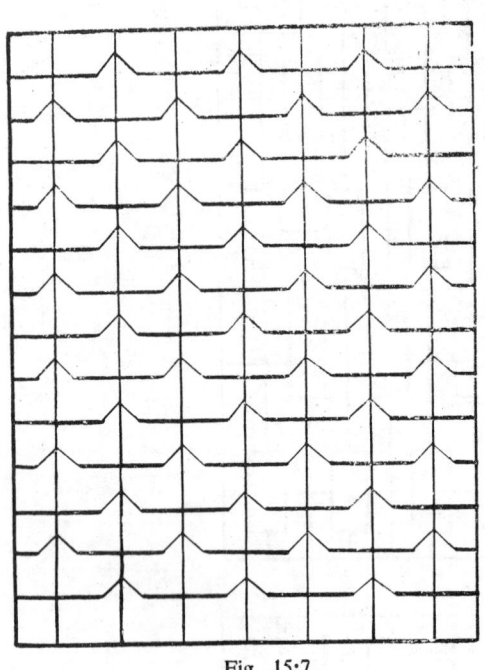

Fig. 15·7.

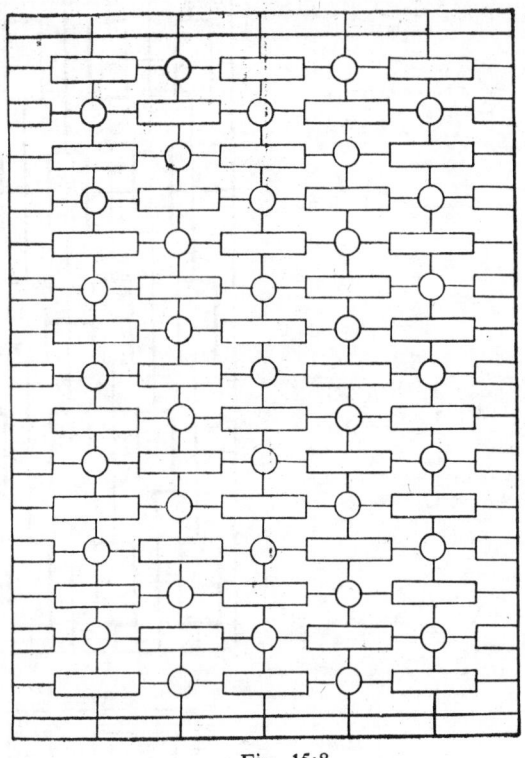

Fig. 15·8.

Fig. 15·9

Fig. 15·10.

Fig. 15·11.

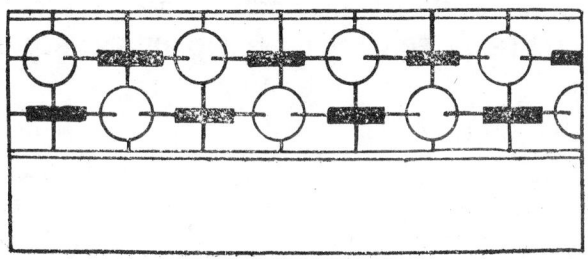

Fig. 15·12.

Gates are provided at the entry to a building or to a campus surrounded by a compound wall. Gates are used for safety and security by restricting the easy access of any person. Grilled

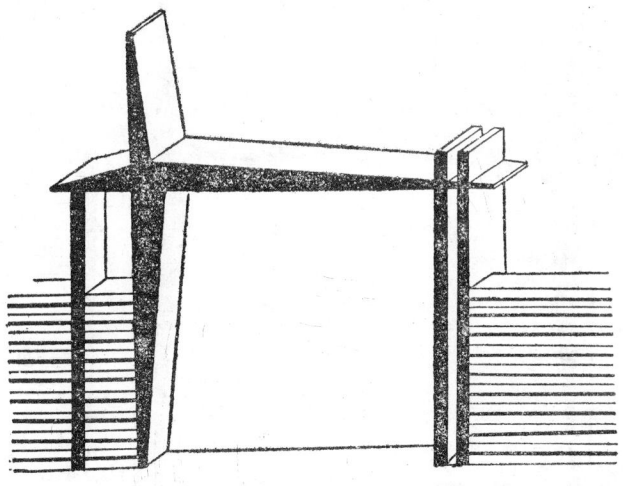

A GATE (ENTRANCE TO A PUBLIC BUILDING)

Fig. 15·13

gates are preferued for clear visibility. But, in restricued buildings, factories, etc, gates with covering sheath are used. The shape and size of a gate depend upon the functional character of a building or campus. Thus, a gate for a public building will differ from that for a domestic building and a gate for an exhibition or expo will differ from a gate for a public building.

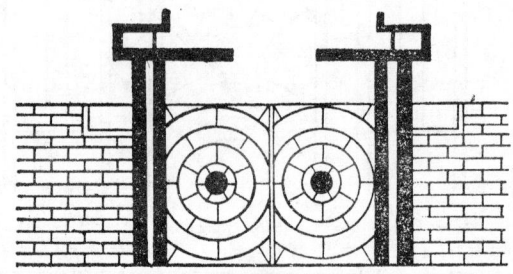

(A GATE ENTRANCE TO A DOMESTIC BUILDING)

Fig. 15·14.

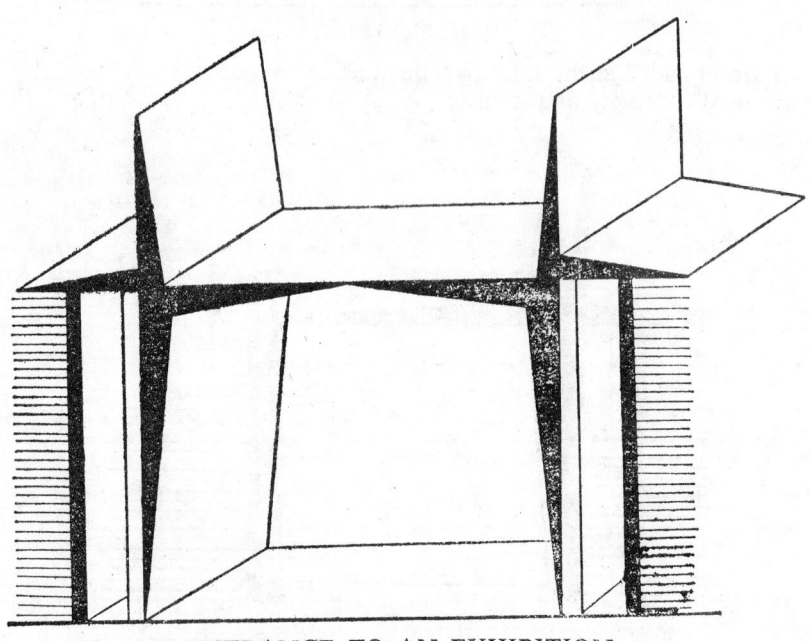

ENTRANCE TO AN EXHIBITION

Fig. 15·15.

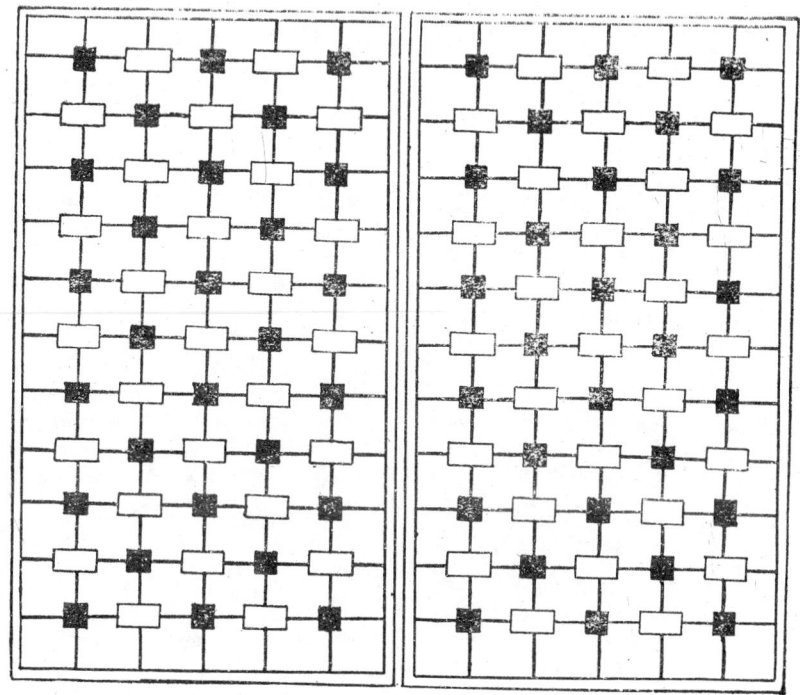

A Gilled Gate
Fig. 15·16.

16

Structural Steel Connections

Stuctural steel connections are needed for steel framed buildings, towers, bridges etc. For highrise buildings, skyscrapers, lofty towers and bridges, construction in steelwork is found economical, safe and durable. First, the required sections are fabricated from good quality mild steel sections and a framework is formed and then erected at site. Various M.S. sections commonly used are plates, R.S.J., Angle sections, Tee sections, and channel sections to form the required steel framework. The steel members after being cut, drilled and arranged in positon are fastened together by means of rivets, bolts-nuts or by welding. The modern techinque is to go for welding, because it is simple and fast working and the jointss become rigid.

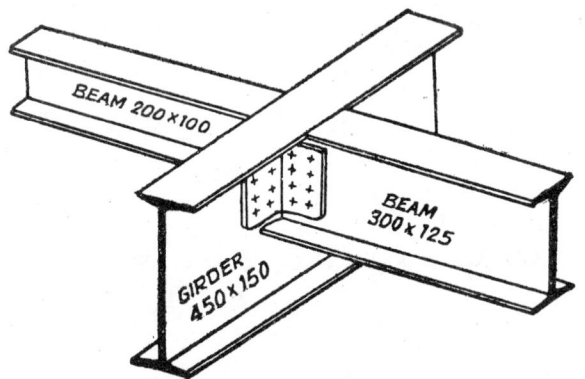

UNEQUAL BEAMS TO GIRDER CONNECTION (Pictorial view)

Fig. 16·1.

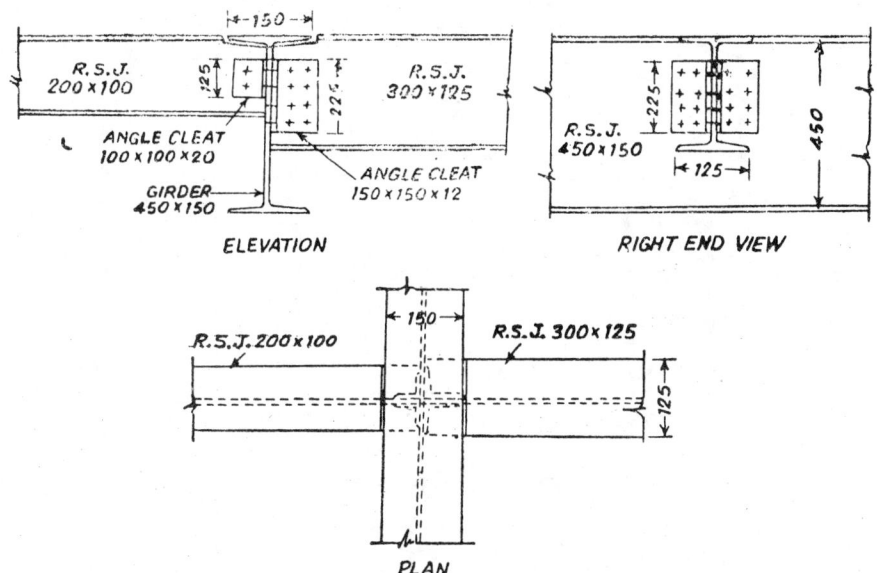

UNEQUAL BEAMS TO GIRDER CONNECTION

Fig. 16·2.

COLMN TO COLUMN CONNECTION

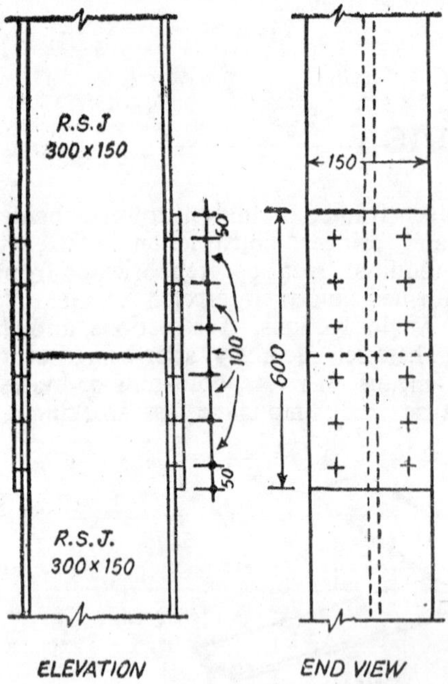

R.S.J
300×150

←150→

↕ 50
↕ 100
↕ 50

600

R.S.J.
300×150

ELEVATION   END VIEW

Fig. 16·3.

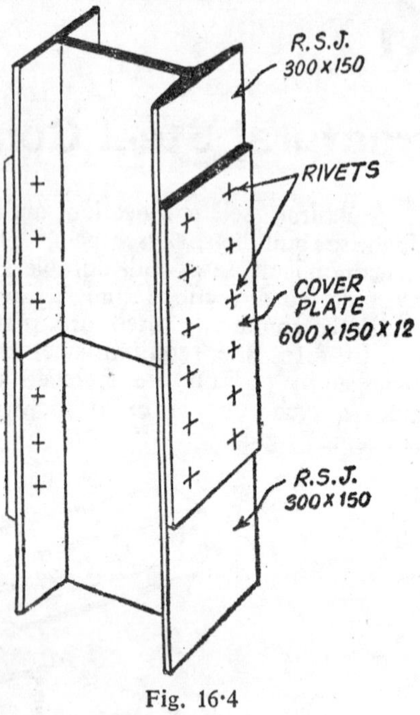

COLUMN SPLICING (EQUAL SECTIONS)

R.S.J.
300×150

RIVETS

COVER
PLATE
600×150×12

R.S.J.
300×150

Fig. 16·4

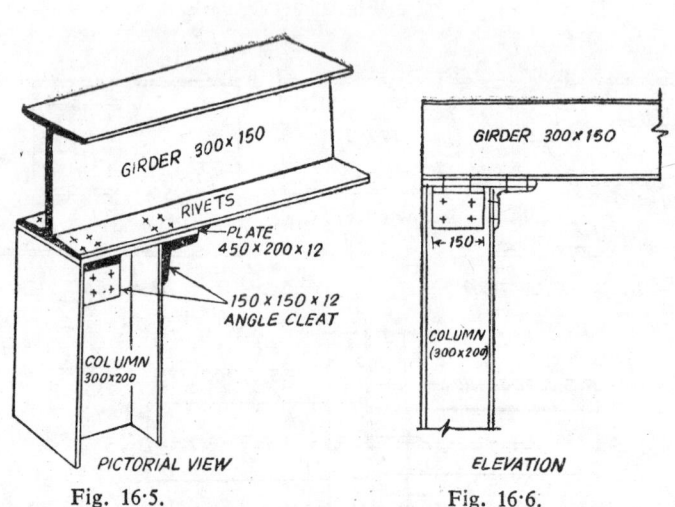

GIRDER 300×150

RIVETS

PLATE
450×200×12

150×150×12
ANGLE CLEAT

COLUMN
300×200

PICTORIAL VIEW

Fig. 16·5.

GIRDER 300×150

←150→

COLUMN
(300×200)

ELEVATION

Fig. 16·6.

GIRDER TO COLUMN HEAD CONNECTION

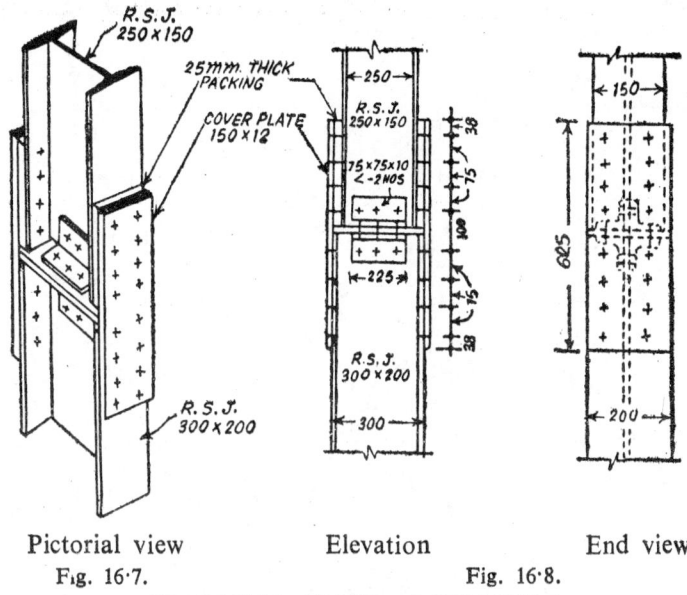

Pictorial view   Elevation   End view

Fig. 16·7.   Fig. 16·8.

UNEQUAL COLUMN SPLICING

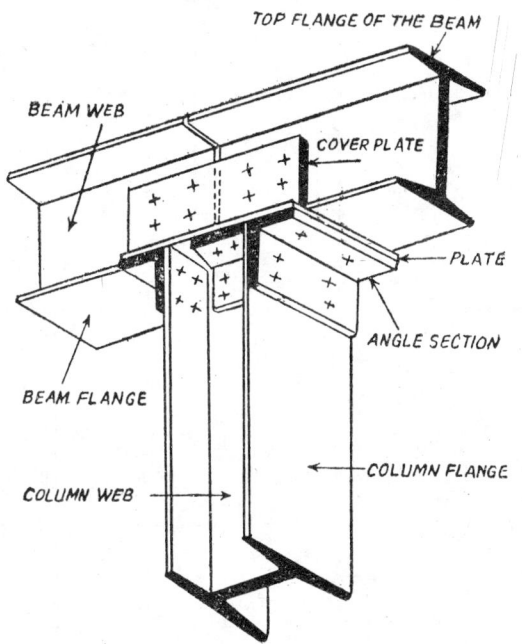

TWO EQUAL BEAM SECTIONS
MEET A COLUMN HEAD

Fig. 16·9.

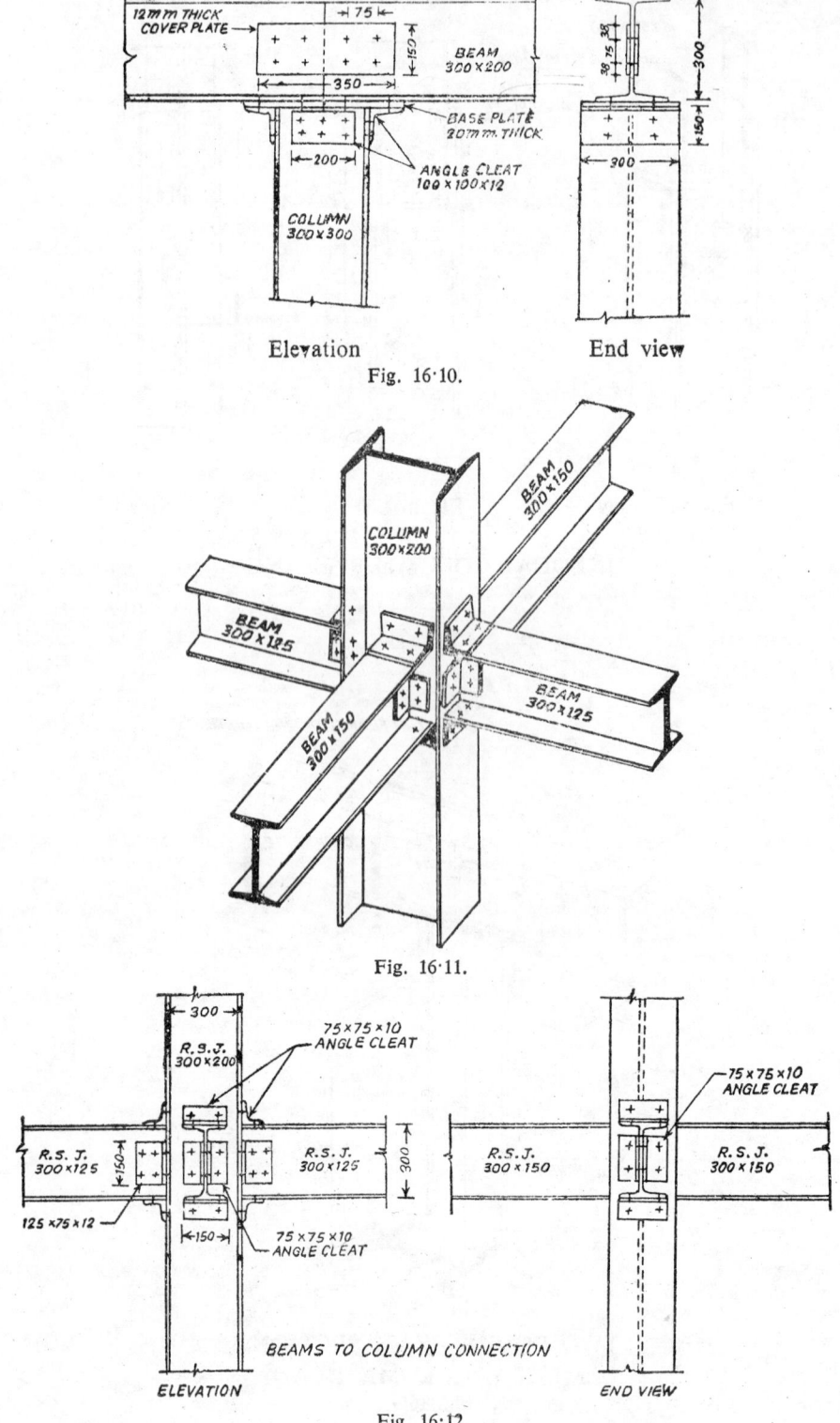

12mm THICK COVER PLATE
75
150
BEAM 300×200
350
BASE PLATE 20mm. THICK
200
ANGLE CLEAT 100×100×12
COLUMN 300×300

300
38 75 38
300
150

Elevation   End view

Fig. 16·10.

COLUMN 300×200
BEAM 300×150
BEAM 300×125
BEAM 300×125
BEAM 300×150

Fig. 16·11.

300
R.S.J. 300×200
75×75×10 ANGLE CLEAT
R.S.J. 300×125
150
R.S.J. 300×125
300
125×75×12
150
75×75×10 ANGLE CLEAT

75×75×10 ANGLE CLEAT
R.S.J. 300×150
R.S.J. 300×150

BEAMS TO COLUMN CONNECTION

ELEVATION   END VIEW

Fig. 16·12.

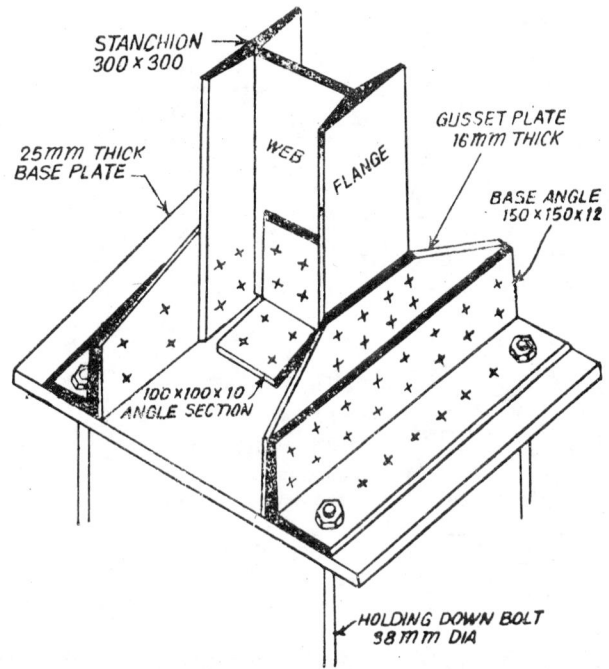

COLUMN BASE CONNECTION (Pictorial view)
Fig. 16·13.

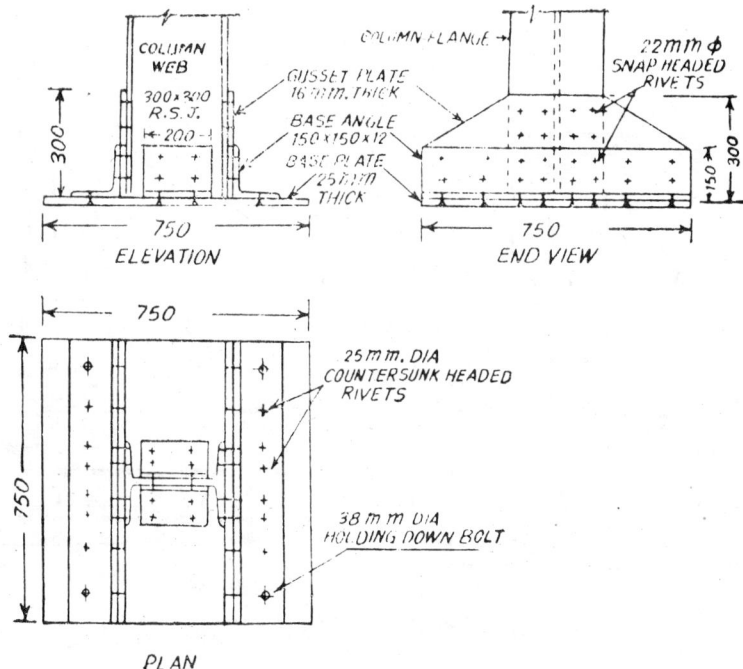

COLUMN BASE CONNECTION
Fig. 16·14.

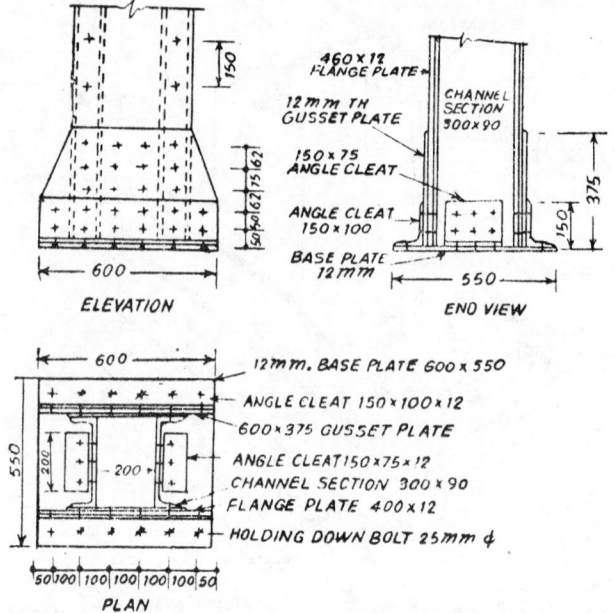

BASE CONNECTION OF A HEAVY STANCHION
Fig. 16·15.

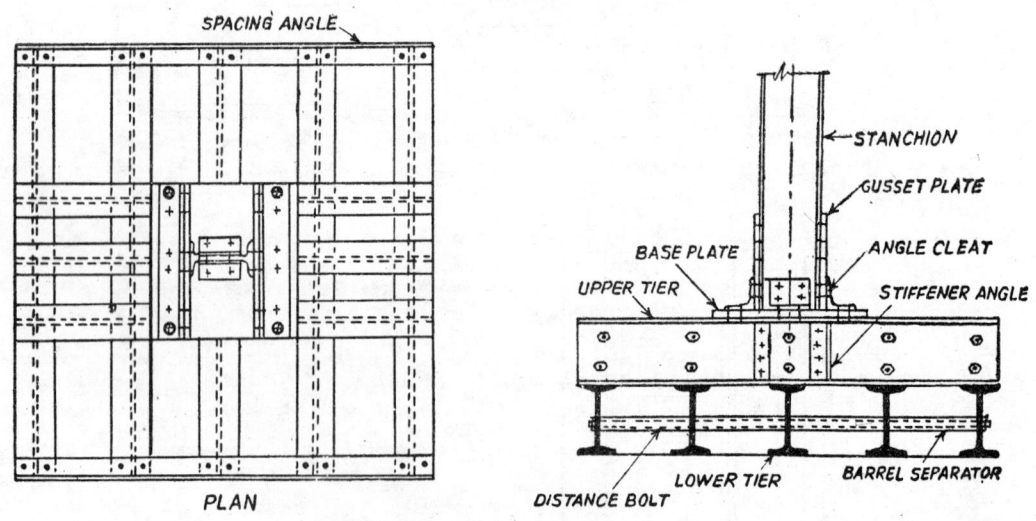

GRILLAGE FOUNDATION
Fig. 16·16.

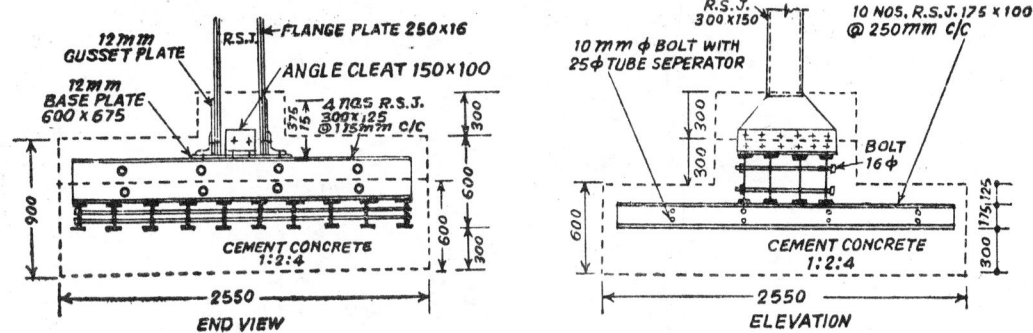

GRILLAGE FOUNDATION
Fig. 16·17.

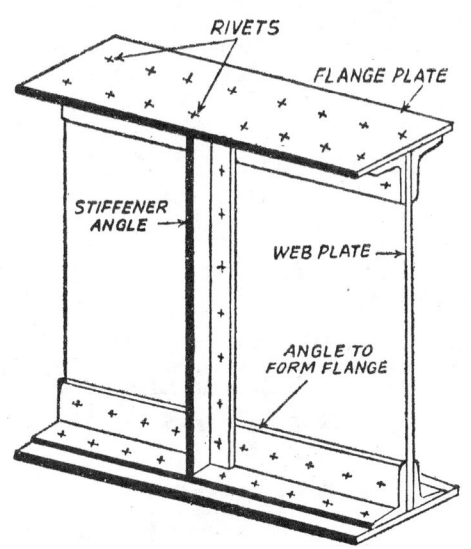

PLATED GIRDER
Fig. 16·18.

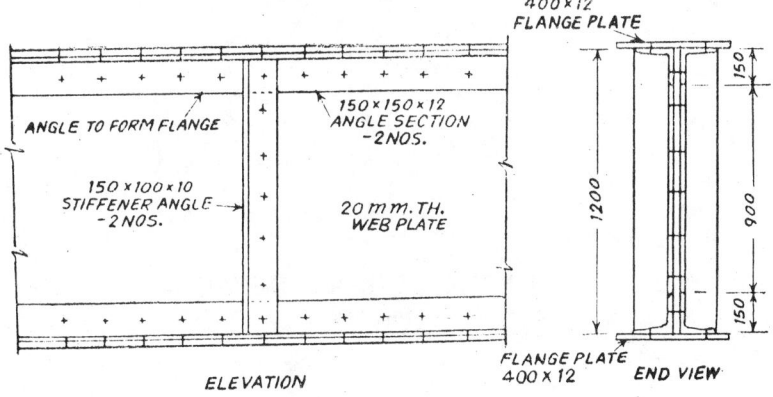

PLATED GIRDER
Fig. 16·19.

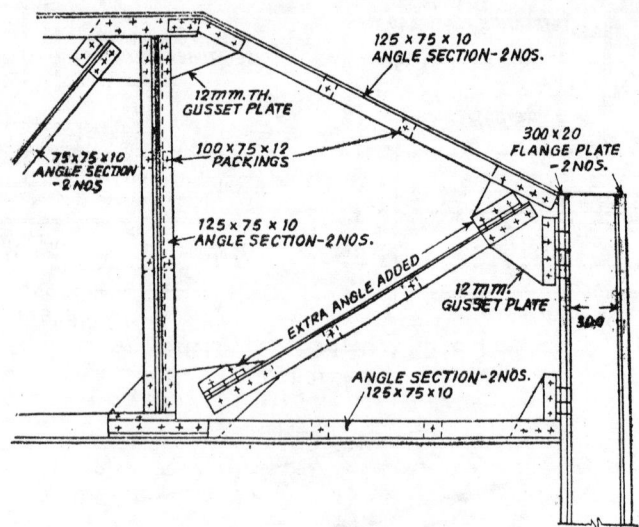

END CONNECTION OF LATTICE GIRDER
Fig. 16·20.

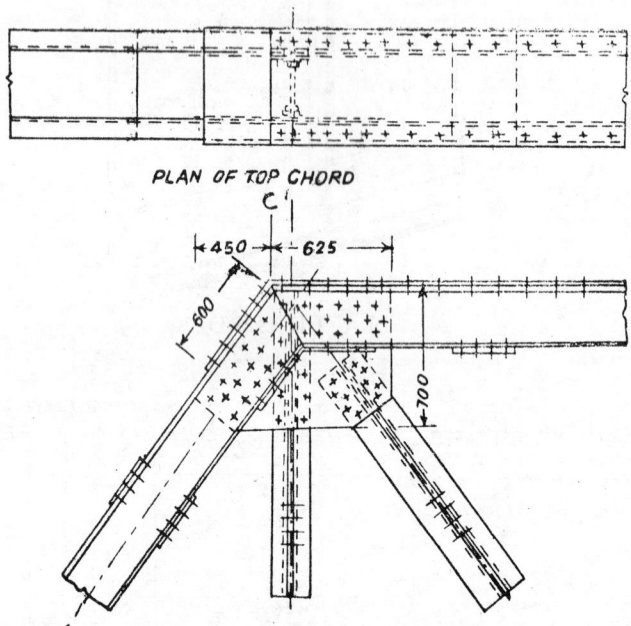

PLAN OF TOP CHORD

LATTICE GIRDER DETAILS
Fig. 16·21.

17

Reinforcement in Structures

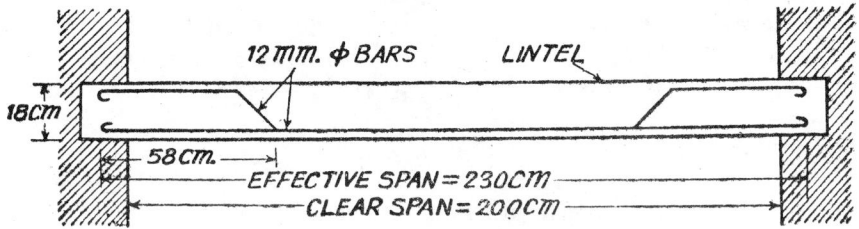

Reinforcement in Lintel
Fig. 17·1.

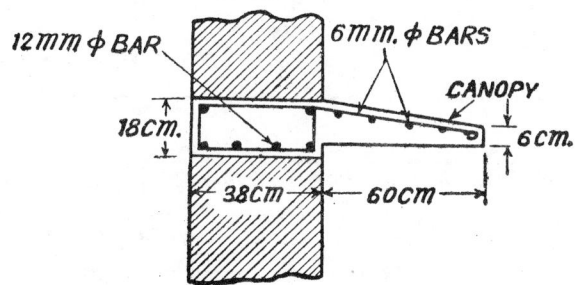

Reinforcement in Lintel with projected canopy
Fig. 17·2.

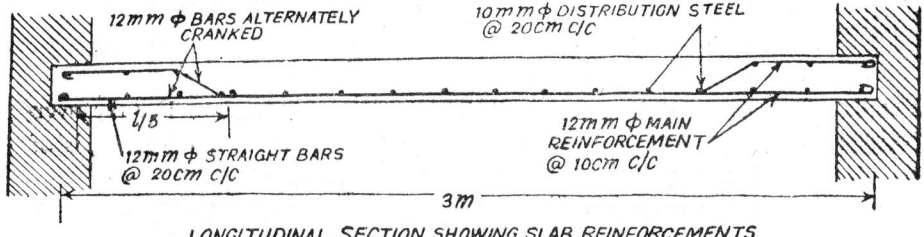

LONGITUDINAL SECTION SHOWING SLAB REINFORCEMENTS

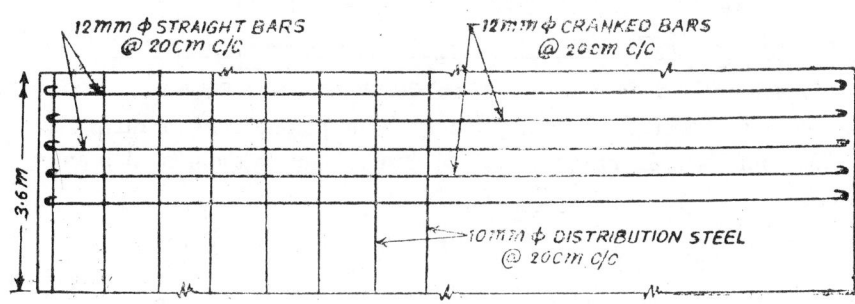

SECTIONAL PLAN SHOWING SLAB REINFORCEMENTS

Fig. 17·3.

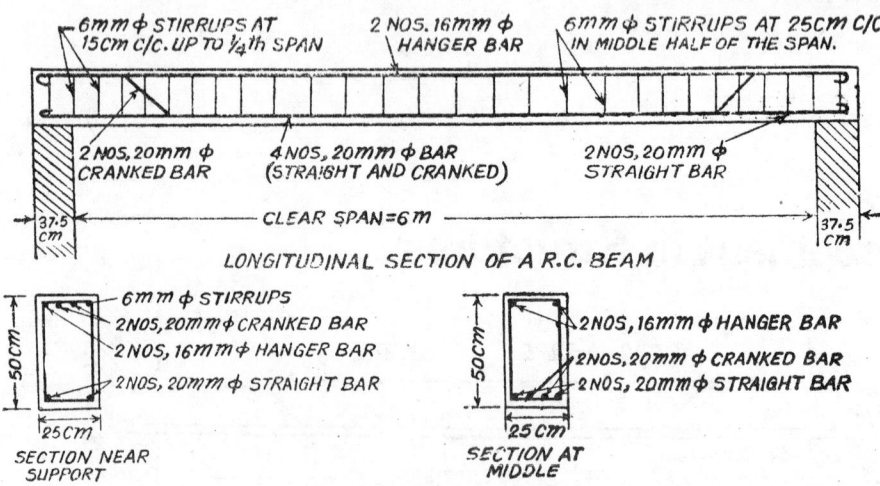

Reinforcement in R.C. Beam
Fig· 17·4.

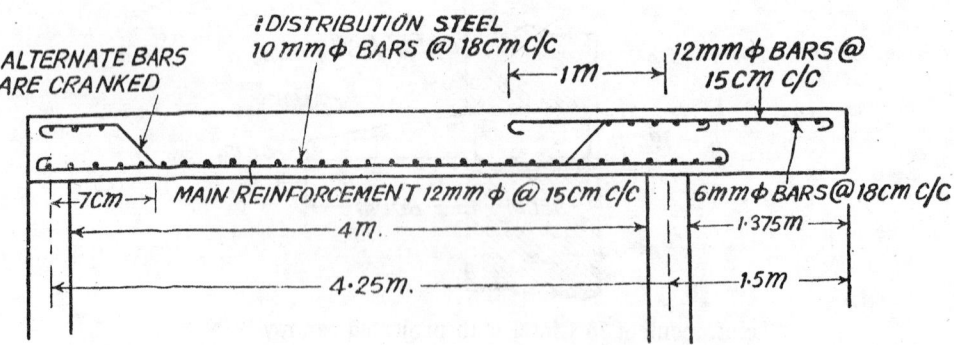

Details of reinforcement steel provided in an overhanging slab. For the over-hanging part, the reinforcement is provided at top, as the bending there, is of hogging nature.

Fig. 17·5.

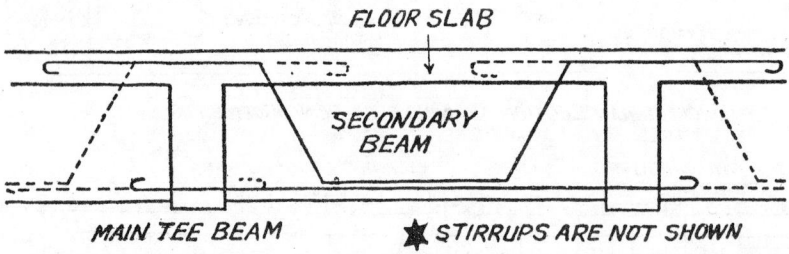

The nature of reinforcement in the secondary beam placed between main Tee beams. The reinforcement in the main Tee beams and in the floor slab are not shown here.

Fig. 17·6.

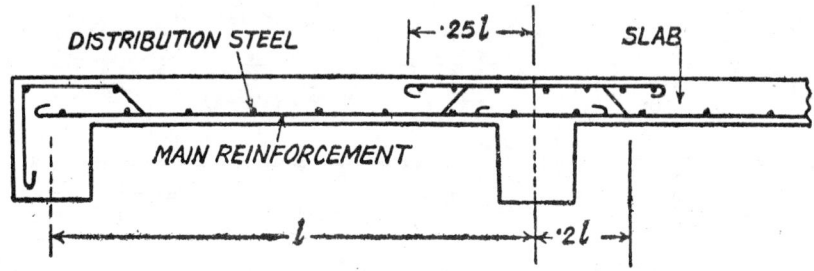

Details of reinforcement in a continuous slab. The slab itslf acts as the compression flange of the Tee Beams. At the end of the slab Ell beams are used. Reinforcements in Tee beam and Ell beam are not shown here.

Fig. 17·1.

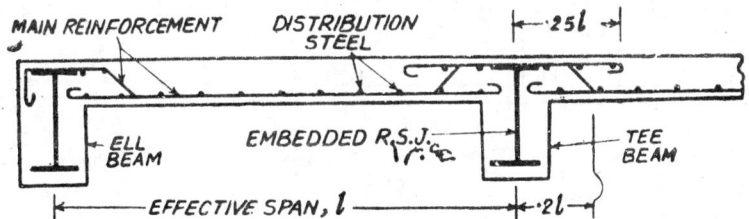

Rolled steel joists are embedded in concrete beams instead of reinforcement steel. Reinforcement bars shown, are for the slab only. Mark carefully, the position and placement of reinforcement bars at the end and intermediate beams.

Fig. 17·8.

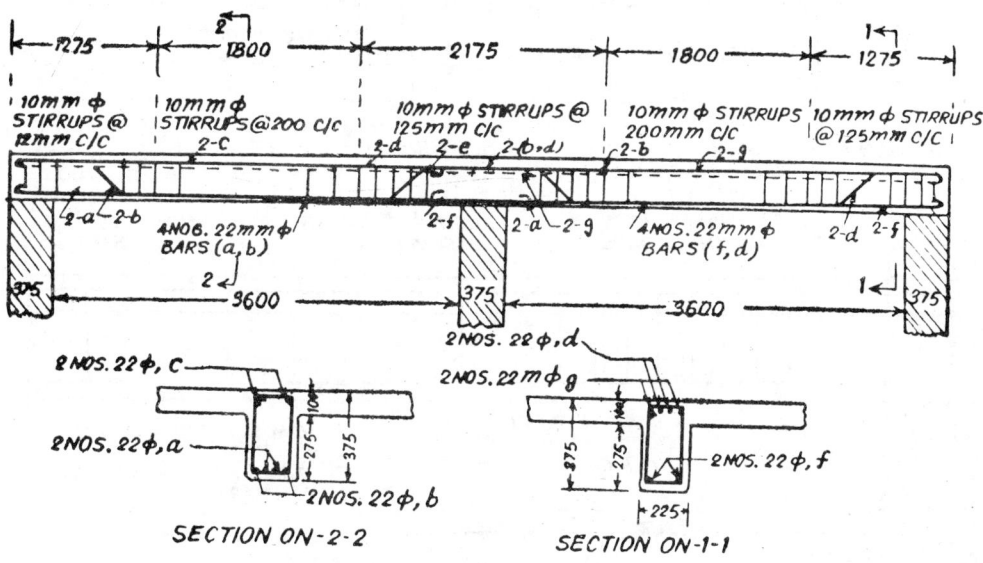

Details of Reinforcement in a Continuous Beam

Fig. 17·9.

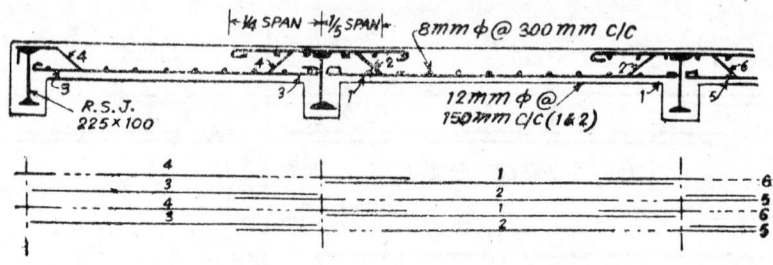

Schedule of reinforcement bars in a continuous slab.
Fig. 17·1₀.

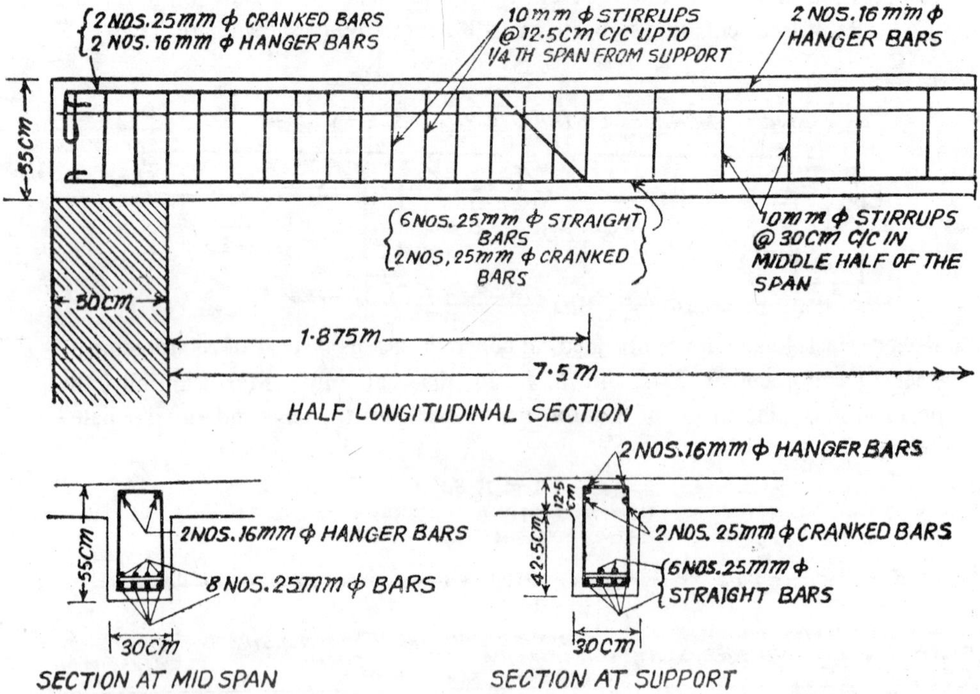

HALF LONGITUDINAL SECTION

SECTION AT MID SPAN   SECTION AT SUPPORT

Fig. 17·11.

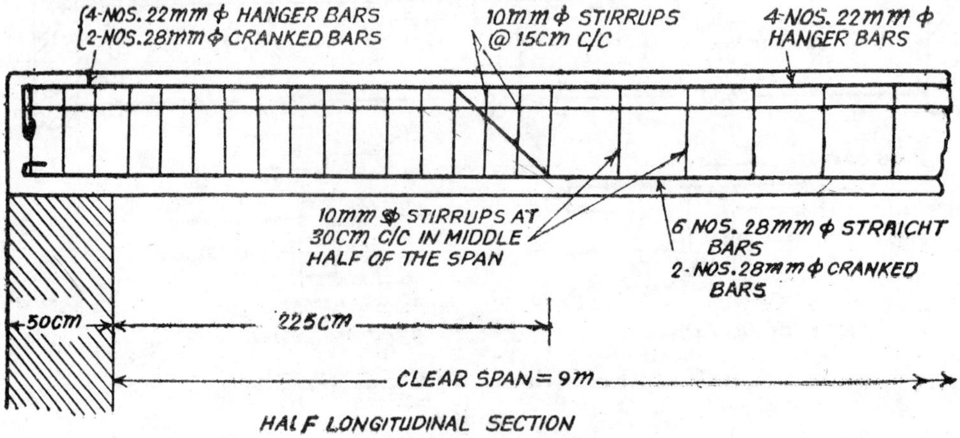

HALF LONGITUDINAL SECTION

Fig. 17·12. (contd)

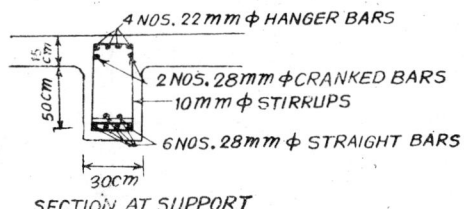

SECTION AT SUPPORT

Fig. 17·12. (contd)

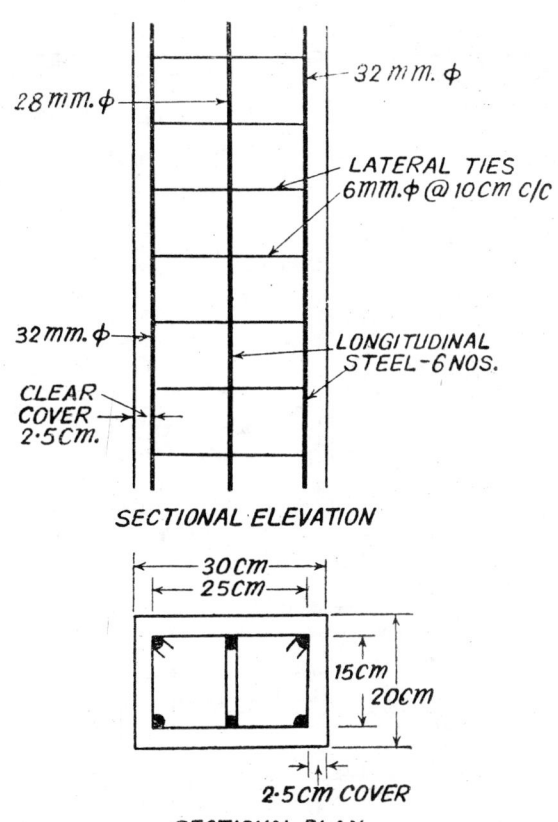

SECTIONAL ELEVATION

SECTIONAL PLAN

Reinforcement in a R.C. column, rectangular in section.
Fig. 17·13.

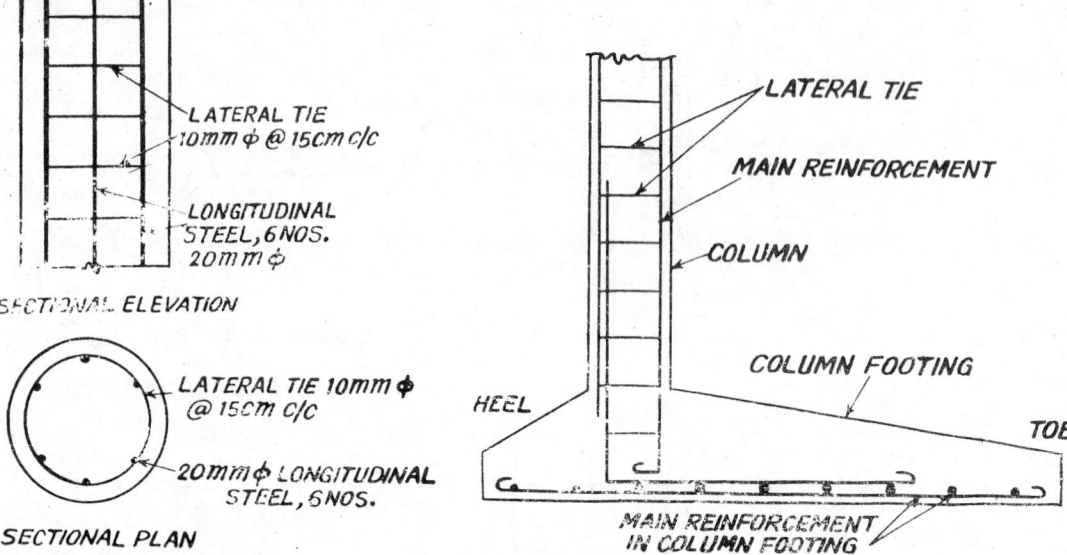

LATERAL TIE
10mm φ @ 15CM c/c

LONGITUDINAL
STEEL, 6 NOS.
20mm φ

SECTIONAL ELEVATION

LATERAL TIE 10mm φ
@ 15CM c/c

20mm φ LONGITUDINAL
STEEL, 6 NOS.

SECTIONAL PLAN

Reinforcement in a circular column.
Fig. 17·14

LATERAL TIE

MAIN REINFORCEMENT

COLUMN

COLUMN FOOTING

HEEL

TOE

MAIN REINFORCEMENT
IN COLUMN FOOTING

Reinforcement in Eccentria colume Footing.
Fig. 17·15

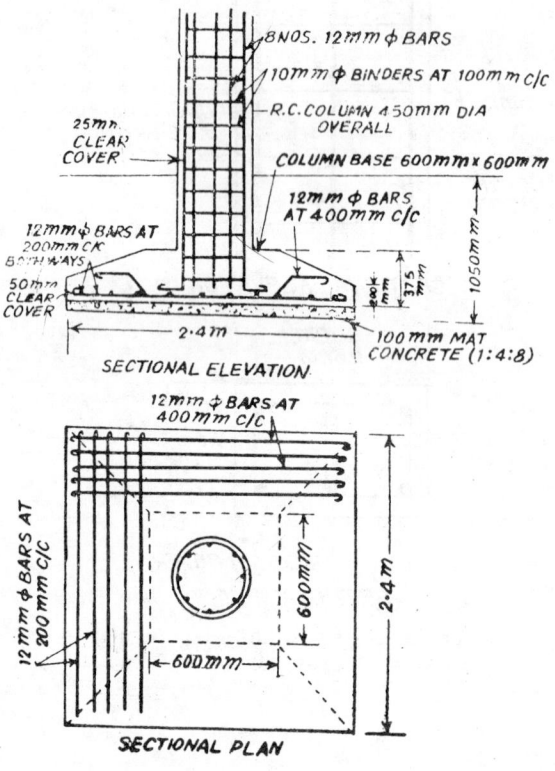

8 NOS. 12mm φ BARS

10mm φ BINDERS AT 100mm c/c

R.C. COLUMN 450mm DIA
OVERALL

COLUMN BASE 600mm x 600mm

25mm.
CLEAR
COVER

12mm φ BARS
AT 400mm c/c

12mm φ BARS AT
200mm C/C
BOTH WAYS

50mm
CLEAR
COVER

375
mm

1050mm

2·4 m

100mm MAT
CONCRETE (1:4:8)

SECTIONAL ELEVATION

12mm φ BARS AT
400mm c/c

12mm φ BARS AT
200mm c/c

600mm

600mm

2·4 m

SECTIONAL PLAN

Fig. 17·16.

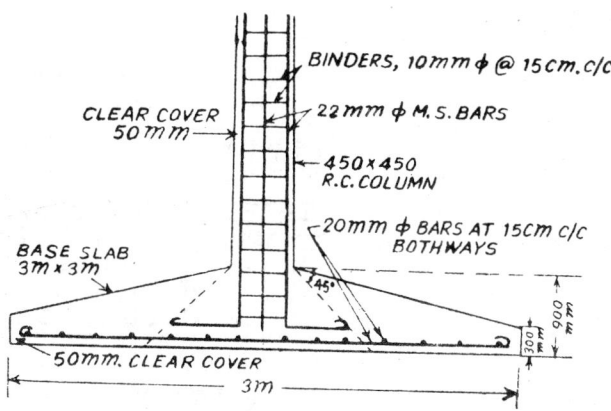

SECTIONAL ELEVATION

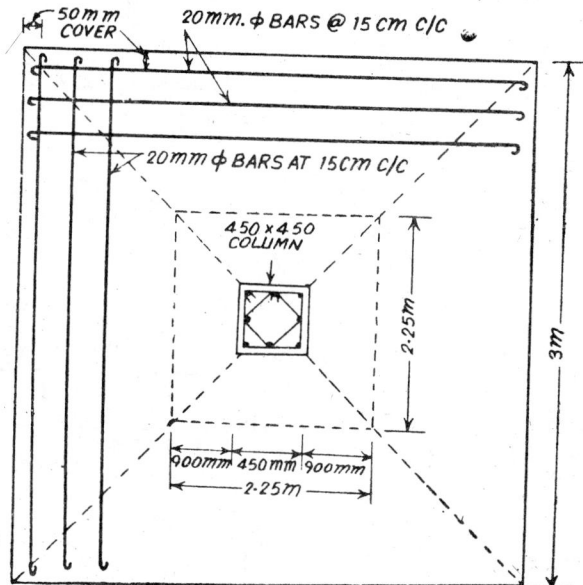

Fig. 17·17.

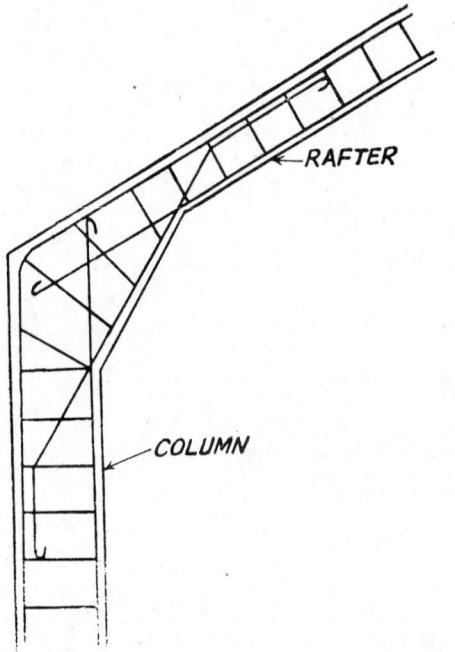

Reinforcement details at the junction of column and rafter for an R.C. framed structure.

Fig. 17·18.

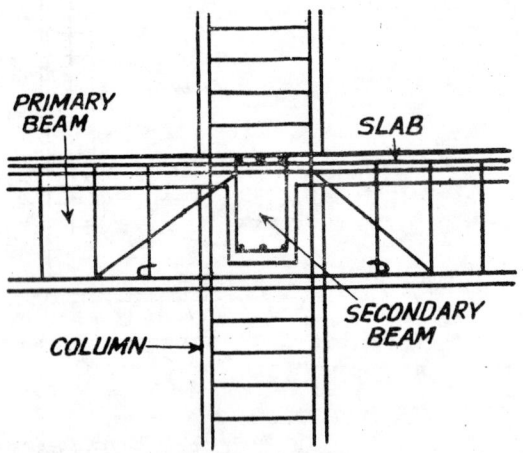

Reinforcement at the junction of primary beam and secondary beam with slab and column.

Fig. 17·19.

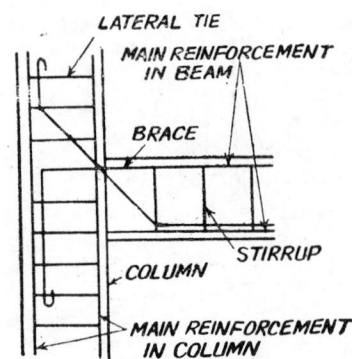

Reinforcement at the junction of column and column brace.

Fig. 17·20.

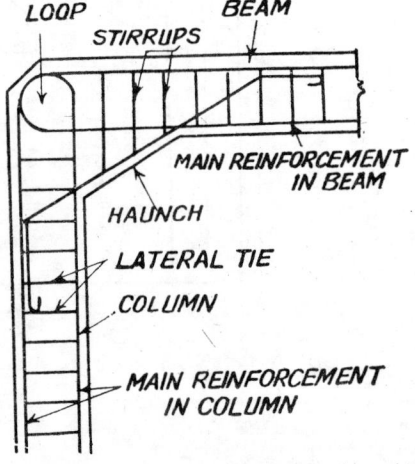

Reinforcement details at the junction of column and beam.

Fig. 17·21.

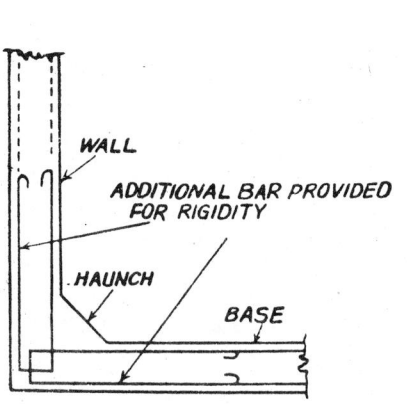

Junction of R.C. wall and the base slab

Fig. 17·22.

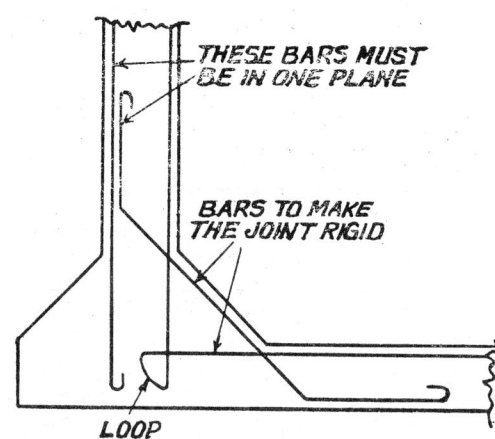

Shows how rigidity is obtained at a junction of wall and Base labs with the help of m.s. bars.

Fig. 17·23.

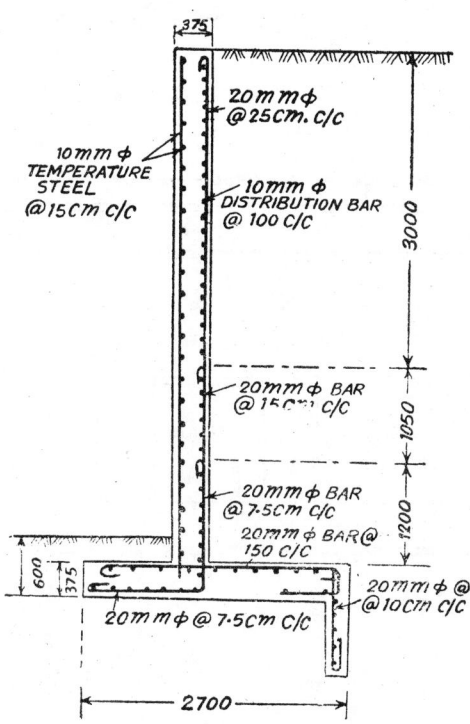

Reinforcement details for a cantilever type Retaining wall.
Fig. 17·24.

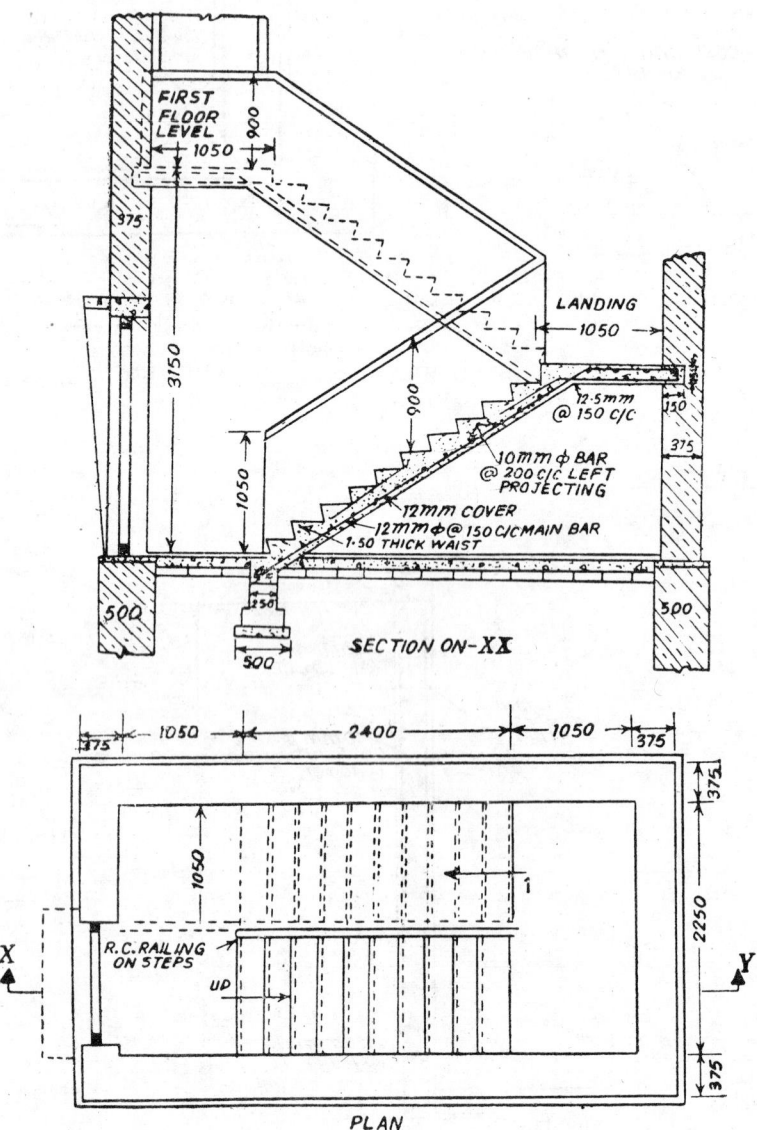

Fig. 17·25.

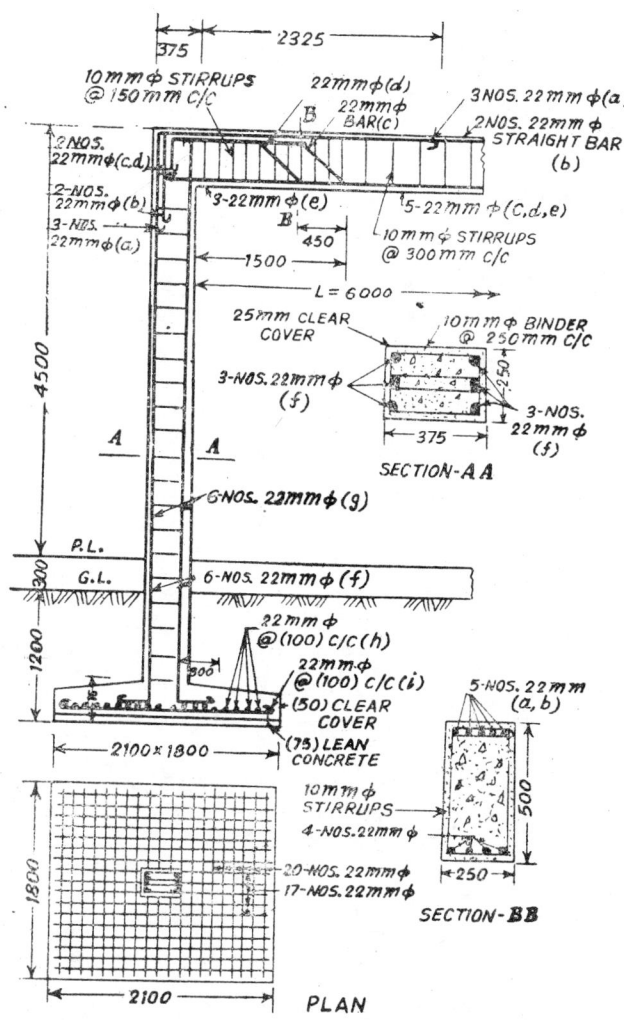

Reinforcement details in a portal frame
Fig. 17 26.

18

Sanitary Fittings and Plumbing Details

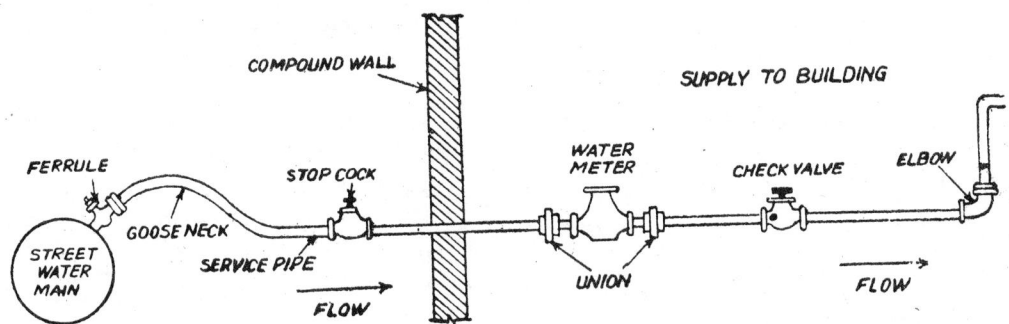

WATER SERVICE CONNECTION FROM STREET MAIN TO PREMISES
Fig. 18·1.

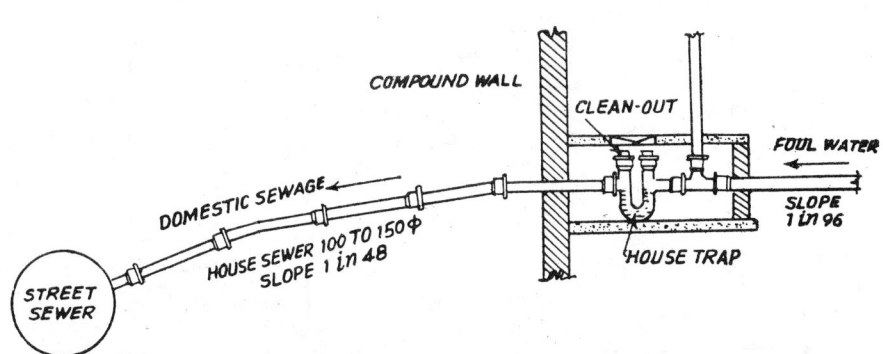

CONNECTION OF HOUSE SEWER TO STREET SEWER
Fig. 18·2.

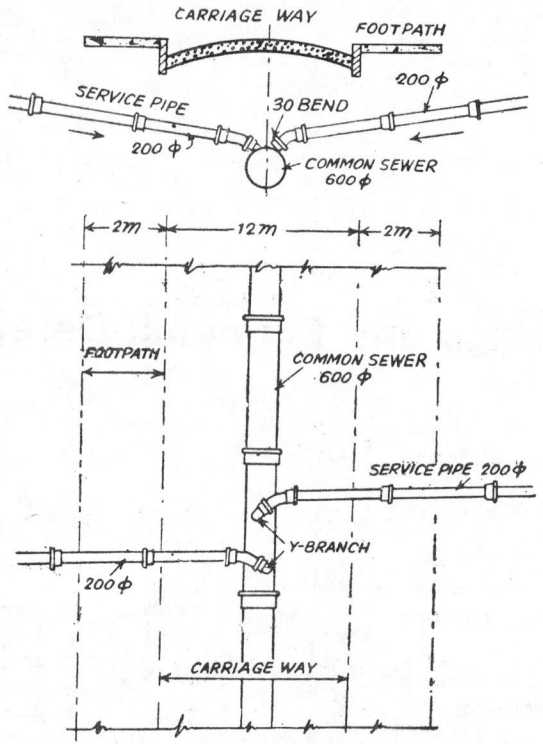

SECTION THROUGH A ROADWAY SHOWING SERVICE CONNECTIONS
Fig. 18·3.

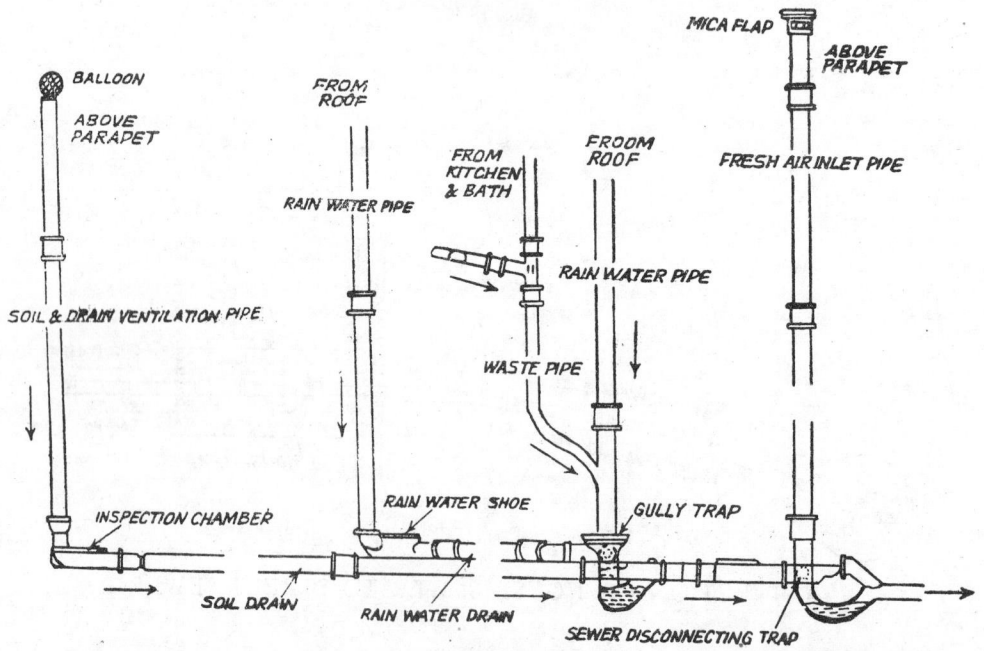

PLUMBING LAYOUT FOR DRAINAGE OF A DWELLING HOUSE
Fig. 18·4.

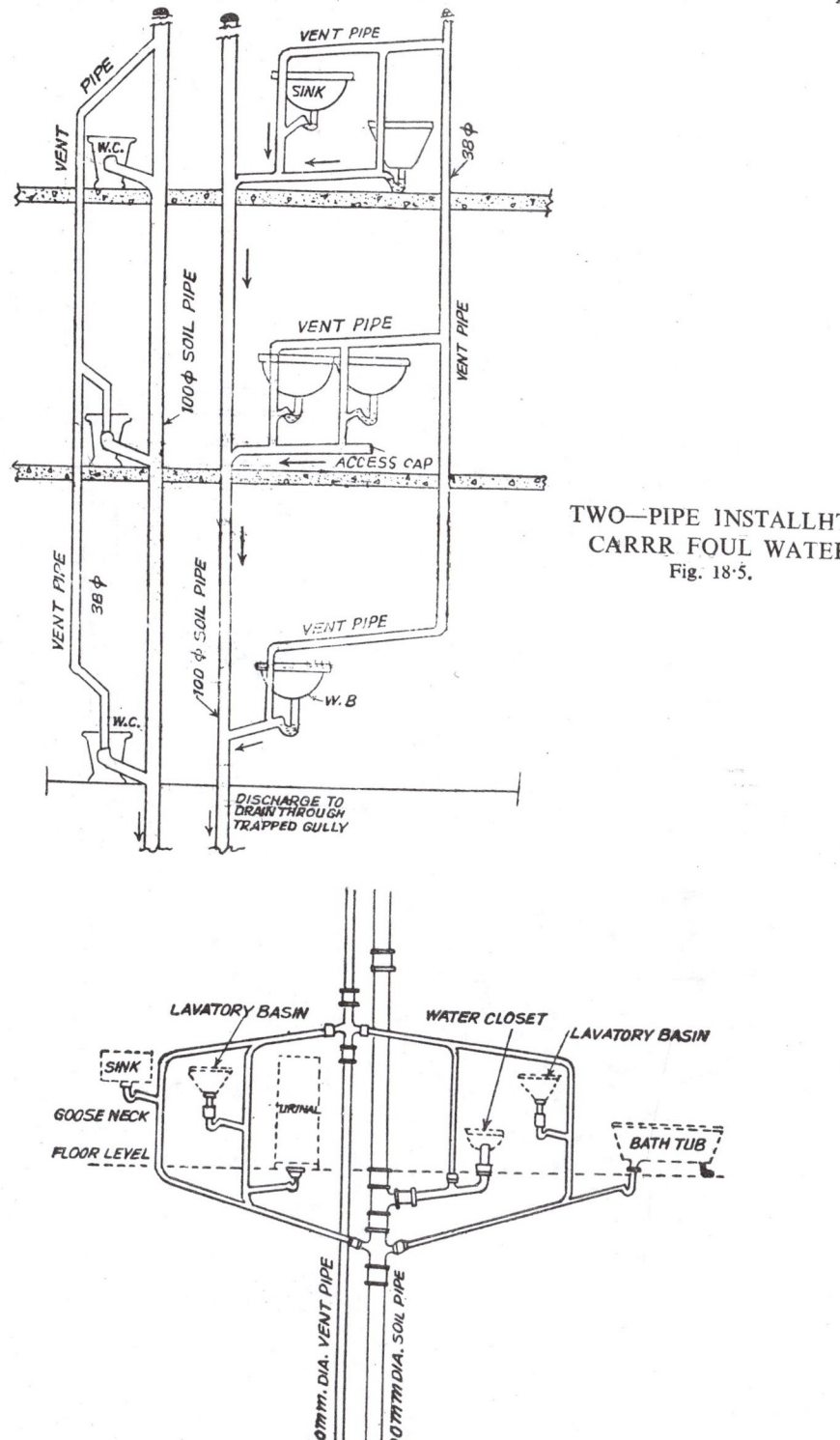

VENT PIPE

SINK

38 φ

W.C.

VENT PIPE

100 φ SOIL PIPE

VENT PIPE

VENT PIPE

ACCESS CAP

**TWO—PIPE INSTALLHTON
CARRR FOUL WATER.**
Fig. 18·5.

VENT PIPE

38 φ

100 φ SOIL PIPE

VENT PIPE

W. B

W.C.

DISCHARGE TO
DRAIN THROUGH
TRAPPED GULLY

LAVATORY BASIN

WATER CLOSET

LAVATORY BASIN

SINK

URINAL

BATH TUB

GOOSE NECK

FLOOR LEVEL

50mm. DIA. VENT PIPE

100mm DIA. SOIL PIPE

ONE-PIPE INSTALLATION TO CARRY FOUL WATER
Fig. 18·6.

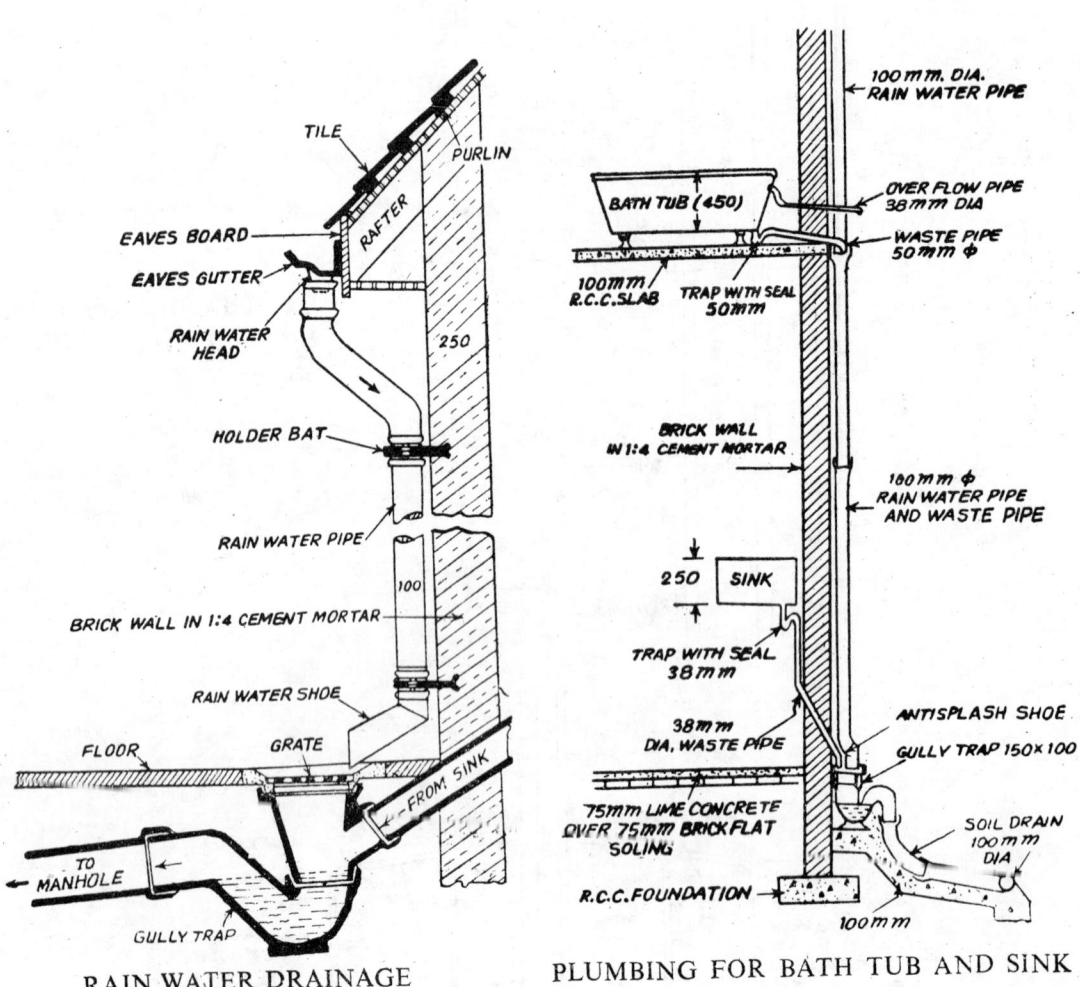

RAIN WATER DRAINAGE
Fig. 18·7.

PLUMBING FOR BATH TUB AND SINK
Fig. 18·8.

Bowl Pattern Urinal :

The two views of a bowl pattern urinal are shown is Fig. 18·15. This is a standing urinal for male. This is flushed by an automatic flushing cistern or manually with the help of a cock after its use each time. The pan is made of white glazed china clay or fire clay.

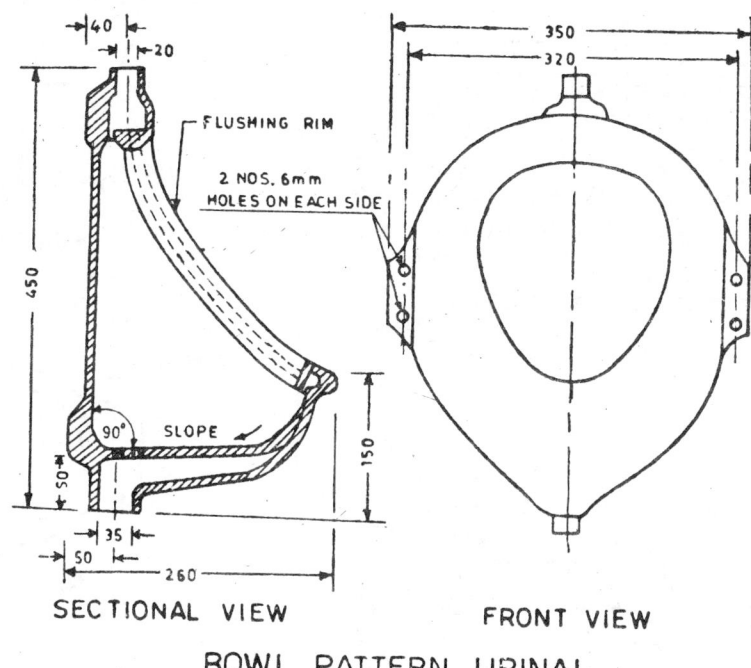

SECTIONAL VIEW   FRONT VIEW

BOWL PATTERN URINAL

Fig. 18·15.

Flushing Cistern :

Fig. 18·16 shows plan and sectional view of a flushing cistern meant for flushing water closet. The flushing capacity of such a cistern is 10 to 15 litres. These are made of cast iron, the minimum thickness being 5 mm. The flushing cistern works on the principle of partial vacuum or syphonic action. When the chain is pulled, the lever raises the bell inside the chamber and the water enters into the stand pipe through the bottom of the bell and the syphonic action starts. A float valve is provided to stop the inflow of water into the cistern when it becomes full. When the float valve fails to operate, the excess water goes out through the overflow pipe. After pulling the chain if it is released suddenly water is forced to come down the flush pipe with a splash.

European Type W.C.

Plan, elevation and sectional view of such a closet are shown in Fig. 18·17. The closet pan, pedestal and trap are cast in one. These are made of glazed stoneware or china clay.

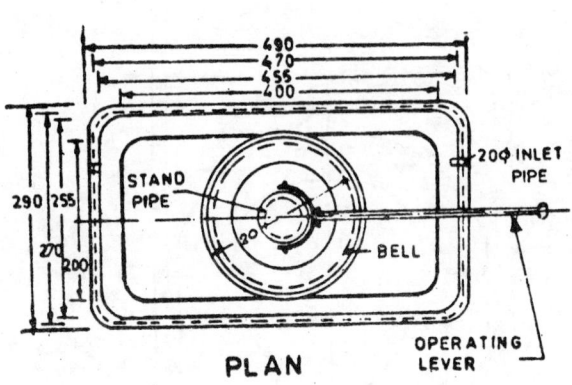

PLAN

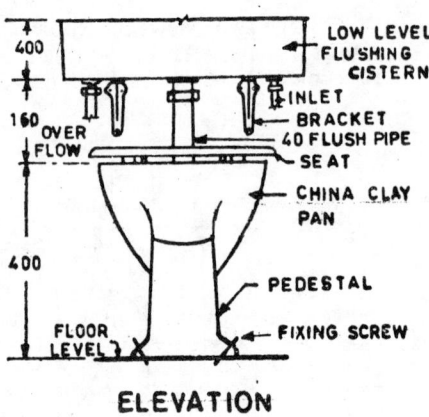

ELEVATION

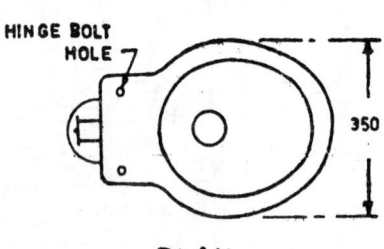

PLAN

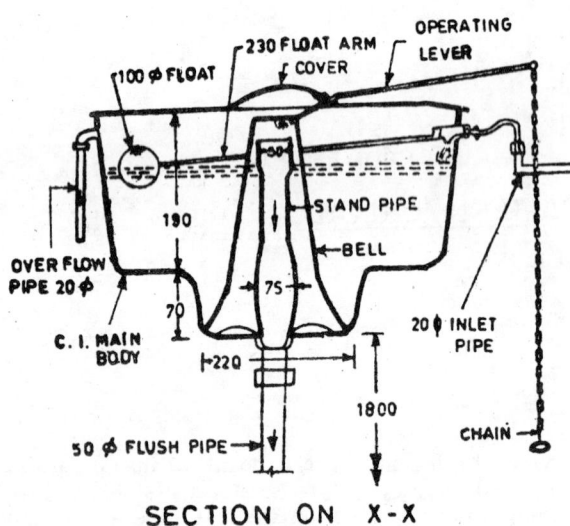

SECTION ON X-X

FLUSHING CISTERN

Fig. 18·16.

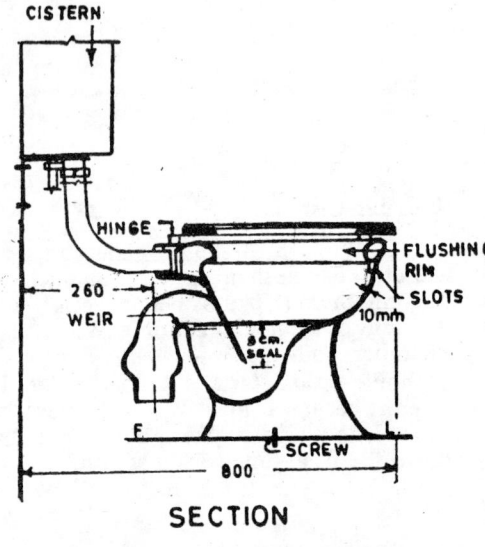

SECTION

EUROPEAN TYPE W.C.

Fig. 18·17.

Bath Tub :

Three views of a bath tub are presented in Fig. 18·18. Usually, these are made of glazed china clay or C.I. coated with plastic or fibre glass sheet. Bath tubs are also made of concrete with mosaic finish.

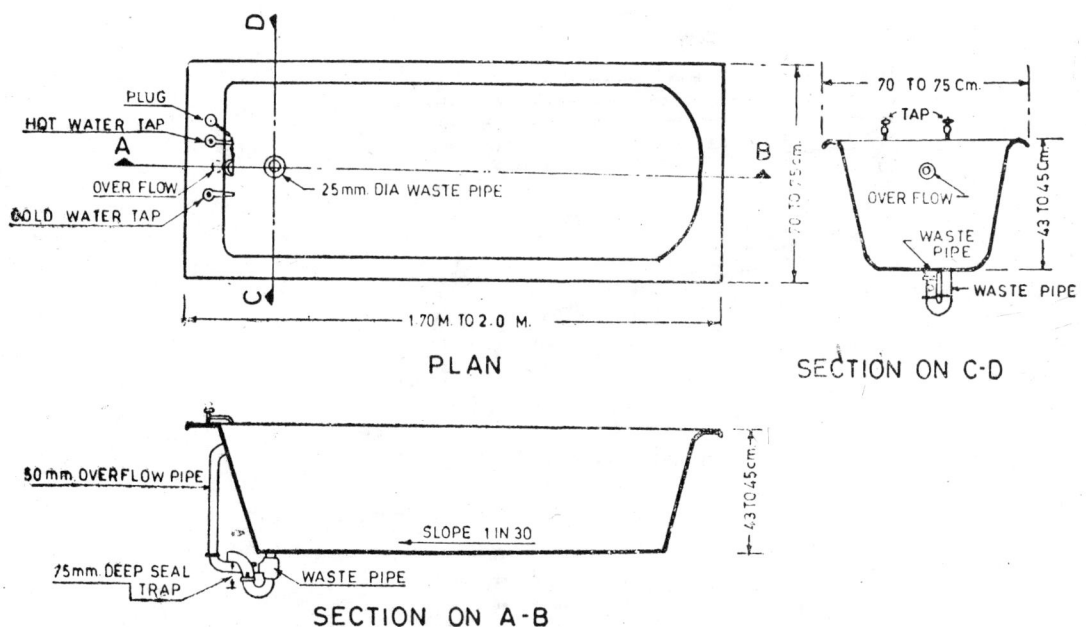

PLAN   SECTION ON C-D

SECTION ON A-B

DETAILS OF BATH TUB

Fig. 18·18.

Squating pan with Trap is shown in Fig. 18·19. This is Indian type water closet pan. The foot rests are either cast with the pan or made separately.

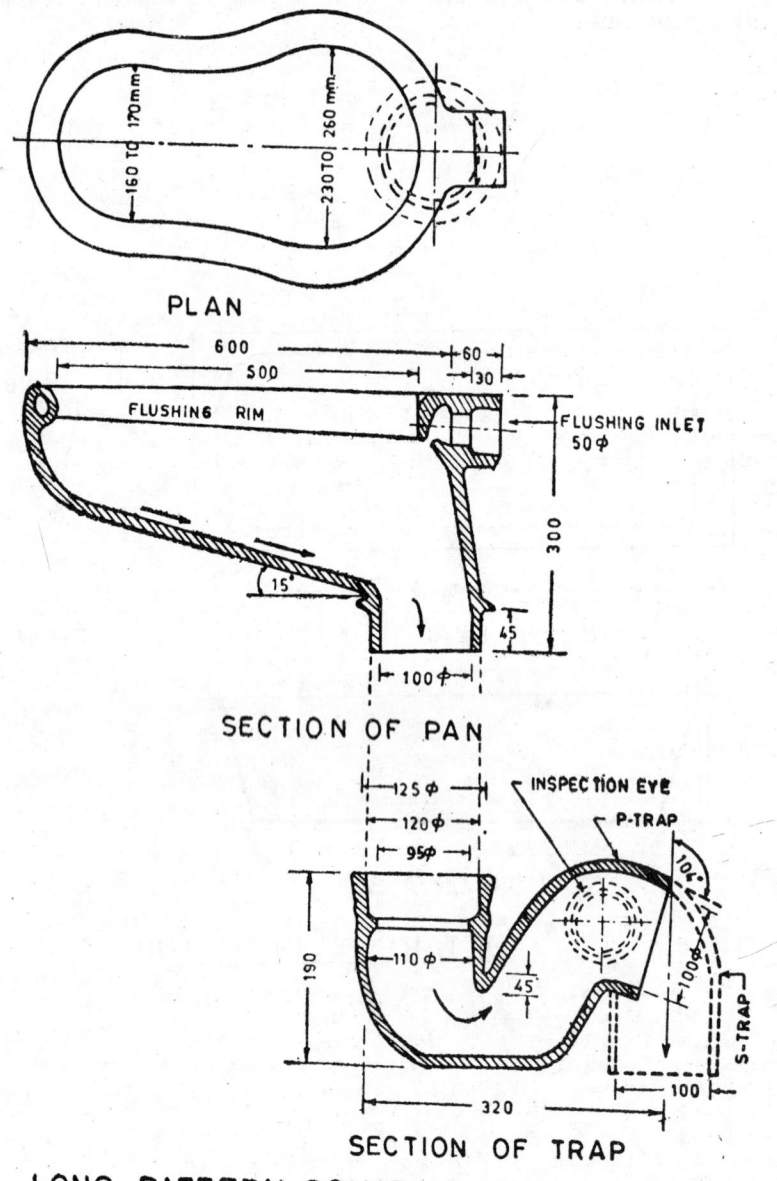

PLAN

SECTION OF PAN

SECTION OF TRAP

LONG PATTERN SQUATING PAN WITH TRAP

Fig. 18·19.

Wash Hand Basin :

Fig. 18·20 shows the isometic view of a wash hand basin from its bottom which reveals the plumbing details.   Plan and two sectional views of the basin are shown in Fig. 18·21.

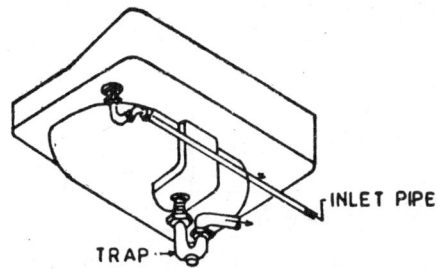

ISOMETRIC VIEW FROM BOTTOM

Fig. 18·20.

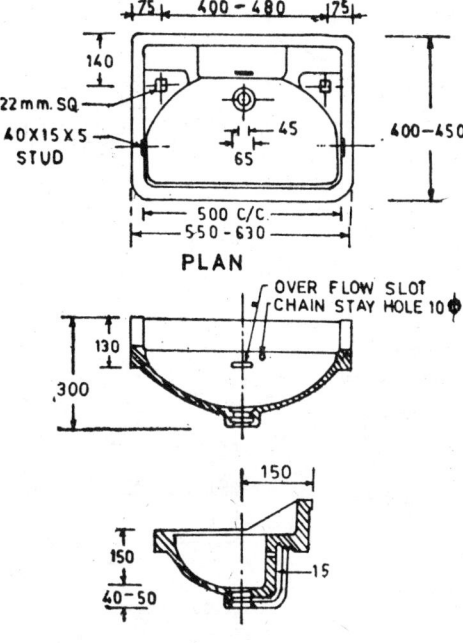

PLAN

SECTIONAL VIEWS

WASH HAND BASIN

Fig. 18·21.

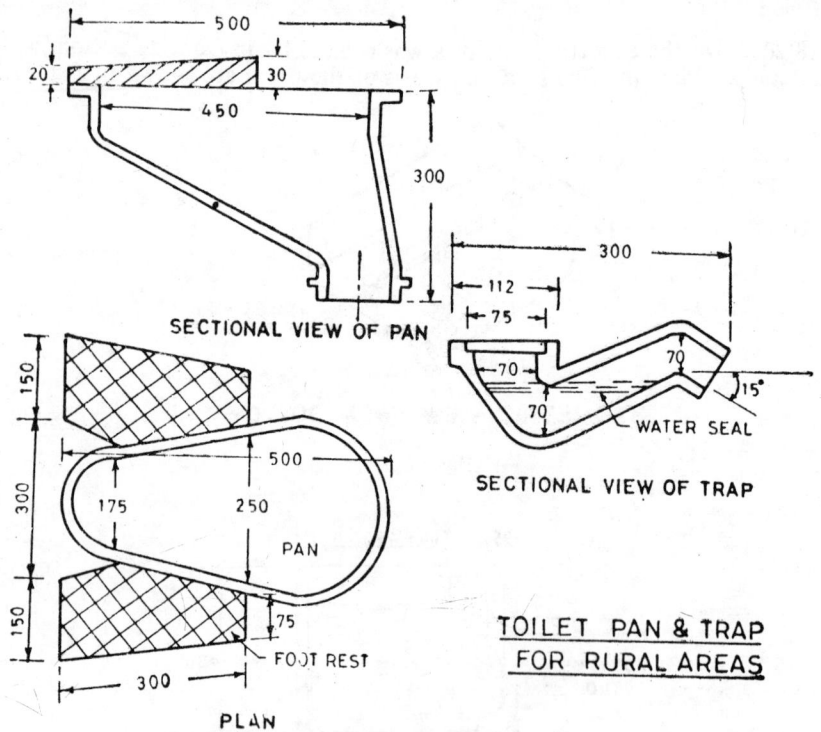

SECTIONAL VIEW OF PAN

SECTIONAL VIEW OF TRAP

PLAN

TOILET PAN & TRAP
FOR RURAL AREAS

Fig. 18·22.

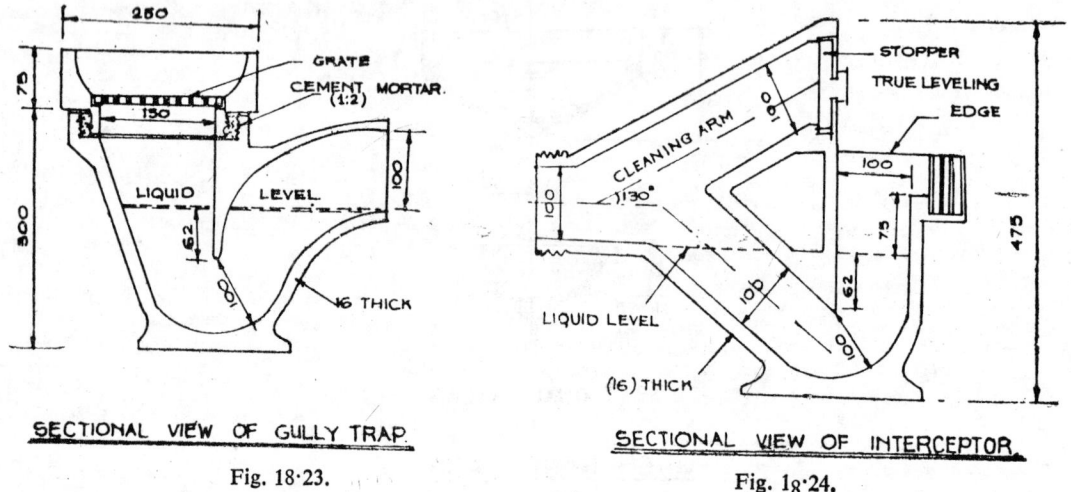

SECTIONAL VIEW OF GULLY TRAP.

Fig. 18·23.

SECTIONAL VIEW OF INTERCEPTOR.

Fig. 1g·24.

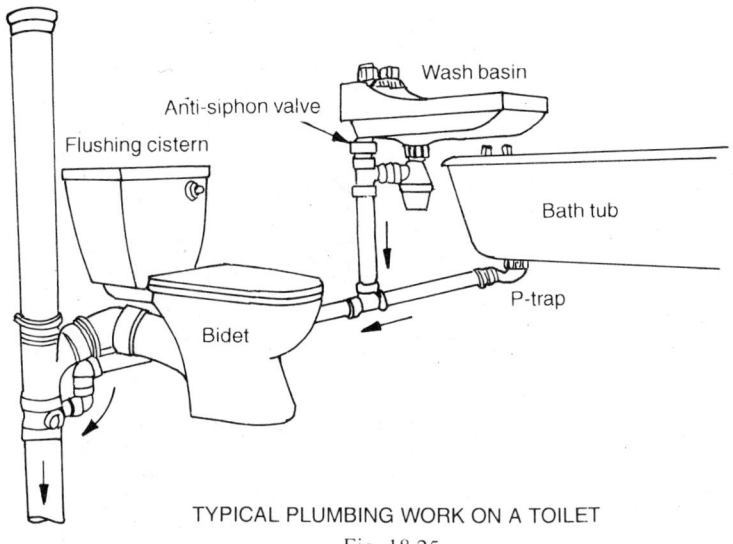

TYPICAL PLUMBING WORK ON A TOILET
Fig. 18.25

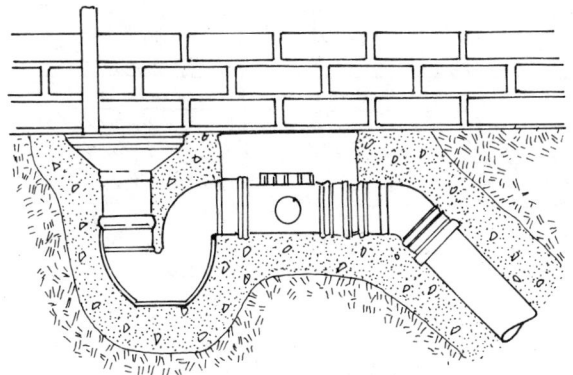

TYPICAL GULLEY TRAP
Fig. 18.26

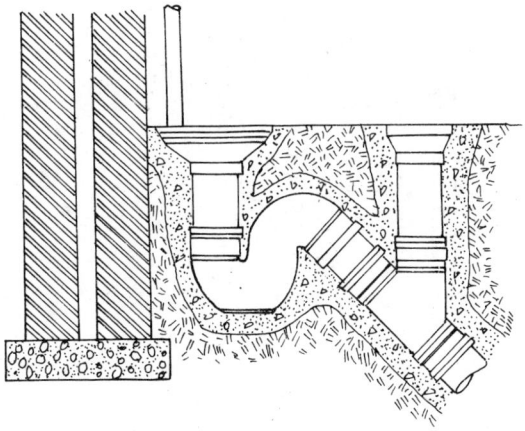

TYPICAL GULLEY TRAP
Fig. 18.27

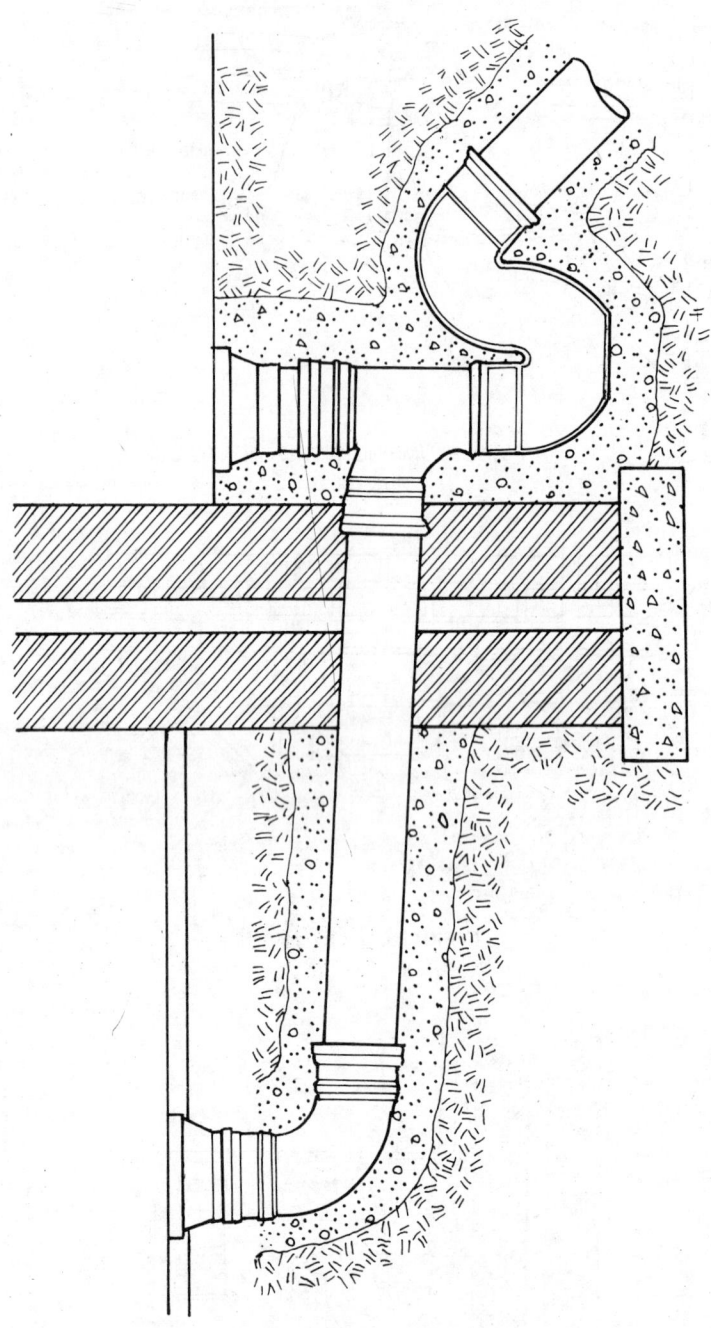

TRANSPORTATION OF FOUL WATER FROM HOUSE TO SEWER MANHOLE THROUGH GULLEY TRAP

Fig. 18.28

19

Design of Buildings

Design of buildings depends on quite a number of factors such as :
*Land Area available ;
*Building character i.e., nature of use ;
*Fund Resource available ;
*Availability of materials to be employed ;
*Character of the area where the building will be erected ;
*Amenities to be provided ;
*Scope of future extension, if any ;
*Meteorological status i.e., climatic condition of the locality :
*Requirement of the users of the building.

The Architect and the Civil Engineer will design the building by keeping in mind the points mentioned above. In building design, orientation of rooms and other spaces plays an important role. For proper orientation, the factors to be considered are :

*easy access to all spaces

*adequate ventilation of air and light

*safety, security and privacy

*maximum utilisation of space

*economy.

This chapter will deal with design of residential buildings only. A residential building should essentially consist of bed room, drawing room, kitchen, dining room and toilet. A small residential house for a poor family should have at least the following units to meet the basic requirements : one bed room, one kitchen and one toilet. On the other hand, a rich family will need the following units : a number of bed rooms with attached bath, drawing room, guest room, parlour, kitchen, dining hall, store, box room, staircase, garage, verandah, servant's room, etc. A family of middle income groop will ask for a small two-storied building comprising of at least two bed rooms, one drawing cum dining room, a kitchen with a store toilets (if not attached to bed rooms) in each floor, the ground floor being kept for tenant., The house owner will not prefer a common entry with the tenant.

Every house owner, whether rich or poor, wants a house having a very good appearance from its front, proper ventilation of air and light and easy access with safety and privacy. Planning and design of a residential house should therefore be made by the architect and the engineers to suit the environment and to meet the need of the house owner.

The orientation should be made in such a fashion that the bed rooms get maximum air and light and no smoke nuissance takes place. The bed rooms must have privacy and no bed room should have direct access to the drawing room or guest room. There should not be any door direct from stair landing to bed room. The common verandah should not extend beyond the windows of the adjacent rooms and the bed rooms should not have windows on the corridor side.

Drawing room or sitting room should be located at the front side of the building with an access to Dining Hall through a passage or with an access to staircase. This room should be spacious with adequate air and light. This may be treated as a living room.

Guest room should preferably be placed on one side of the front portion of the building attached to the drawing room, but disconnected from all other rooms. This room should have attached toilet.

Bed rooms should be placed on the two sides of the building exposed to the outer space for ventilation and light. These rooms should preferably be placed on the side of the prevailing wind and provided with attached toilet.

Dining hall should be approachable from all the rooms and it should be close to the kitchen, staircase (if any) and the drawing room. This space should be properly ventilated.

Kitchen should be placed at one of the rear corners of the building and it should be close to the dining hall. It should be directly approachable from all the rooms. The kitchen should be provided with chimney.

Store room and **pantry** should be very close to the kitchen. These two small rooms should be well-ventilated and should have a number of shelves and cupboards.

Toilet or **bath** and **W.C.** should preferably be attached to the bed rooms. In modern practice, bath and W.C. are combined in one room. This room should have bath tub, shower, wash-hand basin, soap keeper, toilet racks, hangers, brackets and W.C.

Dressing room should be placed in between the bed room and the toilet. The toilet floor level should be made about 100 to 150 mm. lower than the floor level of the dressing room and bed room.

Verandahs may be provided at the front and at two sides of the building.

Servant's quarter should preferably be located in the back portion of the building.

Box room may be placed in the space left between two rooms or two end walls or at the end of a passage. Box room to be used as a strong room should be made concealed.

Lobby which is an entrance hall is sometimes provided in a residential building.

Parlour should be located in a central place of a building or close to the staircase. This is a meeting place of the family members.

Study room should be placed at any corner of a building preferably away from disturbances.

Garage seould be located very close to the roadway.

Roof should be made accessible and it should be provided with a nice garden, where the land is too costly to make a nice lawn with a garden.

Staircase should be located at such a place in a building that it has an easy access from all corners of the building. The stair should be well-ventilated with adequate light. The width of each flight of the stair should be 0·75 m. to 1·25 m. It is preferable to have 1 m. width of stair in each flight. Clear headway should be 2 m. minimum. The relation between rise and tread should be as :

$$\text{Rise} = \frac{66}{\text{Tread}} \; ; \quad \text{when rise and tread are in inches}$$

$$\text{Rise} = \frac{400}{\text{Tread}} \; ; \quad \text{when rise and tread are in cm.}$$

Unitwise floor areas are given below :

Name of Unit	Floor Area in m²
Bed room	8—25
Drawing room	12—30
Dining room	12—30
Guest room	8—12
Kitchen	7—10
Store room/Pantry	6—8
Toilet (Bath & W.C.)	4—6
Bath	2—4
W.C.	0·4—1
Box room	2—4
Dressing room	4—6
Servant's room	7—9
Study room	7—9
Garage	14—18
Staircase	8—12

Plinth Height :

The plinth height should be 300 mm—600 mm. For water logging prone areas and important buildings, the minimum plinth height is 600 mm.

Minimum Room Heights :

Main rooms—3M (Preferably 3·5 M)

Kitchen Store, Pantry, Servant's room and verandah—2·5 M (Preferably 2·7 M).

Plinth Area. Wally cover about $\frac{1}{5}$ th of the total floor area and the verandahs cover about $\frac{1}{10}$ th of the floor area. Thus, if floor area is x, the plinth area is $1·2\ x$.

$$\text{Plinth area rate of construction} = \frac{\text{Total cost of construction}}{\text{Plinth area}}$$

$$\therefore \quad \text{Total plinth area} = \frac{\text{Total cost of construction}}{\text{Plinth area rate}}$$

Ventilation of Building :

For proper ventilation of a room, the area covered by doors and windows should not be less than $\frac{1}{8}$ th of the floor area of the room. Area covered for ventilation should be 4% of the floor area of the room.

Land Area to be kept open :

One-third of the land area should be kept open and specifically 10'—0" width at the back and 4'—0" width on both sides and 4'—0" width at front should be kept open. This is the minimum requirement of open space.

Building Plan :

A building plan should essentially consist of :

      (*i*) a site plan showing the location of the proposed building ;

      (*ii*) Plan of the building (Ground floor, First floor, etc) ;

      (*iii*) Front elevation of the building ;

      (*iv*) End elevation may or may not be shown ;

      (*v*) Longitudinal section of the building ;

      (*vi*) Cross-section of the building ;

      (*vii*) Structural details including foundation details of the building.

The plan of the building will show the orientation of rooms with maximum utilisation of space in each floor.

Front or end elevation of the building will speak of the appearance and beauty of the building.

The sectional views of the building (Longitudinal and Cross sections) will reveal the dimensions of each unit with their appendages, if any. One of the sections must show the staircase and the plumbing details of the toilet.

Structural details should show the sections of walls, beams, pillars, slabs, lintels, chujjas, etc. with reinforcement details. Foundation details should also be shown on the drawing.

Scale of each and individual drawing should be specifically mentioned and North line must be shown on the site plan.

The site plan should show the physical features of the surrounding areas, viz. buildings, road, trees, nullah or drain, sewer line, waterline, etc. etc. and the location of the proposed building.

Shape of a building. The shape of a building is usually guided by the shape and size of the plot where the building is to be erected. However, a good architect/civil Engineer can give a good shape of a building irrespective of the shape of the plot. It is needless to mention that the length of walls of a square plan is about 10 to 20% less than the length of wall of a rectangular plan of same built-up area.

Building Cost :

The cost of a building of compact plan is less than the cost of a building of widespread units *i.e.* a spacious plan. More opening in a building will call for more cost. More numbers of wall will involve more cost. For large rooms there will be an additional cost for provision of beams and extra reinforcement in roof slab. It is seen that cost of a building is directly proportional to the use of brick, cement, steel and timber. The cost of a building can be reduced by proper planning with optimum use of the above said materials. It is obvious that maximum utilisation of space comes under planning.

A Small Building Plan :

A simple building plan for a residential house with two bed rooms, toilets, drawing cum dining room, kitchen and a staircase is shown in Fig. 19·1. The plan is a rectangular one. This type of plan is suitable for a family of middle income group.

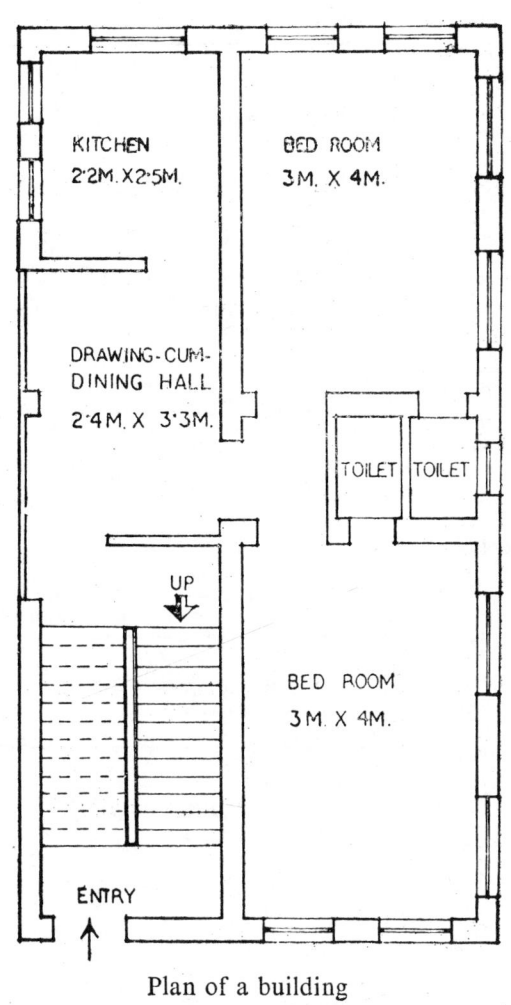

Plan of a building
Fig. 19·1.

Developing a Plan :

The line plan of a square building for a family of middle income group is shown in Fig. 19·2. The dimensions given are all internal dimensions. The developed plan of the building showing wall thickness, stair, doors and windows is presented in Fig. 19·3. Doors are 1·1 m ×2·1 m and most of the windows are 1·1 m ×1·25 m. Windows for toiet and staircase are 0·7 m.×1 m. The toilet doors are 0·8 m×2 m. All walls are 250 mm. thick.

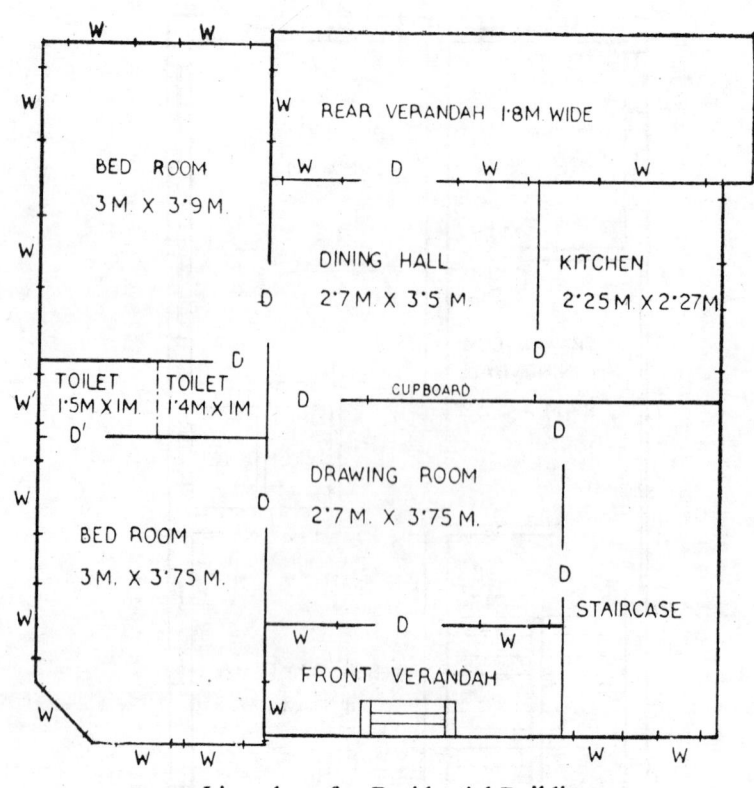

Line plan of a Residential Building
Fig. 19·2.

Total Floor Area=68 sq. m (excluding staircase).

In each floor there are :

two bed rooms with toilet (attached bath and privy.)

drawing room

dining hall

kitchen and verandahs (Front & Rear)

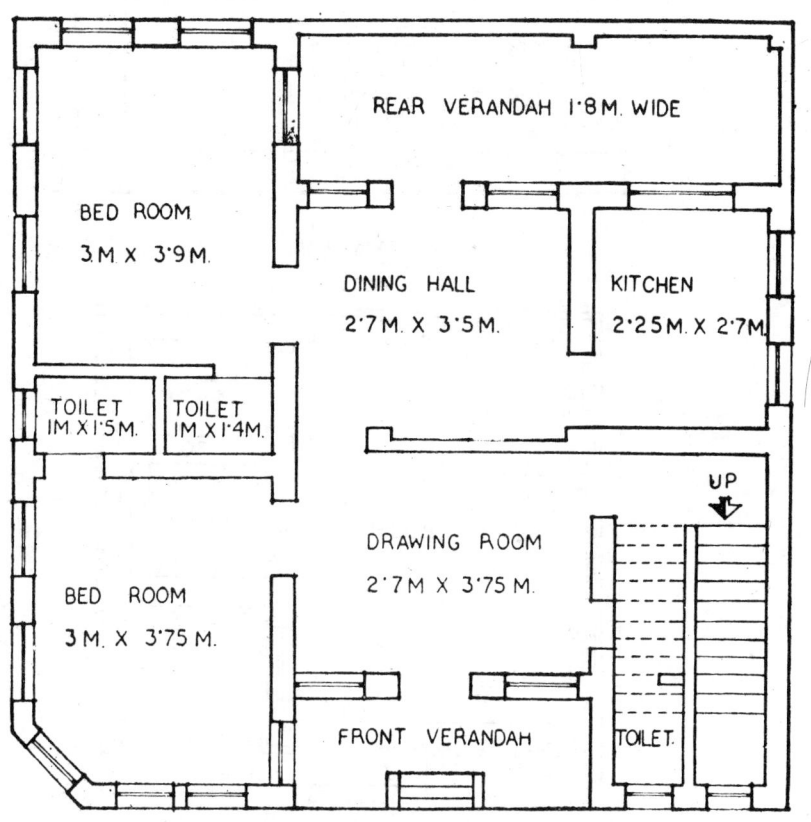

Building plan developed from the line plan
Fig. 19·3.

A Darwan's Quarter :

It is a small house comprising of a room with kitchen and toilet. The plan and elevation of such a quarter are shown in Fig. 19·4. This is a single-storied small house having a room of 3 m×4 m with a kitchen 1·2 m×1·8 m and a toilet 1 m×1·8 m size. The walls are 30 cm. thick including plastering. The room height is 3 m, the headroom for kitchen and toilet being 2·5. m. For proper ventilation and visibility of all corners from the room quite a number of windows are provided. The total floor area is about 16 sq. m. only. Chujjas are provided to protect the room from sun and rain.

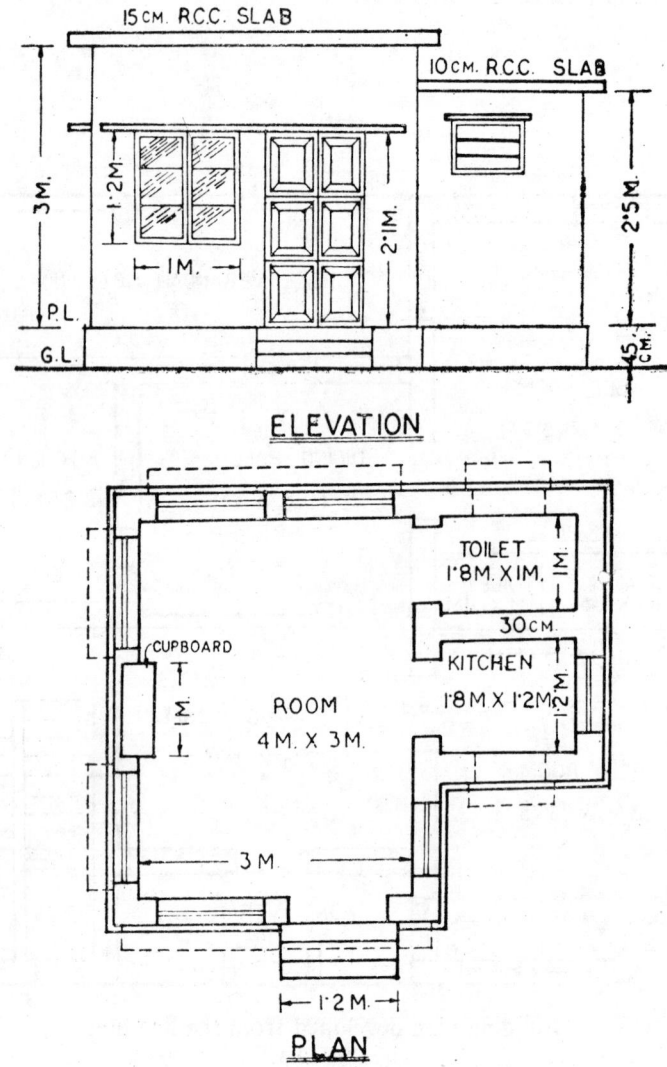

Fig. 19·4.

Low cost house for a family of low income group (LIG) or Economically Weaker Section (EWS) of people is shown in Fig. 19·5. This is a single-storied small house comprising of :

     (*i*) a living room 3·2 m ×4·2 m
     (*ii*) a bed room 3 m×4 m
     (*iii*) a kitchen 2·6 m ×3 m.
     (*iv*) a lavatory 1 m ×1·8 m and a bath 1·2 m ×1·8 m.

A small passage of 1·2 m. width has been kept in between living room, kitchen, toilet and bath. Minimum numbers of doors and windows have been provided, but there will be adequacy of natural air and light. In planning, care has been taken for maximum utilisation of space with common walls and doors. The rear door with steps may be omitted, if required. For economy, roofing may be done with country tiles over a stractural frame of bamboo and timber derived from sawn palm tree which is of low cost. Locally available low cost trees may be used for the purpose of framing the roof. The walls may be built with mud mortar and plastered with lime-sand mortar. For doors and windows, timber of low cost but durable should be used. For flooring, bricks laid on flat with lime-sand mortar may be used. For such a building use of cement and steel may not be required at all. The total floor area is about 40 sq. m. only. Built-up area is 54 sq. m. By keeping 1/3 rd. of the plot vacant, the required land area is 80 sq. m. i.e., about 860 sft. or 1·2 cottah. The construction cost of this building with materials specified above should be within Rs. 15,000/- only. The land cost should not exceed Rs. 5000/-.

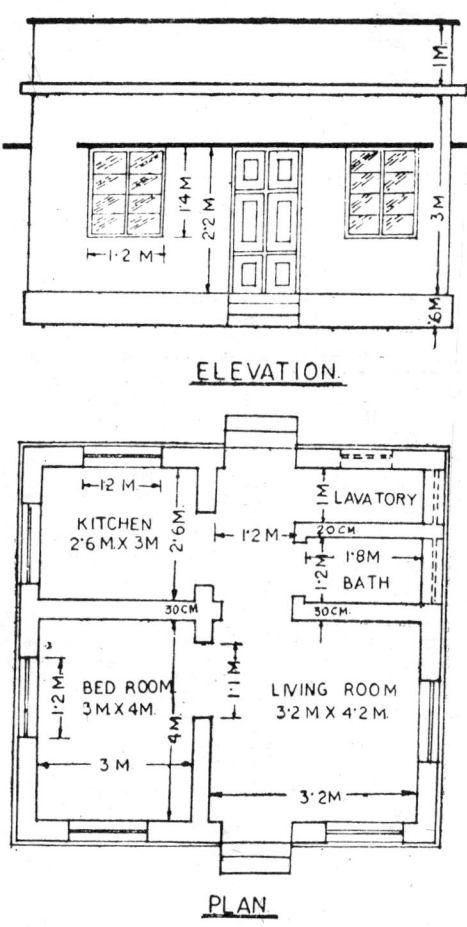

Fig. 19.5.

A Primary Health Centre :

Design of a single-storied Rural Primary Health Centre with the following units is shown in Fig. 19·6.

Waiting Hall	4 m × 6 m.
Doctor's Room	2·8 m × 4 m.
Dispesing Room	2 m × 2·5 m.
Medicine Store	2 m. × 2·2 m.
Toilet	1·7 m. × 2·8 m.

The total floor area required is about 50 sq. m.

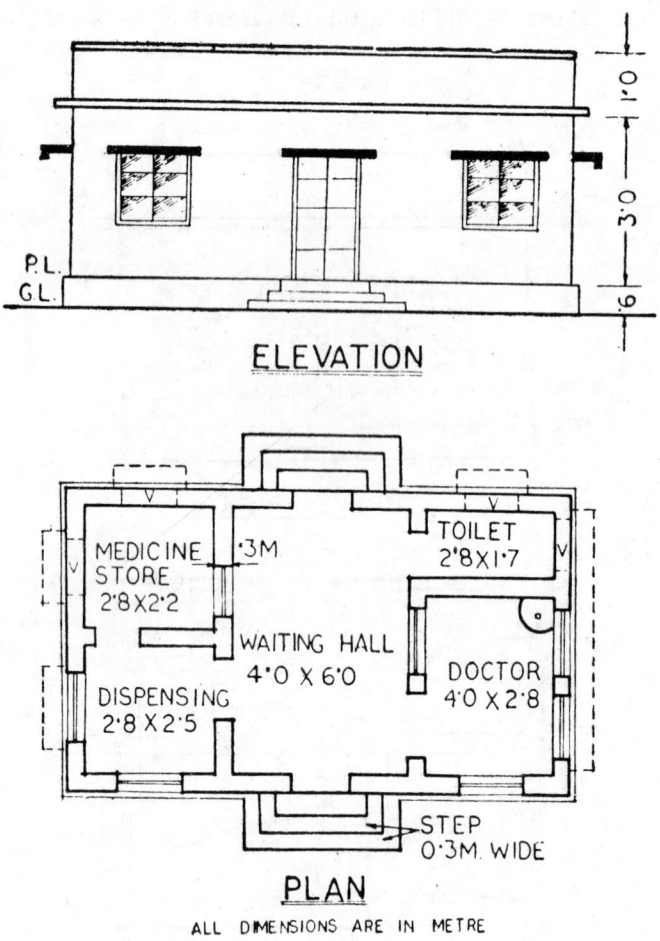

ELEVATION

PLAN

ALL DIMENSIONS ARE IN METRE

Fig. 19·6.

Residential Quarters for families of Low Middle Income Group may be designed single-storied as shown in Fig. 19.7 and 19.8.

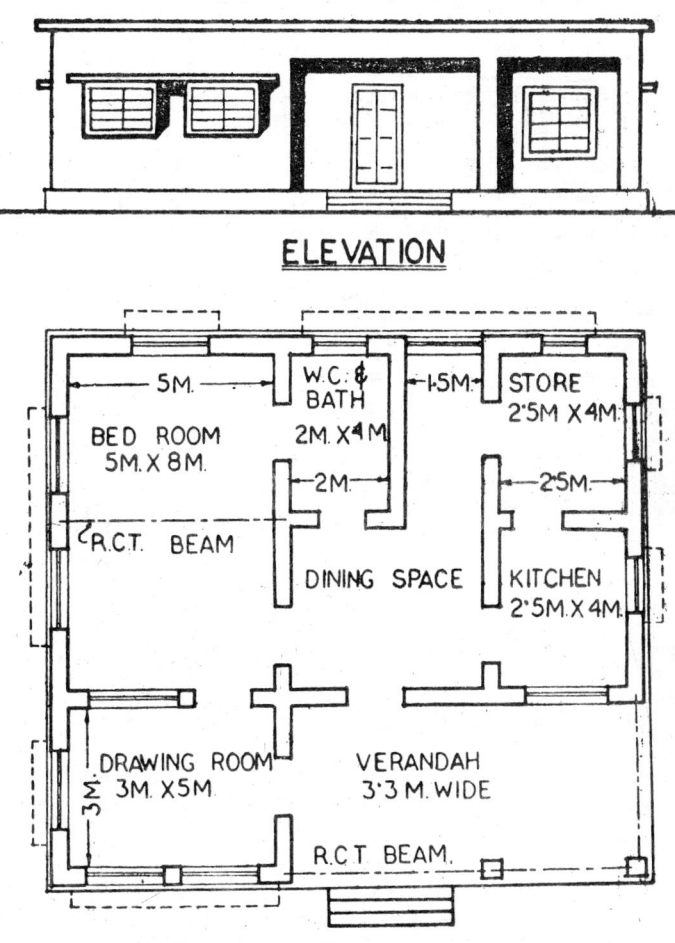

ELEVATION

PLAN

Fig. 19.7.

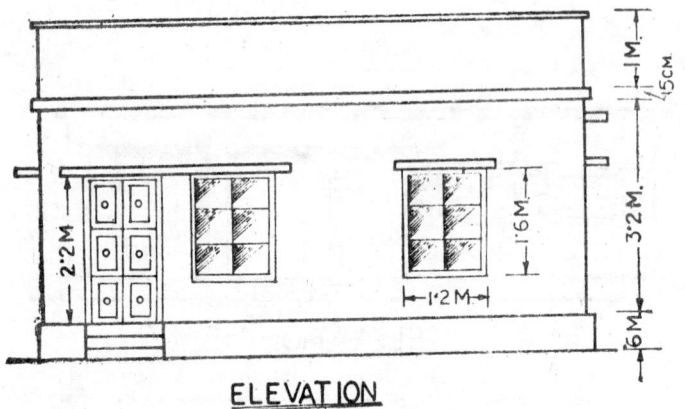

ELEVATION

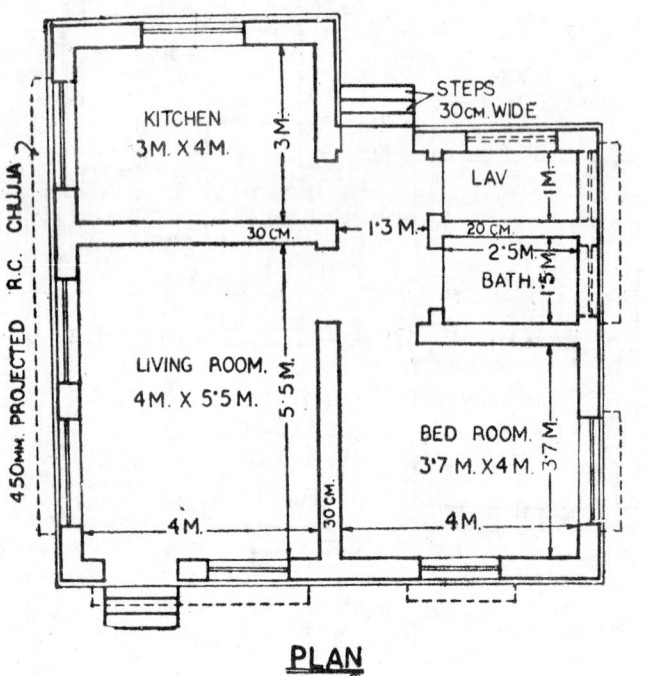

PLAN

Fig. 19.8₉

A **Residential Quarter** for a family of Middle Income Group should essentially consist of at least two bed rooms with a verandah and a dining space close to kitchen. Bath and W.C. may be common to all, if they are not attached to the bedrooms.

Fig. 19.9 shows the Ground floor plan and elevation of such a building. A staircase may be provided for future extension of floors. The depth of the building is about two times its width because of the rectangular plot having a road frontage of 10·0 M only. About 1·25 M width of land is kept open on either side of the building.

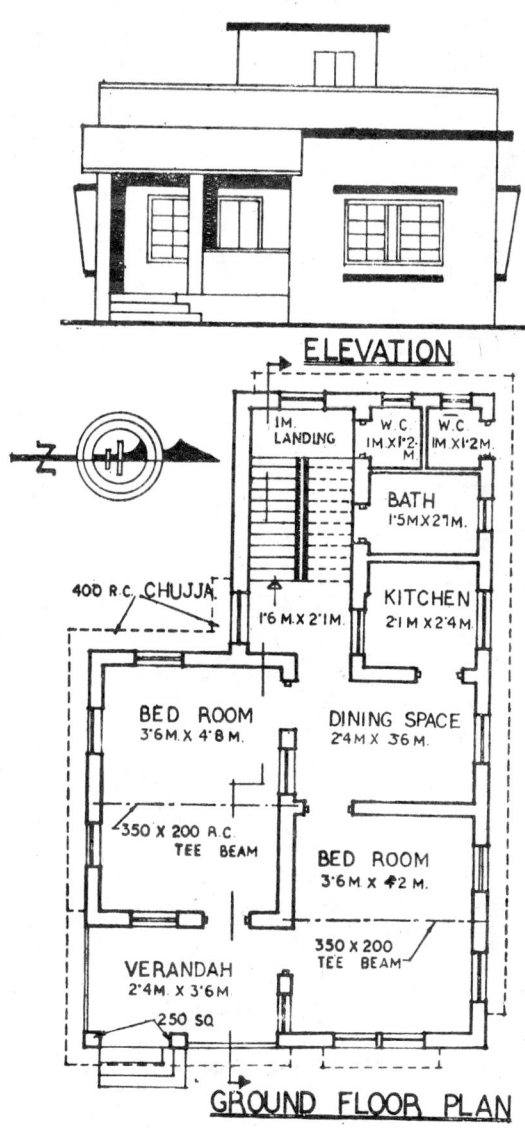

Fig. 19·9.

A Small Building (Fig. 19·10) on a corner plot of 66 sq. m. only i.e., about 1 cottah of land. The built-up area is about 36 sq. m. Each floor has the provision of one bed room with verandah, kitchen, bath and W.C. within such a small space.

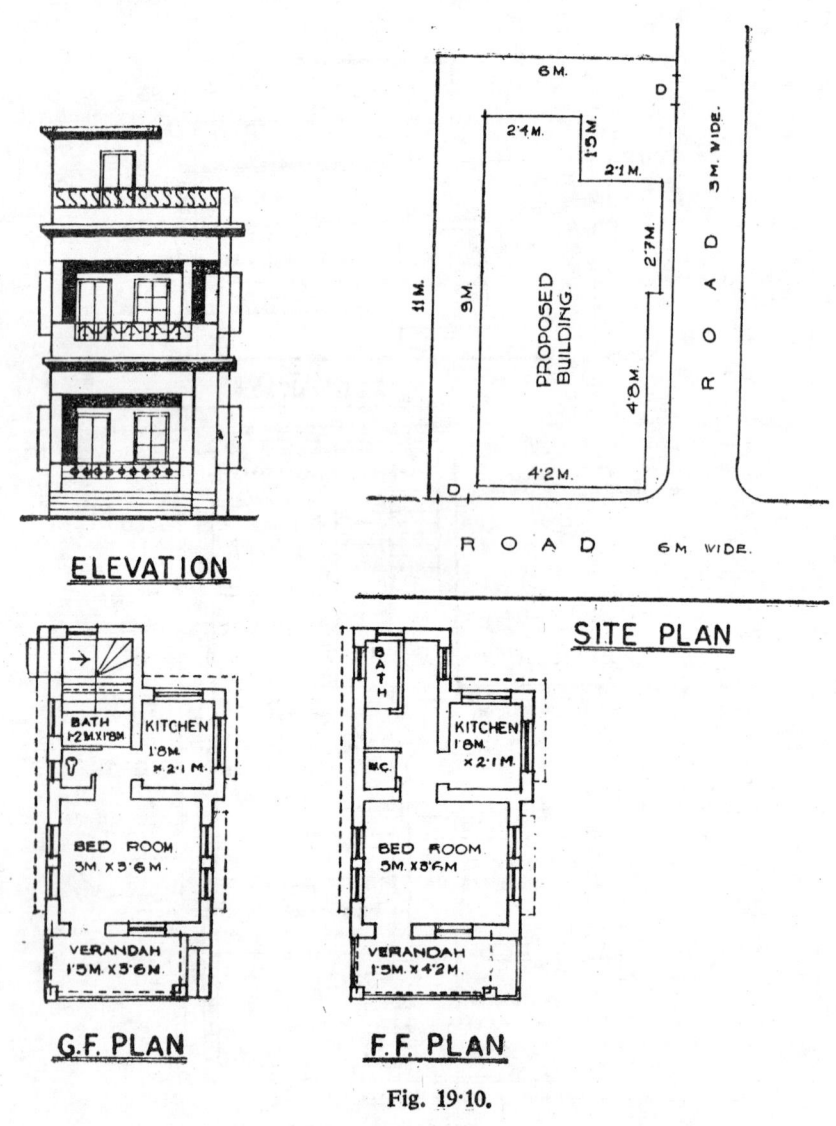

ELEVATION

SITE PLAN

G.F. PLAN   F.F. PLAN

Fig. 19·10.

Planning of a building on similar such a small plot of land is shown in Fig. 19.11.

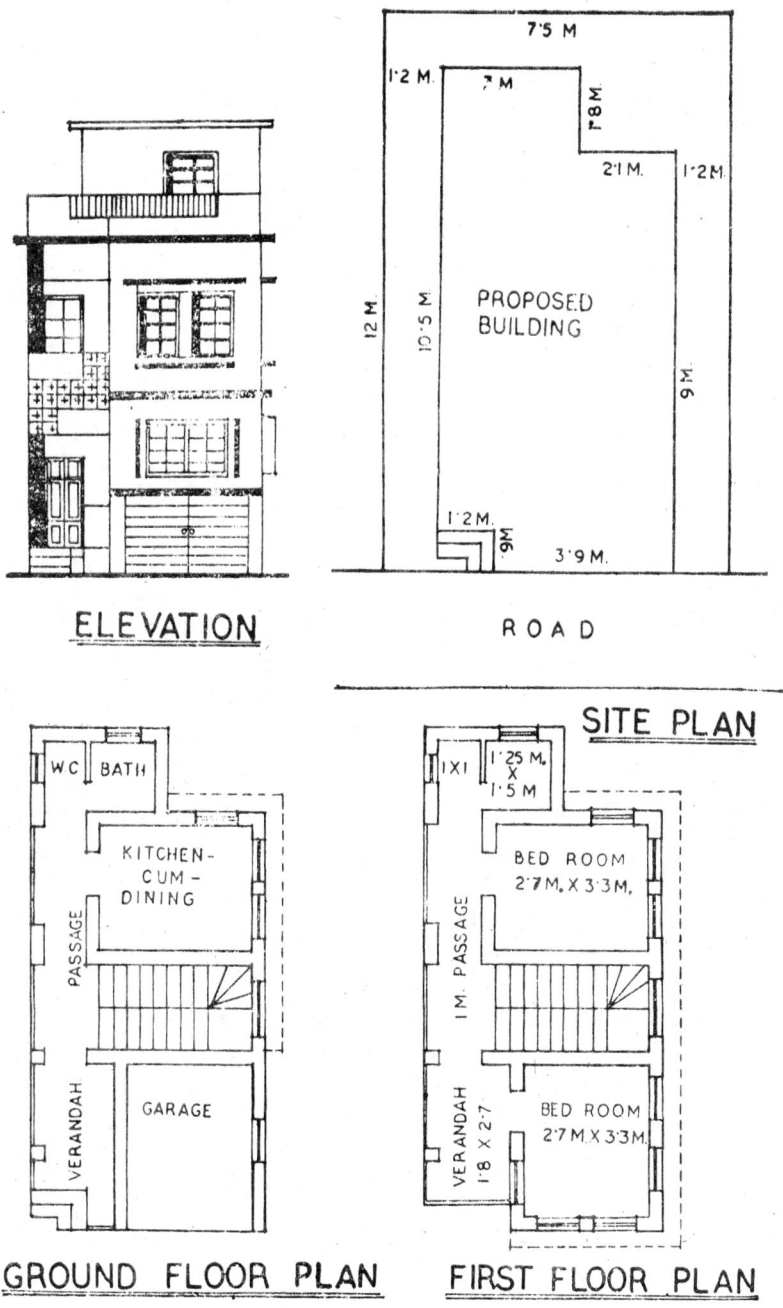

ELEVATION   R O A D

SITE PLAN

GROUND FLOOR PLAN   FIRST FLOOR PLAN

Fig. 19.11.

A Residential Building for a family of High Income Group (H.I.G.) is shown in Fig. 19'12 and 19'13. The Ground floor plan and Top floor plans are shown in Fig. 19'12. The front elevation of the building and the site plan are shown in Fig. 19'13.

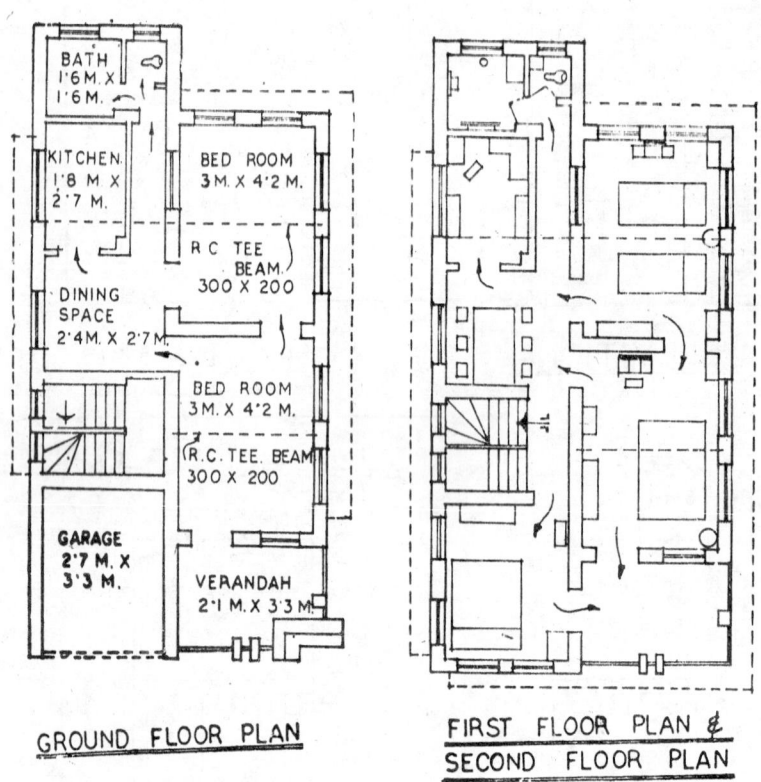

GROUND FLOOR PLAN

FIRST FLOOR PLAN &
SECOND FLOOR PLAN

Fig. 19·12.

The land area is 135 sq. m. i.e., about 2 cottahs of land with a road frontage of 9 m. only i.e., about 30 ft. The orientation of units with maximum utilisation of space is to be noticed. Such type of planning should be made where land is too costly.

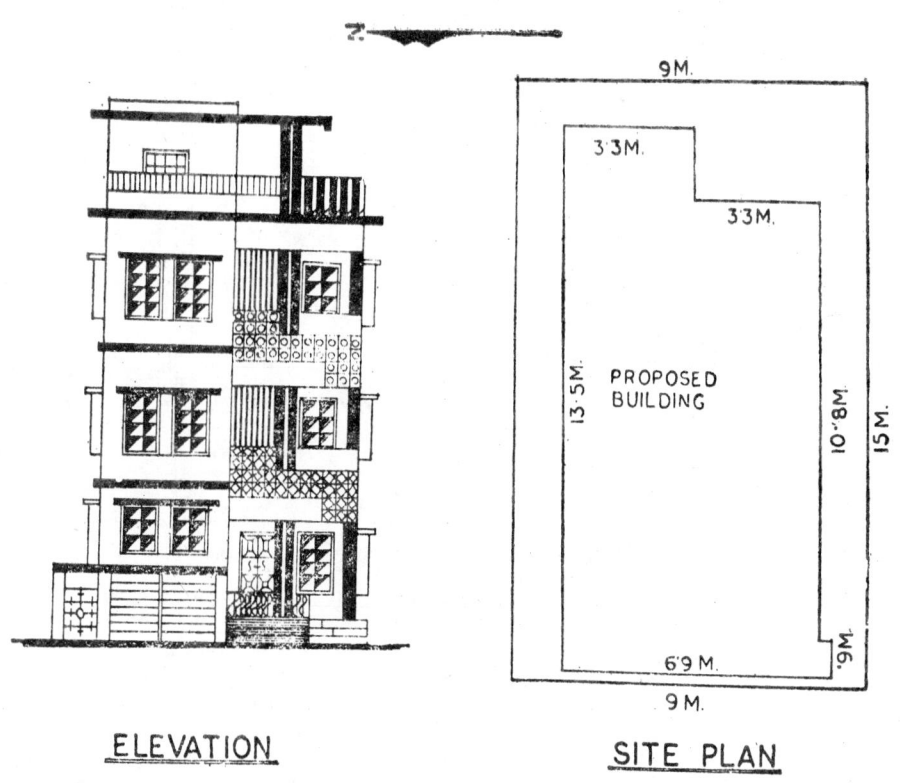

ELEVATION   SITE PLAN

Fig. 19·13.

   A Residential Building for a family of HIG or high MIG may be planned as shown in Fig. 19·14, 19·15 and 19·16 depending upon the land availability and shape of land. The dining space and the drawing room seem to be very small. When the width and depth of land are restricted, the sizes of the dining space and drawing room can be increased simply by shifting the right hand bed room to the back *i.e.*, making identical to the left hand bed

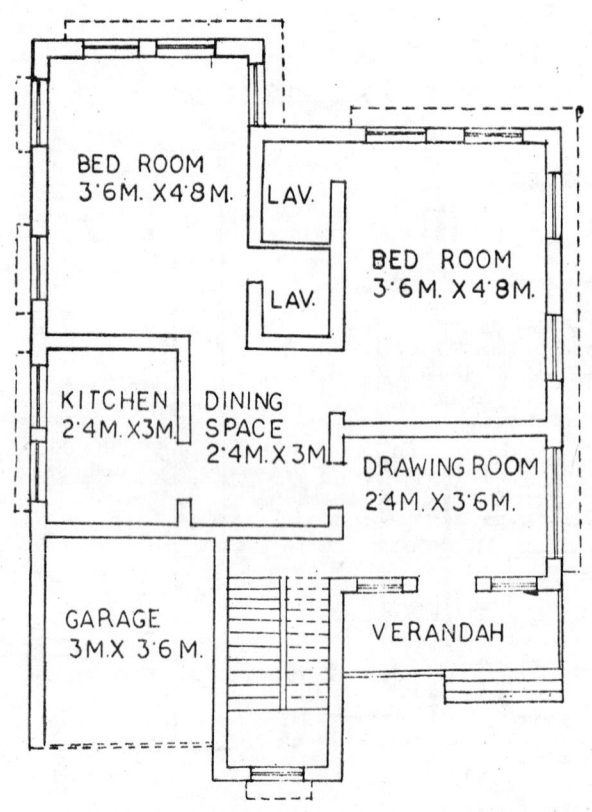

GROUND FLOOR PLAN

Fig. 19·14.

room, advancing the drawing room to the front and by curtailing the front verandah. Here, the planning has been done in such a manner that the ground floor can be used by a tenant and the house owner will use the first floor by keeping the drawing room, verandah and staircase at ground floor for his own use. The entire ground floor can be given to a tenant,

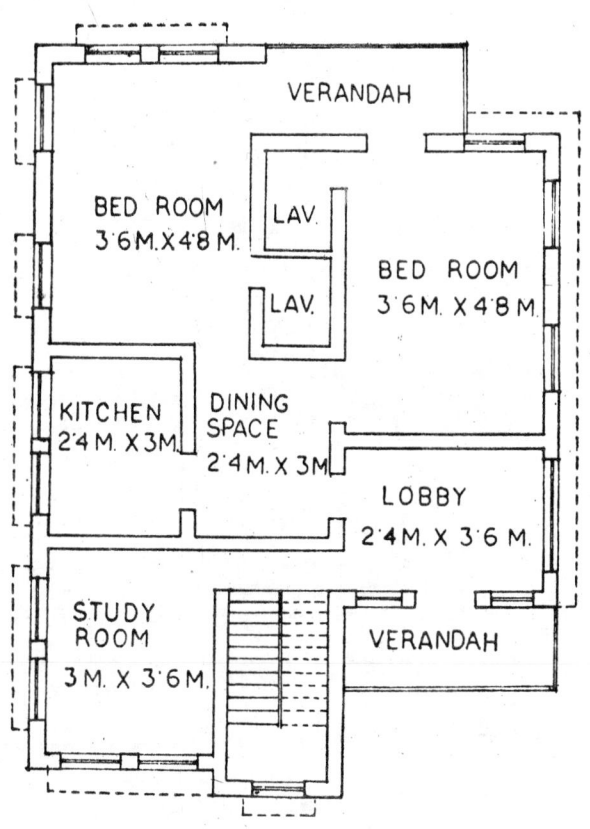

FIRST FLOOR PLAN

Fig. 19·15.

if entry of the house owner is made through staircase by closing the access to staircase from the drawing room. The study room as shown in first floor plan may be utilised as a guest room and the lobby may be converted into a drawing room of the house owner. The elevation shown in Fig. 19'16 can be improved by providing ornamental wall texture, jafriwork and ornamental chujjas or hoods having geometical shapes for windows. Also, the parapet wall can be made ornamental.

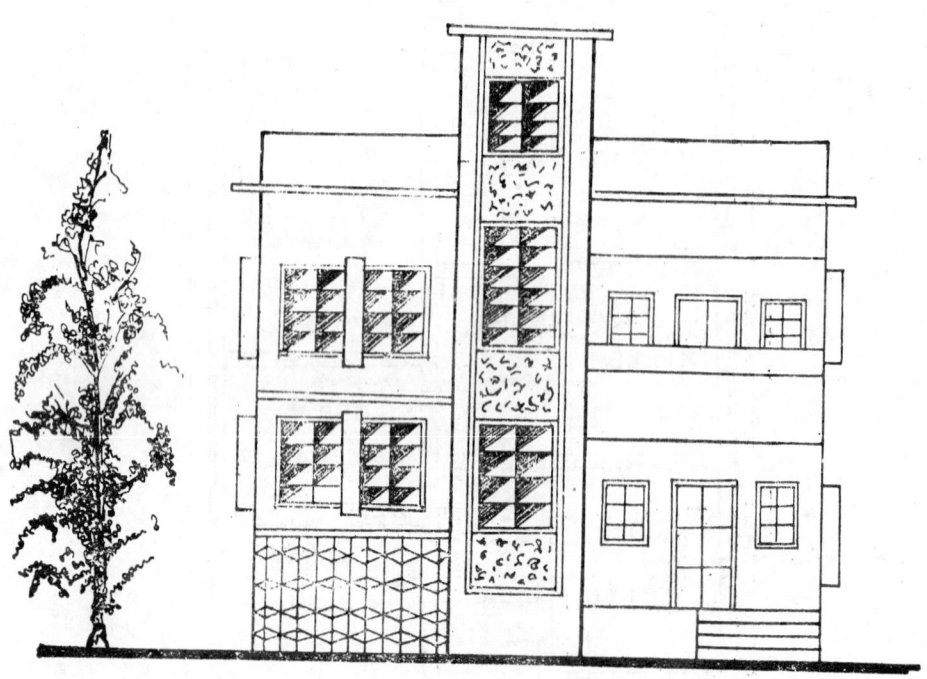

ELEVATION

Fig. 19·16.

A **Residential Flat Building** in an area of heavy rainfall or snowfall can be planned as shown in Fig. 19·17. Each floor is having two residential flats with a common stair located centrally.

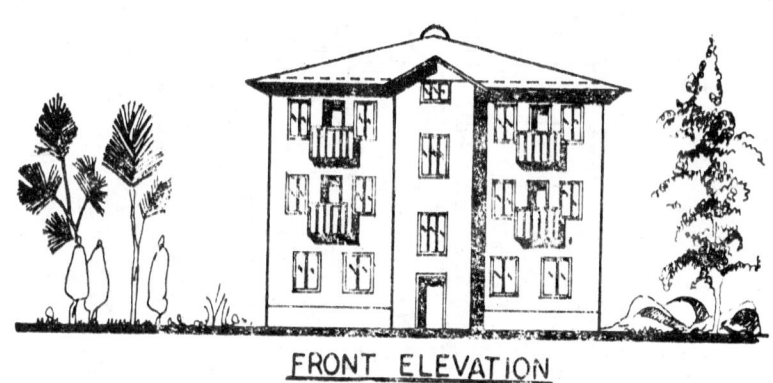

FRONT ELEVATION

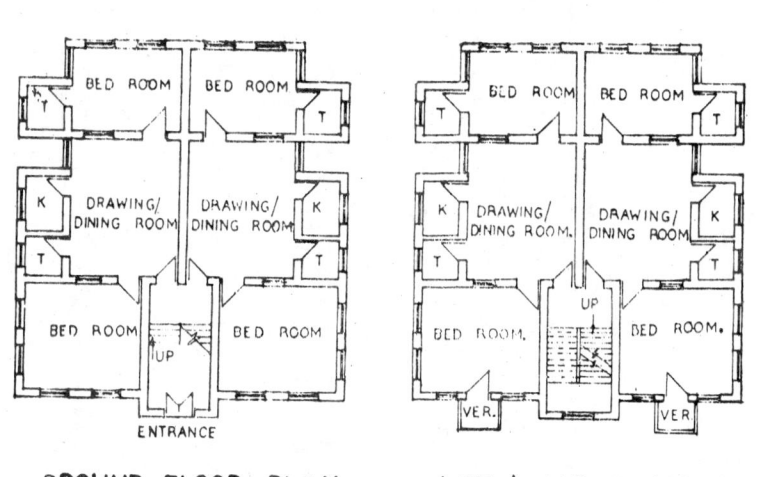

GROUND FLOOR PLAN   1 ST. & 2ND. FLOOR PLAN

Fig. 19·17.

A **Hostal Building** can be planned as shown in Fig. 19·18 through 19·21. The sun breakers provided here imparts additional beauty to the building.

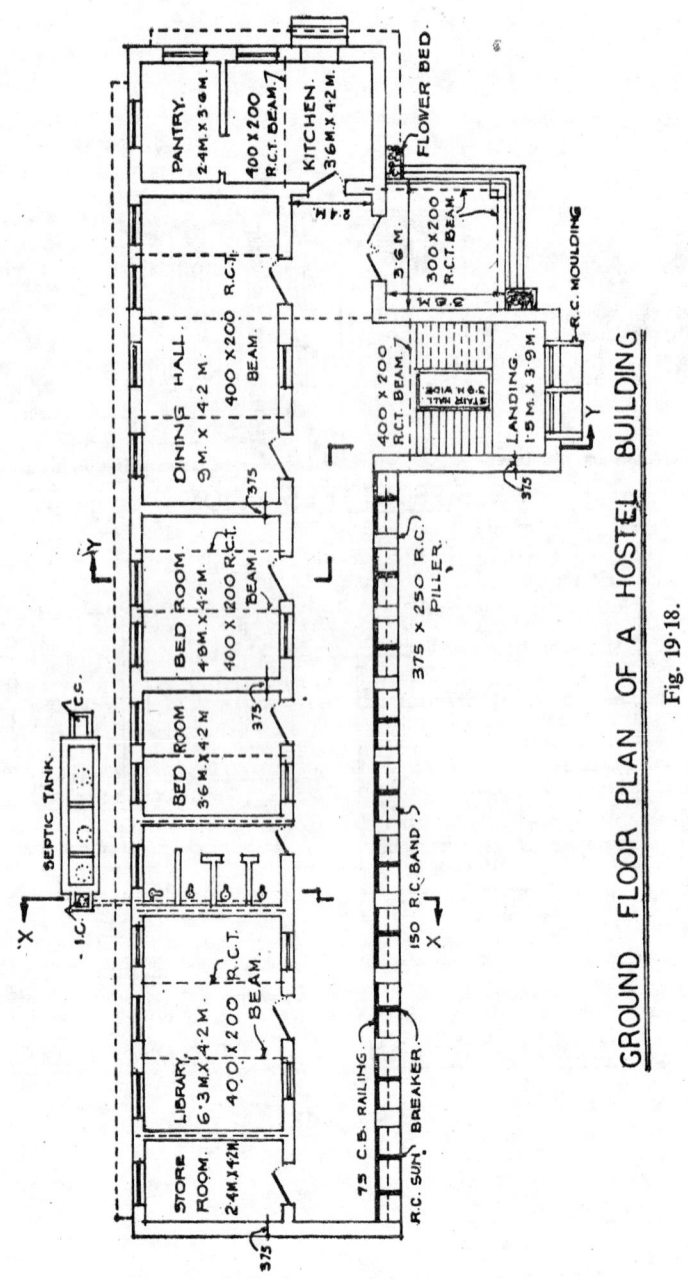

GROUND FLOOR PLAN OF A HOSTEL BUILDING

Fig. 19·18.

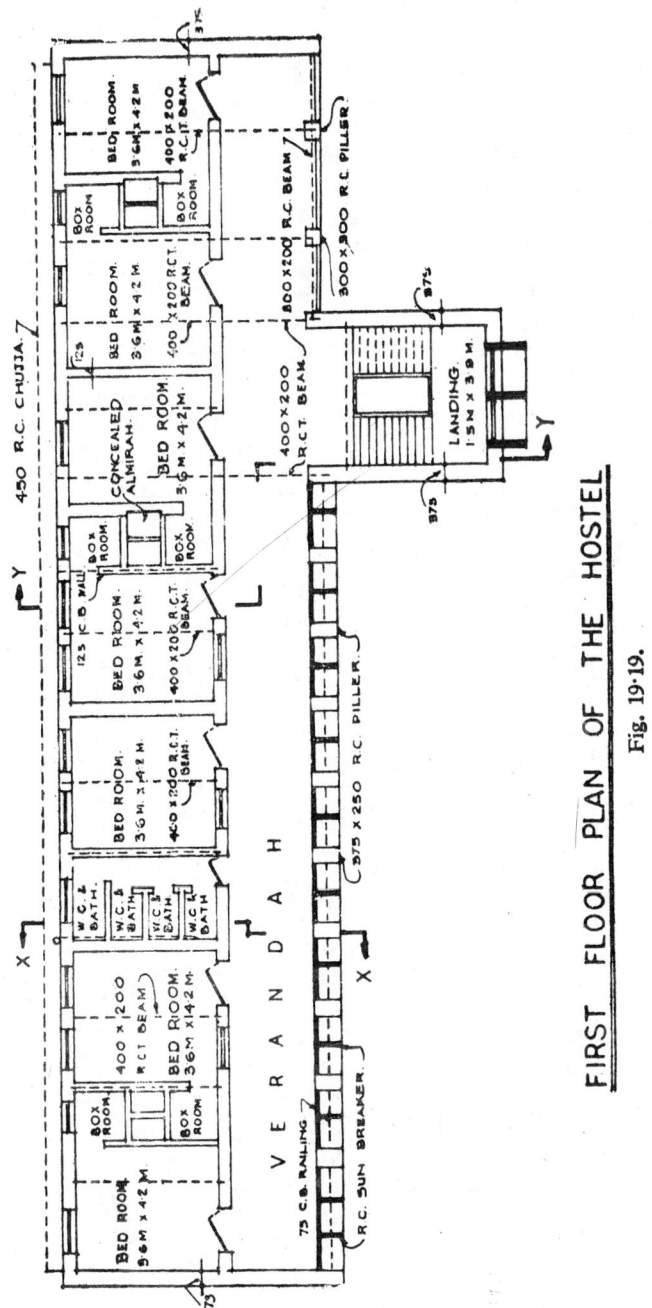

FIRST FLOOR PLAN OF THE HOSTEL

Fig. 19·19.

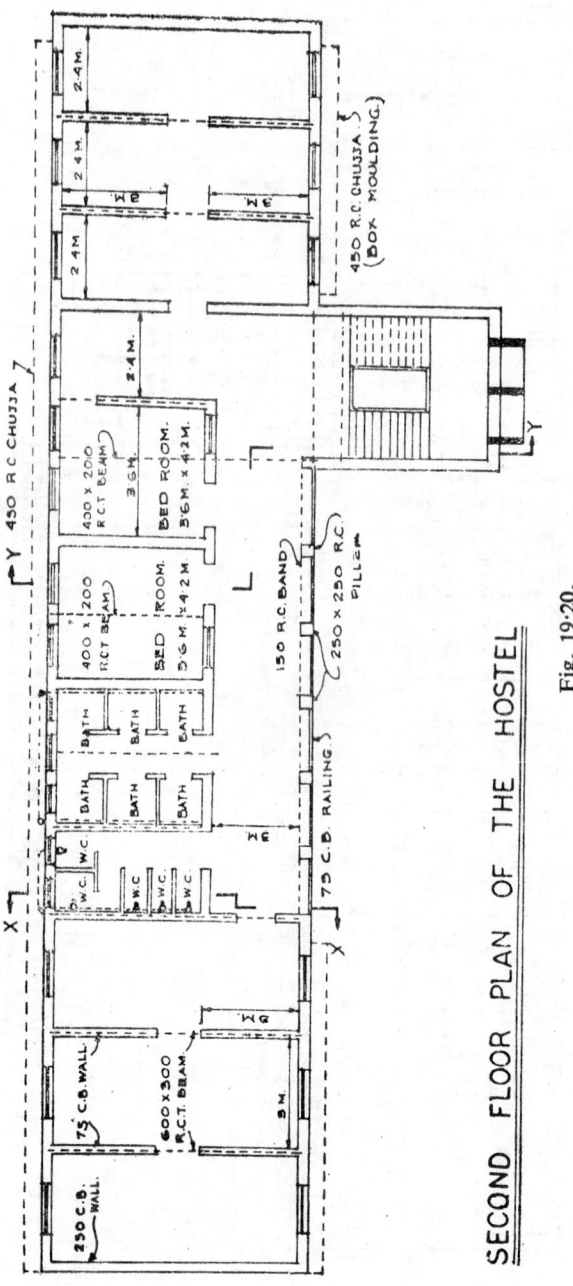

SECOND FLOOR PLAN OF THE HOSTEL

Fig. 19.20.

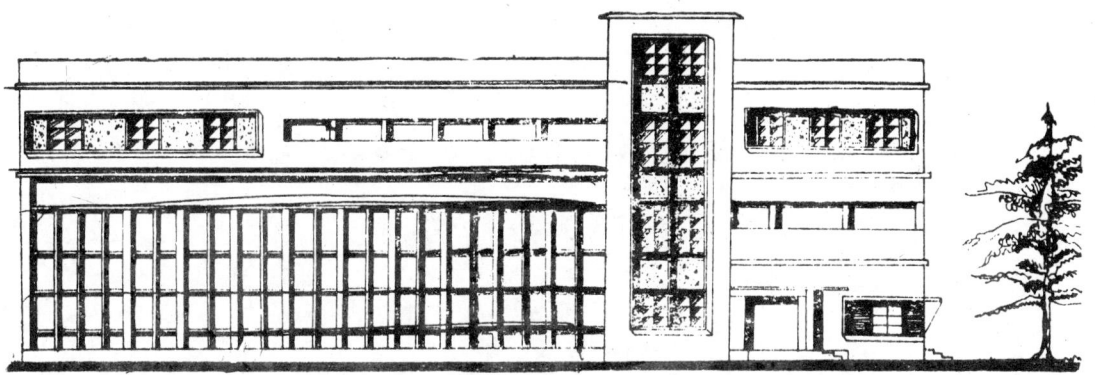

FRONT ELEVATION OF THE HOSTEL
(ARCHITECTURAL COMPOSITIONS ARE SHOWN IN OBLIQUE PROJECTION)

Fig. 19·21.

20

Septic Tank and Latrines

A septic tank is a water tight chamber, usually built underground. This is connected to a latrine or toilet with bath. This tank receives human excreta and other foul matters with flush water. The retention time within the tank being long, gassification, liquefaction and sedimentation of suspended solids take place within the chamber. Also, partial digestion of settled sludge takes place within the chamber. The B.O.D. removal efficiency of a septic tank is not appreciable. About 30 to 40% B.O.D. is removed in a septic tank and therefore the B.O.D. of septic tank effluent is very high. This effluent should be discharged to a soak pit through a vat (for application of bleaching powder). But, it is hardly practised. In most cases, the septic tank effluent is discharged directly to the surface drain which leads to a nullah. This practice is very unhygienic. Septic tanks are adopted as a sanitary measure for disposal of human excreta in absence of water-carriage system *i.e.*, sewerage system. The sectional view of a septic tank is shown in Fig. 20·1.

For Indian conditions, the surface area of a septic tank should be 0·92 m² for every 10 lpm peak flow rate *i.e.*, for 5 persons. The length should be 2 to 4 times the breadth. A minimum depth of sedimentation of about 25 to 30 cm is required. The per capita suspended solids entering the septic tank may be taken as 70 gms/day.

Volume of fresh sludge —0·00083 m³/capita/day.

Volume of digested sludge —0·0002 m³/capita/day.

For determination of volume of digestion zone, the average volume of mixture of fresh and digested sludge is taken as 0·000515 m³/cap./day. Capacity or space required for sludge digestion is 0·032 m³/capita.

For an interval of cleaning of one year sludge, storage capacity required is 0·07 m³/capita. The total tank capacity required per person may be computed as follows :

Surface area for probable peak flow	=0·184 m²
Providing a depth of 30 cm, volume	=0·184×0·3
Volume required for sedimentation	=0·55 m³
Volume required for digestion	=0·032 m³
Volume required for storage of sludge	=0·073 m³

Total=0·16 m³

Free board including 0·25 m³ for seed sludge=0·184×0·3   =0·055 m³

∴ Tank capacity required per person   =0·215 m³

A septic tank usually has two compartments separated by baffle walls, but holes are provided in the baffles for flow of fluid as well as flow of gas, from one compartment to the other. For large septic tanks, three compartments are provided. Each compartment must have a manhole with cover for cleaning the digested sludge. Inside surfaces of walls and floors of the tank should be plastered and lined with neat cement in order to prevent percolation of sewage. The inlet and outlet pipes should be provided with a bend or tee as shown so that the scum and froth formed on the liquid surface is not disturbed. When two or more latrines are

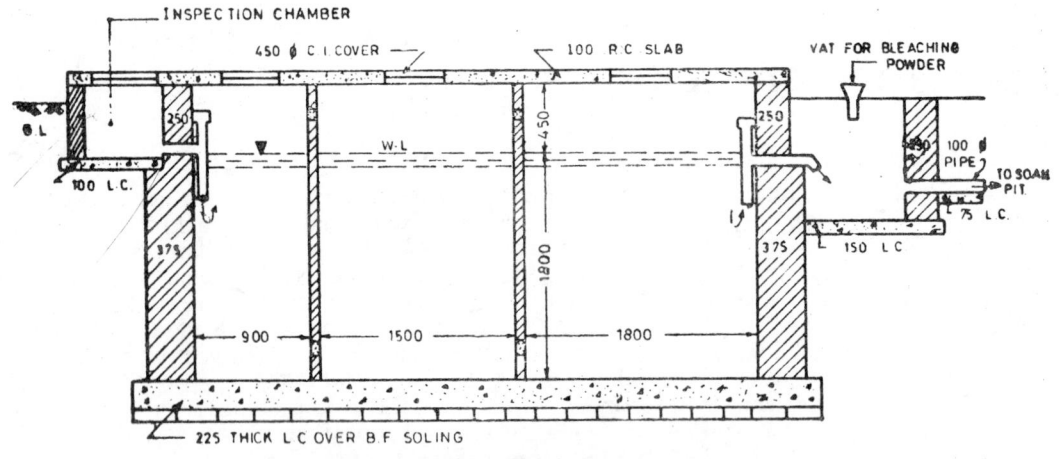

SECTIONAL VIEW OF SEPTIC TANK

Fig. 20·1.

are connected to a single large septic tank, an Inspection Pit is constructed at the Junction of the inlet pipes. A free board of 30 cm should always be kept in the tank. A vent pipe should be provided for disposing off gases produced. The tentative sizes of septic tank for a cleaning period of 2 years are given below :

No. of users	Lengdth in m	Breath in m	Liquid depth (m)	Liquid capacity (m³)	Sludge to be removed (m³)
10	2·0	1·0	1·60	2·50	1·45
20	2·3	1·2	2·00	4·50	2·90
30	2·5	1·4	2·00	7·00	5·00
40	3·0	1·5	2·00	8·50	7·00
50	4·0	1·6	2·00	11·50	7·20
100	5·0	2·0	2·50	23·00	14·00

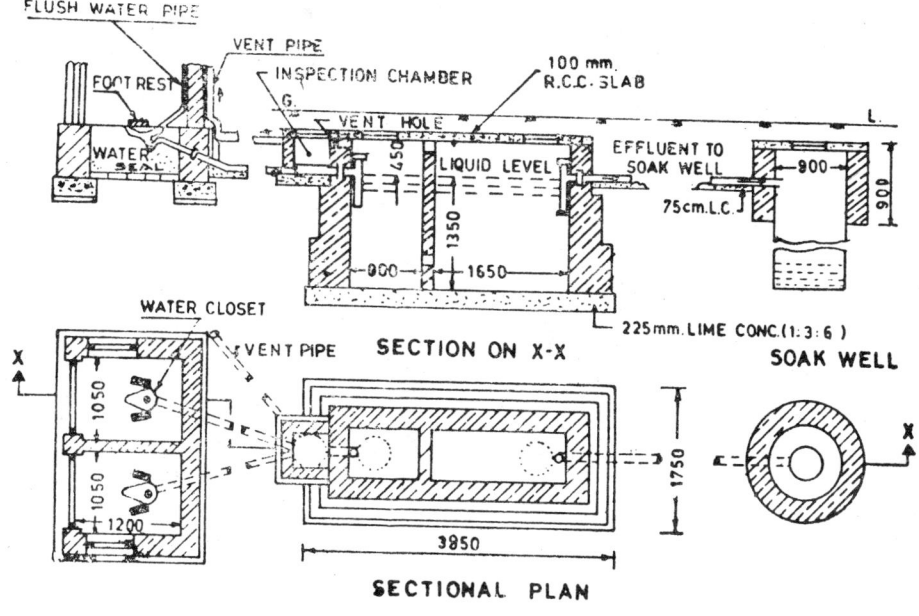

Fig. 20·2.

Details of plumbing from toilet to septic tank to soak pit with details of septic tank and soak pit are shown in Fig. 20˙2. The soak pit must not be constructed within a distance of 15 m from a source of water. The depth of soak pit will vary depending upon the nature of soil and its strata. Sandy soil will approve more hydraulic loading compared to clayey soil. The entire depth of a soak pit should be filled with brick bats, ballasts, etc.

Fig. 20˙3 shows plan and sectional elevation of a cesspool. The type of cesspool shown here is made water-tight. The function of a cesspool is similar to that of a septic tank, but its cost of construction is low compared to that of a septic tank. Sometimes, upper portion of a cesspoal wall is made with open joints in order to permit the seepage of cesspool liquid.

In absence of sewerage system in municipal towns and non-municipal urban areas, disposal of human excreta by head load from service privy tubs and simply by dumping in a trenching ground without any covers, which is the usual practice in India, is very unhygienic. The construction cost of a septic tank or a cesspool are not within the affordability of the common people living in municipal and non-municipal areas. Therefore, for safe disposal of human excreta, low cost sanitation for all should be provided by the municipal authorities with social education for awareness of the people as regard evil effects of pollution.

A few Indian cities are sewered. In most of the cities and towns, septic tanks are constructed for disposal of human excreta. But, in semi-urban and rural areas, use of service privy and open-air defaection are practised. Both are very unhygienic. In some of the semi-urban areas, dug well latrine, bore hole latrine and pit latrine are found to be in use. But, in rural areas, open-air defaecation has become the pleasure of the community. As an outcoming result of this open-air defaecation, the people enjoy various diseases and adore death at an early stage of life.

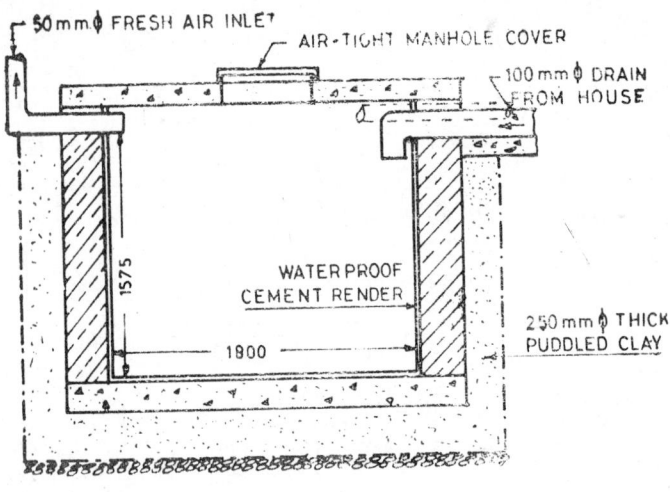

SEC.- A-A

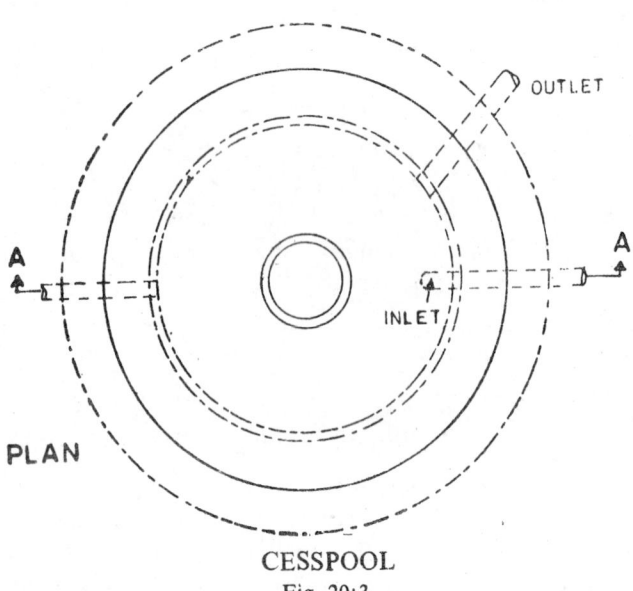

PLAN

CESSPOOL
Fig. 20·3.

As a low cost measure for rural sanitation, any of the latrines mentioned above may be used in rural areas. Any of these latrines should be built at least 15 m away from the water source, in order to avoid pollution of water by leaching action of the sewage. Details of dug well latrine, bore hole latrine and pit latrine are shown in Figs. 20·4, 20·5 and 20·6 respectively.

Of all the latrines mentioned above, pit latrine is less expensive. This can be constructed by the user himself with the locally available low cost materials. A pit latrine may be either dry type or wet type. One of the very important aspects of the pit is its useful life. The longer a pit privy serves a family, the more certain is the health protection which it can assure and therefore it has more value to the family as well as the community. With the increased life of pit privy, the annual per capita cost of installation reduces to a grear extent.

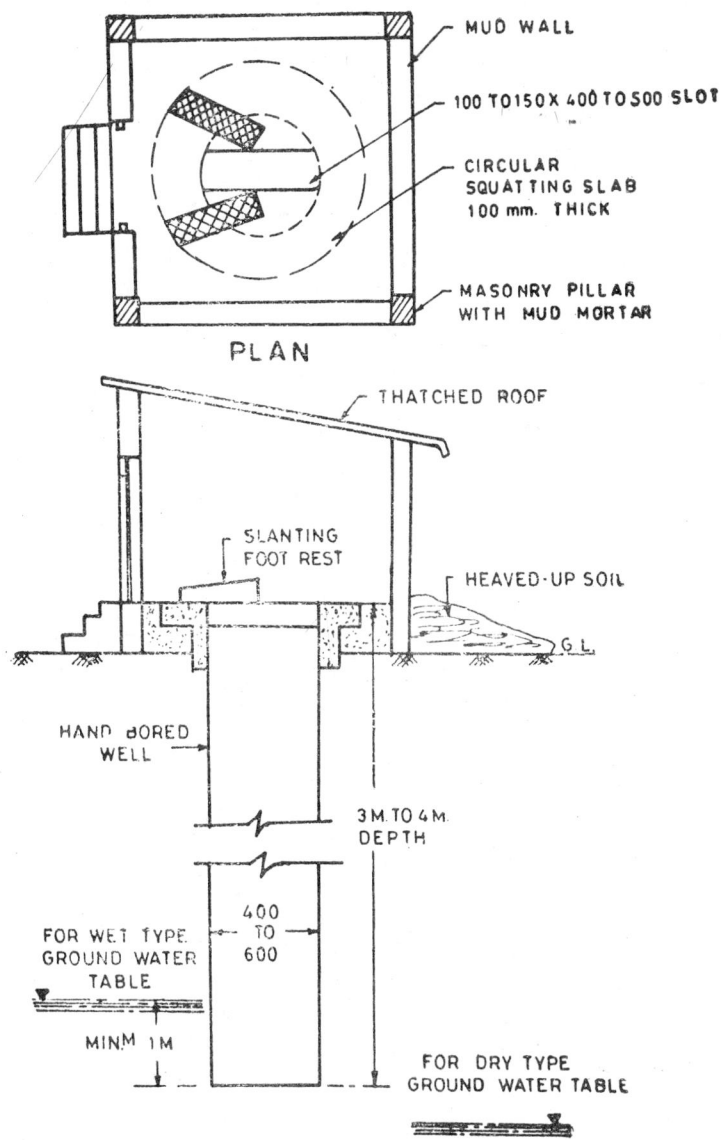

MUD WALL

100 TO 150 X 400 TO 500 SLOT

CIRCULAR
SQUATTING SLAB
100 mm. THICK

MASONRY PILLAR
WITH MUD MORTAR

PLAN

THATCHED ROOF

SLANTING
FOOT REST

HEAVED-UP SOIL

G.L.

HAND BORED
WELL

3 M. TO 4 M.
DEPTH

400
TO
600

FOR WET TYPE
GROUND WATER
TABLE

MIN^M 1 M

FOR DRY TYPE
GROUND WATER TABLE

SECTIONAL VIEW
DUG-WELL LATRINE
Fig. 20·4.

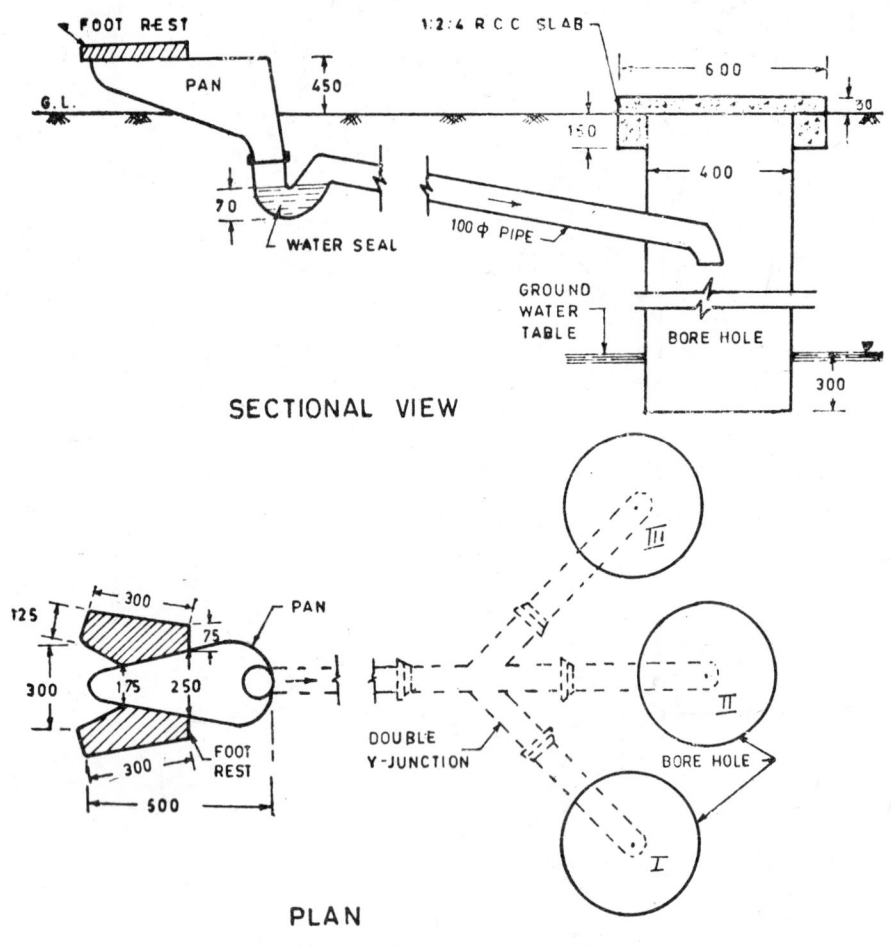

SECTIONAL VIEW

PLAN

BORE HOLE LATRINE

Fig. 20·5.

Actual observations on wet pit latrines where ablution water is used, give a figure of 0·03 cu m of sludge per person per year. The actual volume of material in a wet pit may be reduced in time to about 10 to 20% of the total waste (excreta) deposited. Wet pits should have a maximum depth of 2 m. The pits should be designed at least for a period of 5 years and preferably for eight years. A pit privy with largest possible volume is best. From economic point of view, a deep pit initially of high cost will prove to be a profitable investment.

When the level of excreta comes to within 50 cm of the ground surface, the pit should be closed by filling it with garbage and earth. After closing, the pit should be left for one year by which time the excreta will be converted into humus. This material may be removed and used as a good soil conditioner (fertiliser). The old pit may be reused after two years.

In dry pits, no ablution water should enter the pit and a handfull of dry loose earth should be thrown inside the pit after its use each time.

To prevent mosquito breeding in wet pits, every week a cupfull of kerosene oil should be thrown into the pit.

For a family of five members, a few information on pit latrines, having 0.82 sq. m cross-sectional area, are given below in tabular form. To arrive at overall depth of pit, 50 to. 60 cm. should be added to the effective depth.

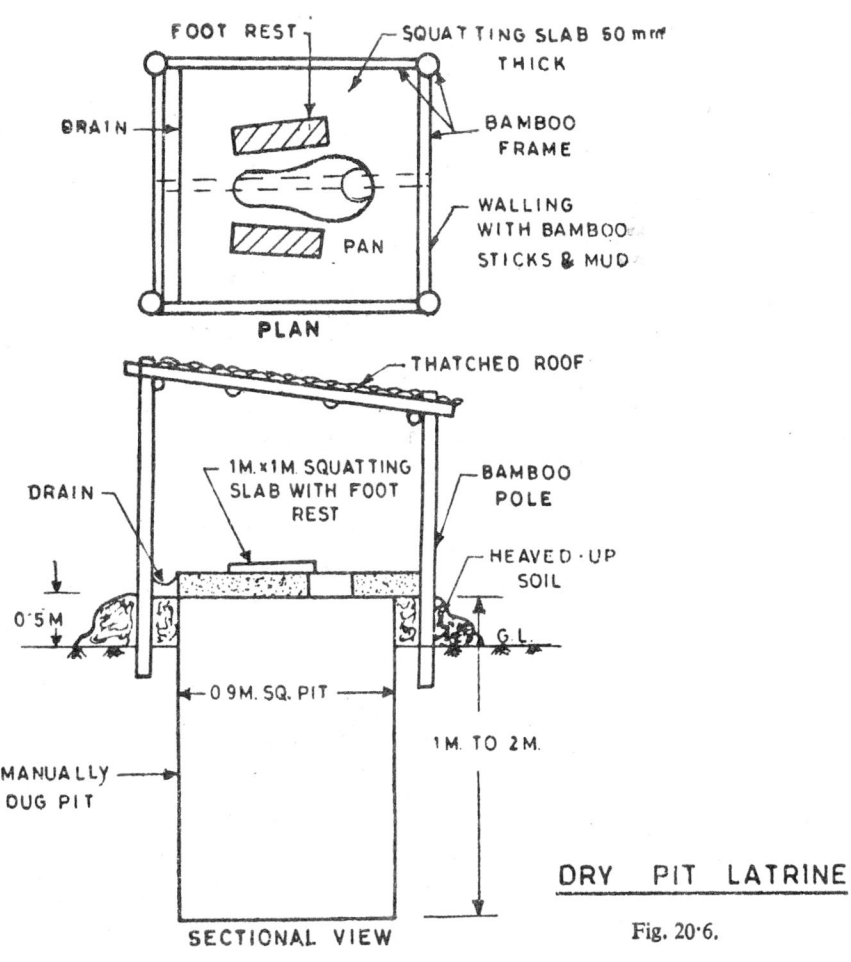

DRY PIT LATRINE

Fig. 20.6.

Wet Pit Latrine

Service life	Pit volume	Liquid depth (effective depth)
4 years	0.74 m³	0.91 m
8 years	1.48 m³	1.82 m

Dry Pit Latrine

Service life	Pit volume	Effective depth of pit
4 years	1.13 m³	1.37 m
8 years	2.26 m³	2.74 m

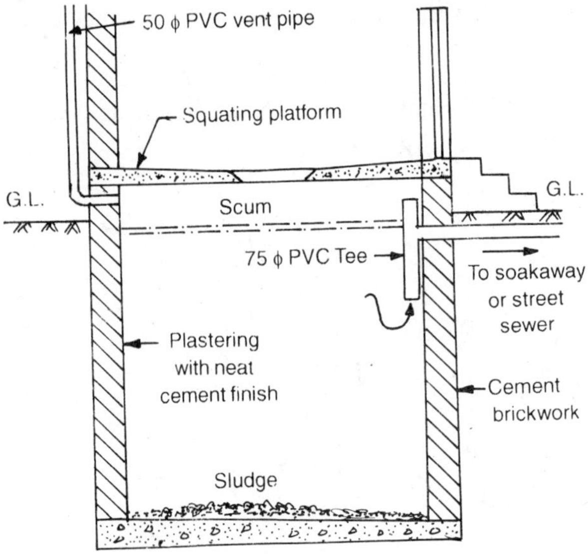

50 φ PVC vent pipe

Squating platform

G.L.

Scum

G.L.

75 φ PVC Tee →

To soakaway
or street
sewer

← Plastering
with neat
cement finish

Cement
brickwork

Sludge

CONVENTIONAL AQUAPRIVY
Fig. 20.7

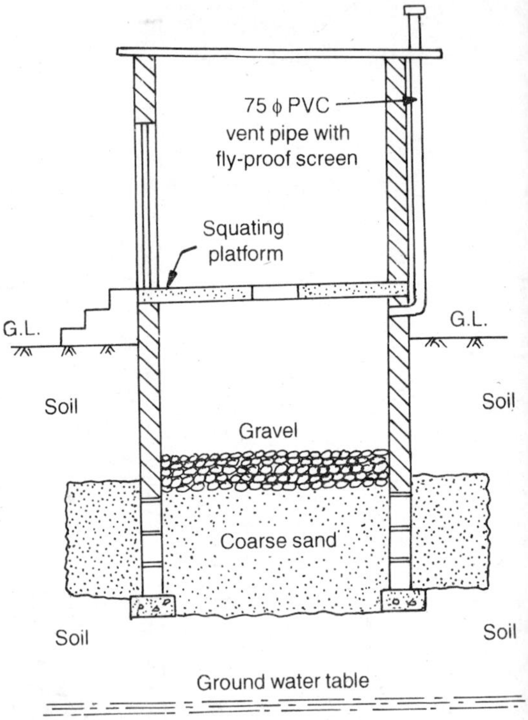

75 φ PVC
vent pipe with
fly-proof screen

Squating
platform

G.L.

G.L.

Soil

Soil

Gravel

Coarse sand

Soil

Soil

Ground water table

VENTILATED IMPROVED LATRINE WITH FILTER
Fig. 20.8

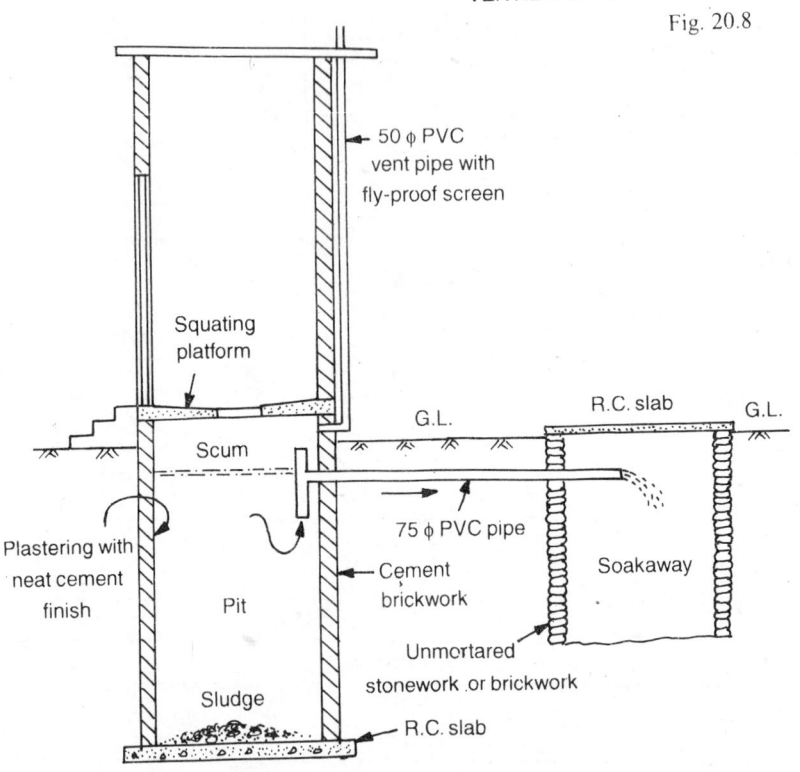

← 50 φ PVC
vent pipe with
fly-proof screen

Squating
platform

G.L.

R.C. slab

G.L.

Scum

Plastering with
neat cement
finish

75 φ PVC pipe

Soakaway

Cement
brickwork

Pit

Unmortared
stonework or brickwork

Sludge

R.C. slab

VENTILATED IMPROVED PIT LATRINE WITH SOAKAWAY

Fig. 20.9

21

Drains and Sewers

Drains and sewers are required to carry the foul water from a community. Underground drains and sewers are either pipes or ducts of required shape and size. Where underground drains and sewers are constructed, street gutters of V-shape or saucer type, street inlets and gully pits are constructed to receive the water (storm water and street wash water) from surface into the drain or sewer. The sanitary sewage is collected from individual house through house connections to sewer line. The combined sewers carry both storm water and foul water during monsoon and only dry weather flow (D.W.F.) *i.e.*, sanitary sewage during the rest part of the year. Owing to this, large size combined sewers have special sections.

Underground drainage or sewerage is very costly. But, sanitary sewers must be laid underground. Therefore, in most of the city outskists, towns and municipal areas, roadside open surface drains either kutcha or pucca are built. Various types of pucca surface drains are shown in Fig. 21.1. These drains are mostly built in city outskists. Rectangular and

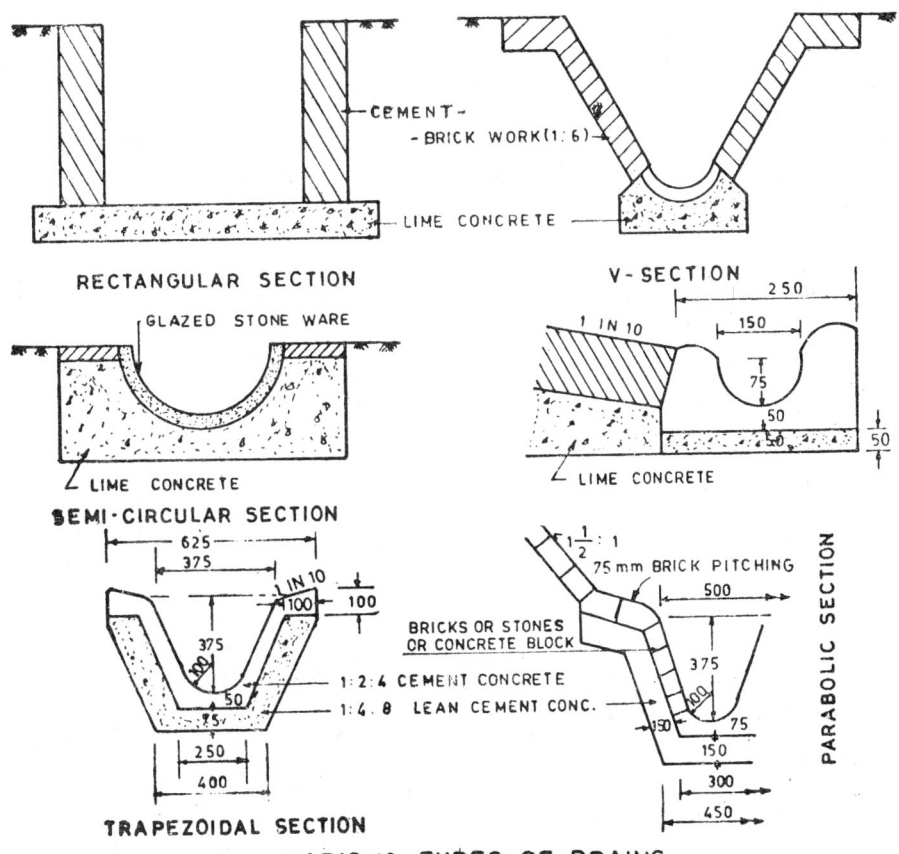

VARIOUS TYPES OF DRAINS

Fig. 21·1.

parabolic sections of open surface drains of different dimensions are shown in Figs. 21·2 to 21·1 These drains are conveniently used in towns, municipalities and non-municipal urban areas. In rural areas, kutcha drains are built by the roadside. In congested areas of old cities and towns where lanes are too narrow for laying underground sewers. the lanes themselves serve the function of drains during monsoon.

The depth to breadth ratio of drains of rectangrlar section veries from 1·5 to 2·5. Usually, this ratio is kept as 2.

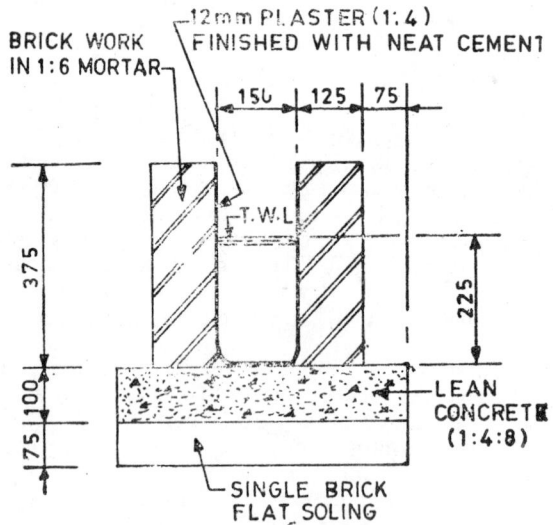

Roadside small Drain

Fig. 21·2.

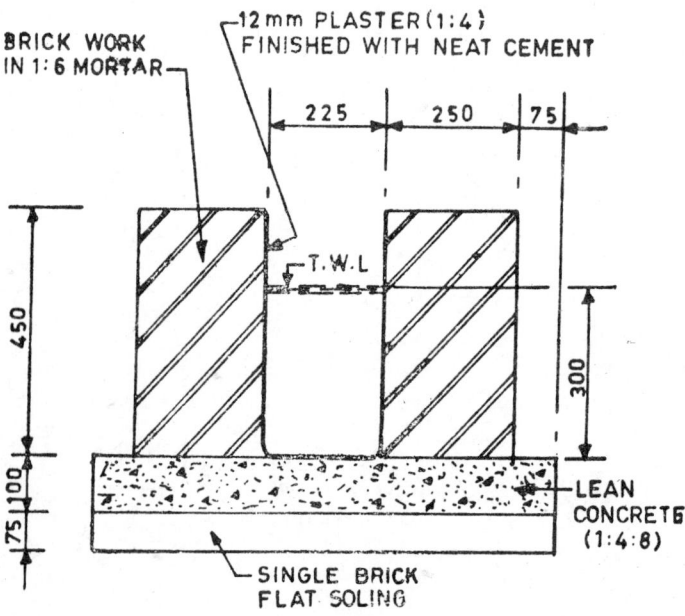

Fig. 21·3.

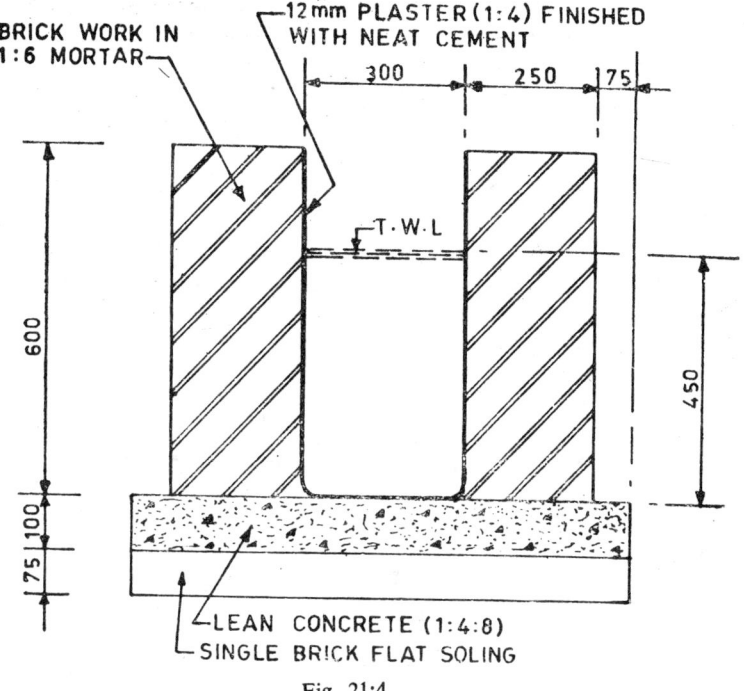

BRICK WORK IN 1:6 MORTAR

12mm PLASTER (1:4) FINISHED WITH NEAT CEMENT

300   250   75

T.W.L

600

450

75 100

LEAN CONCRETE (1:4:8)
SINGLE BRICK FLAT SOLING

Fig. 21·4.

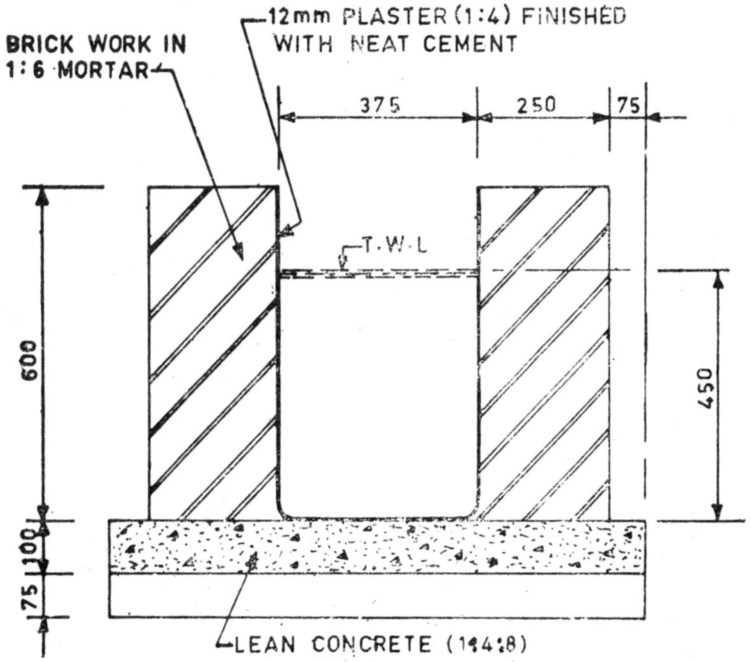

BRICK WORK IN 1:6 MORTAR

12mm PLASTER (1:4) FINISHED WITH NEAT CEMENT

375   250   75

T.W.L

600

450

75 100

LEAN CONCRETE (1:4:8)
Roadside large drain

Fig. 21·5.

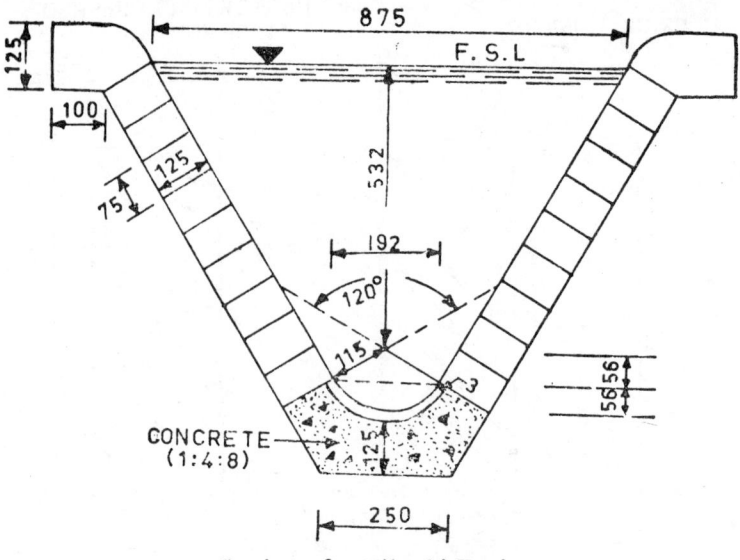

Section of a Nikashi Drain
Fig. 21·6.

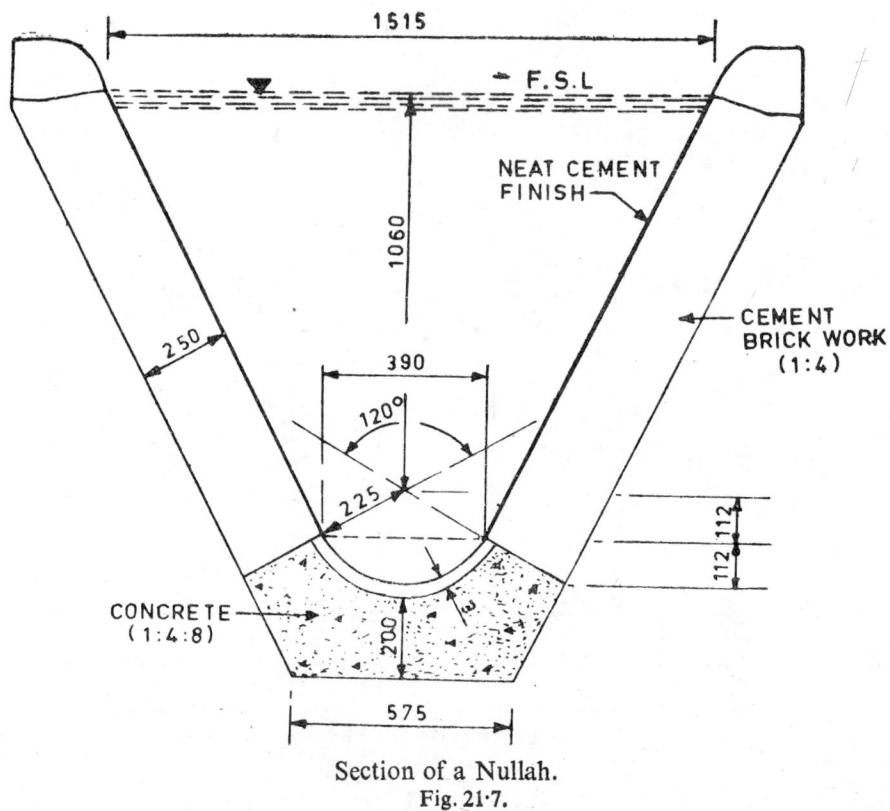

Section of a Nullah.
Fig. 21·7.

The velocity of flow through drains and sewers should be non-silting and non-scouring. Usually, the minimum velocity of flow is kept in a range of 0'75 m/s to 0'9 m/s.

The interior surface of the drains should always be plastered and finished with neat cement. Large size Nullahs and Nikashis are brick lined.

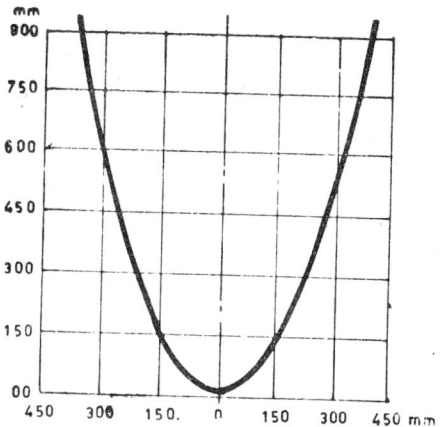

TYPE-A
PARABOLIC SECTION

Fig. 21·2.

EQUATION	SECTIONAL DETAILS IN mm.						
$x^2 = .5Y$	DEPTH	150	300	450	600	750	900
	WETTED PERIMETER	420	750	1080	1380	1710	2040

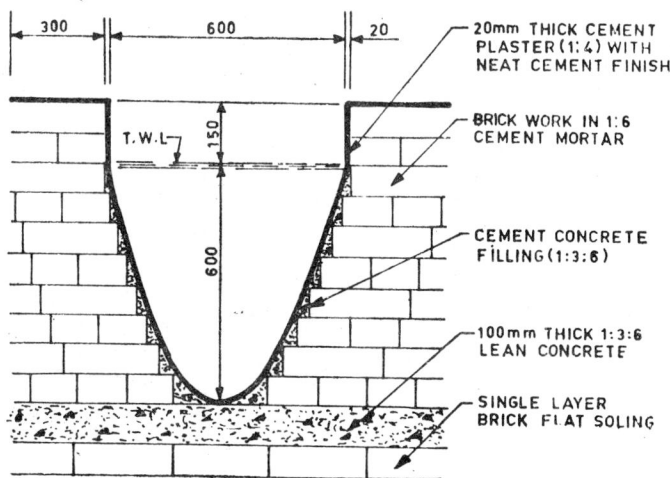

20mm THICK CEMENT PLASTER (1:4) WITH NEAT CEMENT FINISH

BRICK WORK IN 1:6 CEMENT MORTAR

CEMENT CONCRETE FILLING (1:3:6)

100mm THICK 1:3:6 LEAN CONCRETE

SINGLE LAYER BRICK FLAT SOLING

Fig. 21·9.

Four types of drains of parabolic section are shown in Figs. 21·8 to 21·15. The sectional details of each type of drain by using equations $X^2 = 0.5\ Y$, $X^2 = 0.75\ Y$, $X^2 = Y$ and $X^2 = 1.25\ Y$ are given in tabular from. These sections are quite useful in the design of surface drains.

Sewer sactions commonly adopted are : Ci cular, Egg-shaped, Recangular or sequare and Horse-shoe or Moorish.

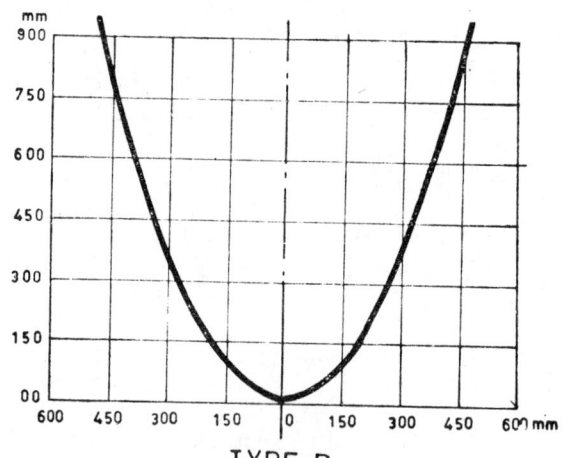

Fig. 21·10.

TYPE-B
PARABOLIC SECTION

EQUATION	SECTIONAL	DETAILS		IN	mm.		
$X^2 = \cdot 75\,Y$	DEPTH	150	300	450	600	750	900
	WETTED PERIMETER	485	860	1160	1460	1760	2060

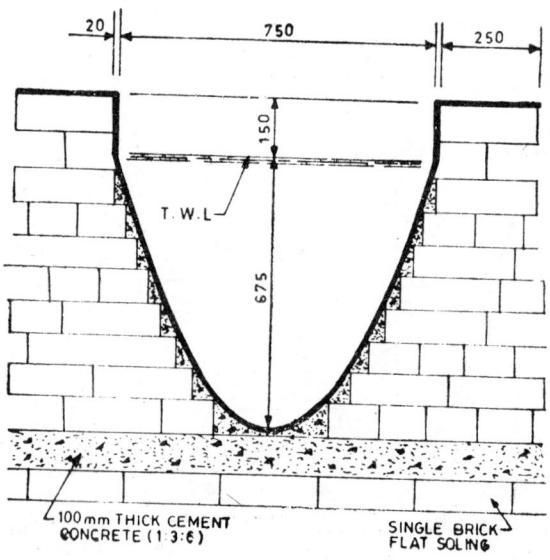

Flg.21·11.

R.C.C. sewers are used upto a diameter of 1800 mm. These are hume pipes (NP-2 and N.P.3). Stone-ware pipe sewers or Glazed china clay sewers are used upto a diameter of 375 mm for branches and laterals. Brick sewers are constructed usually for large sewer sections beyond 1500 mm. diameter.

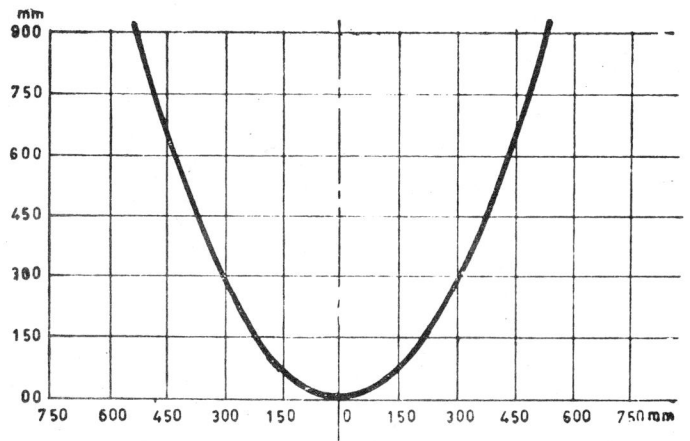

TYPE-C
PARABOLIC SECTION

Fig. 21·12

EQUATION	SECTIONAL DETAILS IN mm					
$x^2 = y$	DEPTH	150	300	450	600	750	900
	WETTED PERIMETER	540	900	1230	1560	1860	2190

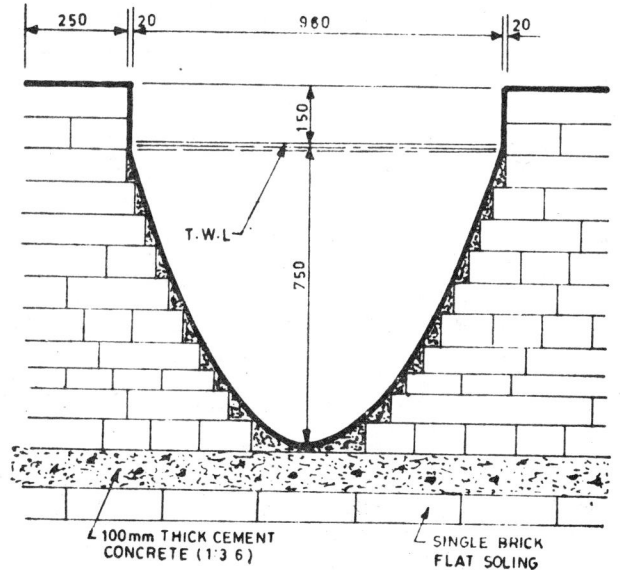

T.W.L

100mm THICK CEMENT CONCRETE (1·3·6)

SINGLE BRICK FLAT SOLING

Fig. 21·13

    The sectional views showing construction of egg-shaped brick sewers are shown in Fig. 21·16. Construction of circular brick sewer is shown in Fig. 21·17. The sectional view of a rectangular masonry sewer topped with a R.C.C. slab is shown in Fig. 21·18.

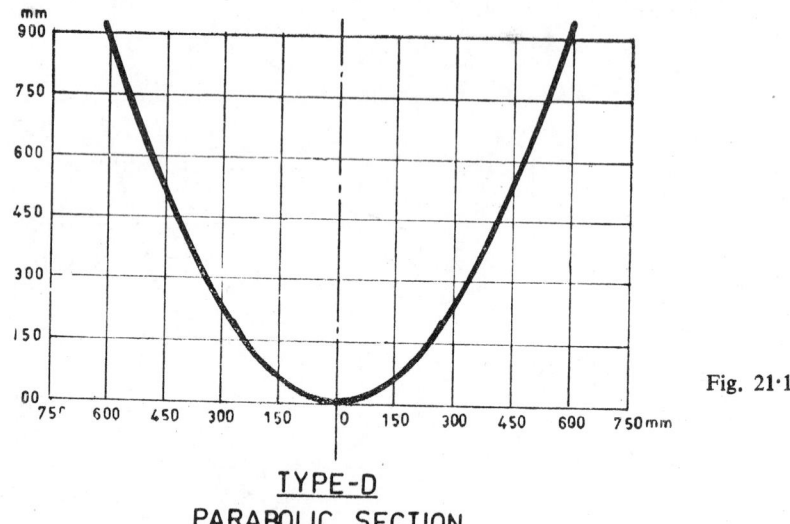

Fig. 21·14

TYPE-D
PARABOLIC SECTION

EQUATION	SECTIONAL	DETAILS		IN	mm	
$x^2 = 1.25Y$	DEPTH	150	300	450	600	750	900
	WETTED PERIMETER	600	935	1275	1610	1950	2250

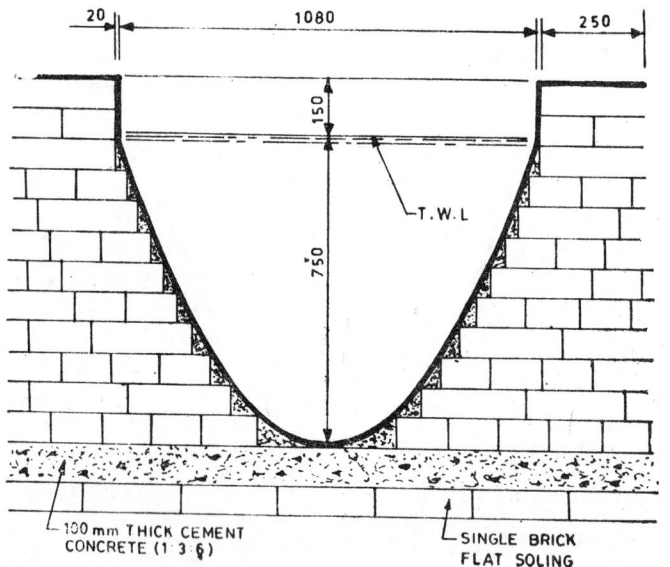

Fig. 21·15

100 mm THICK CEMENT CONCRETE (1:3:6)

SINGLE BRICK FLAT SOLING

T.W.L

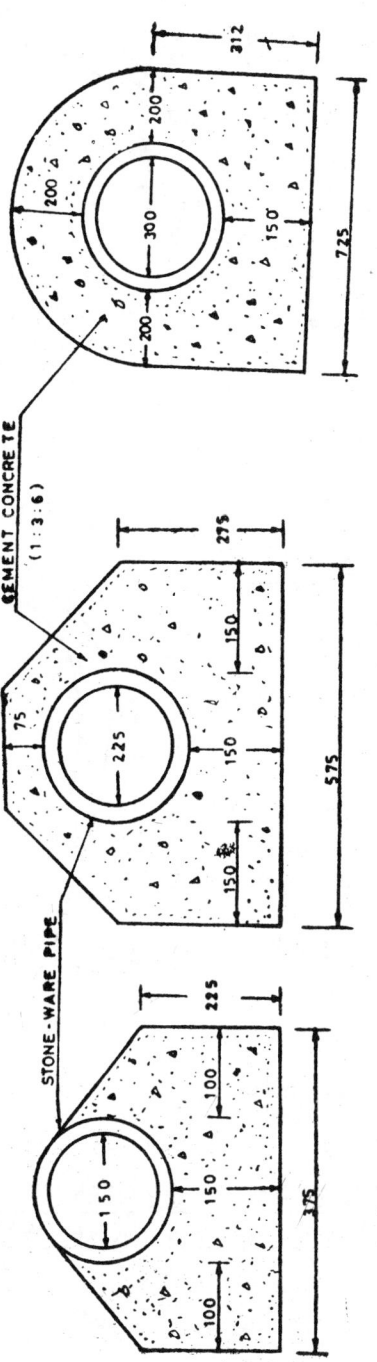

EMBEDDING STONE WARE PIPE SEWER

Fig. 21·19.

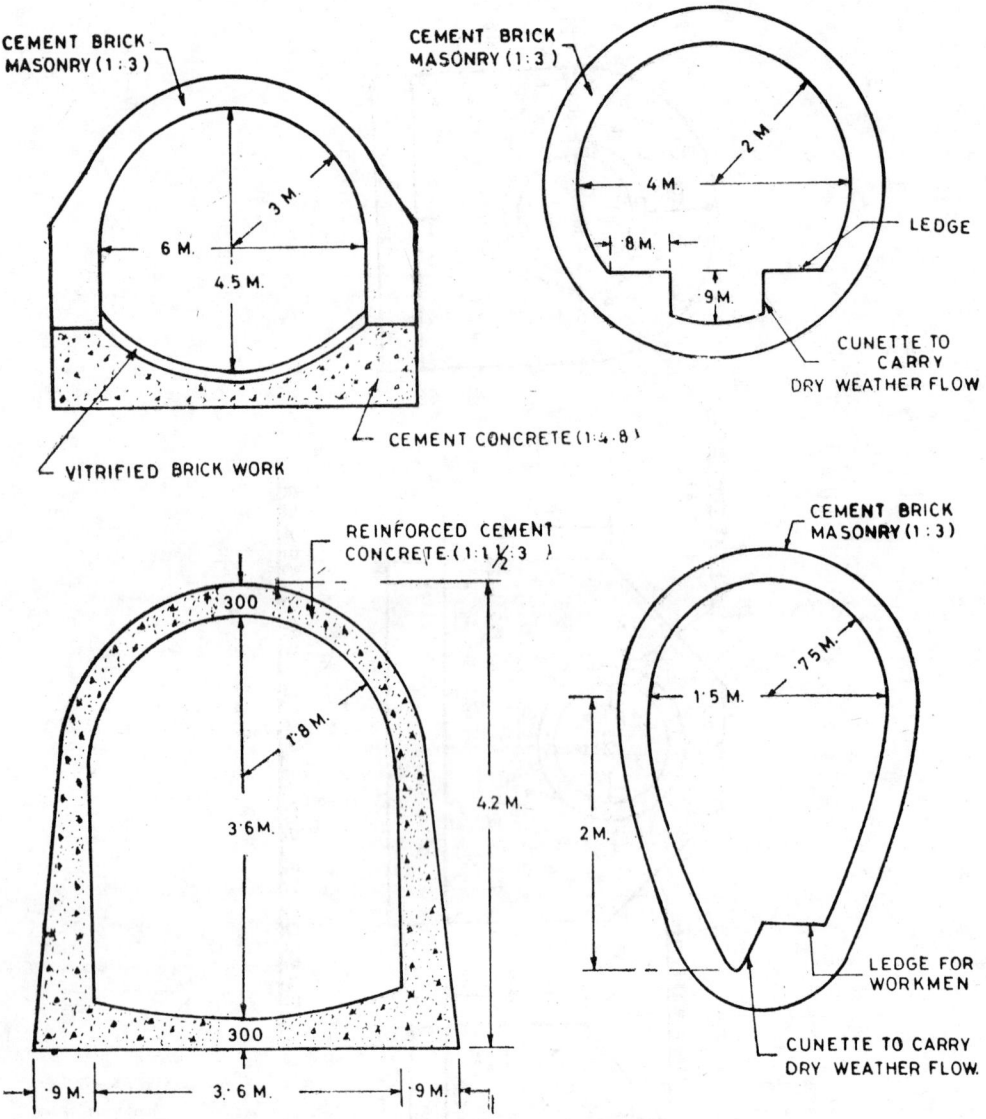

LARGE SIZE SPECIAL TYPE COMBINED SEWERS
(TO CARRY SANITARY SEWAGE & STORM WATER)

Fig. 21·20.

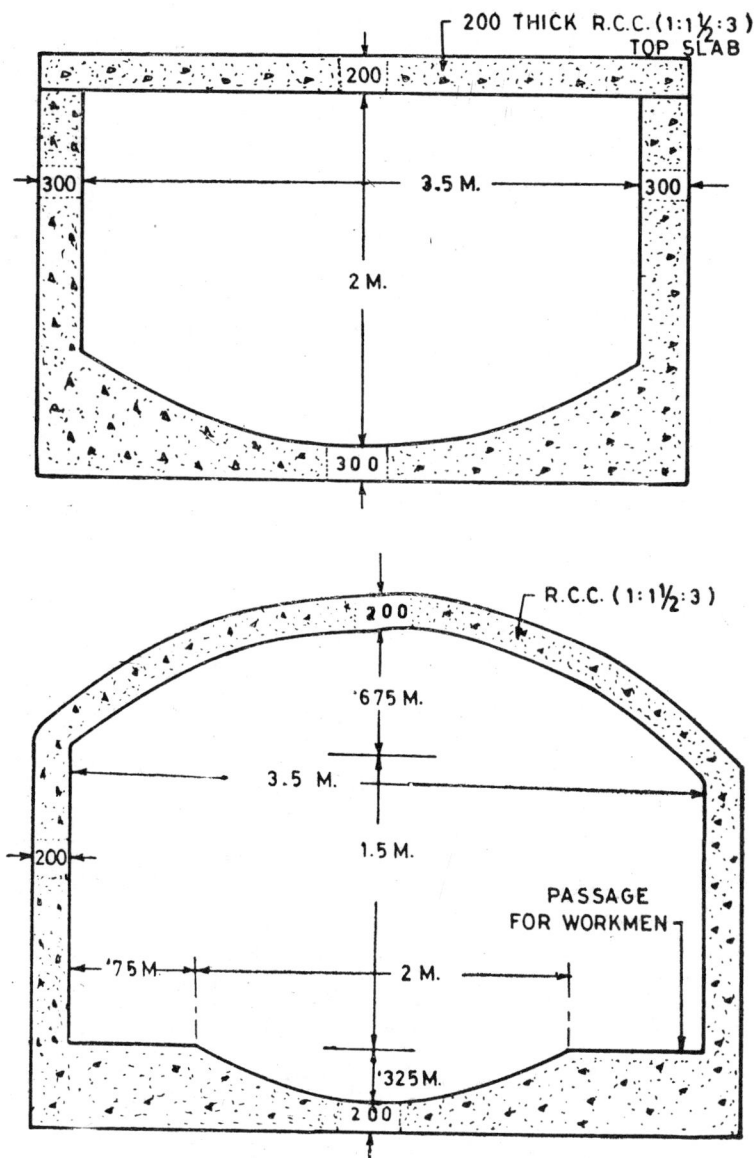

Fig. 20·21 [contd.]

22

Pipe Fittings and Joints

For joining pipes and for branching off, pipe fittings and specials are required. This chapter will deal with various pipe fittings, specials and different joints used for metal pipes, especially cast iron pipes and for stone ware pipes. The spigot and socketed joints and collared joint which are most commonly used for hume pipes are not presented here.

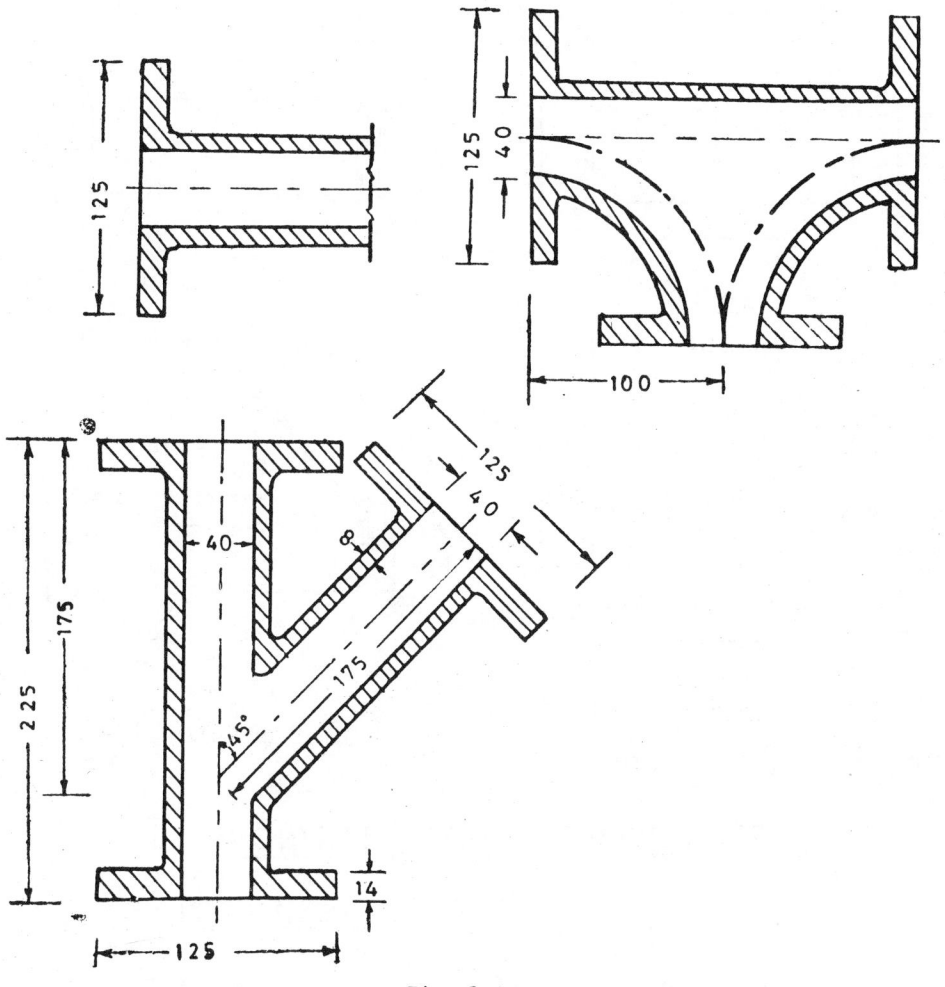

Pipe fittings
Fig. 22·1.

The pipe fittings mostly used for C.I., G.I. and stone-ware pipes are : Elbow, Bend, Tee, Cross, Y, Double Y, Reducer and Enlarger. Fig. 22·1 shows a flanged pipe end, a Tee and a Y-junction used for flanged pipes. Stone-ware pipe fittings are presented in Fig. 22·2. Commonly used cast iron pipe fittings are illustrated in Fig. 22·3.

Most common joints are either flanged or spigot and socketed. Flanged joint may of various types. In flanged joints, the pipe ends with flanges are brought together and bolts and nuts are fitted into the holes of the flanges and tightened. For better tightening and making the joint leak proof, a rubber gasket is inserted in between two flanges. Two views of a flanged joint are shown in Fig. 22·4.

Spigot and socketed joints are used in C.I., stone-ware and cement concrete pipes. For making this joint, the spigot end of one pipe is introduced into the socket end of the other pipe. The annular space between the outer surface of the inserted spigot and the inner surface of the socket is filled with hemp, jute and lead in case of C.I. pipes and cement mortar for stone-ware and hume pipes.

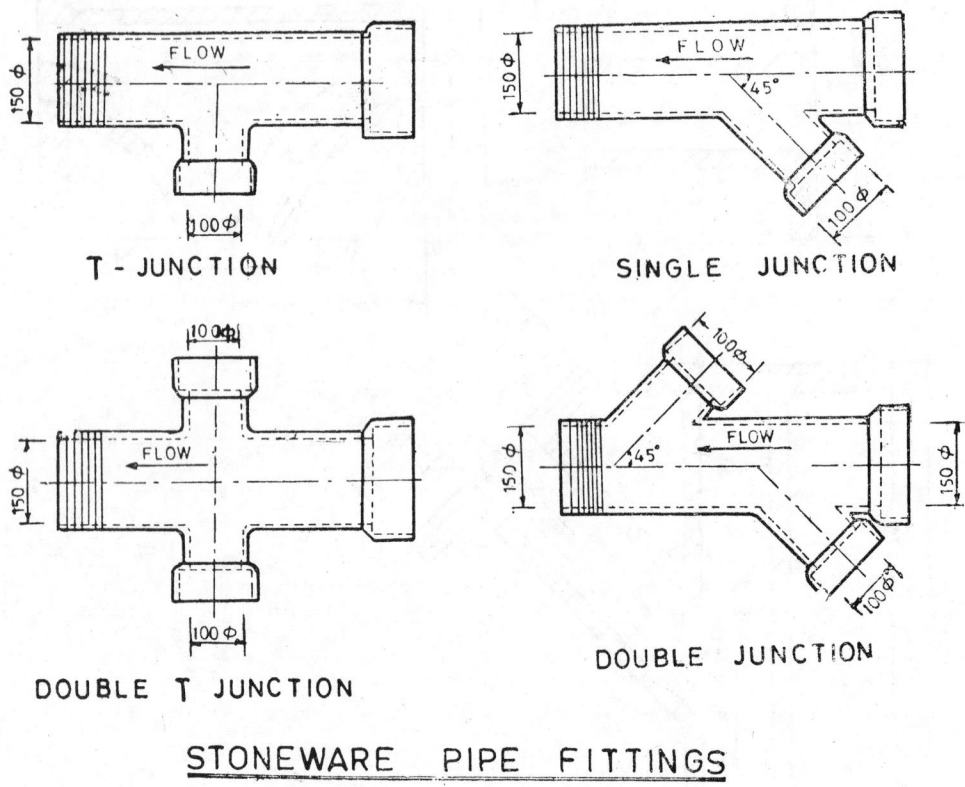

STONEWARE   PIPE   FITTINGS

Fig. 22·2.

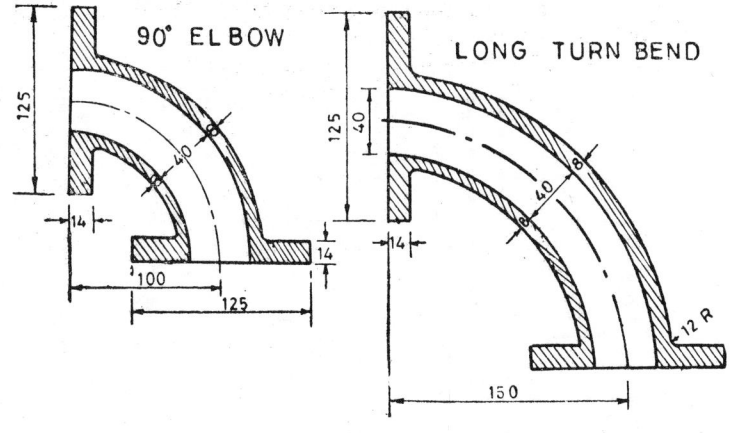

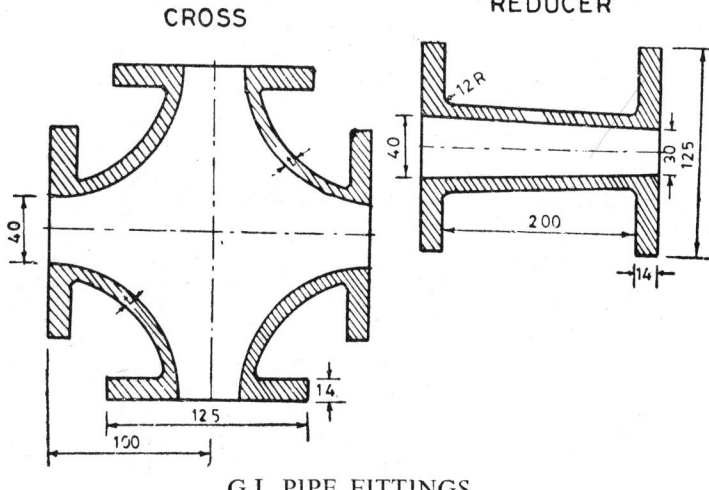

G.I. PIPE FITTINGS
Fig 22.3.

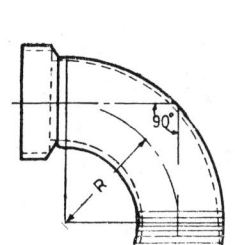

FOR 10Cm. DIA, R=9,15,21Cm.
FOR 15Cm. DIA, R=15,19,22.5Cm.

RADIUS BENDS

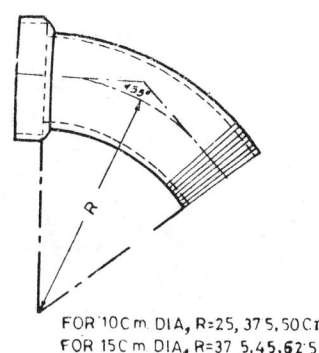

FOR 10Cm. DIA, R=25, 37.5, 50Cm.
FOR 15Cm. DIA, R=37.5, 45, 62.5Cm.

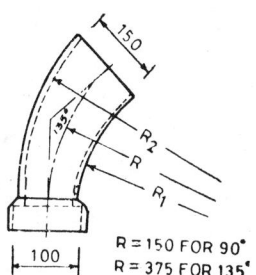

R=150 FOR 90°
R = 375 FOR 135°

TAPER BEND

Fig. 22.3. (Contd.)

Fig. 22·5 shows half-sectional view of a spigot and socketed joint. The dimensions given are for joining 100 mm diameter pipes.

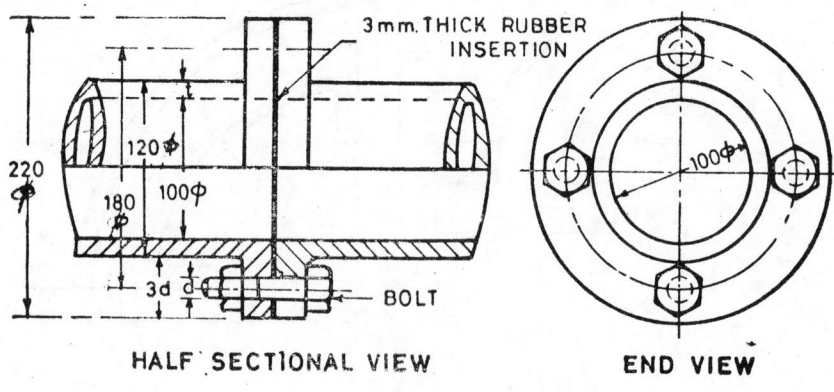

HALF SECTIONAL VIEW   END VIEW

FLANGED JOINT

Fig. 22·4.

Screwed socket joint is mostly used for joining G.I. pipes. For making this joint, the pipe ends are threaded and a separate piece known as socket having internal threads is used to join the two pipes. In this joint, the socket is first screwed at the end of one pipe and then the second pipe is screwed into the socket. White lead, jute or paint is used over the threads of the pipes, before the socket is screwed over it. This type of joint is especially used for small diameter G.I. pipes. Usually large diameter G.I. pipes (100 mm to 200 m diameter) have flanged ends

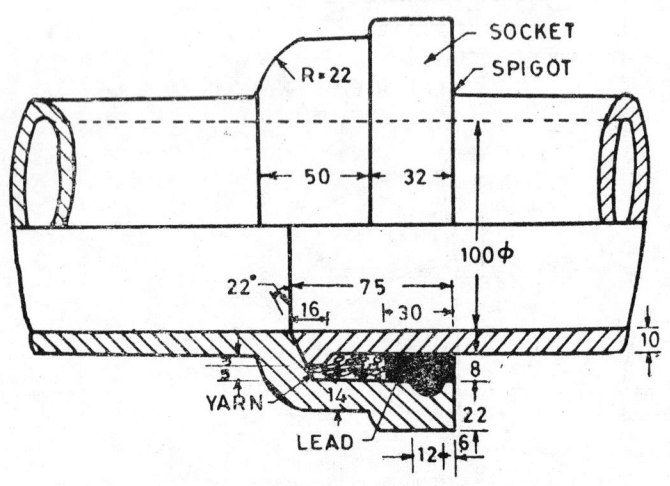

HALF SECTIONAL VIEW
SPIGET & SOCKET JOINT

Fig. 22·5.

with holes in flanges. A union joint for a G.I. pipeline in half-sectional view is shown in Fig. 22·6.

Sectional views of a pipe clamp are shown in Fig. 22·7. Pipe clamps are required to hold a pipe. These are also used to hold the broken parts of a pipeline and to prevent the

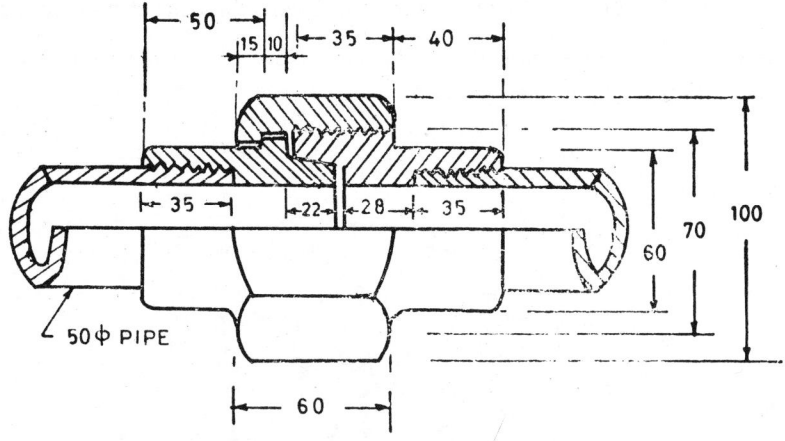

UNION JOINT FOR A PIPELINE
(HALF SECTIONAL VIEW)

Fig. 22·6.

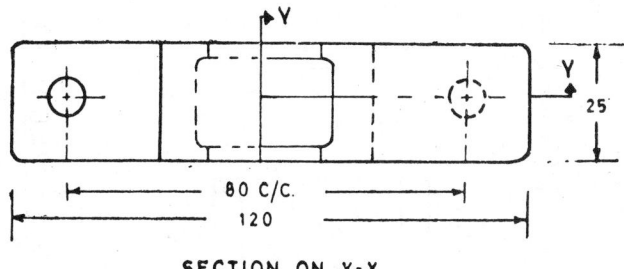

SECTION ON X·X

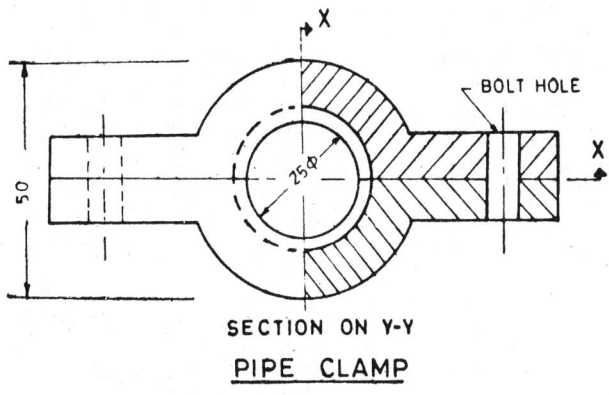

SECTION ON Y-Y

PIPE CLAMP

Fig. 22·7.

leakage from a running line by wrapping a rubber sheet over the broken parts and then by clamping the two pieces together.

Two views of a screwed type flanged joint are shown in Fig. 22·8. Here the pipe ends are threaded and separate threaded flange pieces with holes are screwed onto it. These two flanged ends are then brought together and joined with the help of bolts and nuts.

The half-sectional view and the right end view of a hydraulic joint are shown in Fig. 22·9. This is another type of flanged joint.

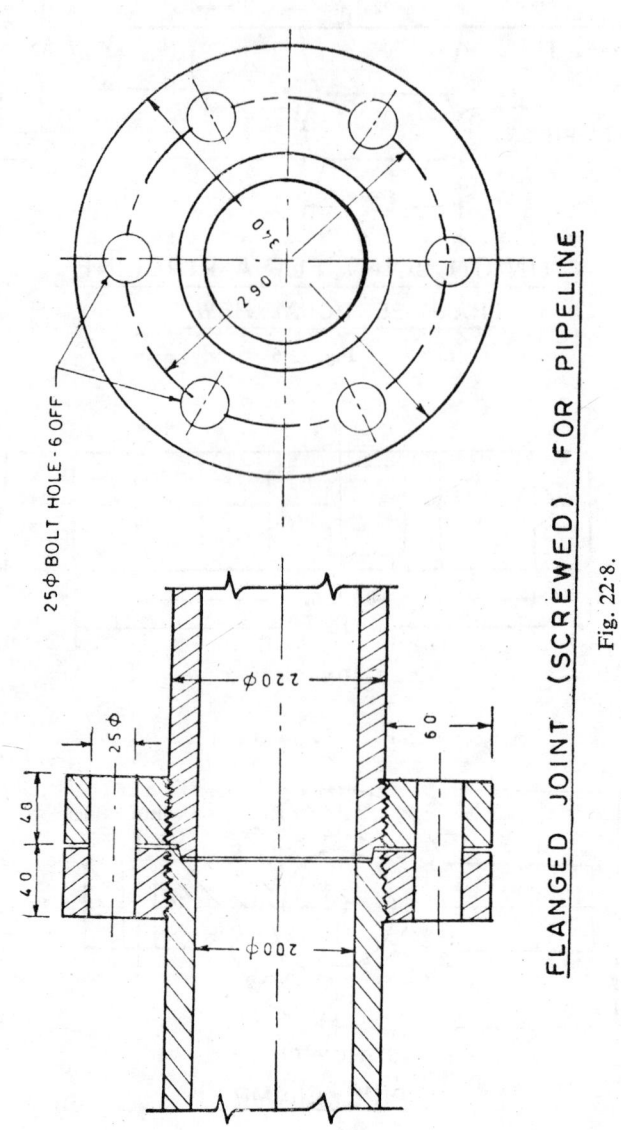

FLANGED JOINT (SCREWED) FOR PIPELINE

Fig. 22·8.

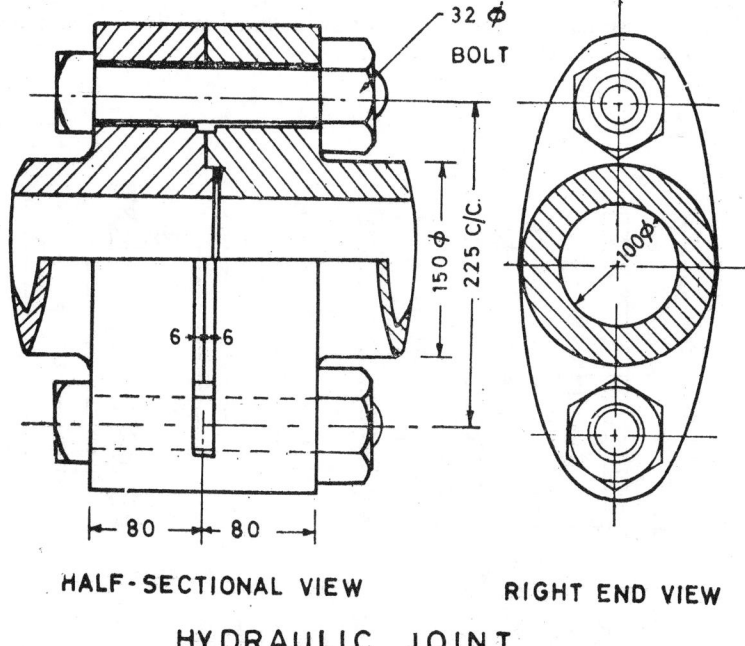

HALF-SECTIONAL VIEW   RIGHT END VIEW

HYDRAULIC JOINT

Fig. 22·9.

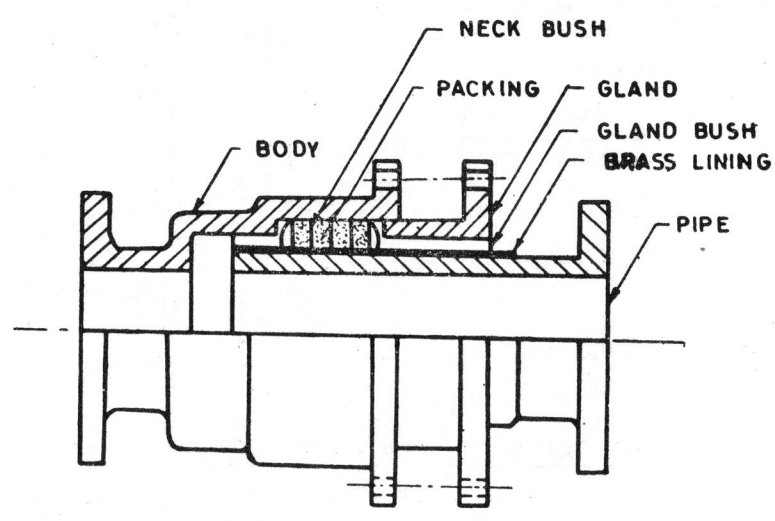

HALF-SECTIONAL VIEW

EXPANSION JOINT

Fig. 22·10.

Fig. 22·10 shows half-sectional view of an expansion joint. This type of joint is used in high pressure lines.

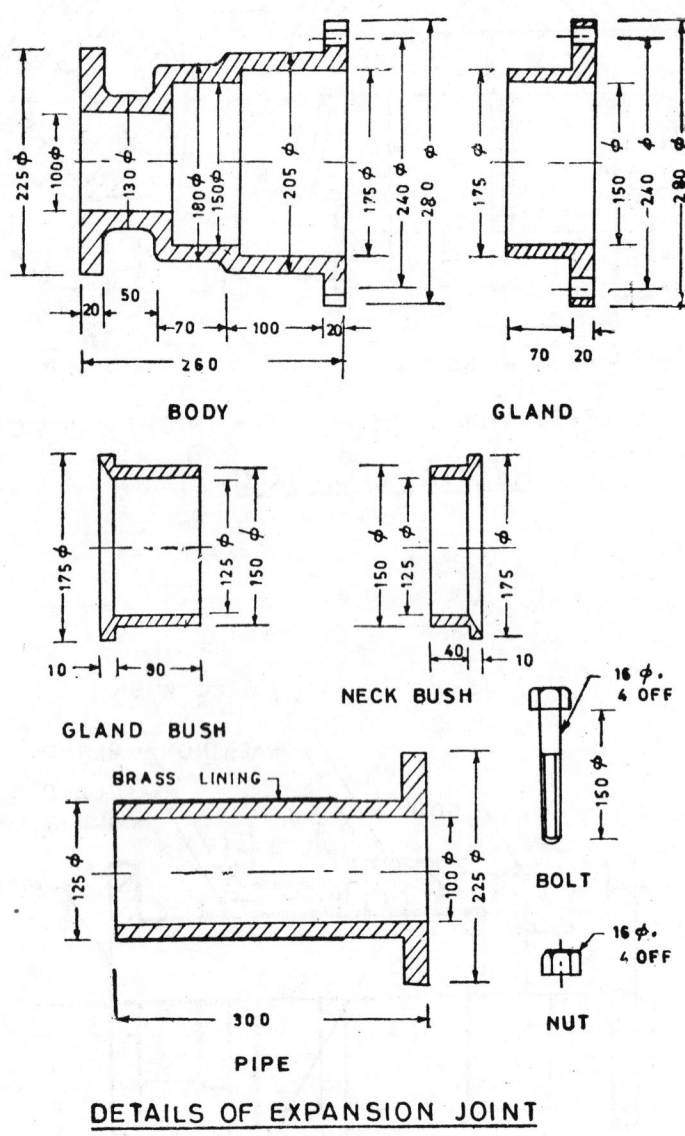

DETAILS OF EXPANSION JOINT

Fig. 22·11.

The details of expansion joint are shown in Fig. 22·11.

Use of a Tee and a Bend in a pipeline is shown in Fig. 22·12. This is an example of a flanged pipe joint. Its pictorial view is presented in Fig. 22·13.

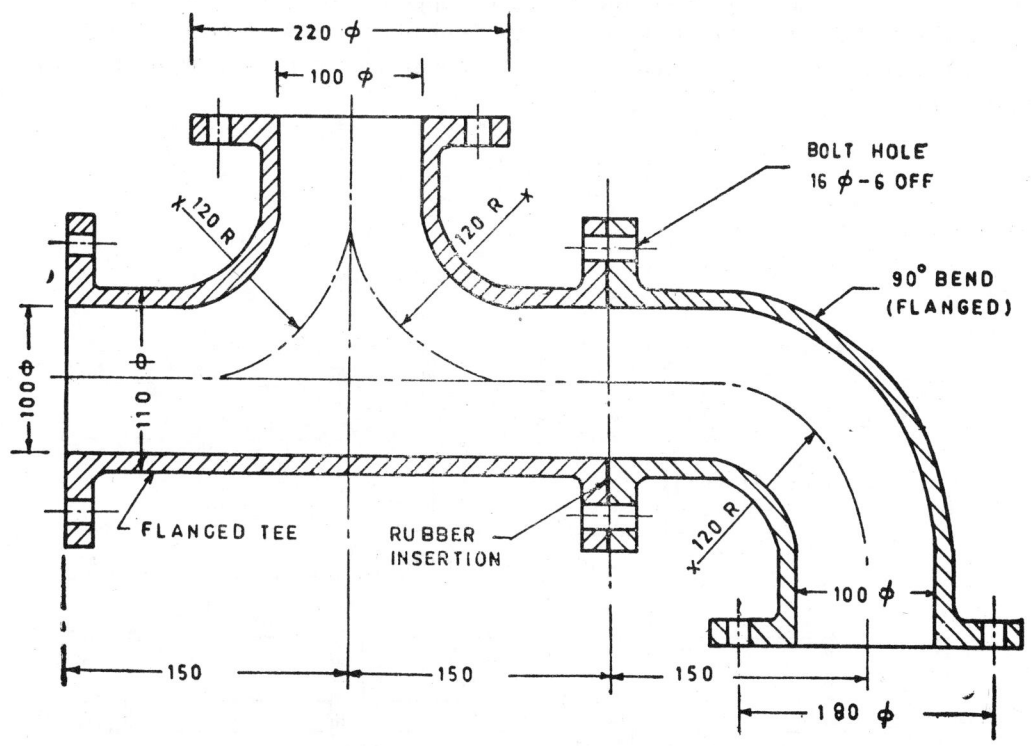

FLANGED PIPE JOINT

Fig. 22·12.

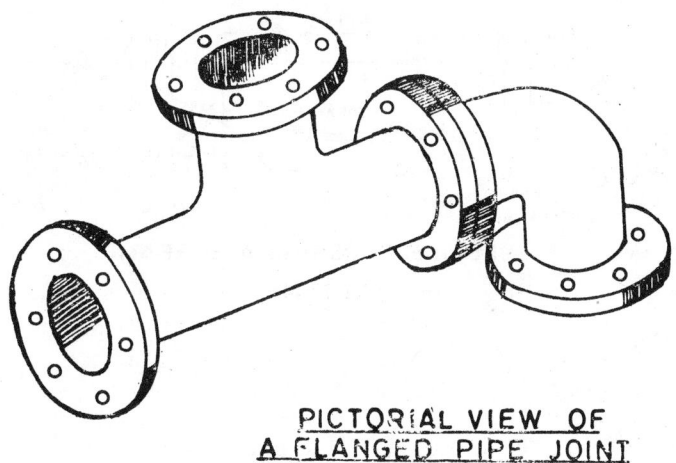

PICTORIAL VIEW OF
A FLANGED PIPE JOINT

Fig. 22·13.

Fig. 22·14 shows how a pipeline should be supported over ground with supports at the spigot and socketed joints and also while crossing a nullah or narrow stream. Raising a pipeline in crossing a stream where needed, as also shown with anchor blocks, required pipe joints and supports.

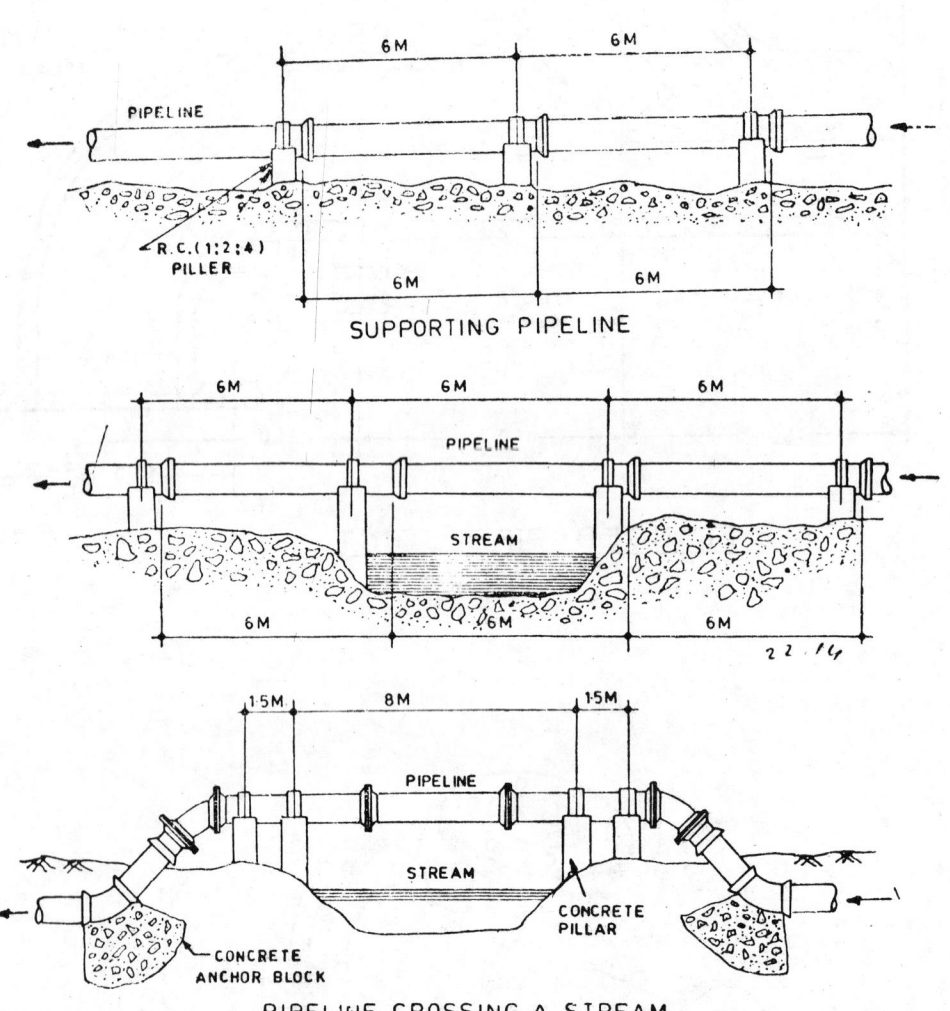

Fig. 22·14.

23

Valves and Valve Chambers

A valve is a controlling device or mechanism used in a pipeline carrying fluid under pressure. With the help of valve the flow of a fluid in a pipeline may be diverted or stopped. There are different types of valves used for different purposes.

In water and sewage lines, Stop valve, Gate valve, Sluice valve, Foot valve, Pressure Release valve, Non-return valve, etc. are used. In steam and gas lines, Stop valve, Relief valve, Safety valve, Feed check valve, etc. are commonly used.

A typical valve connection in a street water main is shown in Fig. 23·1. In order to protect such a valve from external damage, a valve chamber is constructed surrounding the valve and the chamber is topped with a heavy slab.

Gate valves are used to control and to stop the flow of water. Such a valve to shown in Fig. 23·1. These are used in water workes and water distribution lines.

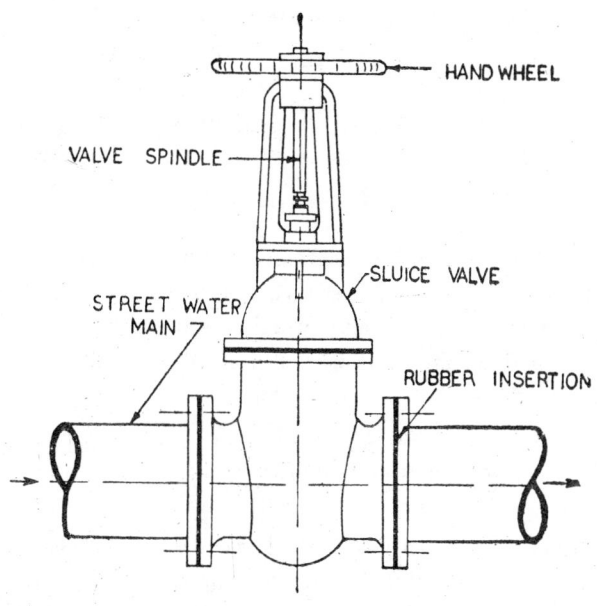

TYPICAL VALVE CONNECTION

Fig. 23·1.

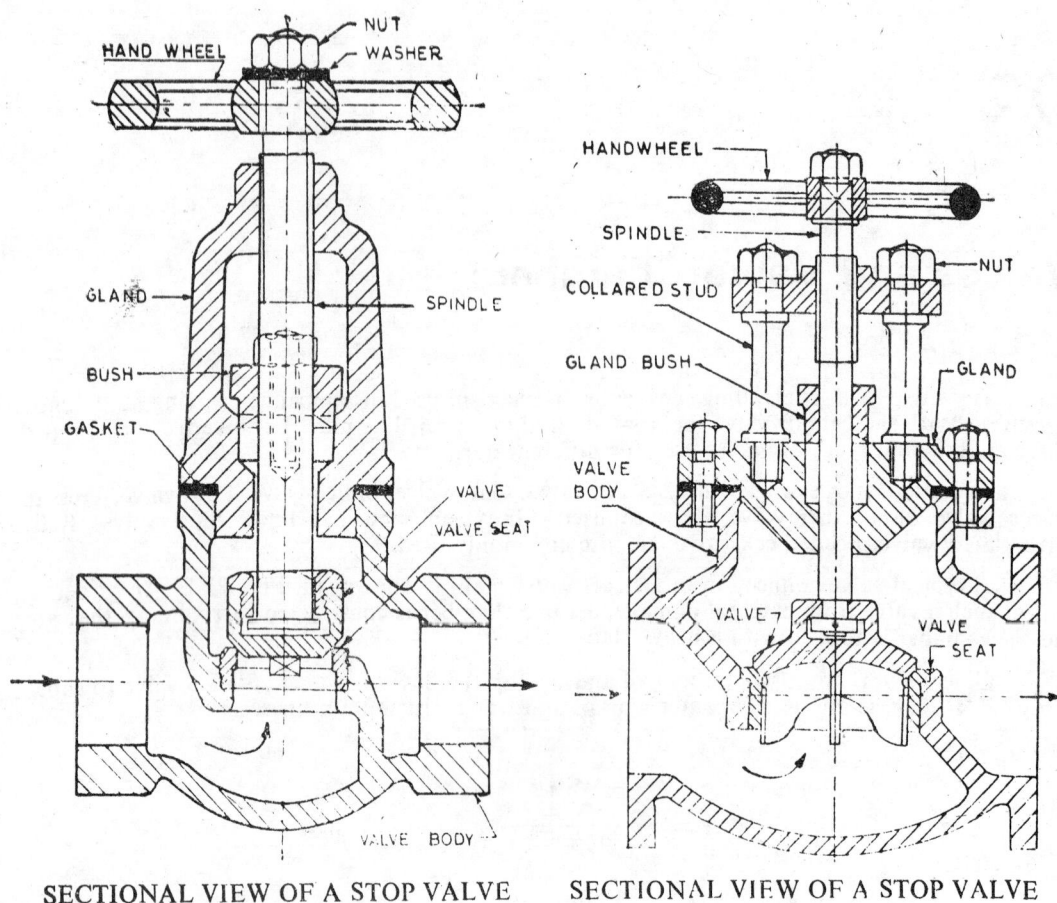

SECTIONAL VIEW OF A STOP VALVE
Fig. 23·2.

SECTIONAL VIEW OF A STOP VALVE
Fig. 23·3.

Sectional views of three types of stop valve used in water, steam and gas lines are shown in Figs. 23·2 to 23·4. These valves and valve seats are usually made of gun metal or brass. In order to guide the valve in raising and bringing down, necessary fittings are provided as shown in the figures. The body of the valve is made of different types depending upon the inlet and outlet arrangement.

Sectional view of a non-return valve used in steam line is shown in Fig. 23·5. It may be seen that the lift of the valve is limited by the projected portion in the cover. This valve will allow unidirectional flow as shown by arrowhead.

A typical sluice valve chamber constructed in a water line is shown in Fig. 23·6. The hand wheel is removed when a sluice valve of rising spindle type is used and it is operated with a long handled key by introducing it through a small hole in the fixed top slab of the chamber. This ensures protection of the valve.

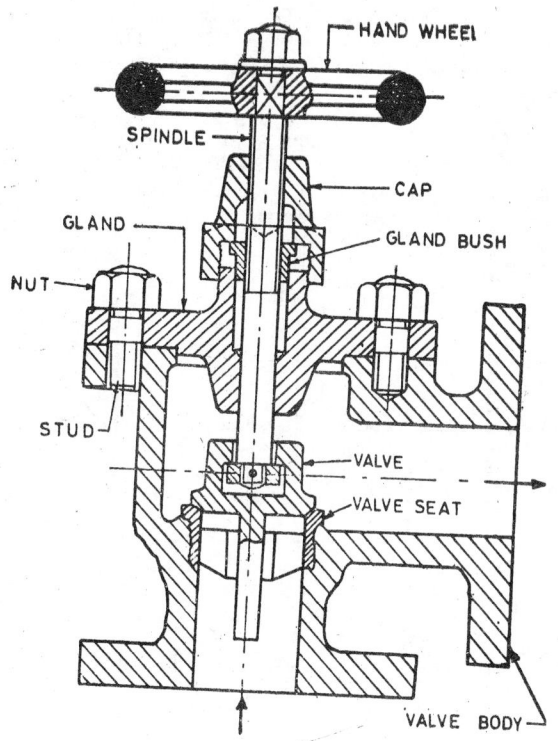

SECTIONAL VIEW OF A STOP VALVE
Fig. 23·4.

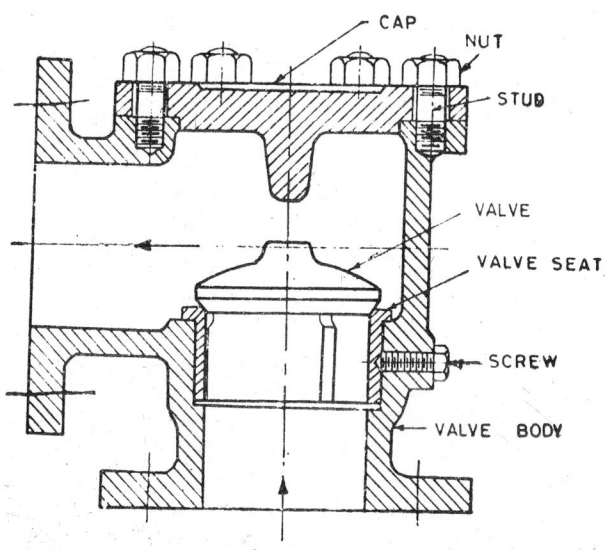

SECTIONAL VIEW OF A NON-RETURN VALVE
Fig. 23·5

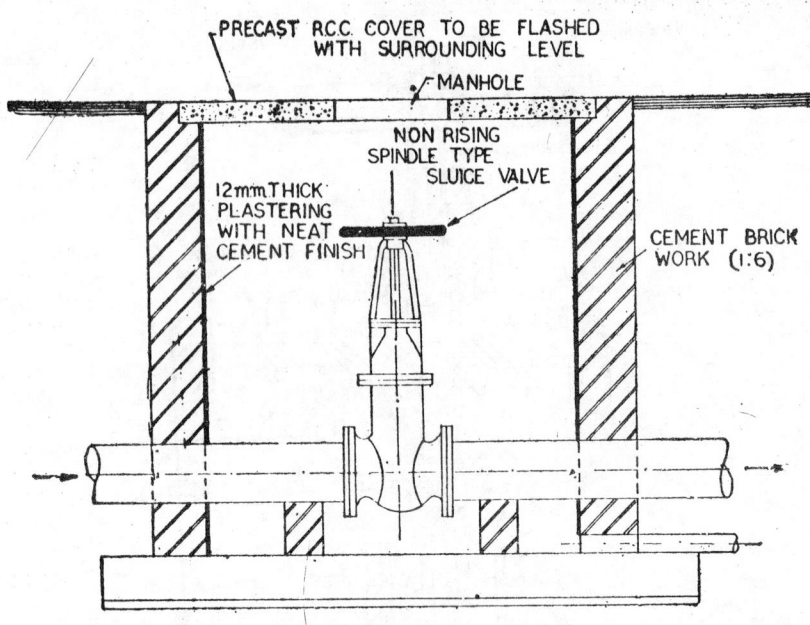

TYPICAL VALVE CHAMBER IN LINE

Fig. 23·6.

Details of Sluice Valve Chamber

Dia. of sluice valve in mm.	Length L in mm	Breadth B in m.	Height in m			Wall thickness in mm	
			H	H_1	H_2	T_1	T_2
200	0·9	0·9	1·10	0·35	0·75	250	250
250	0·9	0·9	1·30	0·40	0·90	250	250
300	0·9	0·9	1·35	0·45	0·90	250	250
450	1·2	0·9	2·10	1·20	0·90	250	375
600	1·50	1·20	2·10	1·20	0·90	250	375
750	1·65	1·35	2·55	1·20	1·35	250	375
900	1·80	1·65	3·00	1·50	1·50	375	500
1050	1·95	2·10	3·45	1·65	1·80	375	500
1200	2·40	2·10	3·75	1·80	1·95	375	500

The detailed dimensions of sluice valve chamber varying with the dimensions of the sluice valve are presented in tabular form. Fig. 23·7 shows the details of such a sluice valve chamber.

Fig. 23·8 shows plan and sectional elevation of a washout valve chamber to be used for 80 to 200 mm diameter waterline.

Plan and sectional view of two types of Gate valve chamber are presented in Fig. 23·9 and 23·10.

Two views of a single Air-Release valve chamber are shown in Fig. 23·11. These valves with valve chambers are needed at summits in a water line, with a view to releasing out the accumulated air in the pipeline.

Fig. 23·12 shows plan and sectional elevation of a washout sump.

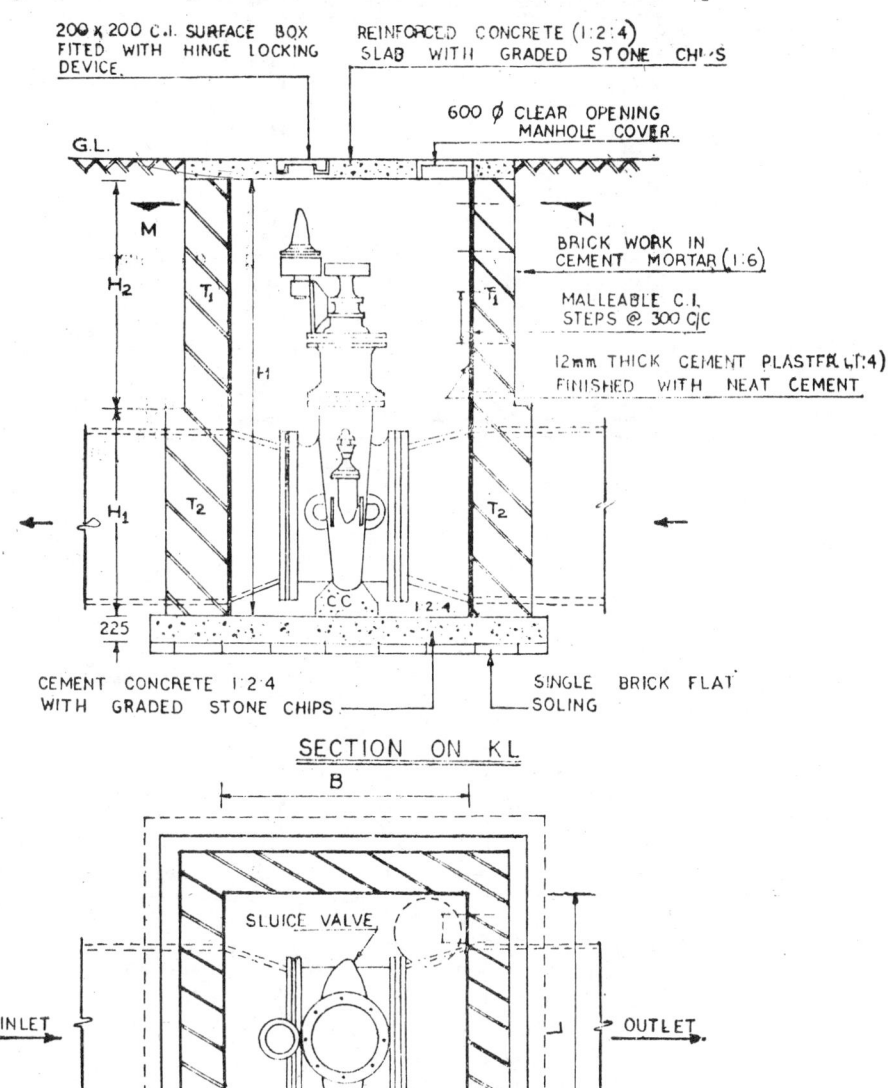

200 x 200 C.I. SURFACE BOX FITED WITH HINGE LOCKING DEVICE.

REINFORCED CONCRETE (1:2:4) SLAB WITH GRADED STONE CHIPS

600 Ø CLEAR OPENING MANHOLE COVER

G.L.

M

H₂

T₁

BRICK WORK IN CEMENT MORTAR (1:6)

MALLEABLE C.I. STEPS @ 300 C/C

12mm THICK CEMENT PLASTER (1:4) FINISHED WITH NEAT CEMENT

H

H₁   T₂   T₂

225   C C   1:2:4

CEMENT CONCRETE 1:2·4 WITH GRADED STONE CHIPS

SINGLE BRICK FLAT SOLING

SECTION ON KL

B

SLUICE VALVE

INLET

OUTLET

K   L

150 Ø BY PASS VALVE

SEC. PLAN ON MN

DETAILS OF SLUICE VALVE CHAMBEʳ

Fig. 23·7.

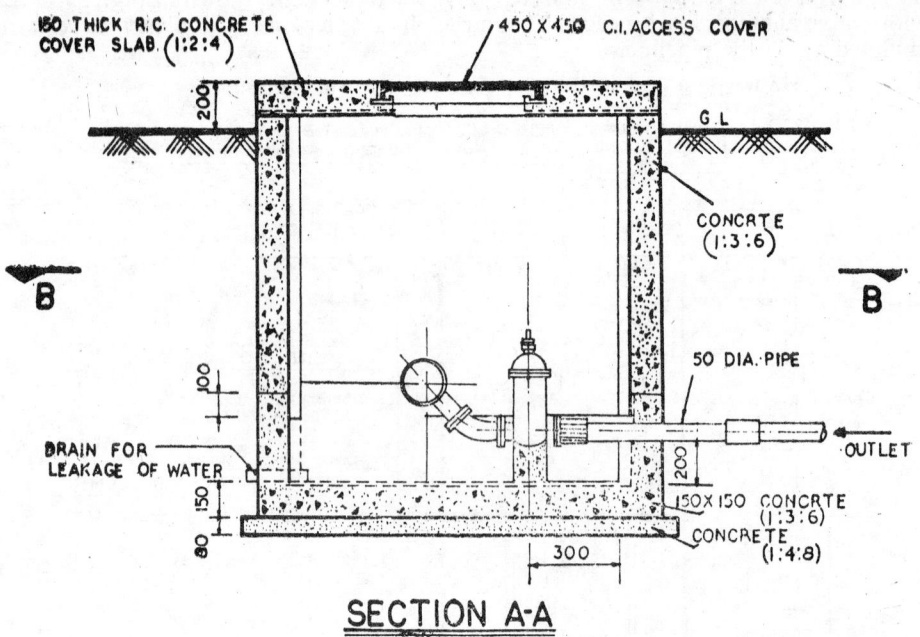

150 THICK R.C. CONCRETE COVER SLAB. (1:2:4)

450 X 450 C.I. ACCESS COVER

200

G.L

CONCRTE (1:3:6)

B   B

100

50 DIA. PIPE

OUTLET

BRAIN FOR LEAKAGE OF WATER

200

150X150 CONCRTE (1:3:6)

150

80

CONCRETE (1:4:8)

300

SECTION A-A

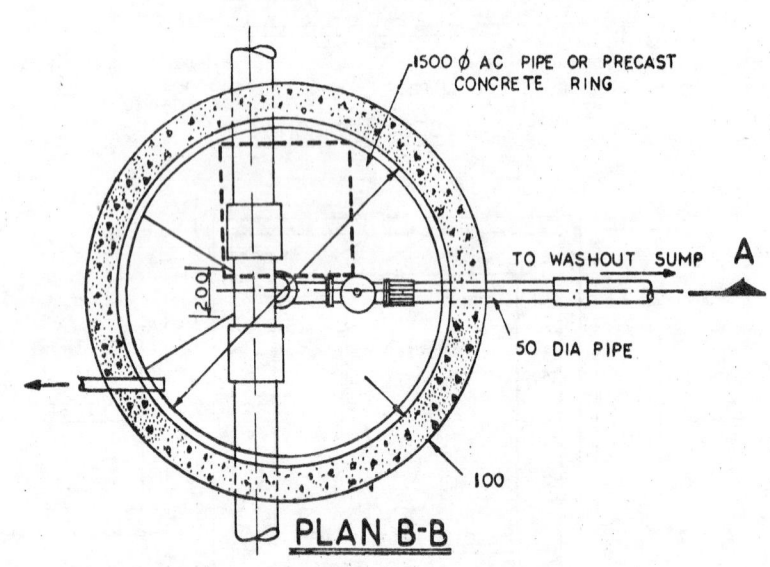

1500 ⌀ AC PIPE OR PRECAST CONCRETE RING

200

TO WASHOUT SUMP   A

50 DIA PIPE

100

PLAN B-B

WASHOUT VALVE CHAMBER
FOR 80-200 MAIN

Fig. 23 8.

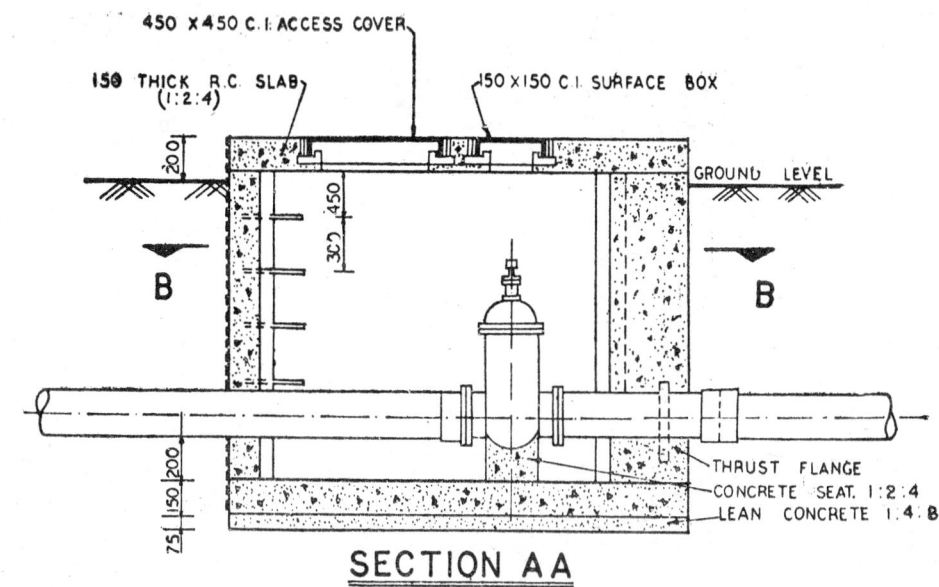

450 X 450 C.I. ACCESS COVER

150 THICK R.C. SLAB (1:2:4)

150 X 150 C.I. SURFACE BOX

GROUND LEVEL

THRUST FLANGE
CONCRETE SEAT. 1:2:4
LEAN CONCRETE 1:4:8

SECTION A A

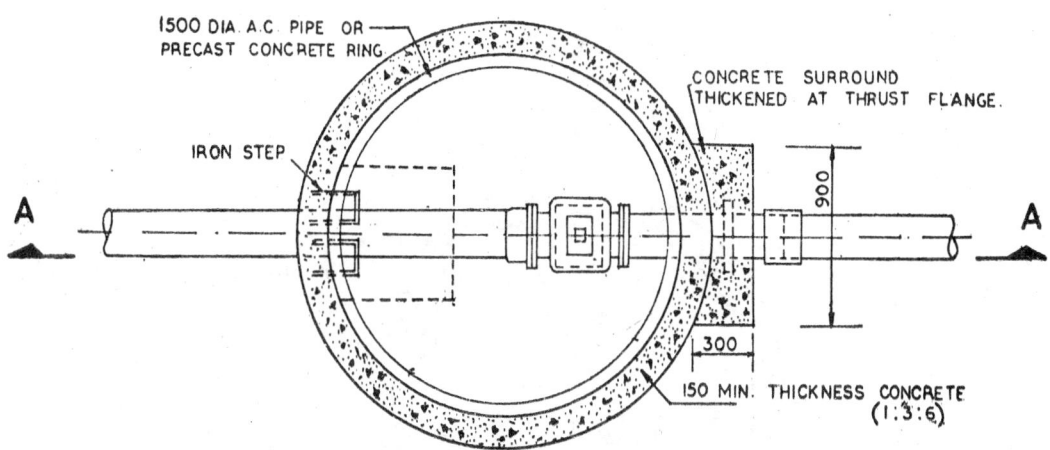

1500 DIA. A.C. PIPE OR PRECAST CONCRETE RING

CONCRETE SURROUND THICKENED AT THRUST FLANGE.

IRON STEP

150 MIN. THICKNESS CONCRETE (1:3:6)

PLAN B B

GATE VALVE CHAMBER
FOR 80-200 DIA. MAIN

Fig. 23·9.

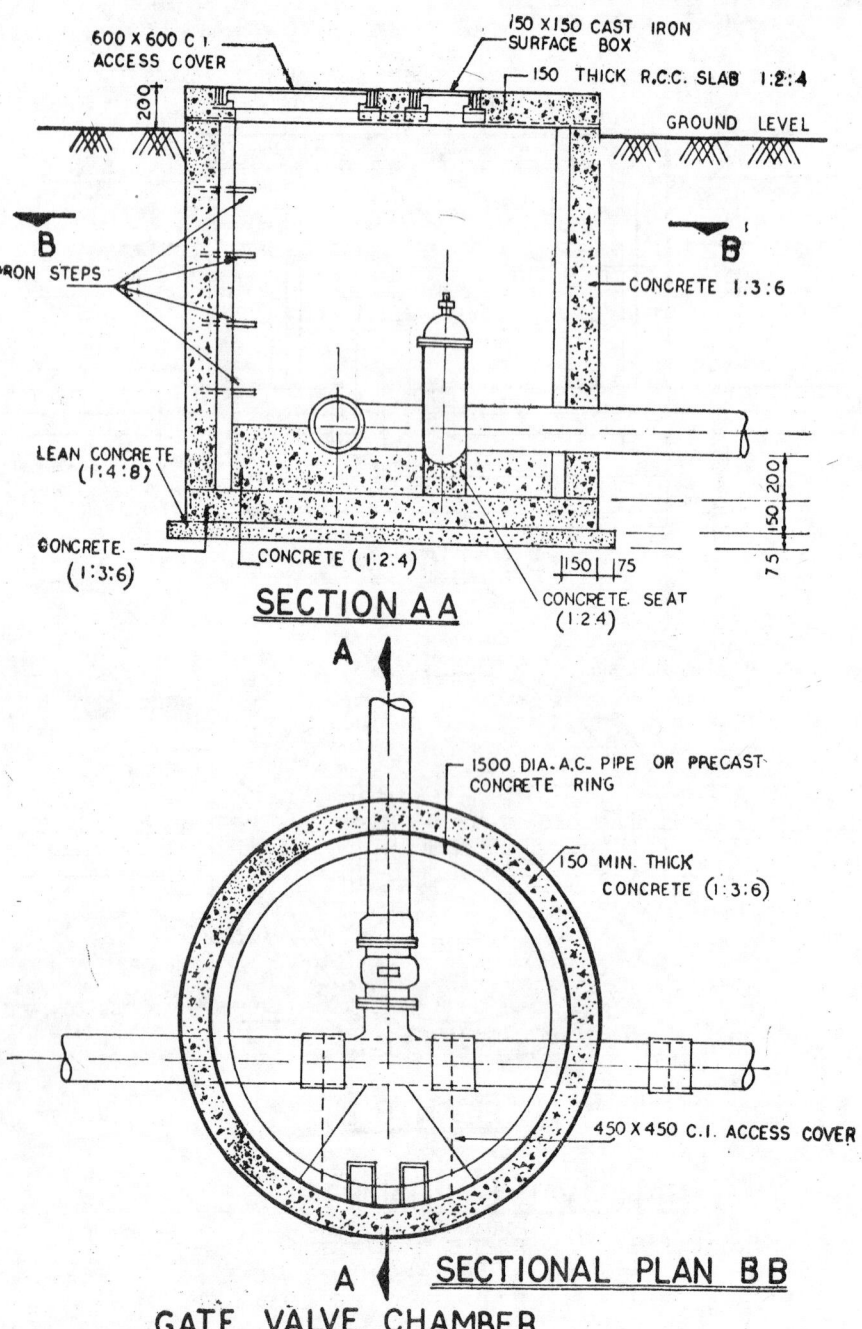

SECTION AA

SECTIONAL PLAN BB
GATE VALVE CHAMBER

Fig. 23·10.

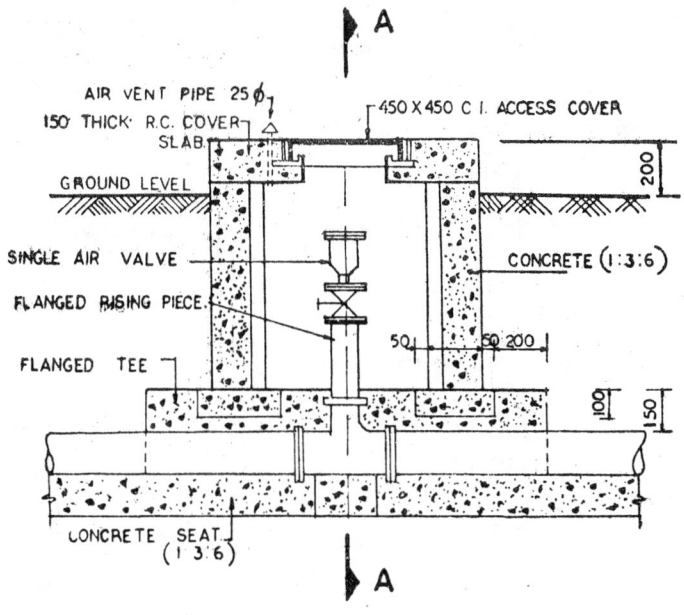

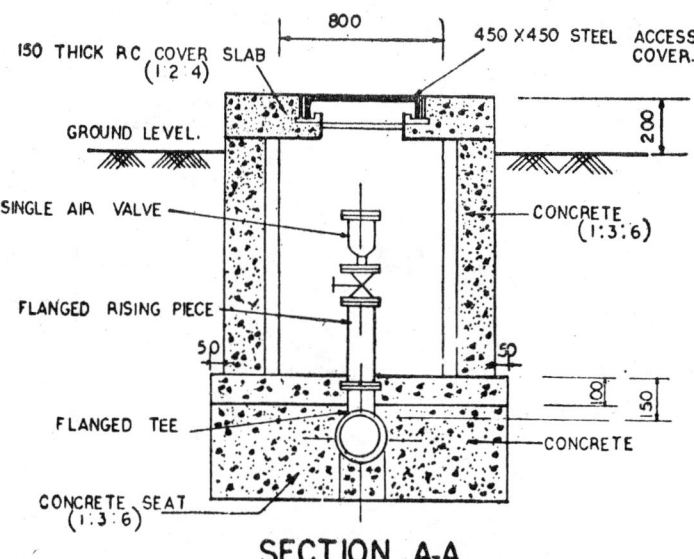

SECTION A-A

SINGLE AIR VALVE CHAMBER

FOR 80-150 DIA. MAIN.

Fig. 25·11.

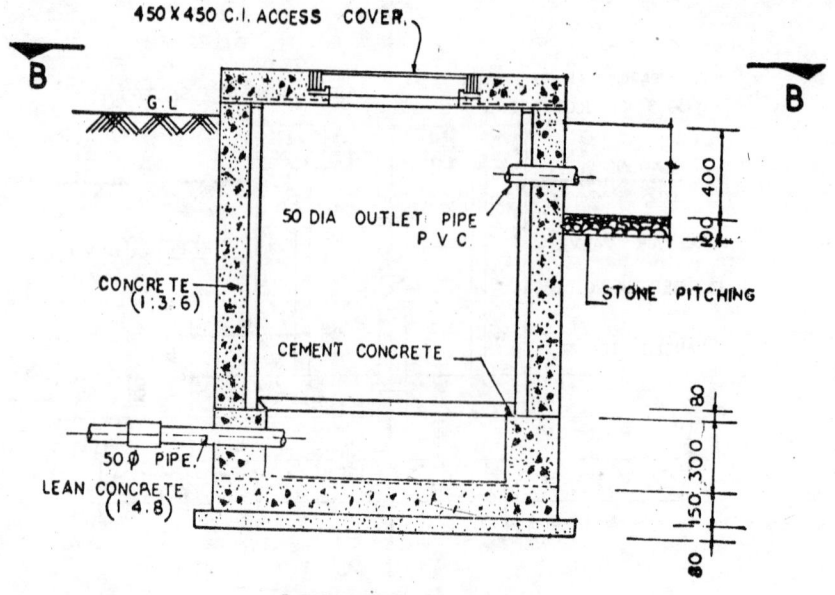

450 X 450 C.I. ACCESS COVER.

B

G.L

50 DIA OUTLET PIPE
P.V.C.

CONCRETE
(1:3:6)

CEMENT CONCRETE

STONE PITCHING

400

50 ø PIPE.

LEAN CONCRETE
(1:4:8)

80
300
150
80

B

SECTION A-A

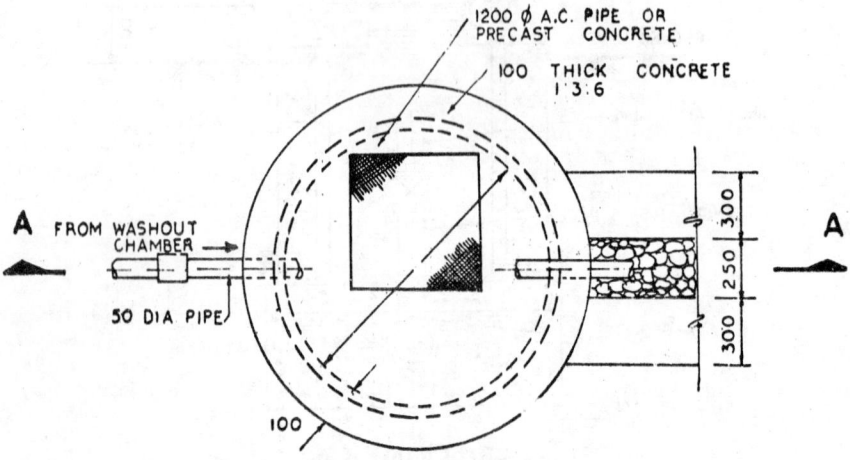

1200 ø A.C. PIPE OR
PRECAST CONCRETE

100 THICK CONCRETE
1:3:6

A   FROM WASHOUT
CHAMBER

50 DIA PIPE

100

300
250
300

A

PLAN B-B

WASH OUT SUMP

Fig. 23·12.

24

Sewer Appurtenances

Sewer appurtenances are the structures and devices related to a sewerline. These are mostly street inlets with gully pits, intercepting chamber, connection pit, manhole, lamphole, flush tank, inverted siphon, etc.

Street inlets are openings on the street gutter or curb designed and spaced at an interval to permit the entry of storm water into the sewer. Thus, street inlets are either gutter inlet or curb inlet, Openings between bars in gutter bottom should retain objects of smallest dimension of 25 mm. The clear height between horizontal bars in the plane of the curb may be 150 mm. A length of 600 mm between supports for horizontal bars is recommended. No bars or grating across the gutter opening should be placed. A rectangular inlet in vertical side of gutter with bottom depressed 75 mm below the bottom of gutter and vertical height of 125 to 150 mm are quite adequate to cause no obstruction. The size of the pipe from the street inlet to the catch basin or sewer is usually 200 to 300 mm in diameter. Cast iron is most commonly used for inlet structures. Three views of a gulley pit with street inlet are shown in Fig. 24·1.

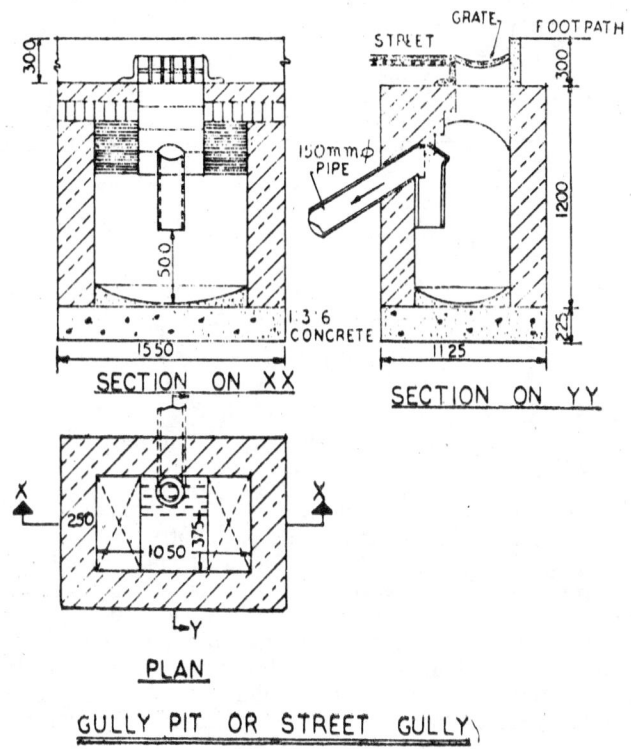

GULLY PIT OR STREET GULLY

Fig. 24·1.

Plan, sectional elevation and sectional end view of an intercepting chamber are shown in Fig. 24·2. This type of chamber was used to be constructed in earlier days for house sewer connections to the street sewer for disposal of sanitary sewage. This chamber is provided with an inscercepting trap and a ventilator pipe. The top slab of the chamber is provided with a manhole which facilitates in cleaning the house sewer in case of blockages. The trap used has a cleaning arm which remains plugged. In recent practice, direct house sewer connections are given and no such chamber is constructed.

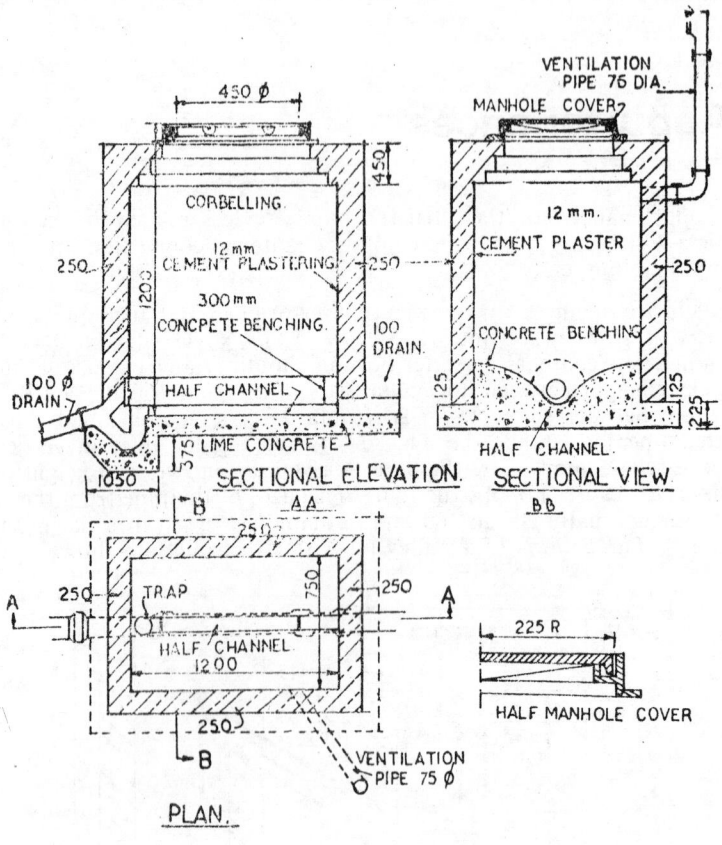

INSPECTING CHAMBER

Fig. 24·2.

Two views of a connection pit are shown in Fig. 24·3. This is also known as junction chamber. This chamber is constructed when two or more sewers from different directions meet at different levels at a point.

A manhole is the most common sewer appurtenance. This is an opening constructed in a sewerline in order to permit entry of a man into the pit for inspection and cleaning of the sewerline. Manholes are mostly square or rectangular in section. A minimum inside dimension of 120 cm × 90 cm is recommended for manholes. But, it should be adequately spacious for easy inspection and cleaning operation without difficulty. For entry of a workman with cleaning equipment into the pit, a minimum clear opening of 500 mm is recommended. Manholes are usually constructed directly over the centreline of the sewer. But, for large sewers the manholes are constructed at a tangent to the side of a sewer for better accessibility. For getting down into the pit, suitable steps preferrably of malleable cast iron bar of U-shape should be firmly set into the wall with a projection of 300 mm and spaced at about 375 mm vertically. The sidewalls are usually made of cement brick work of minimum 250 mm thickness. The inside of the brickwork should be plastered (20 mm. thick) with 1 : 3 cement mortar. Manholes should be spaced at 30 m length for sewers upto 300 mm in diameter and at about 100 m length for sewers above 300 mm diameter.

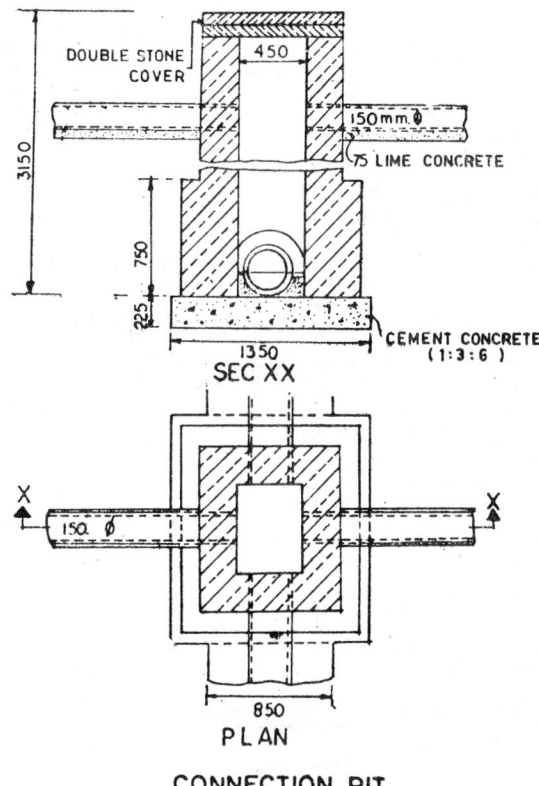

CONNECTION PIT

Fig. 24·3.

A rough expression for the thickness of the walls of a brick manhole more than 12 ft. (3·6 m) deep is given as $t = 2 + \dfrac{D}{2}$; where t is the thickness of wall in inches and D is the depth of manhole in ft. The floor of the manhole is sloped toward the centre, so that the sewage flows in a half round or U-shaped channel. The depth of flow through channel is sometimes made equal to the full diameter of the sewer. The floor should have a pitch towards the sewer of about 1 vertical to 10 horizontal.

Drop manholes are provided when the difference in elevation of the invert levels of the incoming and outgoing sewers of a manhole is more than 600 mm. A vertical drop arrangement with a 90° bend or a double tee junction may be provided. Two views of a drop manhole are shown in Fig. 24·5.

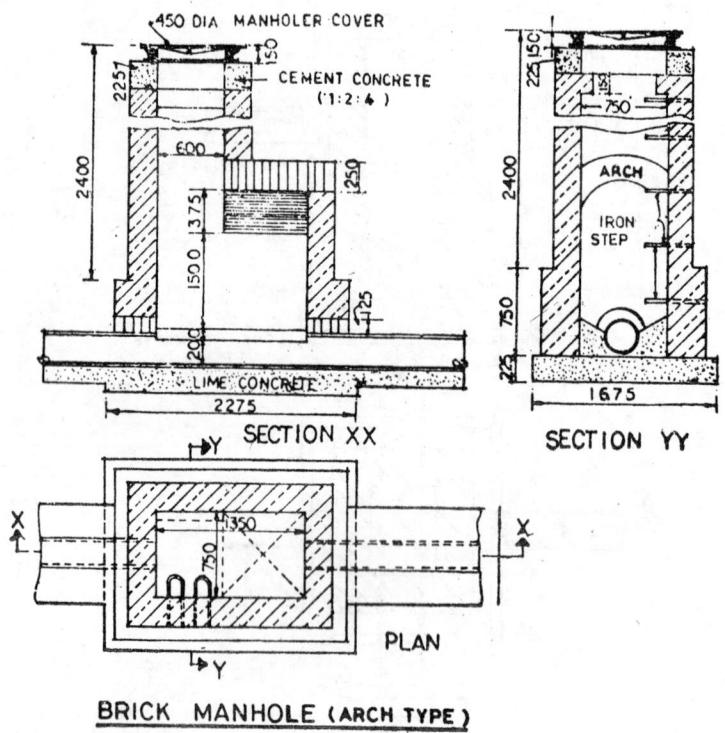

SECTION XX   SECTION YY

PLAN

BRICK MANHOLE (ARCH TYPE)

Fig. 24·4.

The manhole cover and grating frame may be of cast iron or reinforced cement concrete. The manhole should not be less than 160 mm. thick and should be set confirming accurately to the grade of the cavement. Heavy R.C.C. covers with suitable lifting arrangement can also be used in place of C.I. manhole covers.

Plan and sectional views of a single grating frame and a double grating frame for street gulley are shown in Fig. 24·6 and 24·7 respectively.

Gulley pits under carriageway and under footpath are shown in Fig. 24·8 to 24·17.

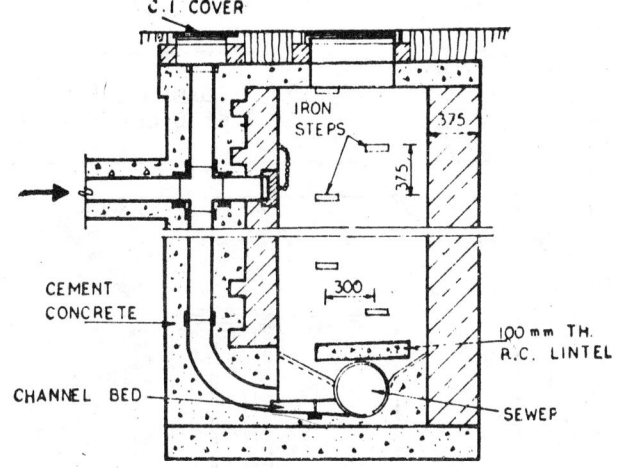

SEC.- A A

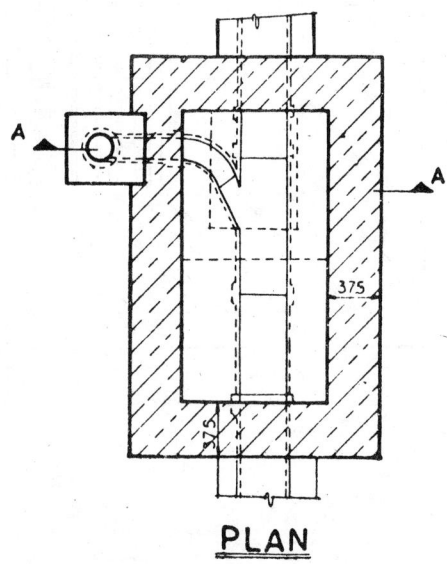

PLAN

DROP MANHOLE
Fig. 24·5.

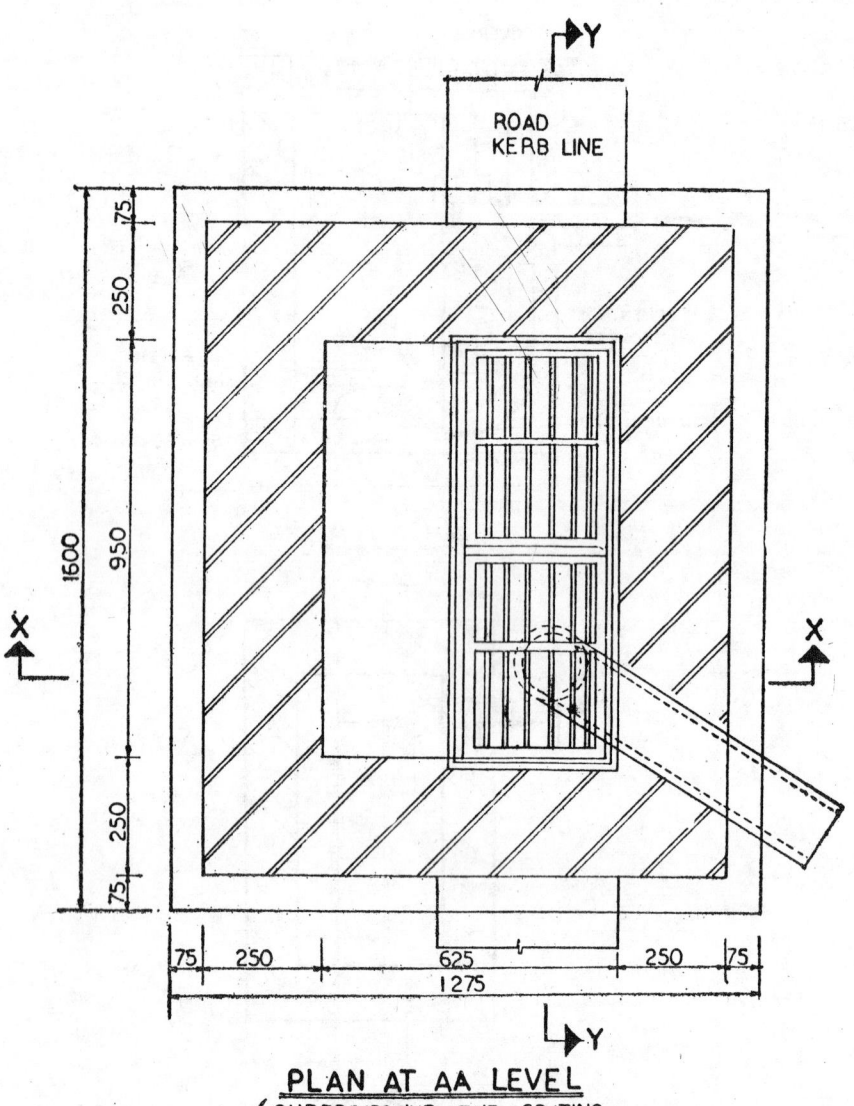

PLAN AT AA LEVEL
(SUPERIMPOSING THE GRATING)

GULLY PIT.

Fig. 42·8.

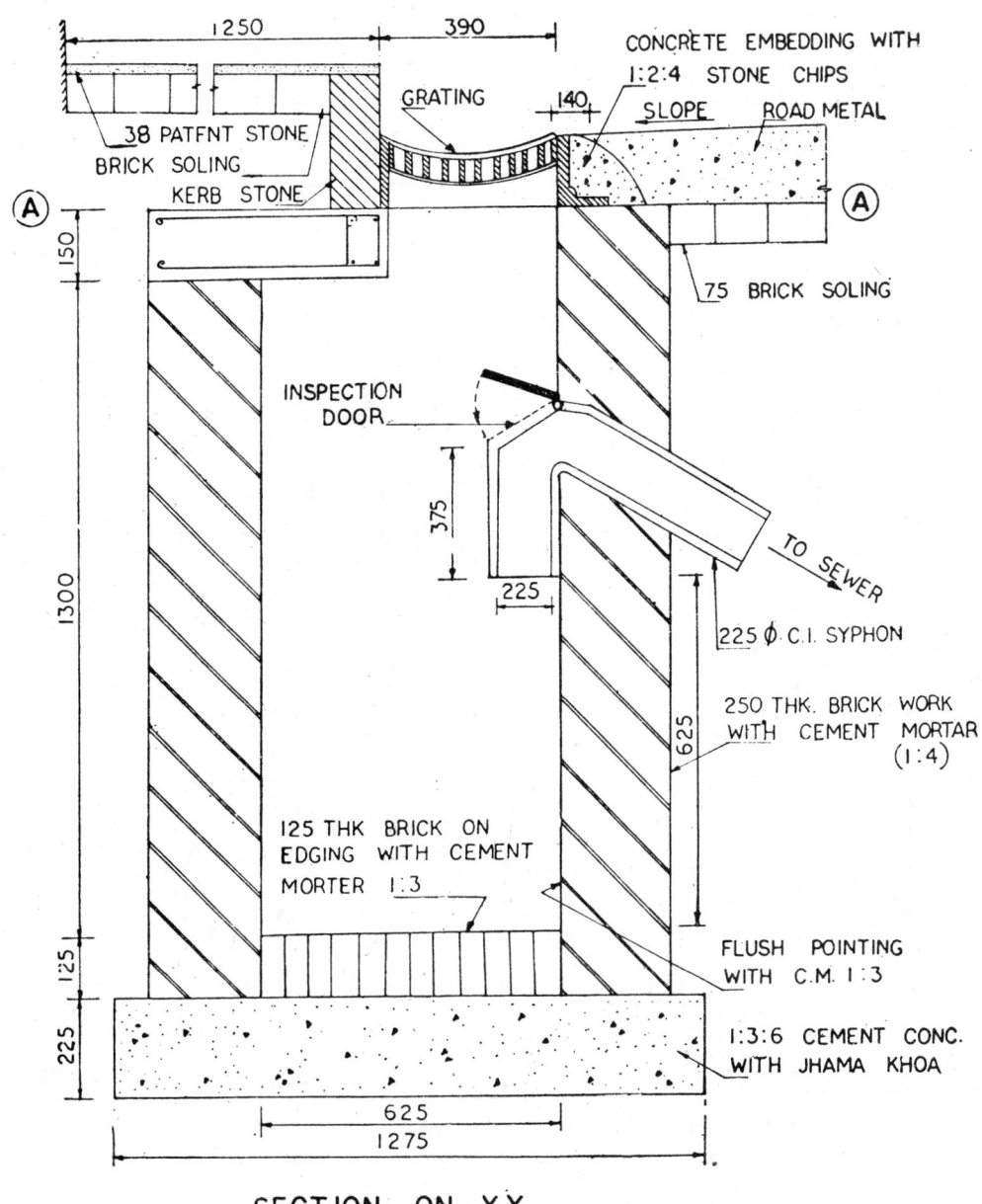

1250
390
CONCRETE EMBEDDING WITH
1:2:4 STONE CHIPS
GRATING
140
SLOPE
ROAD METAL
38 PATENT STONE
BRICK SOLING
KERB STONE
A
A
150
75 BRICK SOLING
INSPECTION DOOR
375
225
TO SEWER
225 Ø C.I. SYPHON
1300
625
250 THK. BRICK WORK
WITH CEMENT MORTAR
(1:4)
125 THK BRICK ON
EDGING WITH CEMENT
MORTER 1:3
125
FLUSH POINTING
WITH C.M. 1:3
225
1:3:6 CEMENT CONC.
WITH JHAMA KHOA
625
1275

SECTION ON - XX

Fig. 24·9.

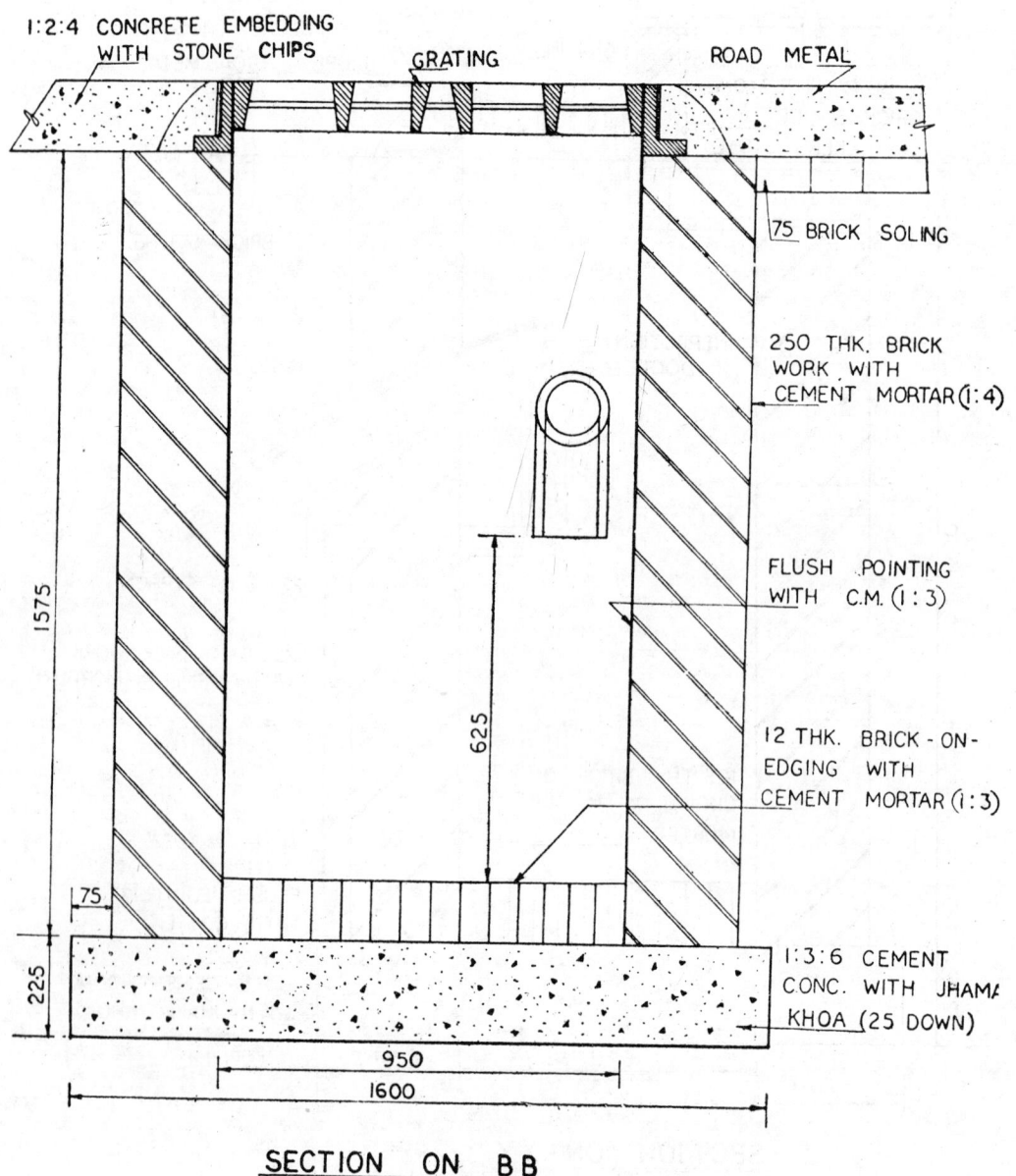

1:2:4 CONCRETE EMBEDDING WITH STONE CHIPS

GRATING

ROAD METAL

75 BRICK SOLING

250 THK. BRICK WORK WITH CEMENT MORTAR (1:4)

FLUSH POINTING WITH C.M. (1:3)

12 THK. BRICK-ON-EDGING WITH CEMENT MORTAR (1:3)

1:3:6 CEMENT CONC. WITH JHAMA KHOA (25 DOWN)

1575

625

75

225

950

1600

SECTION ON BB

Fig. 24·10.

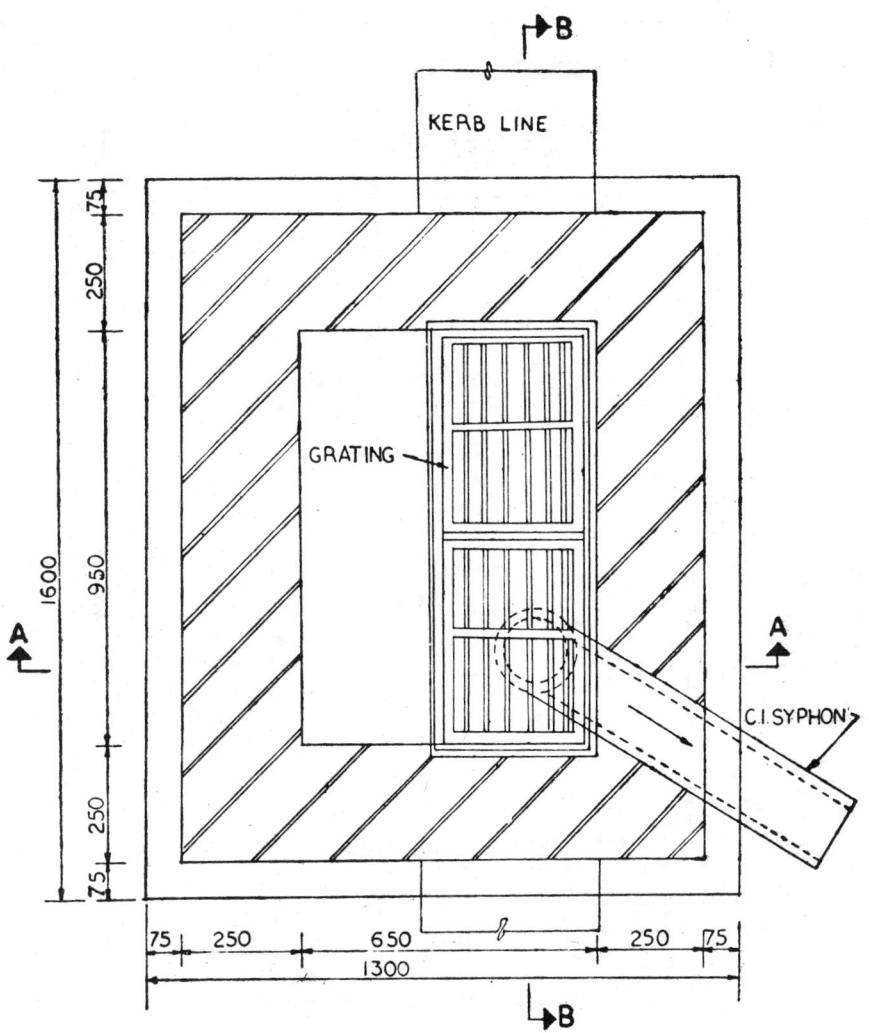

PLAN AT XX LEVEL
(SUPERIMPOSING THE GRATING)

A TYPICAL GULLY PIT

Fig. 24·13.

UNDER FOOTPATH.

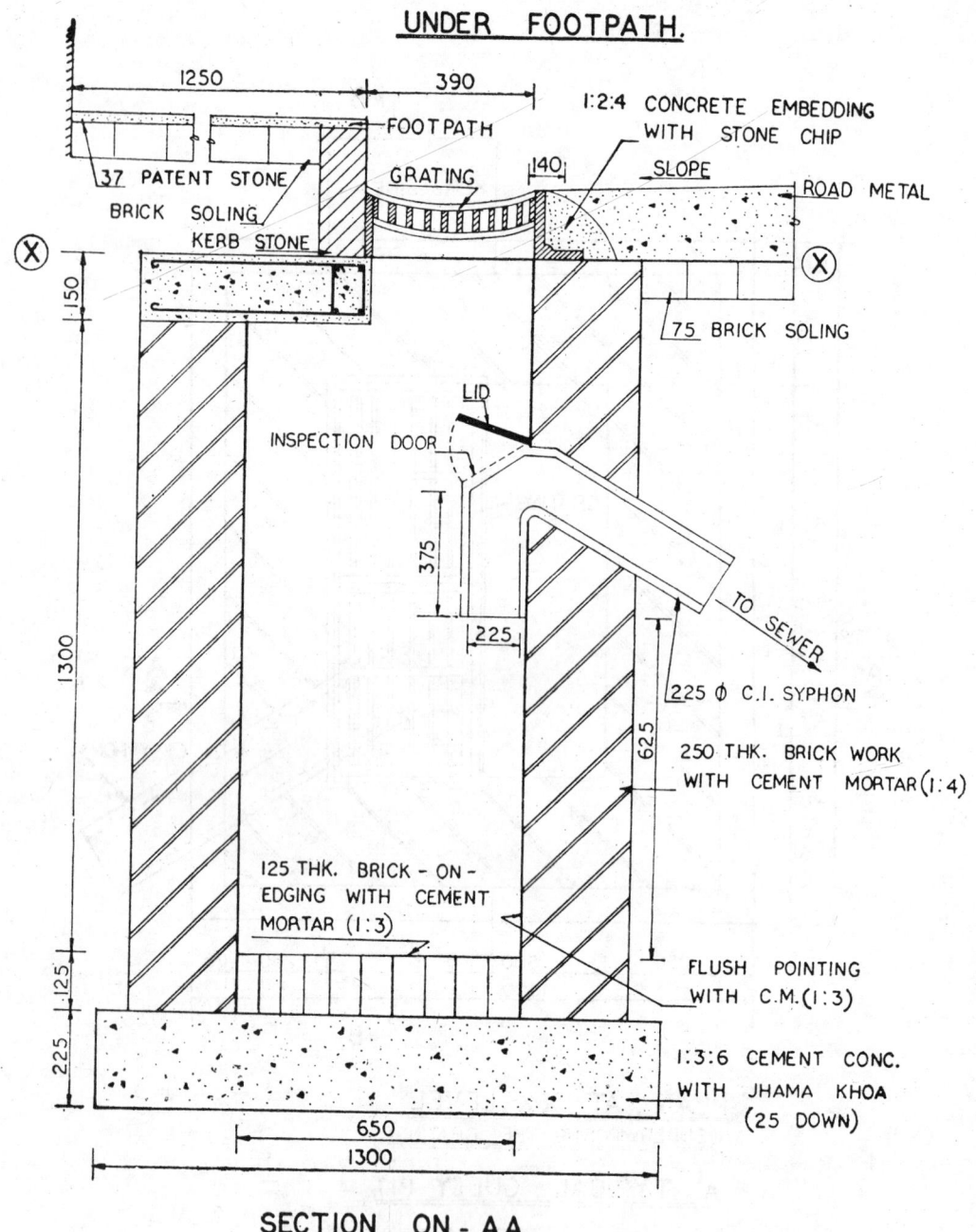

1250

390

1:2:4 CONCRETE EMBEDDING WITH STONE CHIP

FOOTPATH

GRATING

SLOPE

140

ROAD METAL

37 PATENT STONE

BRICK SOLING

KERB STONE

75 BRICK SOLING

150

LID

INSPECTION DOOR

375

TO SEWER

225

225 Ø C.I. SYPHON

1300

625

250 THK. BRICK WORK WITH CEMENT MORTAR (1:4)

125 THK. BRICK - ON - EDGING WITH CEMENT MORTAR (1:3)

FLUSH POINTING WITH C.M.(1:3)

125

225

1:3:6 CEMENT CONC. WITH JHAMA KHOA (25 DOWN)

650

1300

SECTION ON - AA

Fig. 24·14.

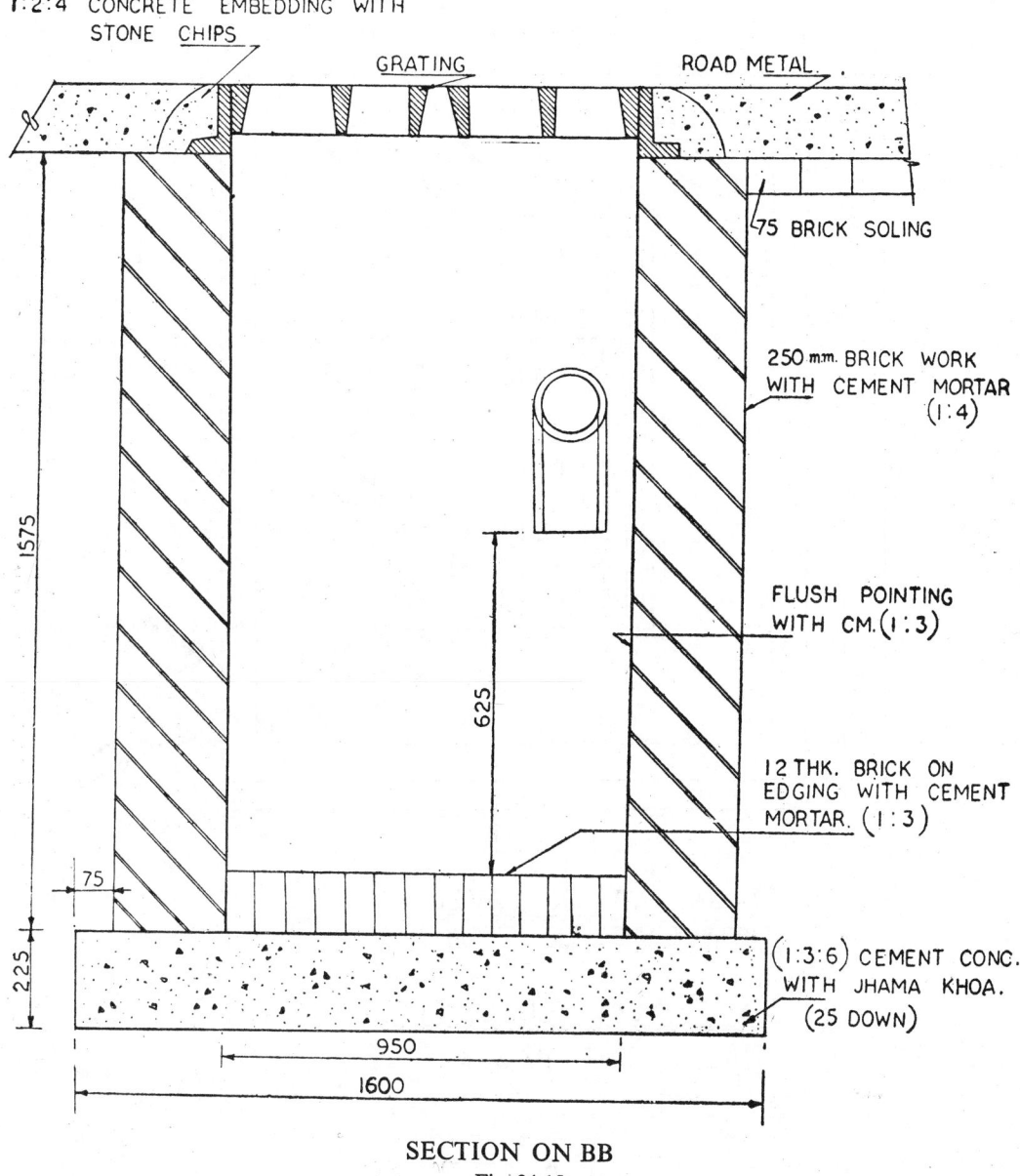

1:2:4 CONCRETE EMBEDDING WITH STONE CHIPS

GRATING

ROAD METAL

75 BRICK SOLING

250 mm. BRICK WORK WITH CEMENT MORTAR (1:4)

FLUSH POINTING WITH CM. (1:3)

12 THK. BRICK ON EDGING WITH CEMENT MORTAR. (1:3)

(1:3:6) CEMENT CONC. WITH JHAMA KHOA. (25 DOWN)

1575

625

75

225

950

1600

SECTION ON BB
Fig. 24·15.

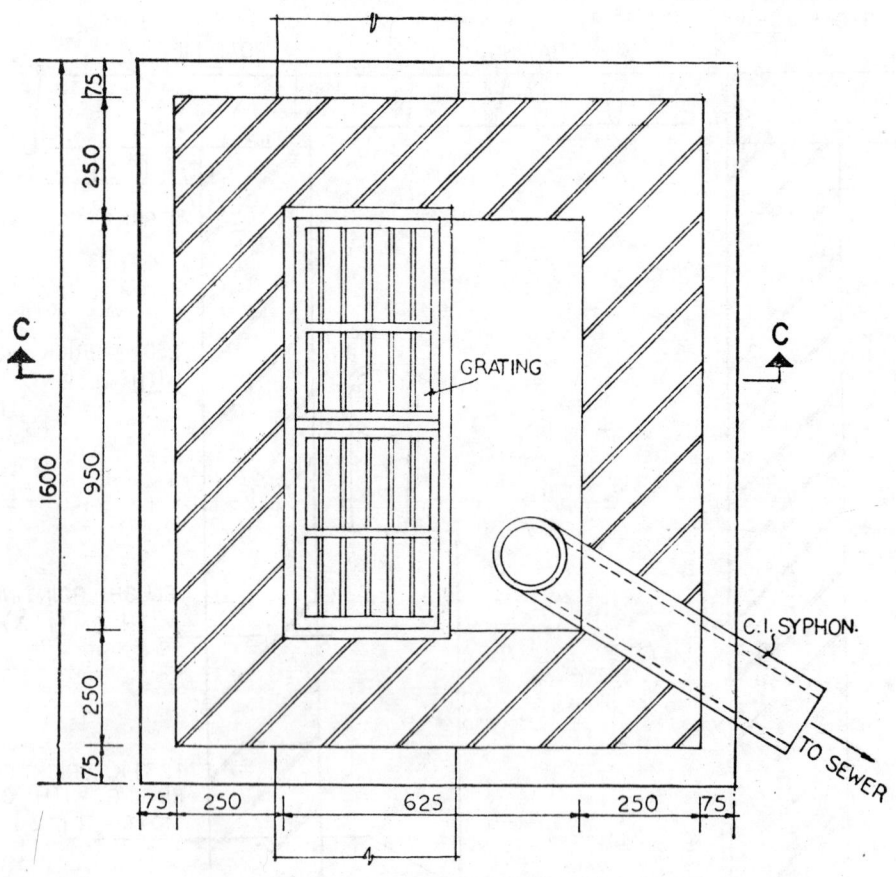

GRATING

C.I. SYPHON.

TO SEWER

75
250
1600   950
250
75

75   250   625   250   75

PLAN AT XX LEVEL
(SUPERIMPOSING THE GRATING.)

A TYPICAL GULLY PIT

Fig. 24·16.

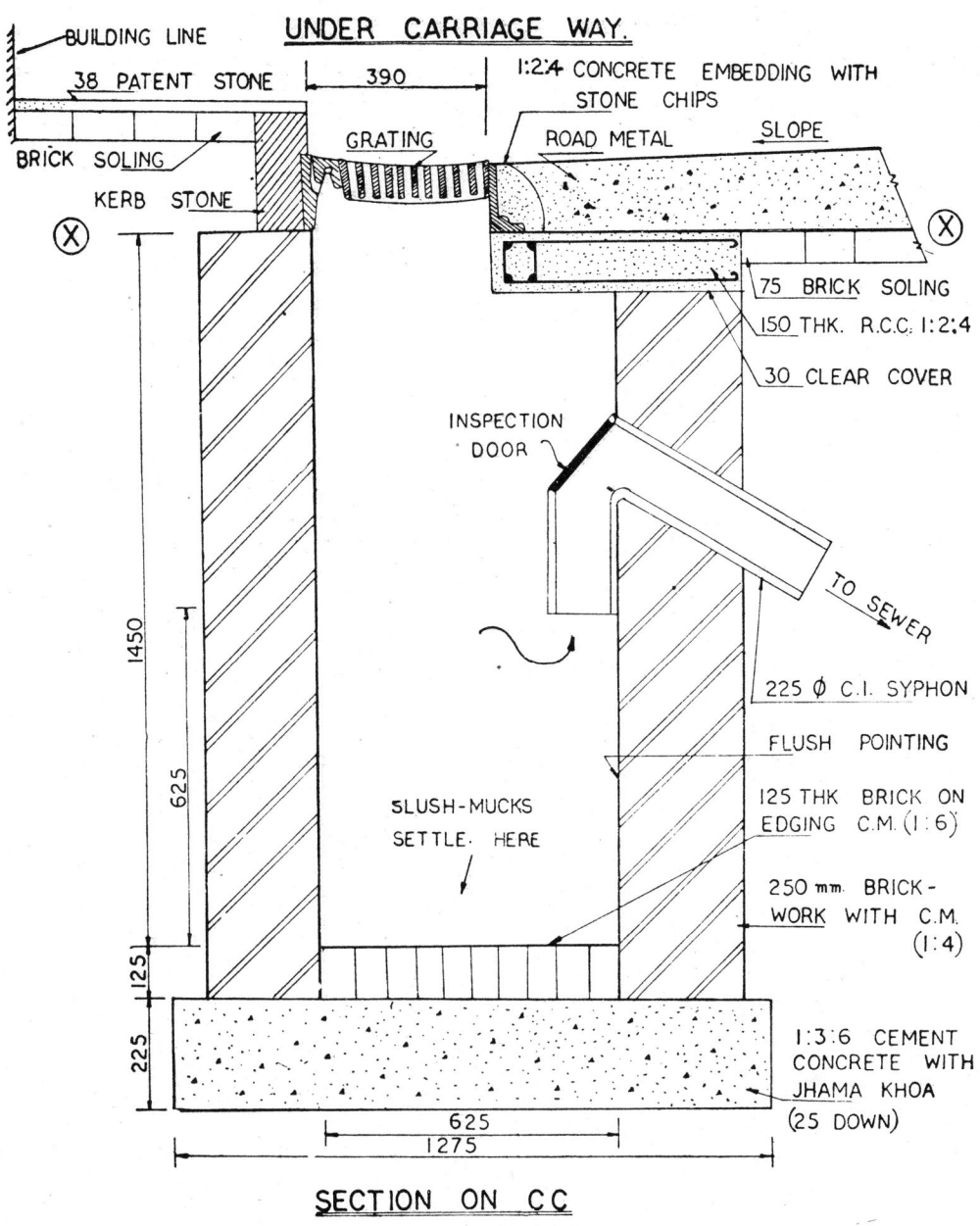

UNDER CARRIAGE WAY.

BUILDING LINE

38 PATENT STONE

390

1:2:4 CONCRETE EMBEDDING WITH STONE CHIPS

BRICK SOLING

GRATING

ROAD METAL

SLOPE

KERB STONE

Ⓧ

Ⓧ

75 BRICK SOLING

150 THK. R.C.C. 1:2:4

30 CLEAR COVER

INSPECTION DOOR

TO SEWER

1450

225 ⌀ C.I. SYPHON

FLUSH POINTING

625

SLUSH-MUCKS SETTLE HERE

125 THK. BRICK ON EDGING C.M. (1:6)

250 mm. BRICK-WORK WITH C.M. (1:4)

125

225

1:3:6 CEMENT CONCRETE WITH JHAMA KHOA (25 DOWN)

625

1275

SECTION ON CC

Fig. 24·17.

25

Fire Places and Chimneys

In regions of extreme cold climate, provision of fire places is made in rooms of a building. In large buildings, especially in public buildings, provision of central heating arrangement is made. Fire places are constructed for heating the rooms. Usually coal or wood is used for firing. In order to prevent spreading of smokes in the room, a vertical duct is constructed above the fire place which drives out the smokes and gases from the room. An

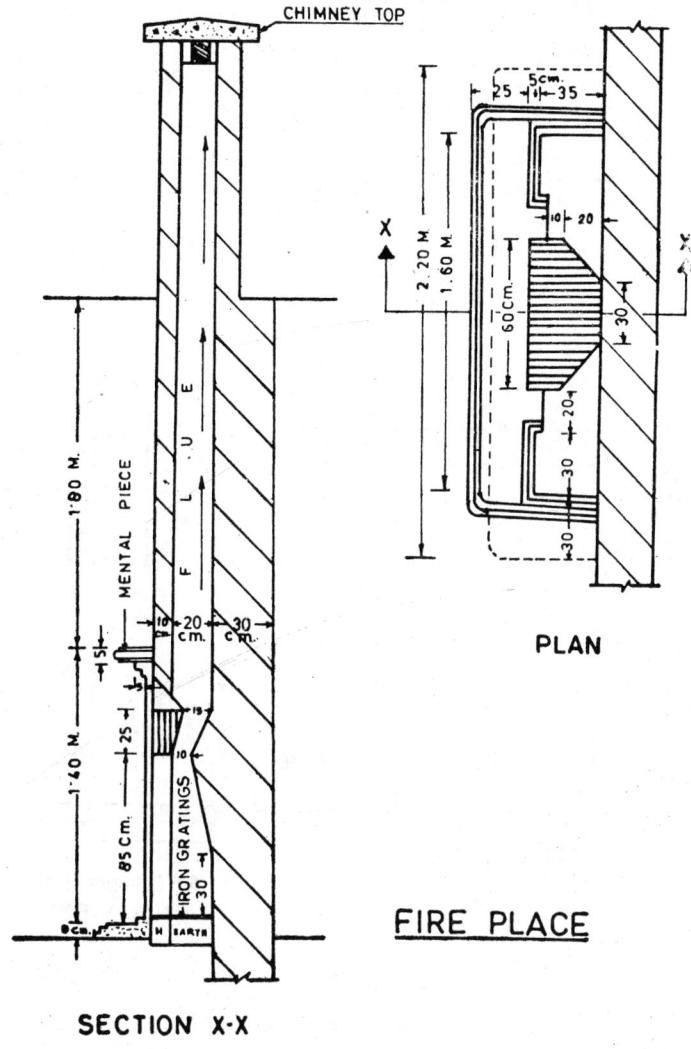

PLAN

FIRE PLACE

SECTION X-X

Fig. 25·1.

iron grating is provided at the fire place for holding the solid fuel. The side walls and the back wall of a fire place should be solid masonry and it is perferred to have fire brick lining. The thickness of these walls should be 20 cm minimum and the hearth should project inside the room by about 50 cm from the chimney. The throat of the fire place should be 20 cm and in any case it should not be less than 10 cm. The plan, elevation and section view of typical fire places are shown in Fig. 25.1 and 25.2.

Cooking ranges are constructed in kitchens. This is similar to fire places. A cooking range essentially consists of a hearth with ash removal arrangemen, a space for coal storage below the cooking platform and a chimney above the hearth for removal of smokes and gases. The plan, elevation and section views of cooking ranges are shawn in Fig. 25.3 and 25.4

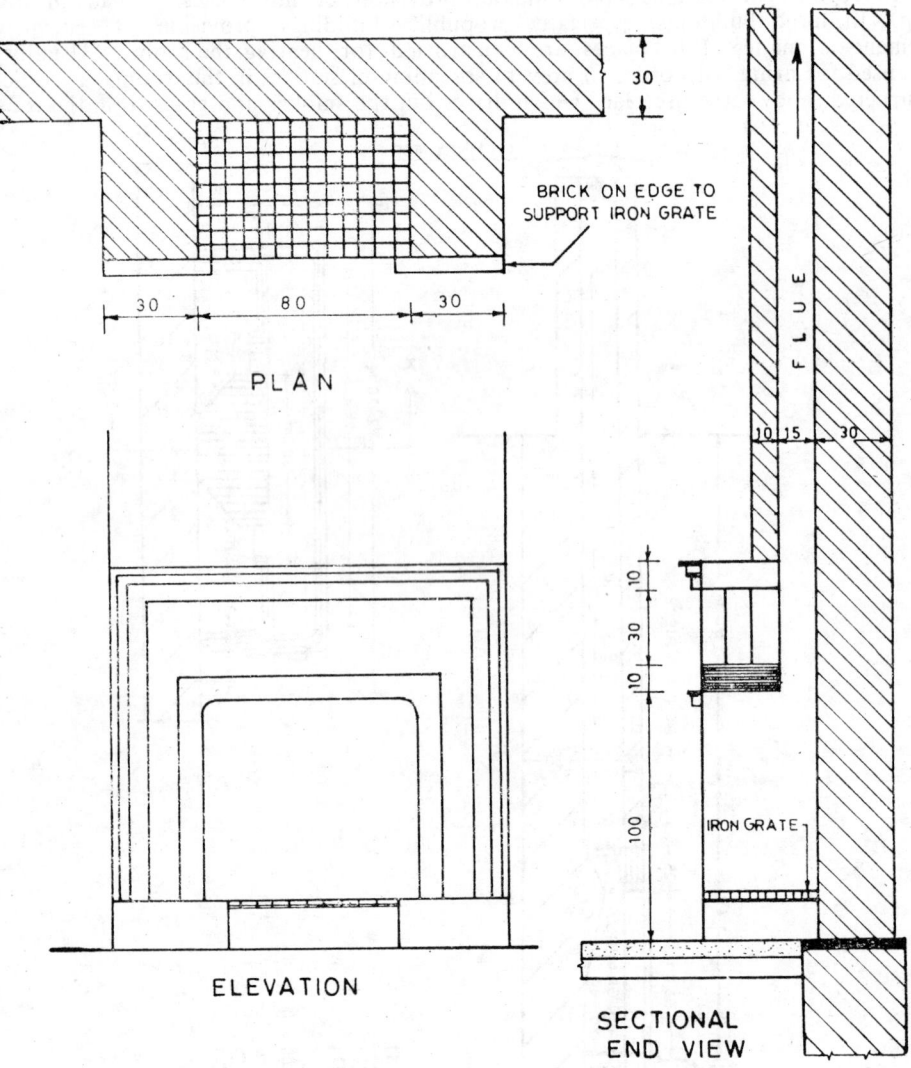

Fig. 25·2.

Construction of chimneys is essential where there are fire places or hearths inside a building or in a factory shade, in order to avoid smoke nuissance. For the removal of smokes from fire places and hearths in a building, chimneys are csnstructed as masonry or concrete ducts. For removal of large quantum of smokes from fire places, hearths and furnaces in factories round the clock, large and fall chimneys are built. These are usually built with iron sheets or are masonry built. Fig. 25'5 shows the sectional view of a 60 M high brick chimney. Fig. 25'6 shows the sectional view of a chimney built with iron sheets. Such a chimney needs support against the action of wind load and for this guy ropes are used as shown in figure.

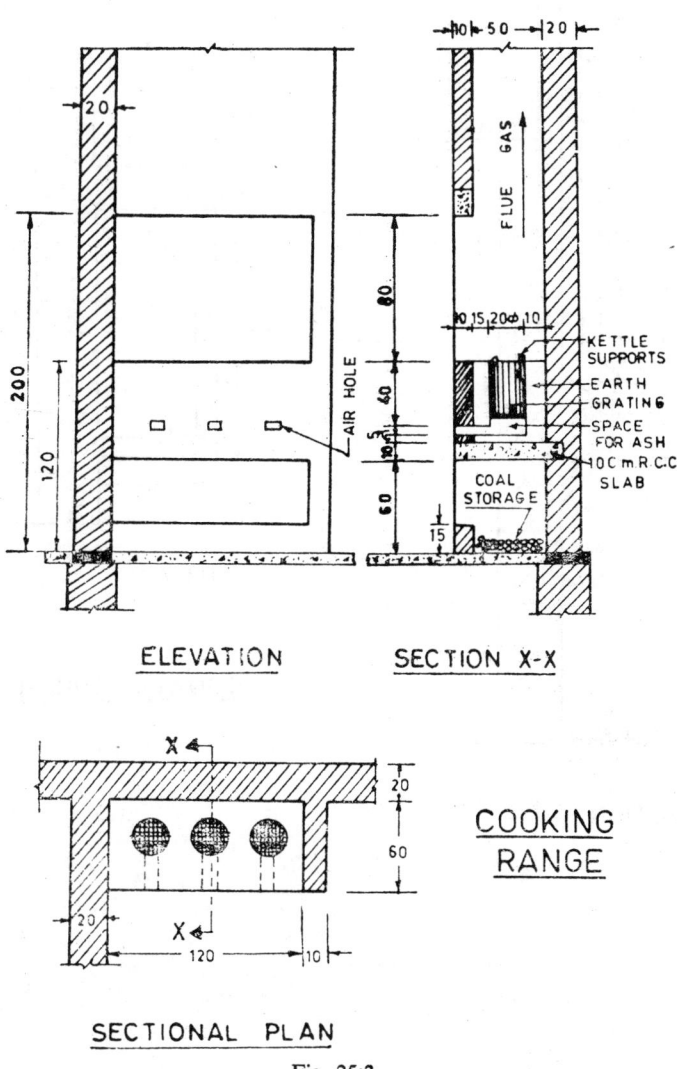

ELEVATION   SECTION X-X

SECTIONAL PLAN

COOKING RANGE

Fig. 25·3.

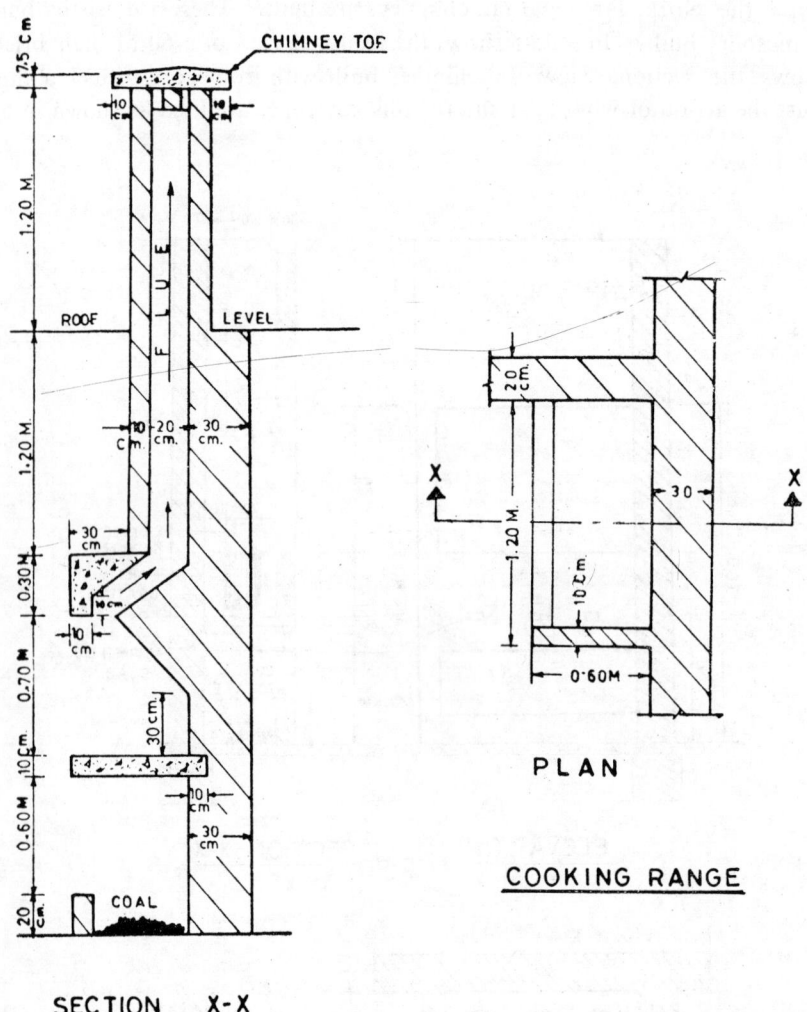

SECTION   X-X

PLAN

COOKING RANGE

Fig. 25·4.

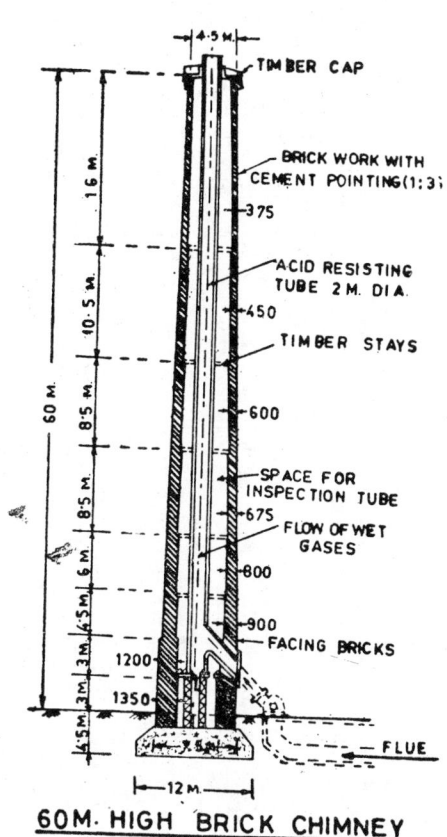

Fig. 25·5.

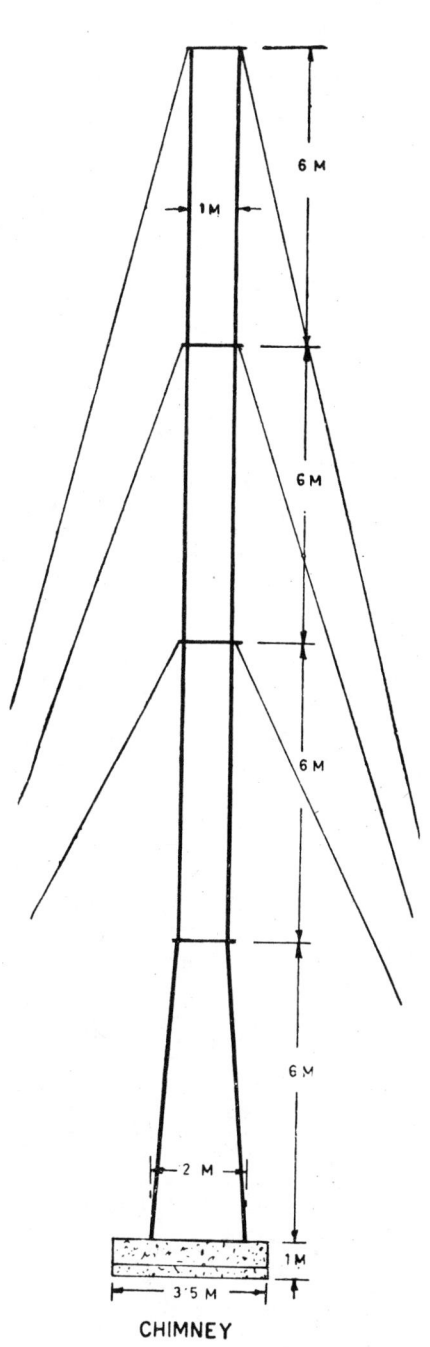

CHIMNEY

Fig. 25·6.

26

Swimming Pools and Water Tanks

Swimming Pools

A swimming pool is a pucca water tank with all sorts of facilities required for the purpose of swimming and diving. Adequate supply of water with filtration, chlorination and pumping arrangement should be ensured in a swimming pool. Filtration, chlorination and pumping become essential where it is required to recirculate the pool water instead of wasting an appreciable quantum of water. However, drainage of overflow water from swimming pool and wash water from bath house should be made properly. The bath house should be provided with showers, lockers for keeping dresses and lavatory. The same swimming pool may be used by both male and female members at different times and for which separate bath house should be constructed in a swimming pool.

The diving end of a swimming pool has a greater depth of water than the other end. The bed of the pool should have a gentle slope of 1 in 15 or 1 in 20 starting from the shallow end. The diving end will have a greater bed slope. The minimum free board should be kept as 300 mm.

The dimensions for learner's swimming pool are given below :

Length in m	Breadth in m	Depth in mm	
		At Deep end	At Shallow end
12 to 16	7 to 9	1·70	1·00
20	7·5 to 9	3·00	1·00
25	9	3·50	1·00

The dimensions of swimming pools for competion purpose are :

Length in m	Breadth in m	Depth in m	
		At deep end	At shallow end
20	12·75	1·70	1·00
36	12·75	3·00	1·00
50	17·00	3·50	1·00

Design of a standard swimming pool for 100 members of a swimming club with diving arrangement and other facilities are presented here. Fig. 26˙1 shows the layout plan of the swimming pool. The line plan and the sectional view of the swimming pool are illustrated in Fig. 26˙2.

No. of Users	—100
Pool Size	—22˙5 m × 7˙5 m
Water Surface Area	—53 sq. m.
Structure	—Concrete walls and floors; walls with bluish glazed tiles finish.
Depth Marks	—Depth of water shall be marked on both sides and ends of the pool at all strategic points.
Lifeguard Chair	—One elevated lifeguard chair.
Life Line	—A life line with coloured floats at not greater than 1˙5 m. spacing shall be provided at the break of grade between the shallow and deep water.
Ladders	—Two ladders shall be provided.
Minium Depth	—Minimum depth of water in shallow area shall be 1 m.
Diving Tower	—Height of diving tower must not exceed 3 m in any case.
Walks	—Walks shall be continuous around the pool with a minimum unobstructed width of 2˙4 m. The pavement should be made of chekered tiles.
Fence	—The pool area shall be enclosed with a 2˙0 m. high wall with barbed wire fencing at top.
Overflow Gutter	—Shall be continuous around the pool. Overflow drains shall be spaced at a maximum of 4˙5 m on centres.
Maximum Pool Capacity	—2, 36, 392 litres.
Pressure Filter	—Diameter-1˙05 m., Filter Area-0˙27 sq. m. Rate of Filtration-218 1pm @ ˙65 1pm/sq. m. Backwash. Water—436 1pm @1˙30 1 pm/sq. m.
Complete turnover of pool capacity in 18 hours. Pipe sizes	—Suction line from pool 38 mm diameter. Filtered water line from filter to the overhead reservoir—38 mm diameter. Filter backwash line—50 mm diameter. Supply line (main)—62 mm diameter.

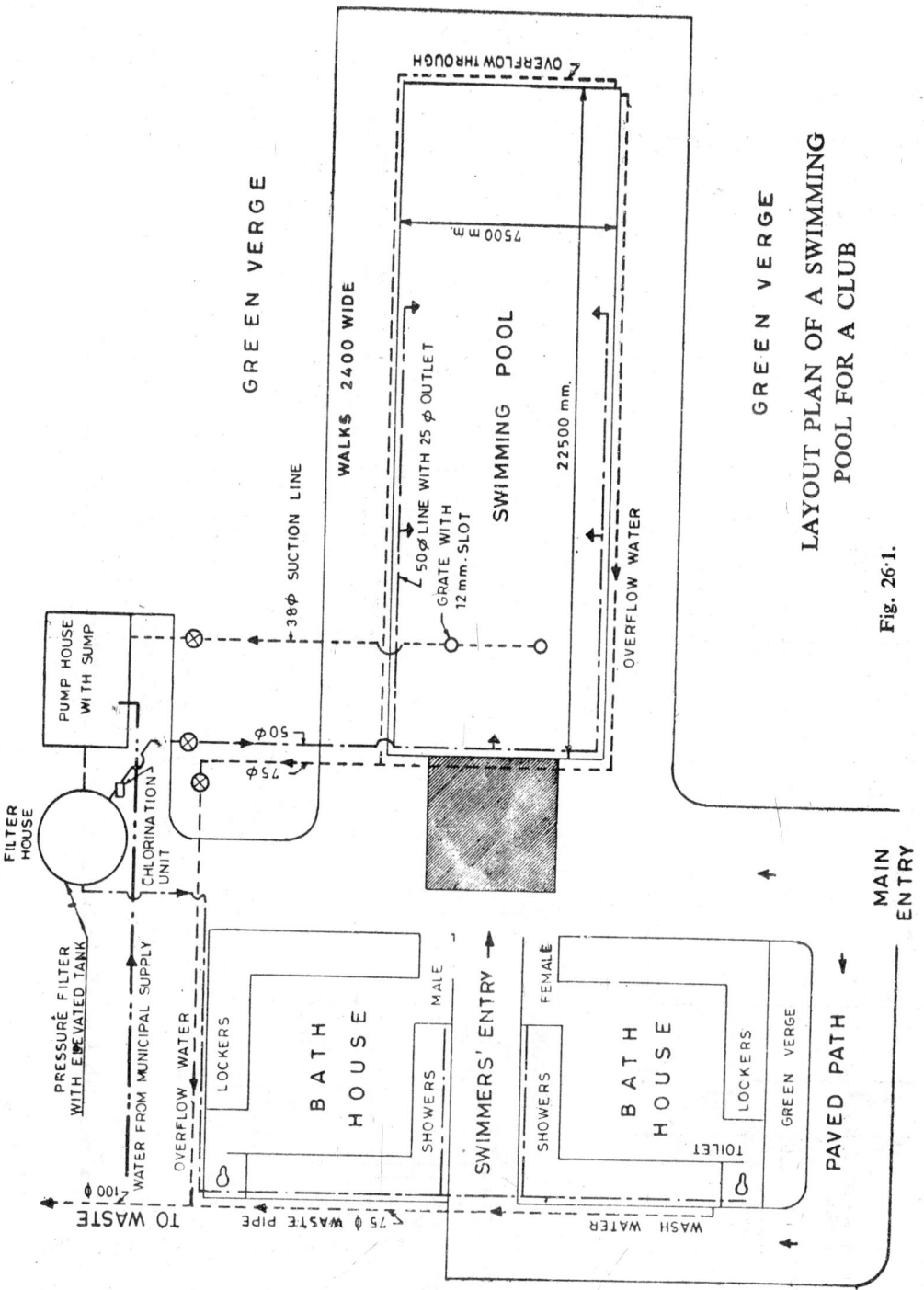

GREEN VERGE

OVERFLOW THROUGH

7500 m m

WALKS 2400 WIDE

SWIMMING POOL

22500 mm.

38φ SUCTION LINE

50φ LINE WITH 25 φ OUTLET

GRATE WITH 12 mm. SLOT

OVERFLOW WATER

GREEN VERGE

PUMP HOUSE WITH SUMP

FILTER HOUSE

CHLORINATION UNIT

50φ

75φ

PRESSURE FILTER WITH ELEVATED TANK

WATER FROM MUNICIPAL SUPPLY

OVERFLOW WATER

φ 100

TO WASTE

75φ WASTE PIPE

WASH WATER

LOCKERS

BATH HOUSE

SHOWERS

MALE

SWIMMERS' ENTRY

FEMALE

SHOWERS

BATH HOUSE

TOILET

LOCKERS

GREEN VERGE

PAVED PATH

MAIN ENTRY

LAYOUT PLAN OF A SWIMMING POOL FOR A CLUB

Fig. 26·1.

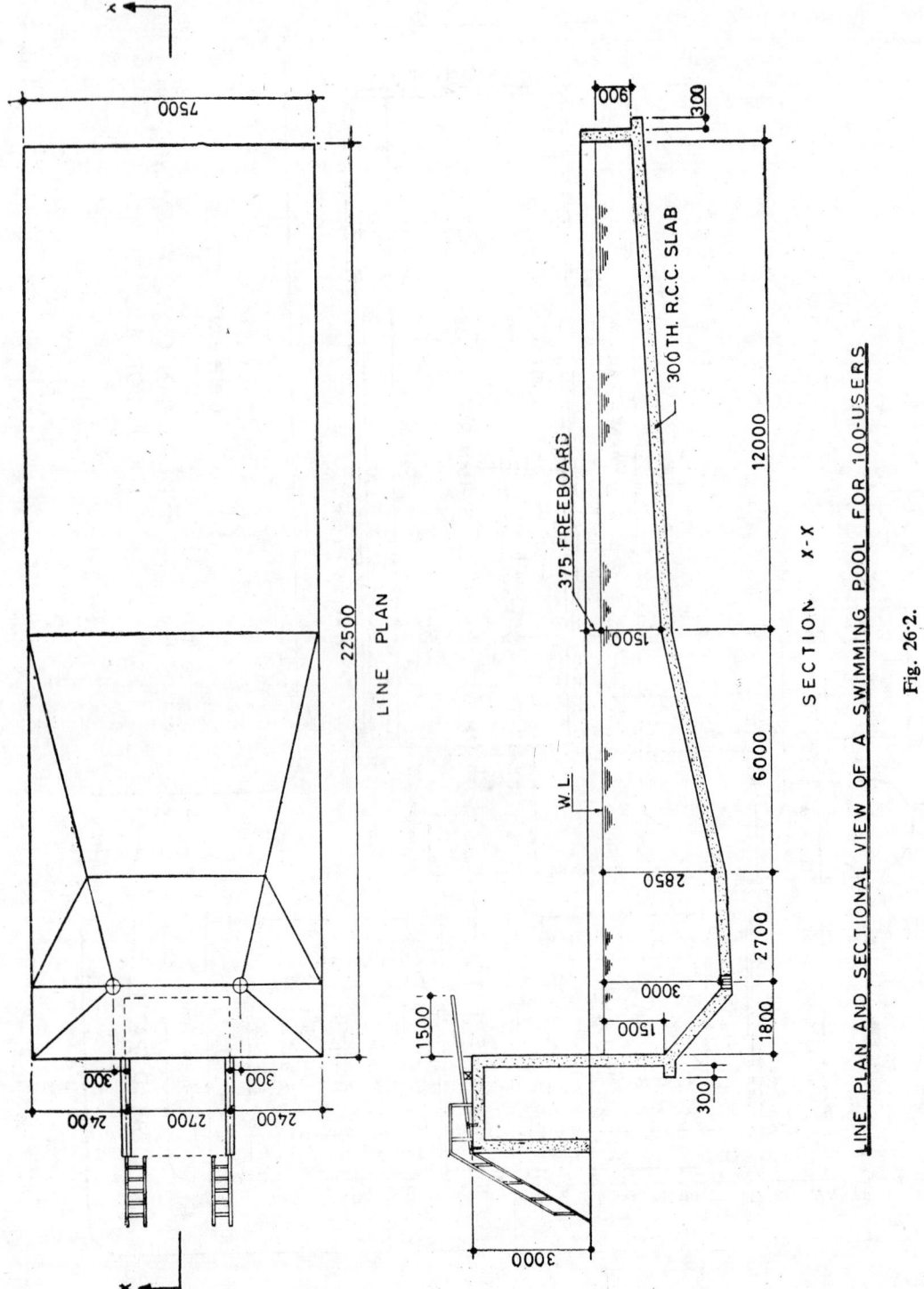

LINE PLAN AND SECTIONAL VIEW OF A SWIMMING POOL FOR 100-USERS

Fig. 26.2.

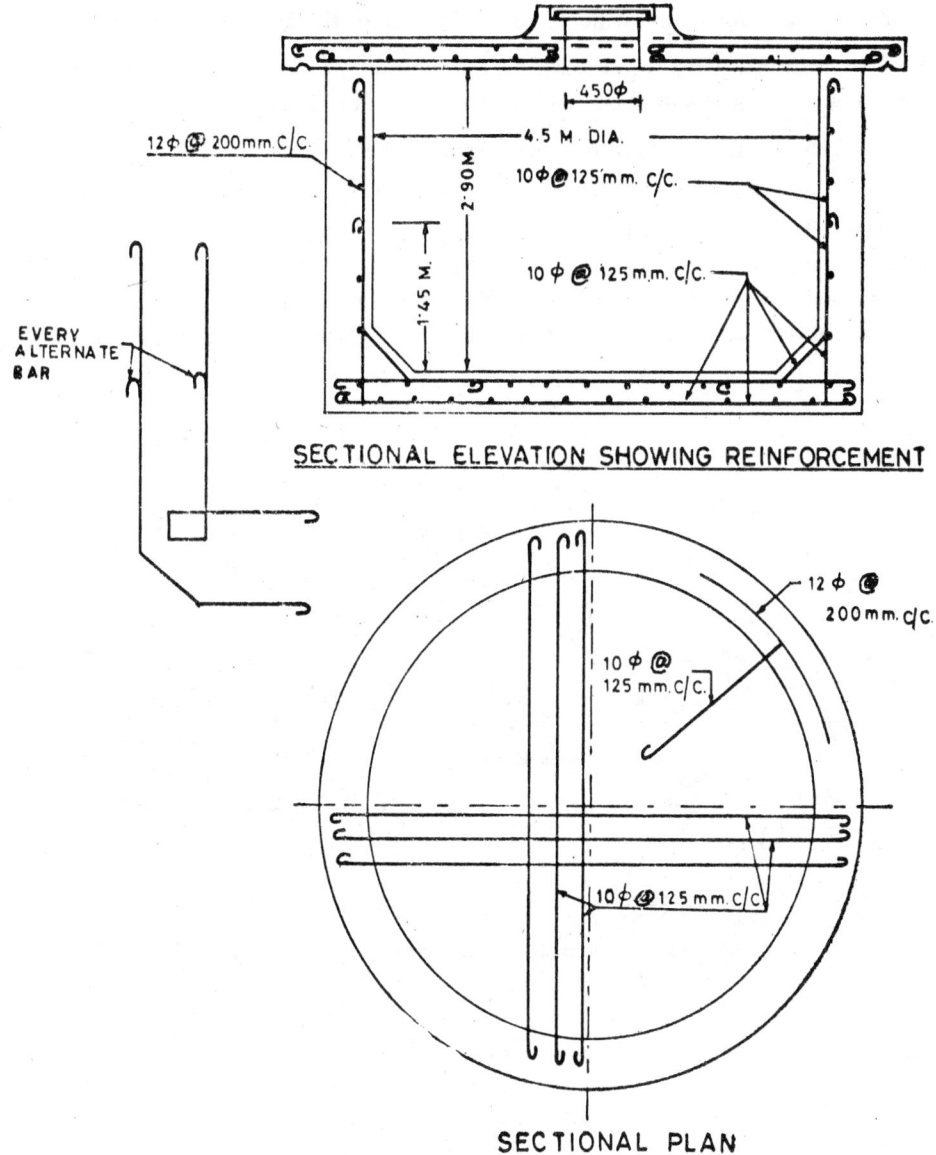

SECTIONAL ELEVATION SHOWING REINFORCEMENT

SECTIONAL PLAN

Fig. 26·3.

Water Tanks

Water tanks or reserviors are constructed for storage of water in a water supply system. These tanks are constructed either undergound or overground on a hill top or on an elevated structure. The tanks are built either of brick masonry or R.C.C When they are underground or on a hill top. Reinforced cement concrete or steel is used in case of elevated water tanks. The structures are made of steel or R.C.C. Mostly, R.C.C. elevated reserviors are constructed.

Design of a circular tank of 45460 litres (10,000 gallons) capacity is shown in Fig. 26·3. The tank is of 4·5 m diameter and 2·9 m depth. The figure shows the reinforcement of the tank in sectional plan and in sectional elevation. The top slab is provided with a manhole of 450 mm diameter for ready inspection and cleaning of the tank as and when required.

The sectional view of a clear water sump is shown in Fig. 26·4 and its part sectional view showing the reinforcement details is presented in Fig. 26·5.

The elevated reserviors of different shapes and sizes for different capacities are illustrated in Fig. 26·6.

Half-sectional elevation of a typical overhaad reservoir having a staging height of 19·5 m is shown in Fig. 26·7. The staging is made of a R.C.C. hollow circular shaft and it is provided with steel ladder and landing facility in order to have accessibility to the overhead reservoir for ready inspection and repair. The internal diameter of the hollow

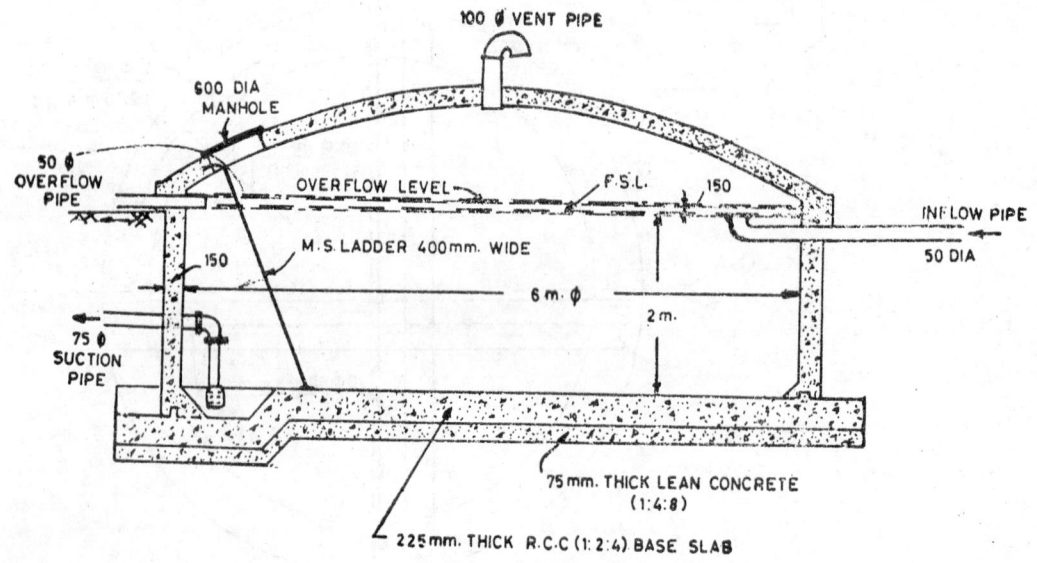

CLEAR WATER STORAGE SUMP

Fig. 26·4.

circular shaft is 5 m and the diameter of the overhead reservoir is 8 m. For other dimensions read the drawing. Fig. 26·8 shows the reinforcement details of the reservoir.

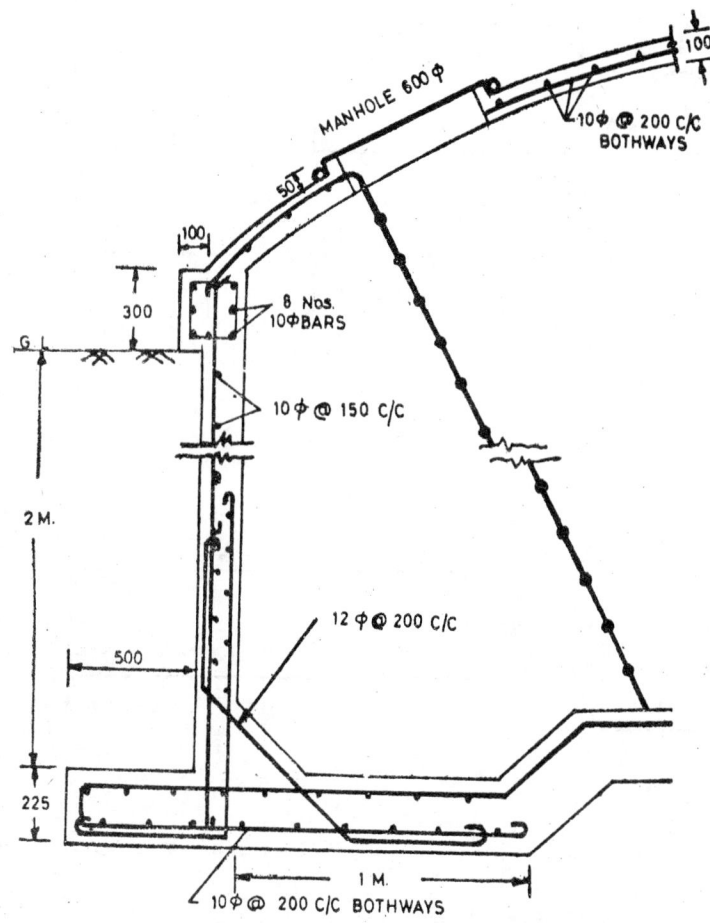

PART SECTIONAL VIEW OF SUMP
SHOWING REINFORCEMENT DETAILS

Fig. 26·5.

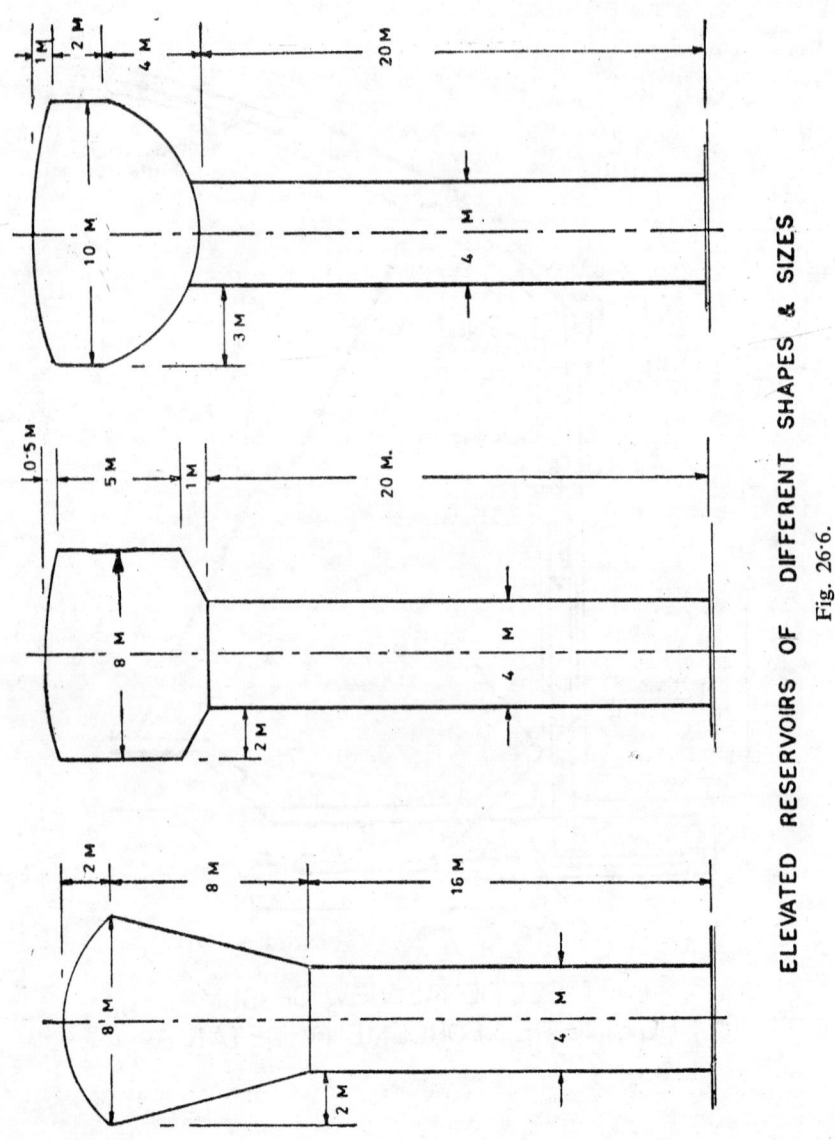

ELEVATED RESERVOIRS OF DIFFERENT SHAPES & SIZES

Fig. 26·6.

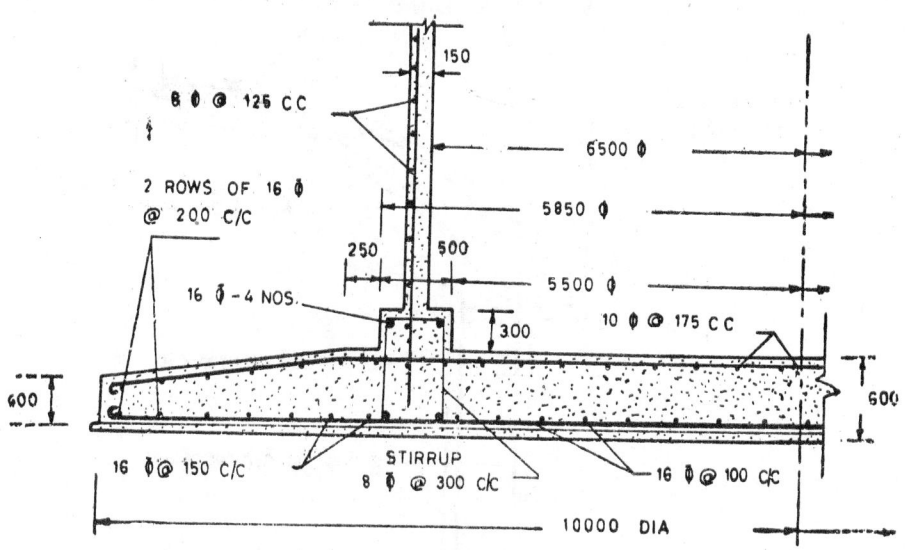

Details of Reinforcement at Base

FOUNDATION SLAB

Fig. 26·9.

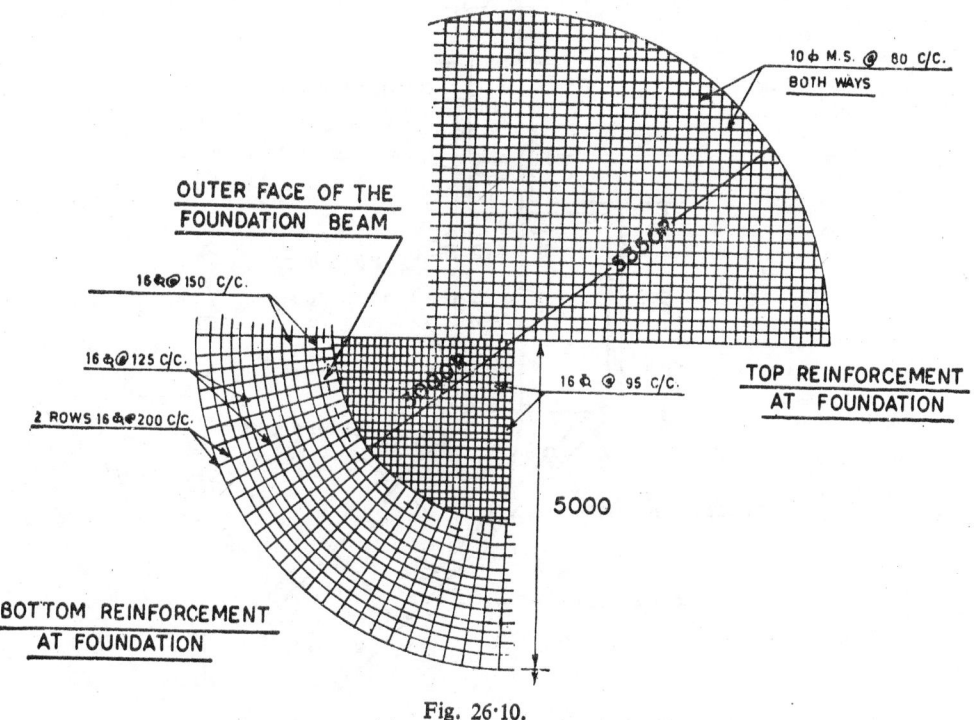

Fig. 26·10.

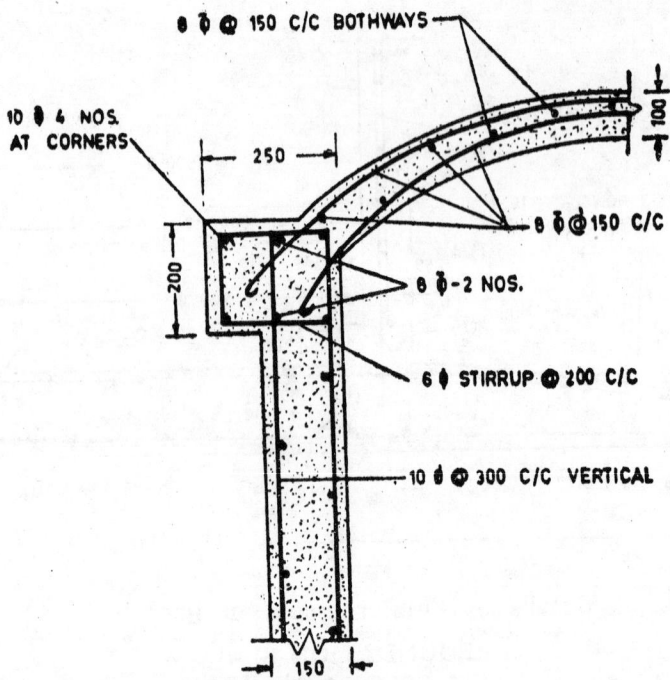

Details at A

Fig. 26·11.

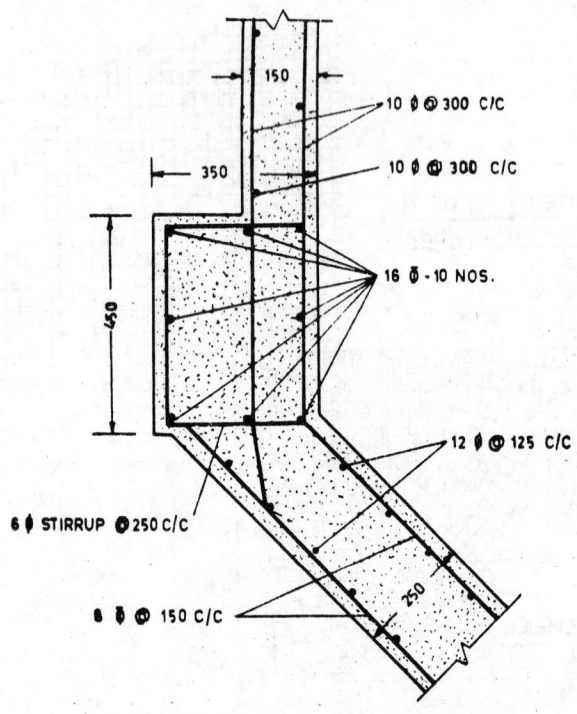

Details a B

Fig. 26·12.

The reinforcement details of the footing foundation *i.e.*, base slab with circular beam are presented in Fig. 26·9. The foundation slab is of 10 m overall diameter. The top reinforcement and bottom reinforcement of the base slab are given in Fig. 26·10. In Fig. 26·8 it may be seen that joints are marked A, B and C. The reinforcement details at joint A are focussed in Fig. 26·11. The details of reinforcement at B and C are illustrated in Figs. 26·12 and 26·13 respectively.

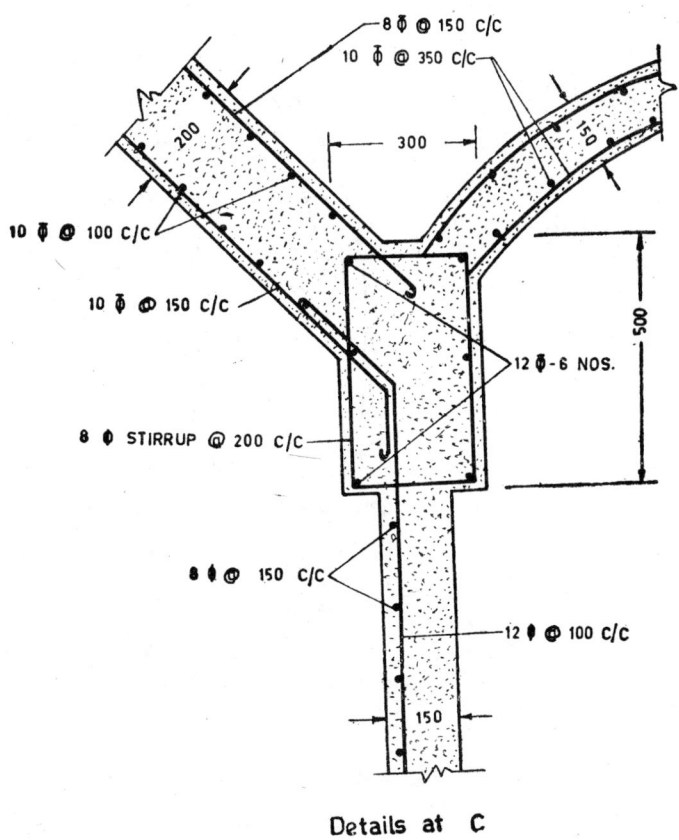

Details at C

Fig. 26·13.

Fig. 26·14 shows the reinforcement details of the vertical shaft and that of the top dome of the reservoir.

PLAN

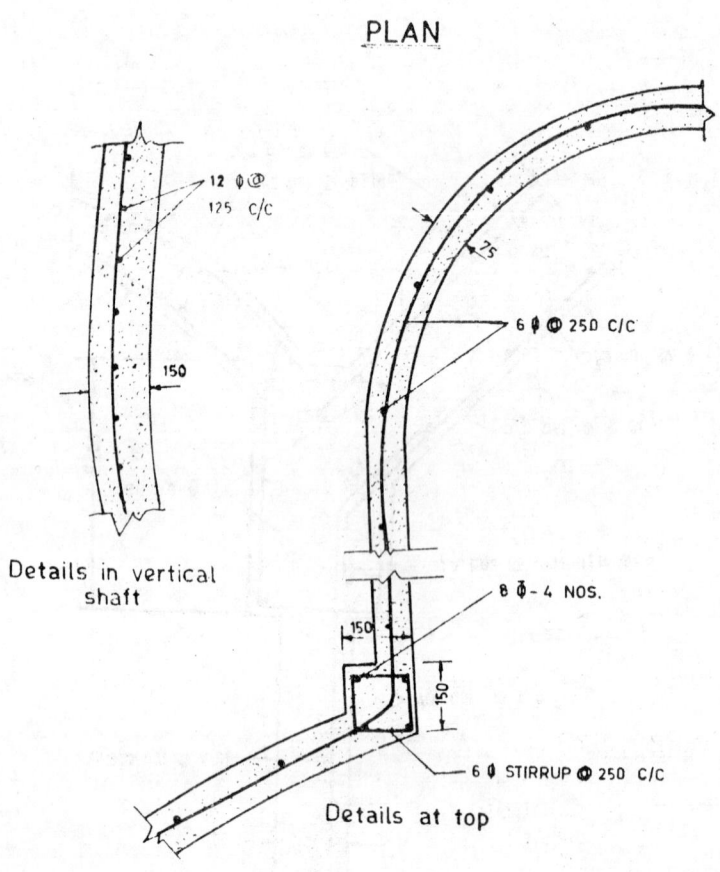

Details in vertical
shaft

Details at top

Fig. 26·14.

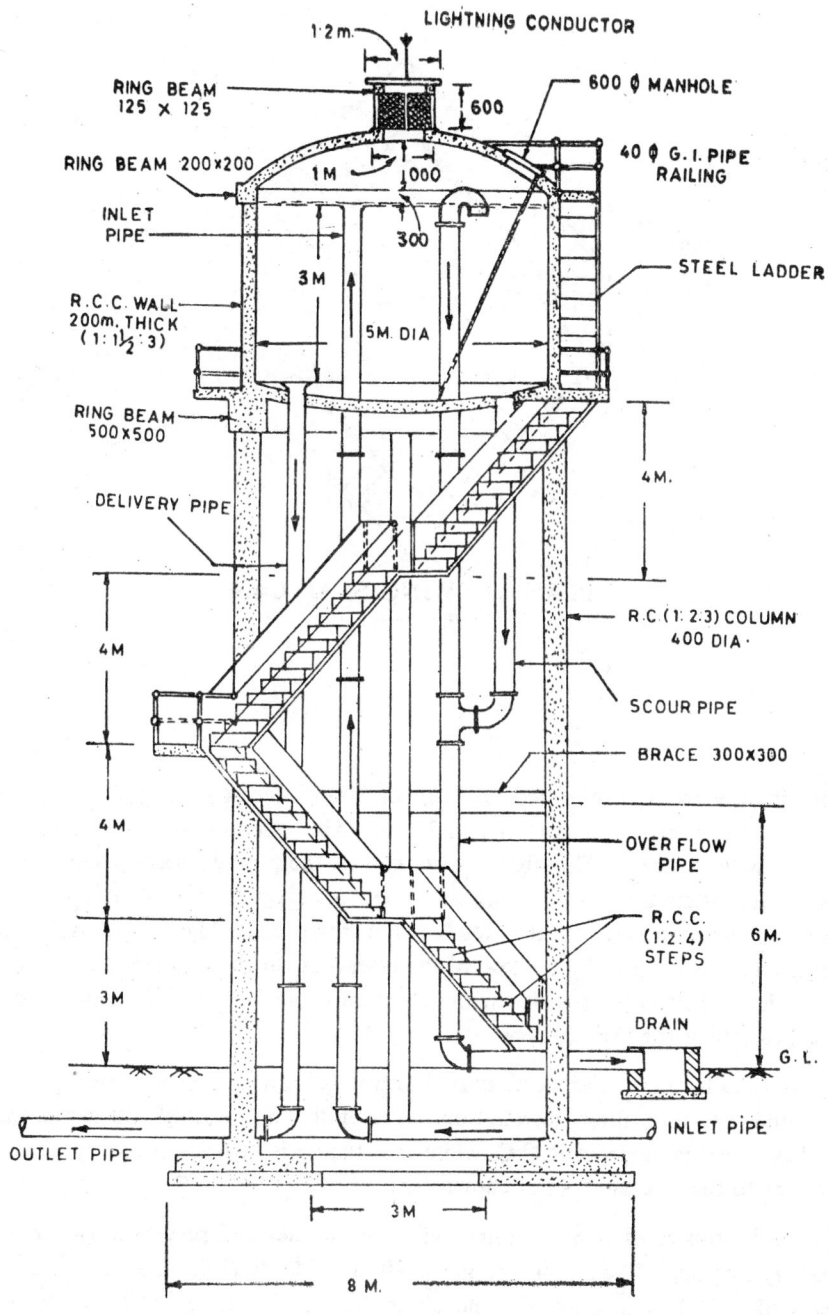

LIGHTNING CONDUCTOR

1·2 m.

RING BEAM
125 × 125

600

600 ∅ MANHOLE

40 ∅ G.I. PIPE
RAILING

RING BEAM 200×200

1M   1000

INLET
PIPE

300

STEEL LADDER

3 M

R.C.C. WALL
200m. THICK
(1:1½:3)

5M. DIA

RING BEAM
500×500

4 M.

DELIVERY PIPE

R.C. (1:2·3) COLUMN
400 DIA·

4 M

SCOUR PIPE

BRACE 300×300

4 M

OVER FLOW
PIPE

R.C.C.
(1:2·4)
STEPS

6 M.

3 M

DRAIN

G. L.

INLET PIPE

OUTLET PIPE

3M

8 M.

SECTIONAL VIEW OF AN ELEVATED RESERVOIR
Fig. 26·15.

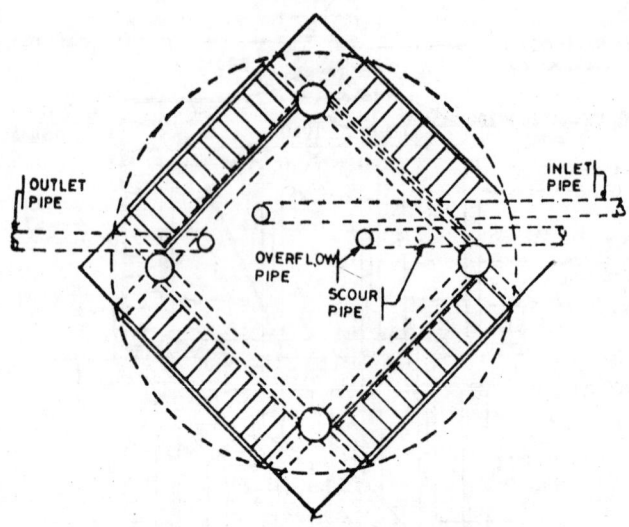

PLAN SHOWING STAIRCASE
Fig. 26·16.

A sectional view of an elevated reservoir of 5 m diameter resting at a staging height of 15 m from ground level is shown in Fig. 26'15. The reservoir is supported by four nos. of R.R.C. pillars with braces. The figure also shows the piping arrangement (for inlet, outlet, overflow and scour), staircase with landing for access to staging height and steel ladder with safety railing for easy access to the top of the reservoir. The dome is provided with a manhole for getting down into the bottom of the reservoir which is necessary for cleaning and repairwork. The dome top is provided with a ventilator and a lightning conductor, Fig. 26'16 shows the staircase plan for the reservoir.

Fig. 26'17 illustrates the sectional elevation of an overhead reservoir with piping arrangement and pressure filter installation. The filter house itself acts as the staging. The water tank (Reservoir) is made of R.C.C. while the staging is of brick masonry. The staircrse or ladder for access to the reservoir is not shown here.

Fig. 26'18 illustrates the sectional view of an elevated service reservoir placed over a circular masoury staging. The staging height is 10 m. The R.C.C. tank is encased in brick-work. The tank is of 3 m diameter and 2'5 m depth. For access to the tank, an angle iron ladder supported by angle iron stays is provided.

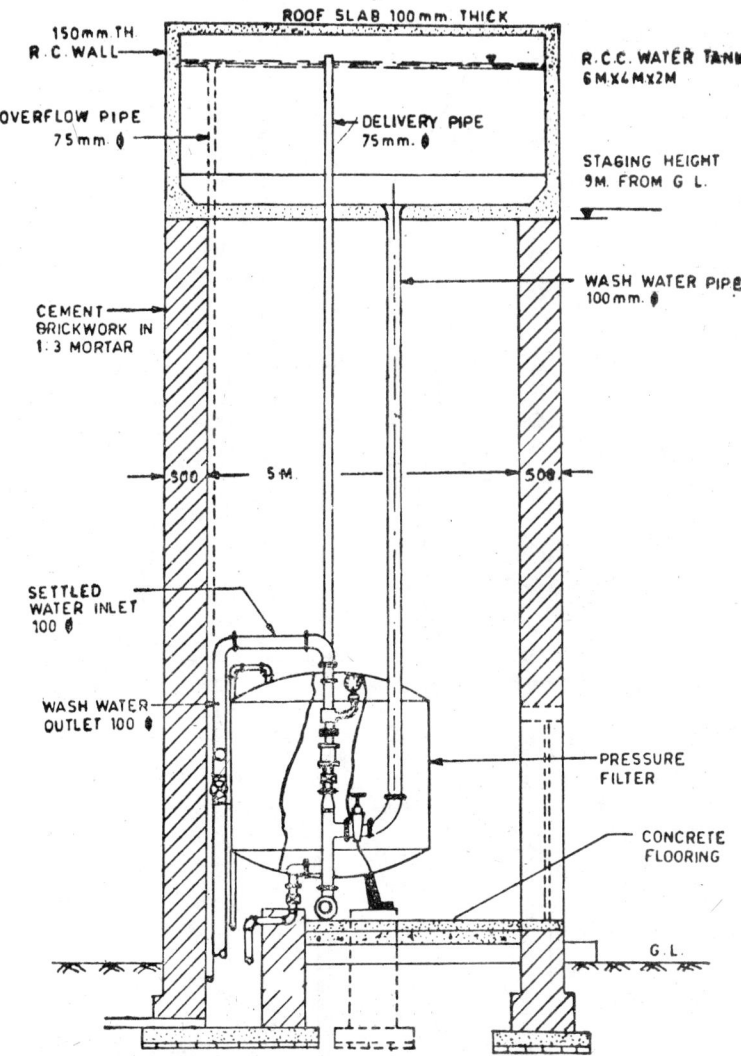

SECTIONAL ELEVATION OF AN OVERHEAD RESERVOIR SHOWING
THE INSTALLATION OF A PRESSURE FILTER
Fig. 26·17.

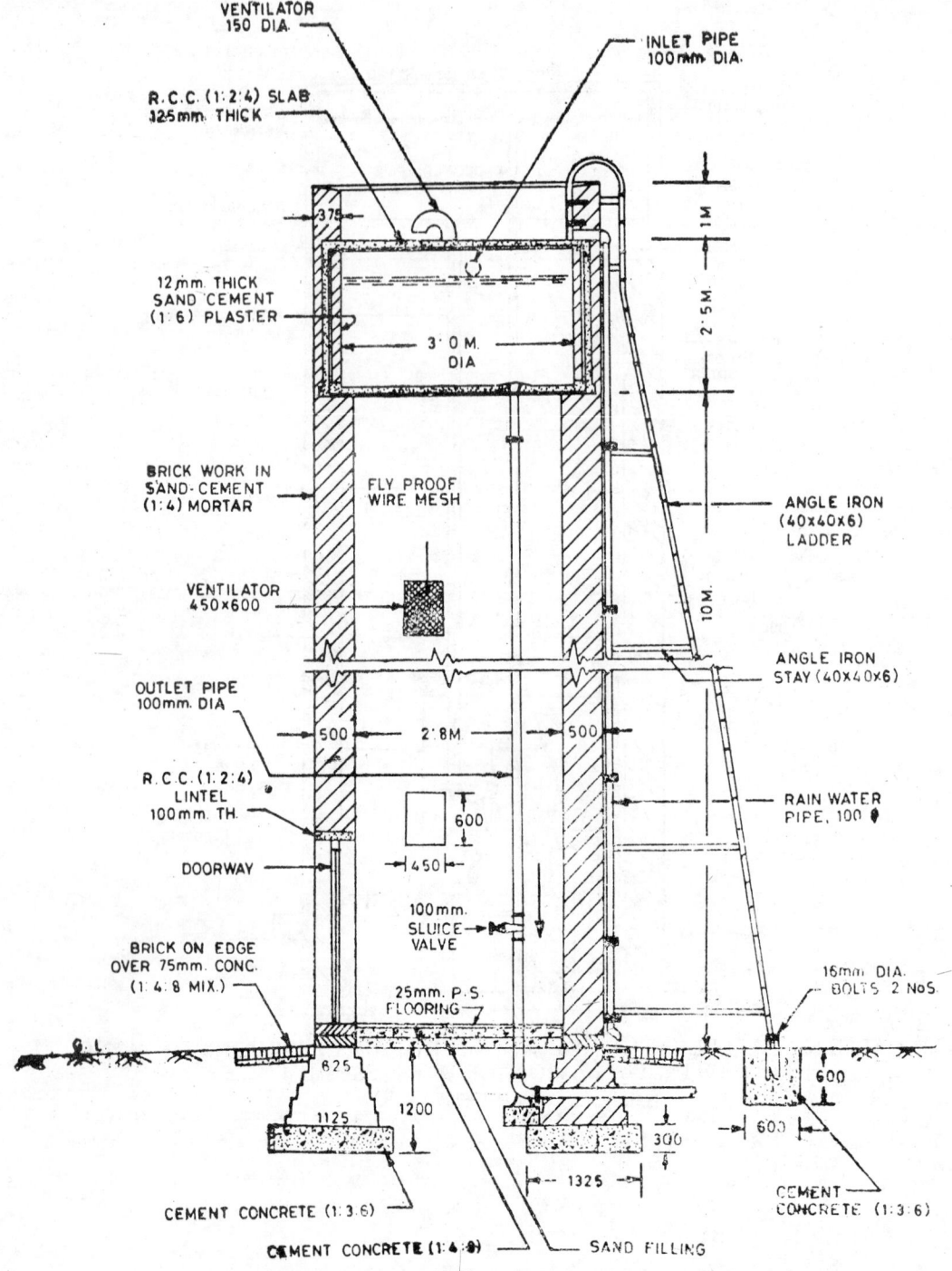

VENTILATOR
150 DIA.

INLET PIPE
100 mm DIA.

R.C.C. (1:2:4) SLAB
125mm THICK

12 mm. THICK
SAND CEMENT
(1:6) PLASTER

3·0 M.
DIA

BRICK WORK IN
SAND-CEMENT
(1:4) MORTAR

FLY PROOF
WIRE MESH

VENTILATOR
450×600

ANGLE IRON
(40×40×6)
LADDER

10 M.

1 M.

2·5 M.

375

OUTLET PIPE
100mm. DIA

ANGLE IRON
STAY (40×40×6)

500   2·8M   500

R.C.C. (1:2:4)
LINTEL
100mm. TH.

600

RAIN WATER
PIPE, 100 Φ

DOORWAY

450

100mm.
SLUICE
VALVE

BRICK ON EDGE
OVER 75mm. CONC.
(1:4:8 MIX.)

16mm. DIA.
BOLTS 2 Nos.

25mm. P.S.
FLOORING

G.L.

625

1200

300

600

1125

600

1325

CEMENT CONCRETE (1:3:6)

CEMENT CONCRETE (1:4:8)   SAND FILLING

CEMENT
CONCRETE (1:3:6)

Fig. 26·18.

27

Water Treatment and Sewage Treatment Units

This chapter deals with layout plans of water treatment plant, sewage treatment plants (employing various sewage treatment processes) and unitwise design and drawing. Detailed level drawings are also furnished.

Although pumps and pumping arrangement do not fall under treatment, they are closely associated with the treatment of water and sewage. Pump house designs with pumping arrangement are therefore included in this chapter.

The designs and drawings provided here will give a clear conception about the units and their actual shape and size used in field. The dimensions of different components and units given here are based on actual design as per standard practice and designs by following different manuals.

The first half of this chapter speaks of water treatment units and the second half deals with sewage treatment units and pump house construction.

WATER TREATMENT UNITS

A layout plan of a water treatment plant using surface water as source is shown in Fig. 27·1. The essential treatment units presented here are :

(1) Cascade Aerator (Raw water fountain)) for step aeration ;

(2) Flash mixer for thorough mixing of chemicals ;

(3) Clariflocculator for flocculation and clarification of turbid water ;

(4) Filters for filtration of water ;

Apart from these, the points of application of heavy chemicals like alum and lime and chlorination point are also shown here which are essential for treatment of water.

The civil structures required for such a treatment plant are :

(i) Cascade aerator.

(ii) Raw water channel.

(iii) Flash mixer.

(iv) Clariflocculator.

(v) Chemical house for storage and preparation of chemical solution.

(vi) Filter house with filter beds and pipe gallery.

(vii) Chlorination chamber with chlorine house.

(viii) Pure water sump and pump house.

(ix) Back wash water tank.

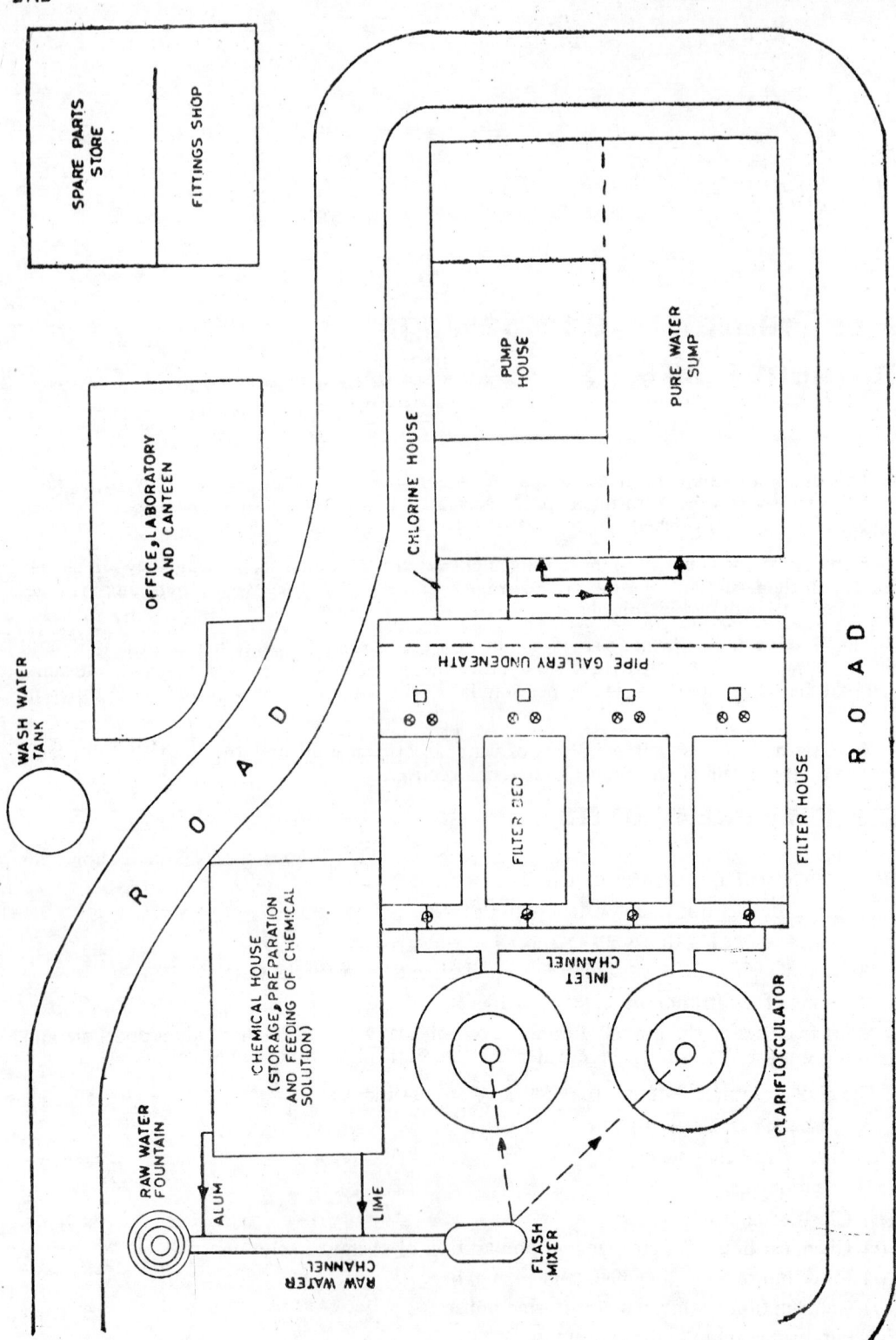

Fig. 27·1. LAYOUT PLAN OF A WATER TREATMENT PLANT.

(*x*) Office building with laboratory, canteen, etc.

(*xi*) Spare parts store and fittings shop.

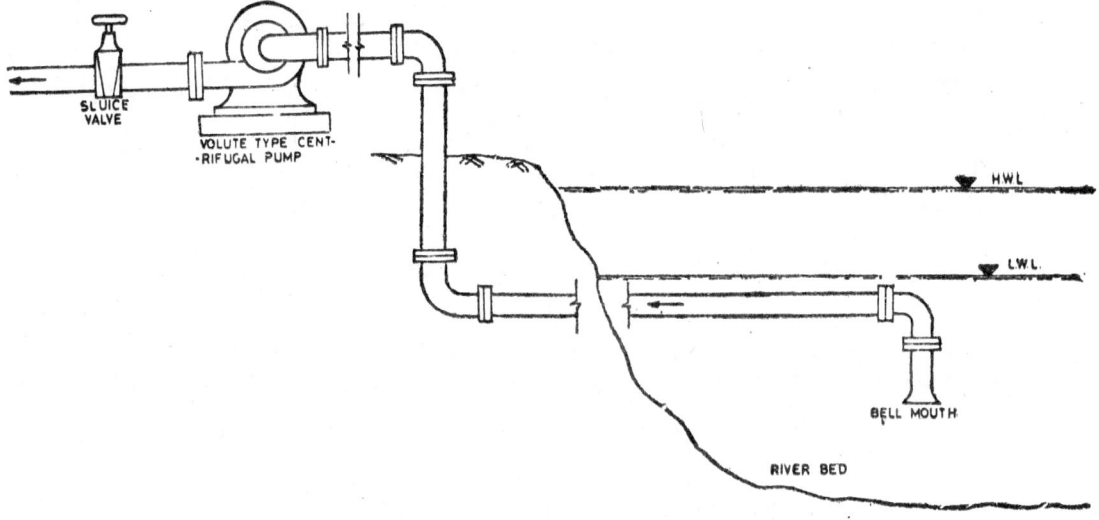

RAW WATER PUMPING

Fig. 27·2.

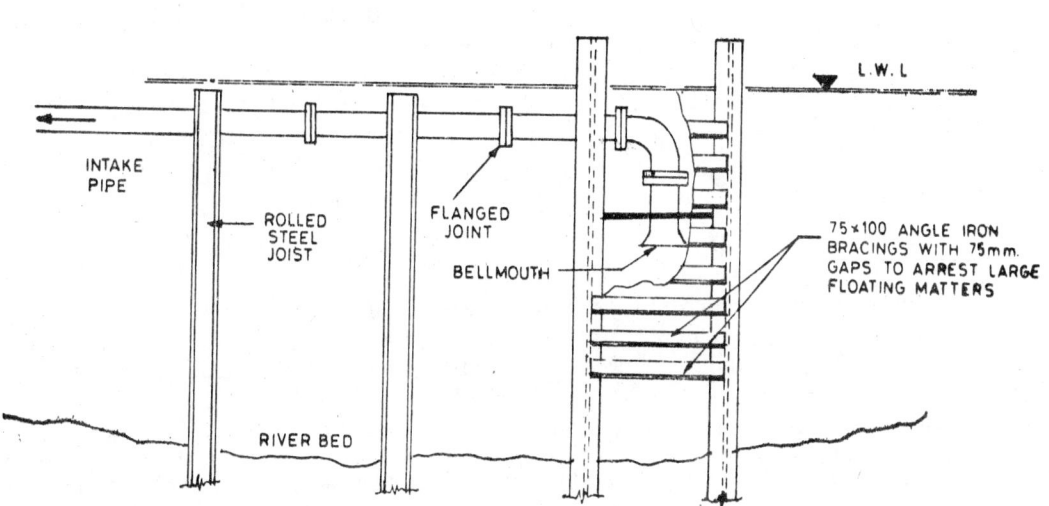

RAW WATER INTAKE STRUCTURE

Fig. 27·3.

INTAKE OF WATER :

Arrangement for raw water pumping from a surface water source is presented in Fig. 27·2. The pump to be employed here is a volute type centrifugal pump. The figure shows direct drawal of raw water from river and no intake well is proposed for construction. The projected pipe with bell mouth should be properly supported.

Fig. 27·3 illustrates the arrangement to support the raw water suction main with bell mouth and the cage made of angle irons to arrest large floating matters.

Fig. 27·4 shows sectional view of a radial collector well with pump house and pumping arrangement. This is similar to infiltration well. This type of intake arrangement should be recommended for drawal of water, when the river is likely to get dried up during summer.

CLARIFLOCCULATOR :

Fig. 27·5 shows the plan of a clariflocculator which is essentially a cylindrical water tight retention reservoir having flocculator and clarification zones. It has a central feed pillar with ports at top. The raw water enters into the flocculation zone through these ports. The flocculation zone is provided with fixed vertical baffles through which another set of vertical baffles passes with the movement of the clariflocculator bridge from which they are suspended. The scraper blades to sweep the sludge from the bottom of the clariflocculator are also hung from the bridge. With the movement of the bridge the scrapers sweep the sludge from the clariflocculator floor and the sludge is collected in the sludge pit from where it is discharged out by opening the sludge discharge valve. The clarified water flows up in the clarification zone and discharges into the peripherial launder through V-notches. A sectional view of the clari-flocculator is shown in Fig. 27·6.

SLOW SAND FILTER :

A slow sand filter is a water-tight basin consisting of a sand layer of 75 to 90 cm. thickness supported on a layer of gravel 20 to 30 cm thick. The gravel layer is underlain by a system of open joint underdrains which lead to a central drain for collection of filtrate. The effective size of sand used is 0·2 to 0·3 mm, the uniformity co-efficient being 2 to 3. The sand is usually placed in two layers, the fine sand being placed at top. The gravel is usually put in four layers for a total depth of 30 cm. The common gradation of gravel is from 2 to 50 mm. For under-drains, concrete pipes or baked clay pipes 30 to 40 cm. long are laid with open joints, the maximum spacing between the two cross drains being 2 M. Alternatively, R.C.C. slabs or flat bricks laid with open joints are used as filter underdrain. The cross drains are connected to a central collecting drain which delivers the filtered water (filtrate).

Slow sand filters are suitable when the raw water turbidity is not to exceed 50 JTU. The filter is usually filled with water to a depth of 1 to 1·5 m above the sand surface and the water is passed through the media at a rate 100 to 150 litres per hour per m² of surface area.

Fig. 27·7 shows the top plan layout of a slow sand filter with inlet and outlet chambers, underdrainage system and piping arrangement for settled water inlet, filtrate outlet, etc. etc.

The sectional view in details of the slow sand filter is presented in Fig. 27·8. The figure is self explanatory.

The alternative filter underdrainage systems are illustrated in Fig. 27·9.

When the filter bed is clogged, 20 to 30 mm. depth of sand is removed from top of the sand layer by scraping manually and new sand to similar specification is placed over it. For making the filter effective there is no backwashing arrangement as it is in a rapid sand filter.

A normal period of operation of a slow sand filter between two cleanings with an average turbidity of 30 JTU is taken as 6 months. A minimum depth of 45 cm of the media should always be maintained. The filter operation should be stopped when the loss of head exceeds 1·3 M.

RAPID SAND FILTER :

A rapid sand filter is a water-chamber built under a shed. The standard rate of filtration through a rapid sand bed is normally 100 to 200 litres/minute/m² of filter bed surface.

For rapid sand filter beds, a maximum area of 100 m² for a single unit with two halves, each of 50 m² area is recommended.

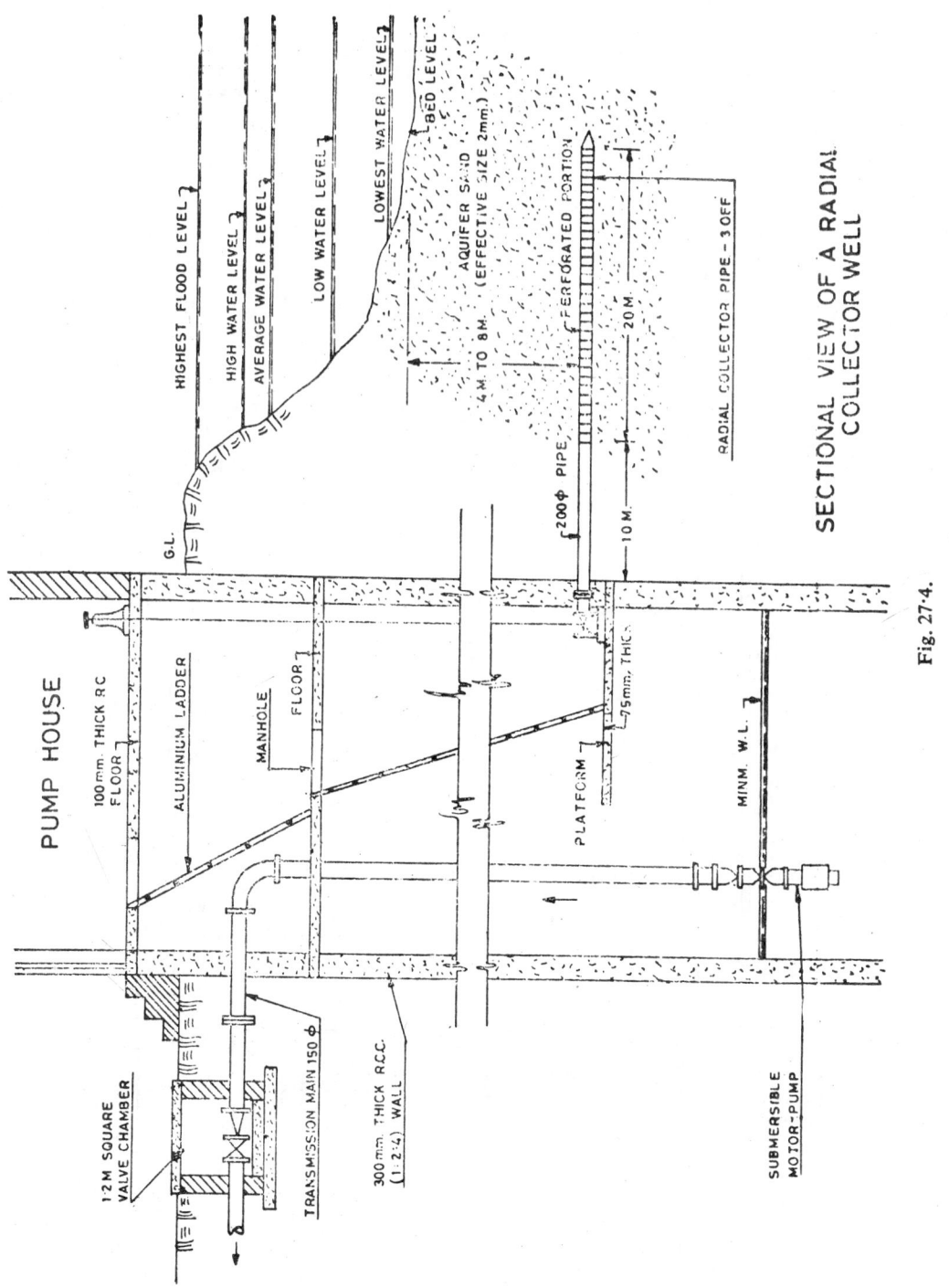

Fig. 27·4.

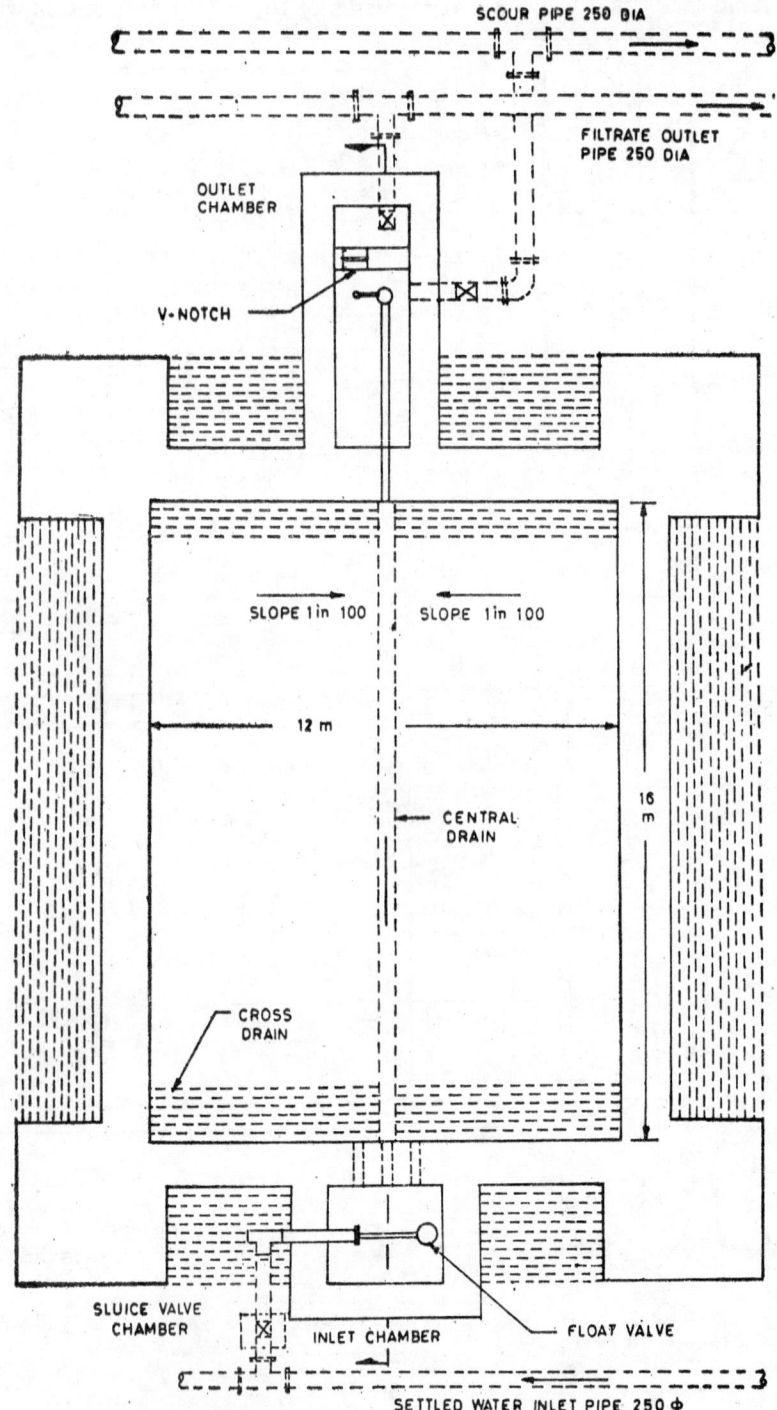

TOP PLAN OF A SLOW SAND FILTER
Fig. 27·7.

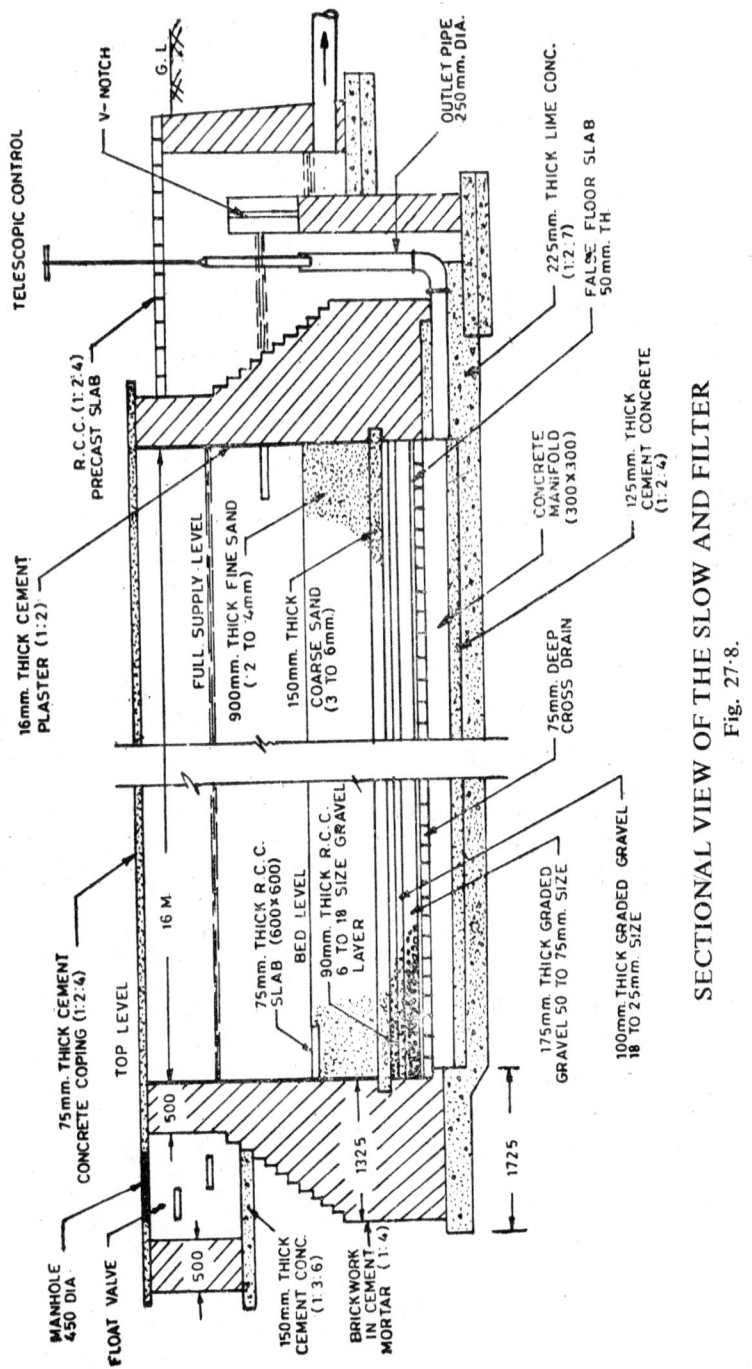

SECTIONAL VIEW OF THE SLOW AND FILTER

Fig. 27·8.

The filter sand should have effective size of 0·45 to 0·70 mm. with uniformity co-efficient ranging from 1·3 to 1·7. The specific gravity of filter sand should be around 2·65. The depth of sand layer should be 60-75 cm.

The filter gravel should be placed between the filter sand and the underdrainage system and the depth should be 45 cm. The sizes of graded gravel should vary from 2 to 5 mm. at top to 50 mm. at bottom.

The standing depth of water over sand surface may vary from 1 to 2 m., a minimum freeboard of 50 cm. being kept above the water level.

The underdrains consist of a manifold with headers and laterals covering the entire filter bottom. For this purpose, usually C.I., concrete or A.C. pipes with perforations are used. The ratio of length to diameter of laterals should be within 60. The spacing of laterals should be approximately 30 cm. The perforations may vary from 5 to 12 mm. in diameter, the spacing of staggered perforations being 8 cm. for 5 mm. dia. and 20 cm. for 12 mm. dia. perforations along the lateral.

The wash water troughs should be designed as free falling weirs, the bottom of the trough or gutter being kept 50 mm. above the expanded sand during backwash. For sand expansion, 50% of the sand depth should be considered.

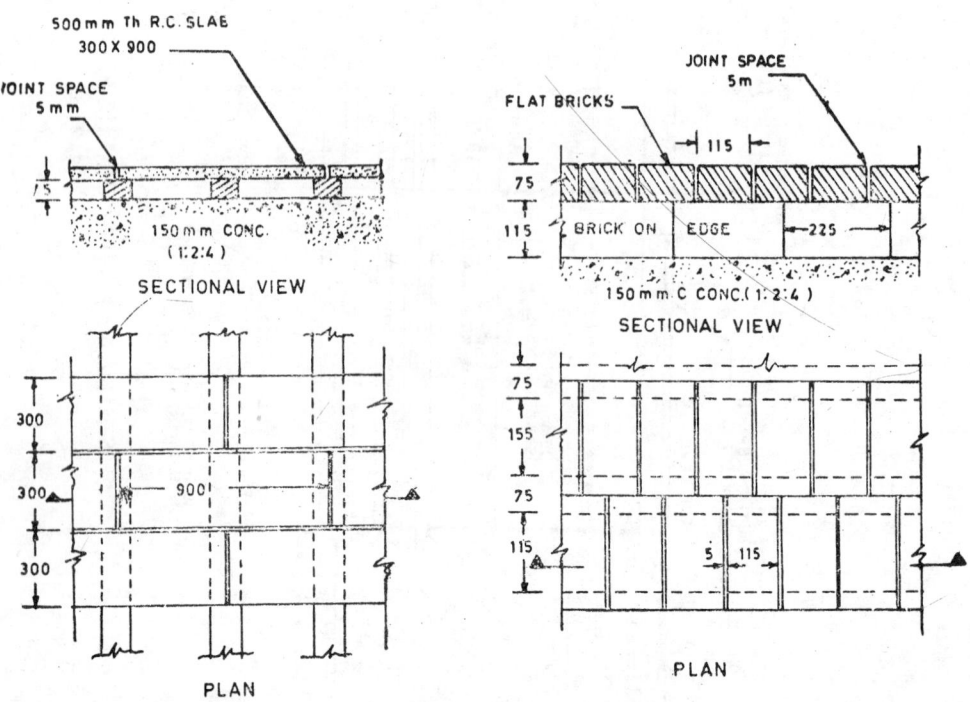

FILTER UNDER DRAIN
Fig. 27·9.

Fig. 27·10 shows a sectional view of a typical rapid sand filter bed and Fig. 27·11 represents the filter bed layout. The sectional views to show the positions of the wash water trough and the gullet with influent pipe are presented in Fig. 27·12.

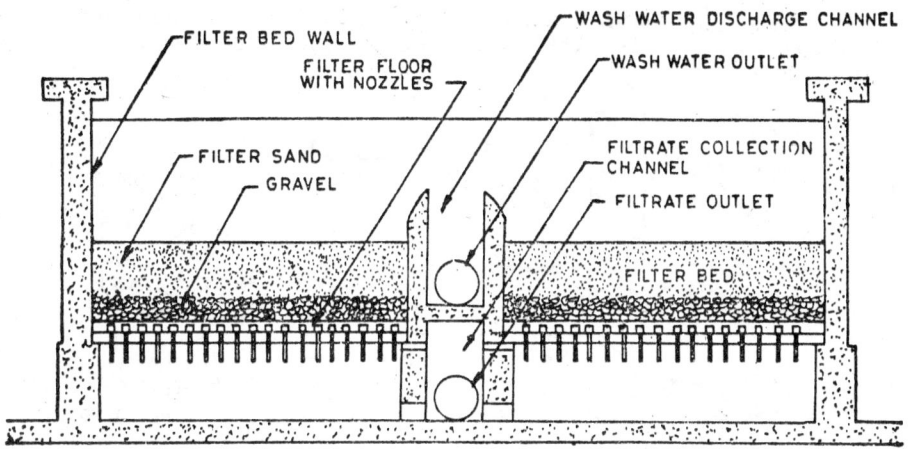

SECTION THROUGH A RAPID SAND FILTER BED

Fig. 27·10.

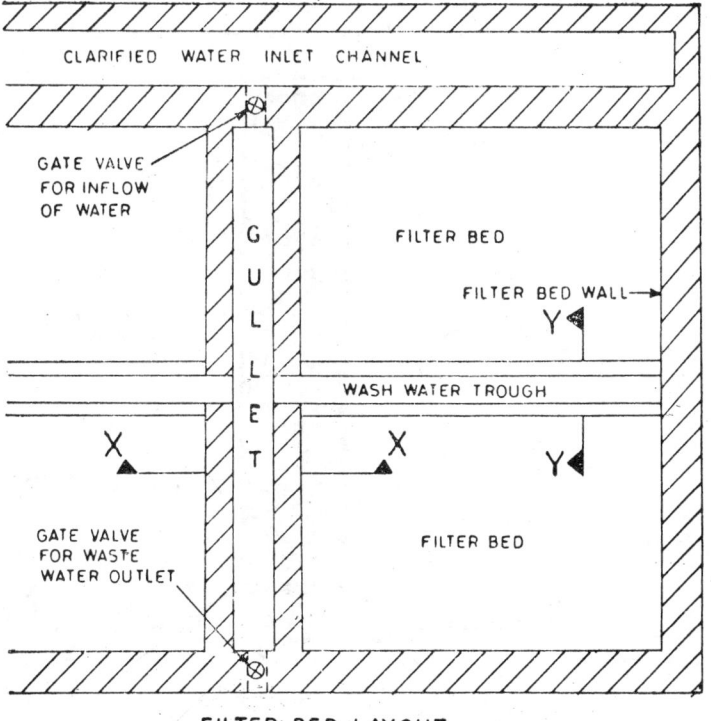

FILTER BED LAYOUT

Fig. 27·11.

The details of chlorine injection tower, which is required for disinfection of filtered water, are shown in Fig. 27·13. The connection of chlorine cylinder with the chloronome board is not shown here.

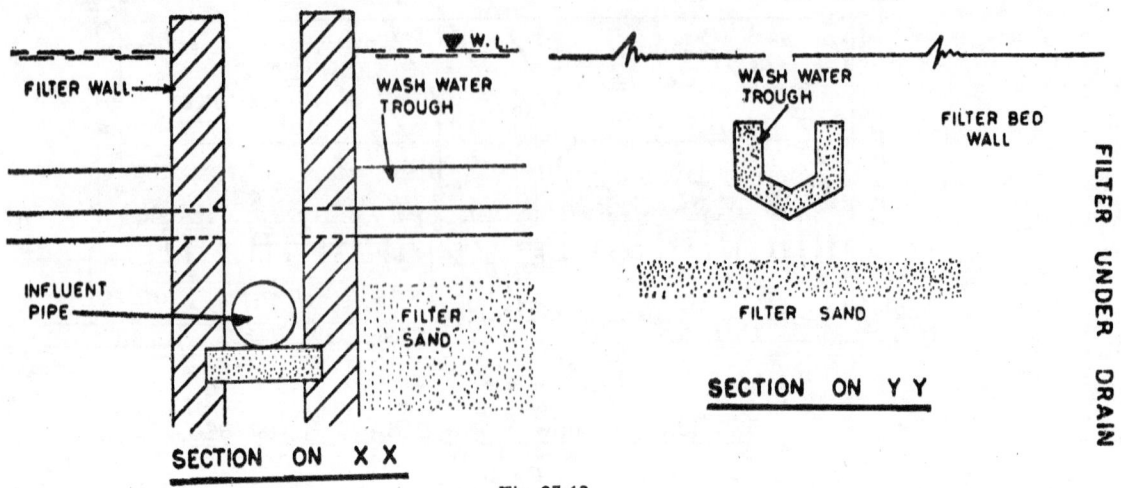

SECTION ON X X

SECTION ON Y Y

FILTER UNDER DRAIN

Fig. 27·12.

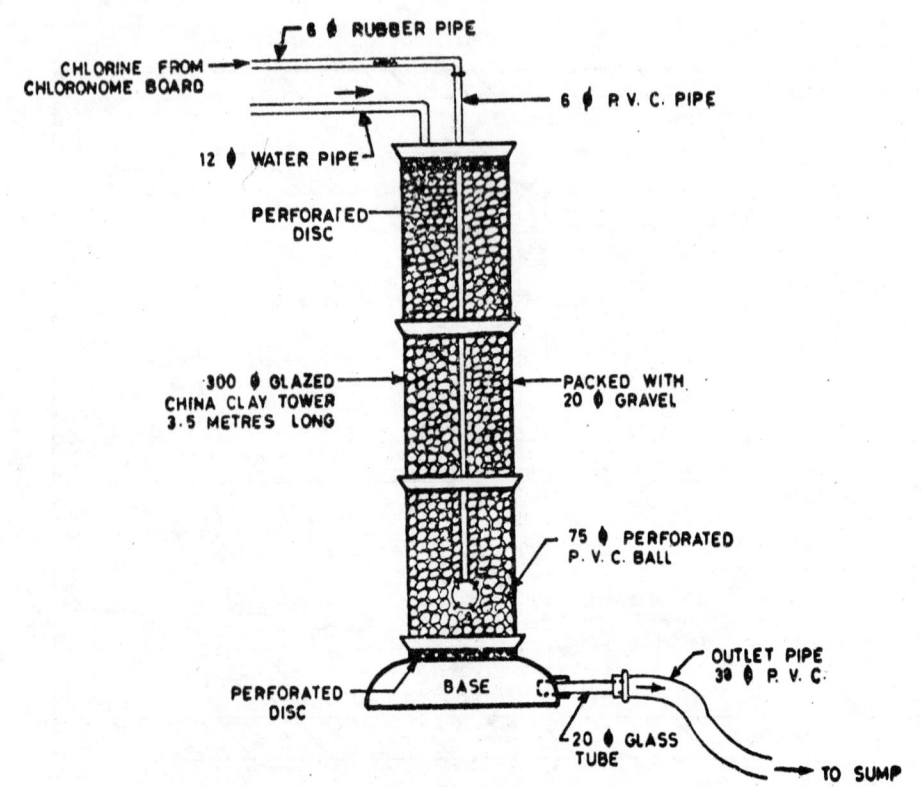

DETAILS OF CHLORINE INJECTION TOWER

Fig. 27·13.

SEWAGE TREATMENT UNITS

Various sewage treatment processes commonly used are :

Activated Sludge Process   Biofiltration (Trickling Filter) Process

Oxidation Ditch   Oxidation (Stabilisation) Pond

Aerated Lagoon

Activated Sludge Plant :

A schematic layout plan of an Activated Sludge Plant is shown in Fig. 27·14. Capacities of such plant s in use range from 5 mld to above 100 mld. The B.O.D. removal in the process is 85 to 95°/₀. The aeration chamber of an activated sludge plant is the heart of the system. The capacity of aeration tank is determined from the F/M ratio and MLSS value selected for the plant. The F/M ratio ranges from 0·4 to 0·2 and MLSS value lies in a range of 1500 to 3000 mg/l. kg. of O_2 per kg. of BOD_5 removal is 0·8 to 1·1.

The inlet and outlet channels of the aeration chamber should be designed to keep a minimum velocity of 0·2 m/s with a view to preventing deposition of solids.

Aeration tanks are designed as long narrow channels, the depth of tank being in the range of 3 to 4·5 m. The width is usually kept between 5 and 10 m.

The width to depth ratio should be between 1·2 and 2·2. The length should not be less than 30 m. and not more than 100 m.

For aeration of sewage, mechanical aerators are employed. The air supply should be at least 0·25 m³/minute per metre length of channel. For diffused air aeration, spacing of the

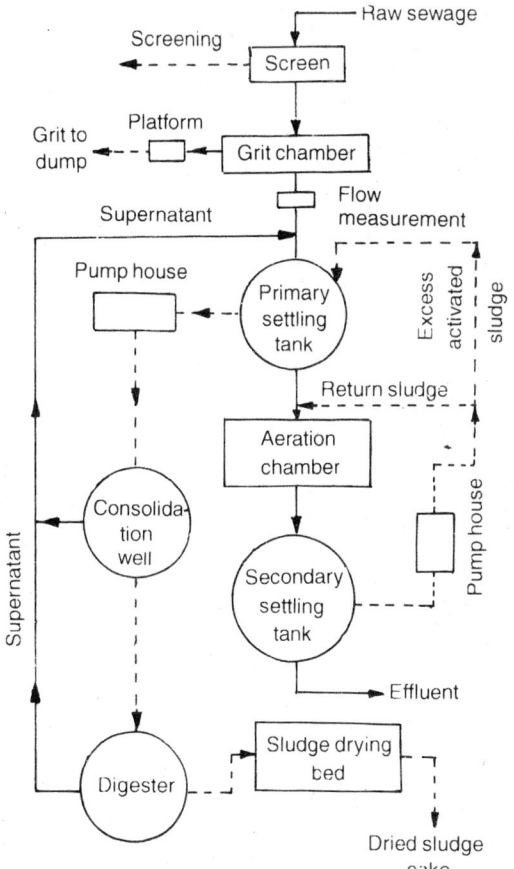

TYPICAL LAYOUT PLANT OF AN ACTIVATED
SLUDGE PLANT

Fig. 27·14.

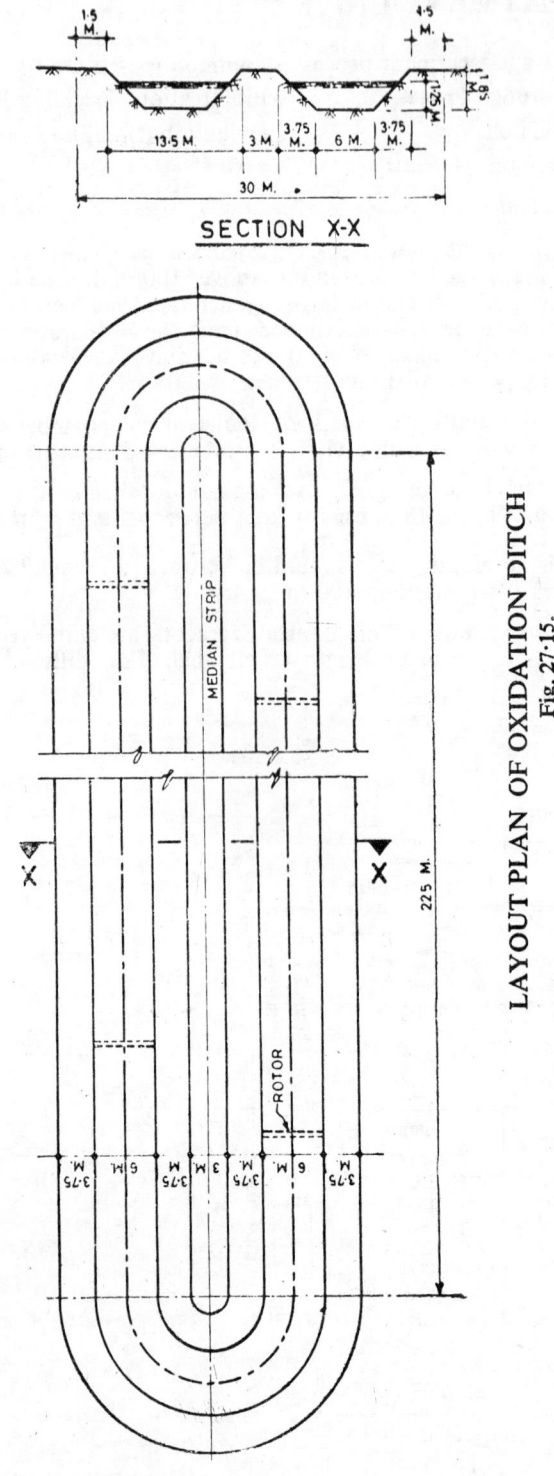

SECTION X-X

LAYOUT PLAN OF OXIDATION DITCH
Fig. 27·15.

diffusers should be 0.6 to 1 m. Standard ceramic plate diffusers of 0.3 m.×0.3 m size having pores of 0.3 mm dia. may be used. Aeration can also be done by introducing air through open ends of vertically submerged pipes of 20 to 40 mm. dia. They are also various types of diffusers and mechanical aerators.

Oxidation Ditch :

Layout plan of an oxidation ditch with dimensions for 1 mgd. (4.55 mld) capacity is presented in Fig. 27.15. This is essentially a process of extended aeration. An oxidation ditch is a long continuous channel, earthen or lined, usually oval in plan, the depth being restricted to 1 to 1.5 m. The circulalting sewage is mechanically aerated by surface aerators placed in slaggered position. The velocity of flow of sewage is 0.3 to 0.4 m/s. The cage rotors for aeration usually have a diameter of 70 cm and a speed of 75 rpm. At 16 cm. depth of immersion, oxygen transfer capacity is about 2.75 kg O_2/m. length of ditch/hour, the power requirement being about 1.35 kw per metre length.

Stabilization Pond :

A typical layout plan of stabilization ponds is shown in Fig. 27.16. Stabilization or oxidation ponds are flow through open earthen reservaoirs or tanks built to treat sewage under a detention period of two to three weeks. Individual pond area should not exceed 40 ha. The maximum length of a pond should not exceed 750 m. and the length should not exceed three times the width of the pond   The optimum depth of an oxidation pond should be in a range of 1 to 1.5 M. The top width of the pond embankment should be at least 1.5 m. for small ponds and 3 m. for large ones. The outer slopes should be 1 to 2.5 and the inner sloped surfaces should be pitched with bricks, stone bonders or concrete blocks in cement mortar. The free board should be 0.5 to 1 m. above the liquid level in the pond. The influent pipeline should extend into the pond at least 15 to 20 m. from the water edge and sewage is dicharged over a splash pad at about 0.3 m below the liquid surface of the pond. The outlet pipe is submerged to a depth of 0.25 m. below the liquid surface and is provided with a tee connection. A series of pond may be interconnected by incerting pipes through the embankments at a depth of 0.25 m below the liquid surface of the ponds.

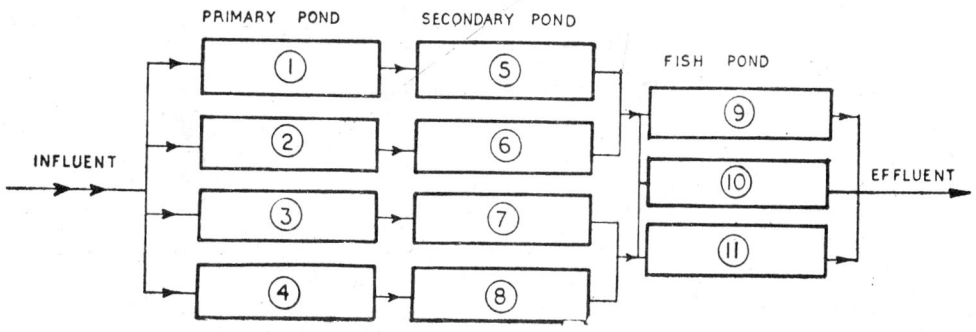

A TYPICAL LAYOUT PLAN OF STABILIZATION POND

Fig. 27.16,

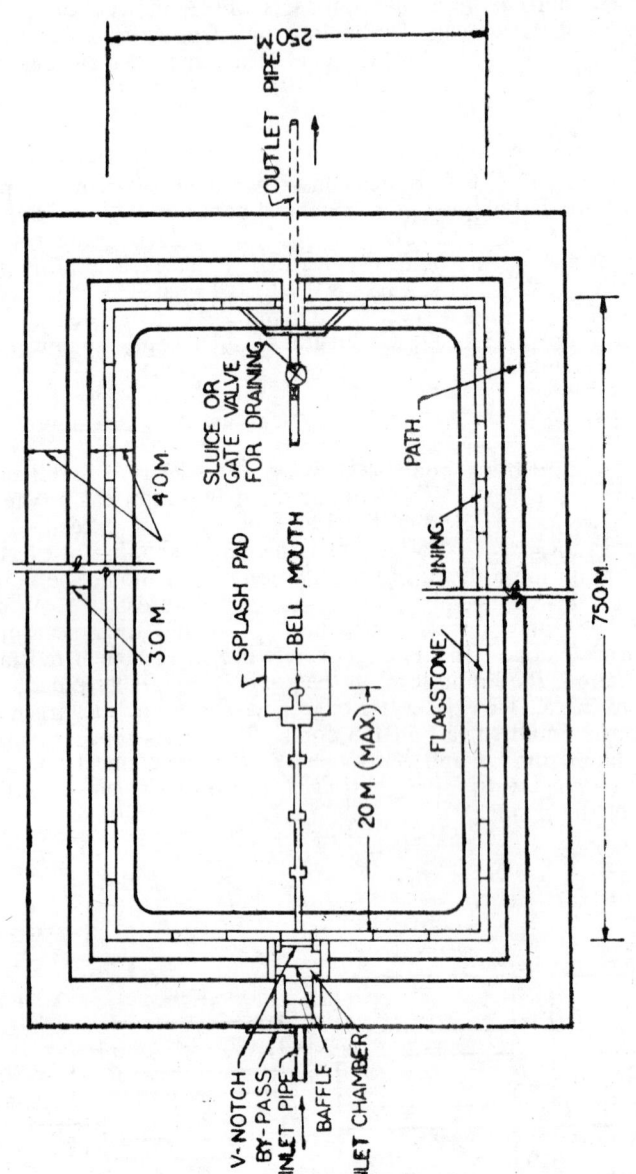

TYPICAL PLAN OF A SINGLE UNIT OF WASTE STABILIZATION POND

Fig. 27·17.

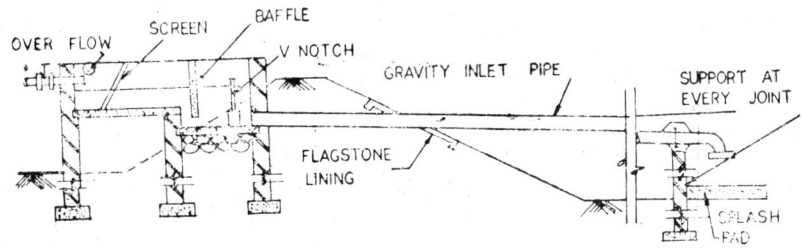

TYPICAL DETAILS OF INLET CHAMBER

Fig. 27·18.

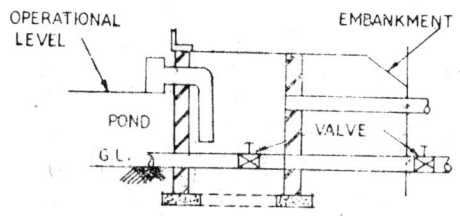

TYPICAL DETAILS OF OUTLET CHAMBER

Fig. 27·19.

A typical plan of a single unit of waste stabilization ponds showing inlet and outlet arrangements with flow measuring device is illustrated in Fig. 27 17.

Typical details of the inlet chamber of an oxidation pond are shown in Fig. 27·18 and those of the outlet chamber are presented in Fig. 27·19.

Biofiltration Plant (Trickling Filter)

A schematic layout of a typical biofiltration plant with series-parallel is presented in Fig. 27·20. This is a two stage high rate biofiltration plant with series-parallel arrangement. There are quite a number of alternative flow diagrams of which the complicated one is shown here. The basic units for a biofiltration plant are :

Screen or classifier ;

Detritns tank or grit chamber with washing arrangement.
Primary clarifier
Secondary clarifier
Sludge diagerter
Sludge drying beds

Pump houses with pumping arrangement are sludge pumping for recirculation of sewage and sludge pumping from secondary clarifier to primary clarifier and there from to the sludge digester.

SCSEEN CHBMBER :

Typical details of screen chamber with placement of screens (hand raked), gates and by pass arrangement for diversion of flow are shown in Fig. 27·21. The sectional view of the screen chamber by taking section at XX is shown in Fig. 27·22 A screen may consist of rods,

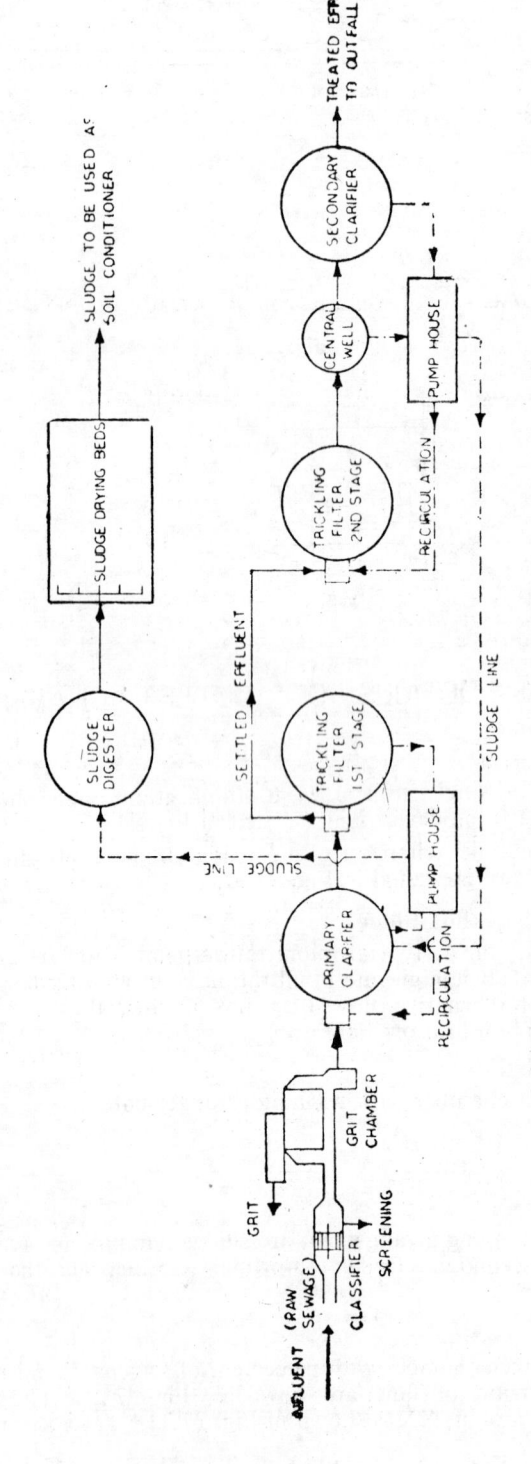

PLAN ON L-L

SCHEMATIC LAYOUT OF A BIOFILTRATION PLANT

Fig. 27-20.

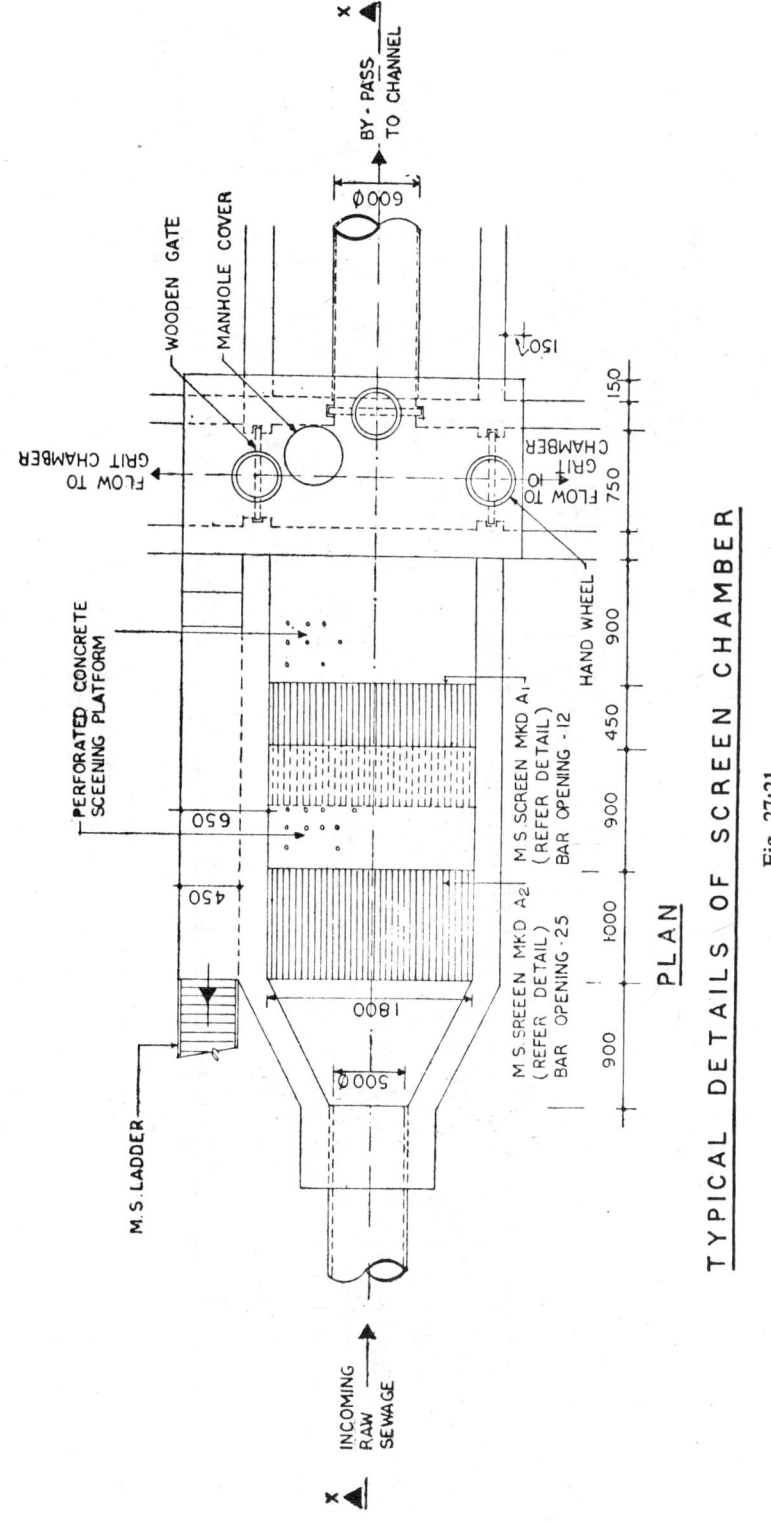

PLAN

TYPICAL DETAILS OF SCREEN CHAMBER

Fig. 27·21.

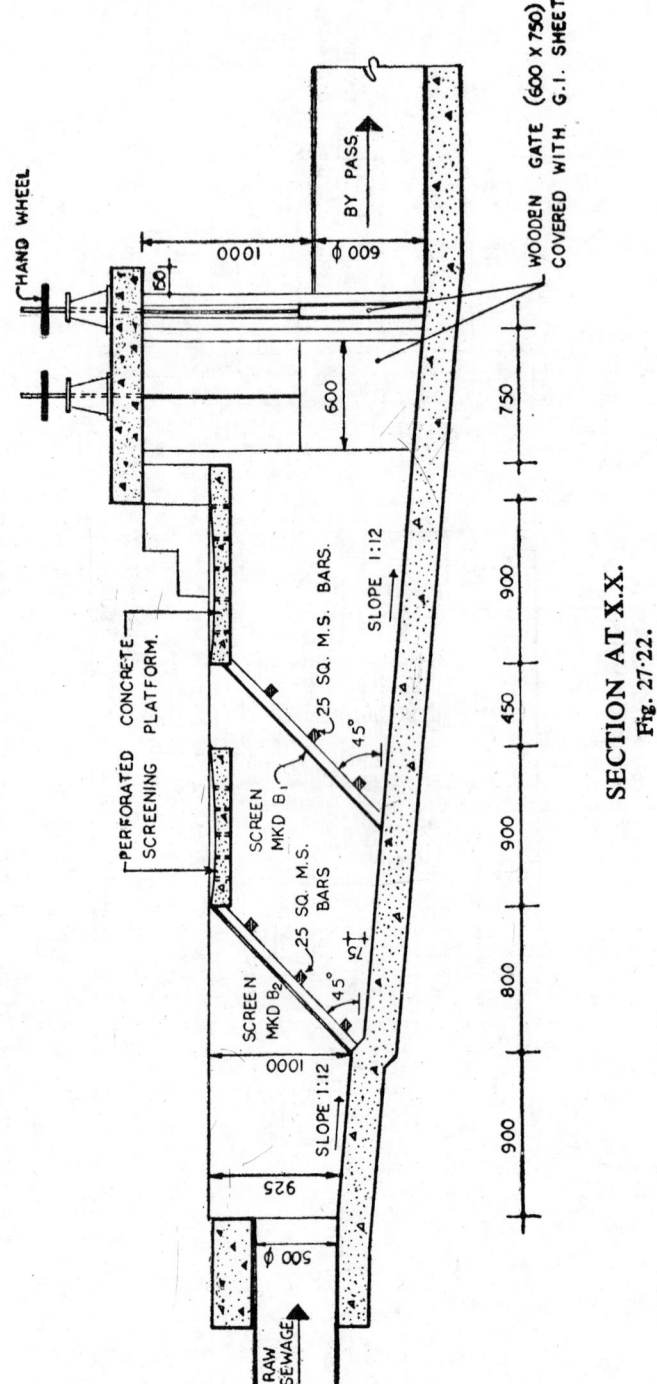

SECTION AT X.X.
Fig. 27·22.

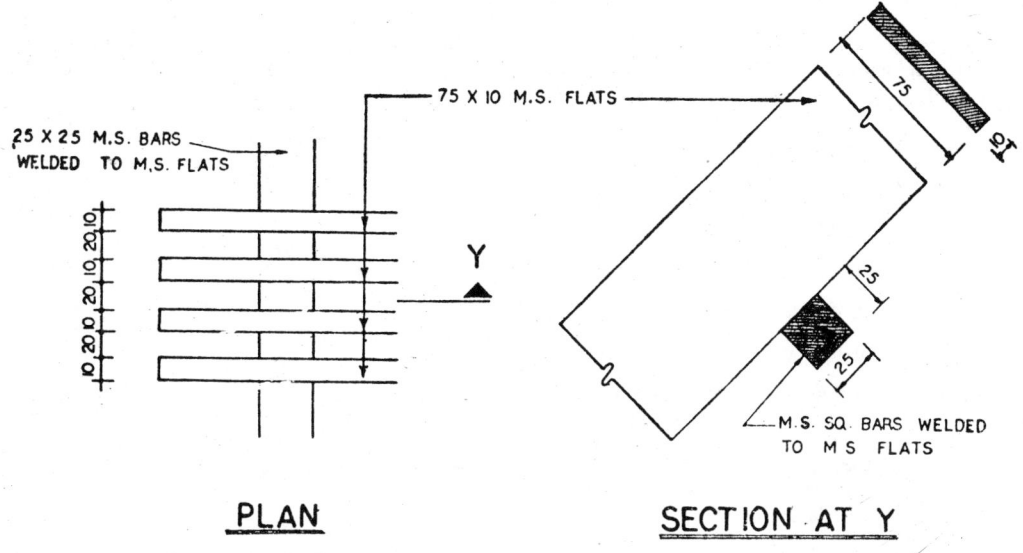

PLAN

SECTION AT Y

DETAILS OF SCREEN MKD. B₁

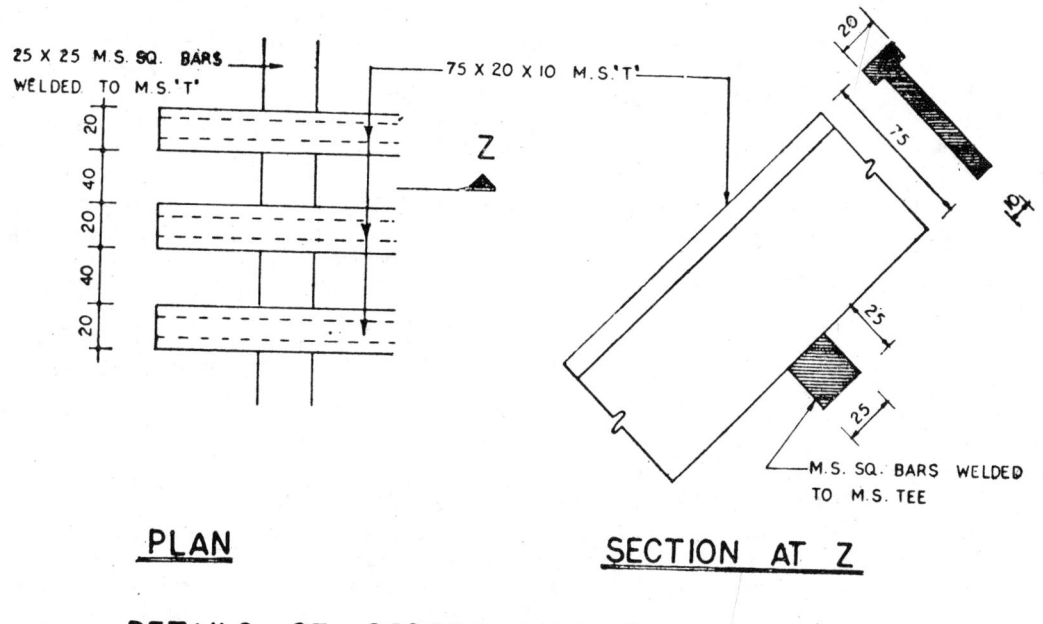

PLAN

SECTION AT Z

DETAILS OF SCREEN MKD B₂

Fig. 27·23.

parallel bars, gratings, wiremesh or perforated plates and medium screens used in a sewage treatment plant are mostly bar screens composed of vertical or inclined bars spaced at equal intervals across a channel through which raw sewage flows in a straight direction. Mechanically cleaned trash racks or screens are mostly vertical. Manually cleaned bar screens should be located such that they are readily accessible for cleaning by hand raking.

The characteristics of hand-cleaned bar racks or screens are given below :

Bar size : Width (6 mm—15 mm)

Depth (25 mm—75 mm)

Spacing of bars : 25 mm to 50 mm.

Inclination of bars : 30° to 60°.

Approch velocity of sewage : 0.3 to 0.6 m/s.

Allowable head loss : 150 mm.

The screen channel invert should be 75 to 150 mm. below the invert of the incoming sewer. This is illustrated in Fig. 27.22. The details of screens are shown in Fig. 27.23.

PRIMARY CLARIFIER :

The plan and sectional view of a primary clarifier are presented in Fig. 27.24. A primary clarifier is a circular up-flow tank used as a sedimentation tank. The diameter of such a tank is commonly in a range of 12 to 30 M, the side water depth varying from 2 to 3.5 M. The tank floor is sloped from periphery to centre @ 10% slope. The sewage enters through the hollow central shaft and is discharged through a number of ports at the top of the shaft. A slow rotating collector arm ploughs settled sludge to the sludge draw off pocket at the centre of the tank bottom. A surface skimmer attached to the arm collects scum from the surface and drops it into a scum box that drains into the scum pit. The detention time for sewage in the tank is kept 2 to 2.5 hr. for settling. The overflow rate should be 35 to 50 m³/day/m² of surface area for average flow of sewage in the tank. The outlet of clarified sewage (settled sewage) is through the peripheral weir. The weir loading should not exceed 100 m³/day/m. length of the weir.

The inlet baffle is placed concentric to the control feed shaft generally with a diameter of 10% to 20% of the tank diameter and extending 1 to 2 m. below the liquid level. The sewage entryports are usually submerged 0.3 to 0.6 m. below the liquid level. A peripheral scum baffle extending 0.20 to 0.30 m. below liquid level is generally provided ahead of the effluent weir. The effluent weir is provivded with shallow V—Notches about 50 mm. deep and spaced 0.15 to 0.30 m. apart. The surface skimmer withe swinging skimmer blades are supported from the trussed rake arm. The total out-flow approachss the effluent discharge pipe from the two sides of the peripheral launder.

TRICKLING FILTER :

Fig. 27.25 shows the plan of a bio-filter (trickling filter) and its sectional view is presented in Fig 27.26. A trickling filter consists of a permeable bed of filter media of brick bats or broken stones. The optimum range of media depth of a high rate trickling filter to 1.5 to 2 m. The size of broken stones is usually in a range of 25 to 75 mm. The media should be packed by hand for at least a height of 30 cm above the underdrains of the filter For distribution of sewage on to the filter, a reaction type rotary distributor is used which essentially consists of a feed column at the centre of the filter, a turn table assembly at the top of the bed, a support for pipe column and hollow radial distributor arms (square or rectanglar in section) with orifices and orifice plates. A section showing central column and distributor arms of a biofilter is presented in Fig 27.27. The distributor arms are fabricated of steel and they may be made gradually tapered towards the end near the periphery of the filter bed. The length of each piece of fabricated arms should not be more than 6 metres. The arms when revolving should clear the surface of the bed by 150 to 200 mm. The distri-

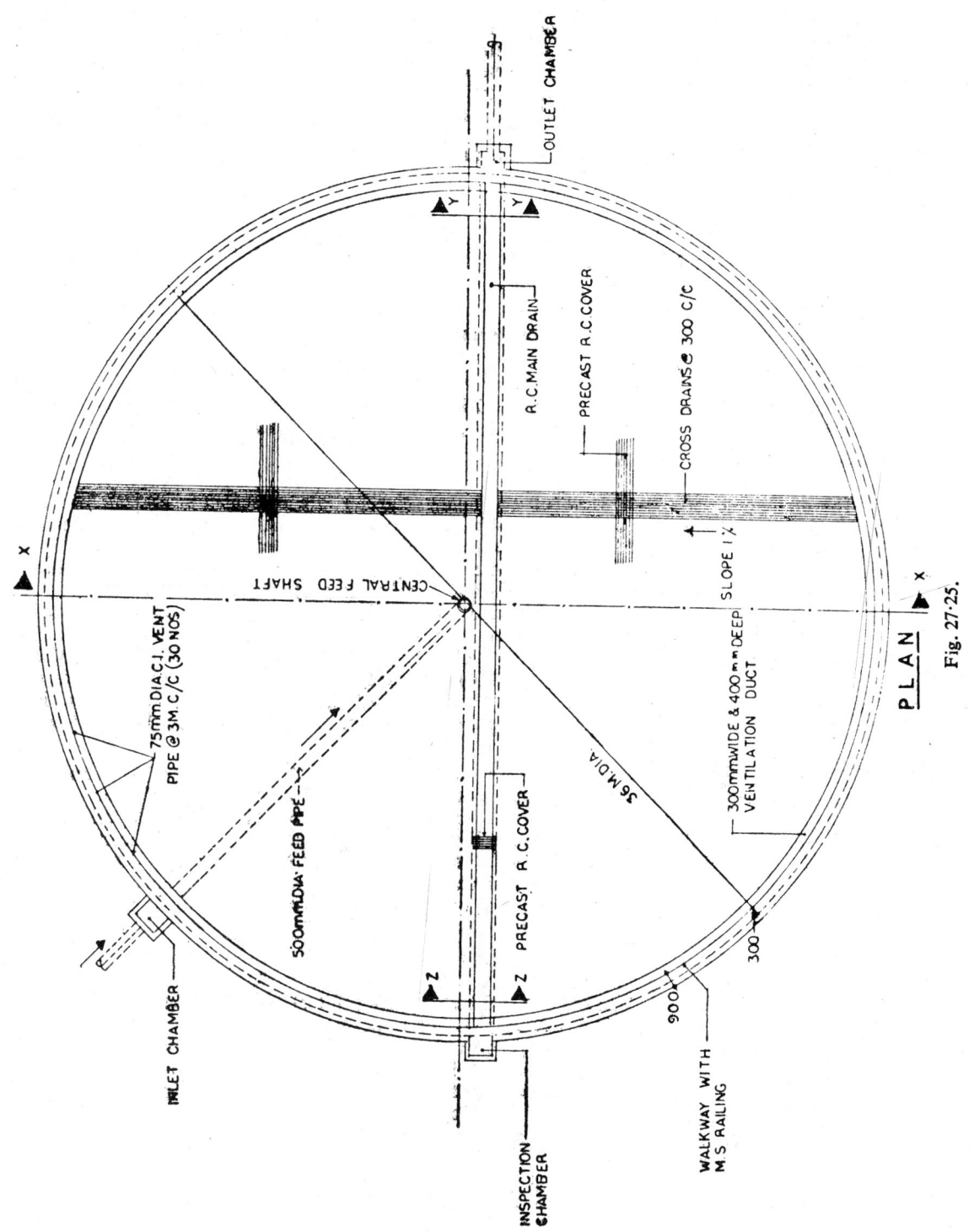

OUTLET CHAMBER

R.C.MAIN DRAIN

PRECAST R.C COVER

CROSS DRAINS @ 300 C/C

SLOPE 1 %

300mm WIDE & 400mm DEEP
VENTILATION DUCT

56 M DIA

CENTRAL FEED SHAFT

75mm DIA.C.I. VENT
PIPE @ 3M C/C (30 NOS)

500mm DIA. FEED PIPE

PRECAST R.C.COVER

300

900

WALKWAY WITH
M S RAILING

INLET CHAMBER

INSPECTION
CHAMBER

PLAN

Fig. 27·25.

butor arms are held in position with the help of guy rods. For safety, a mild steel or G.I. railing is provided along the peripheral wall of the filter bed. The bottom of the filter bed is given a slope of 1% from periphery to the centre. The rectangular slits or orifices in the distributor arms should be provided with spreader plates of deflector plates made of aluminium.

The distributor arms should have gates at the end for flushing them for which at least one end plate should have arrangement for jet impinging on the sidewall. The end gate of a distributor arm is shown in Fig. 27·27. The plan of distributor arms with position of guy rods is shown in Fig. 27·28.

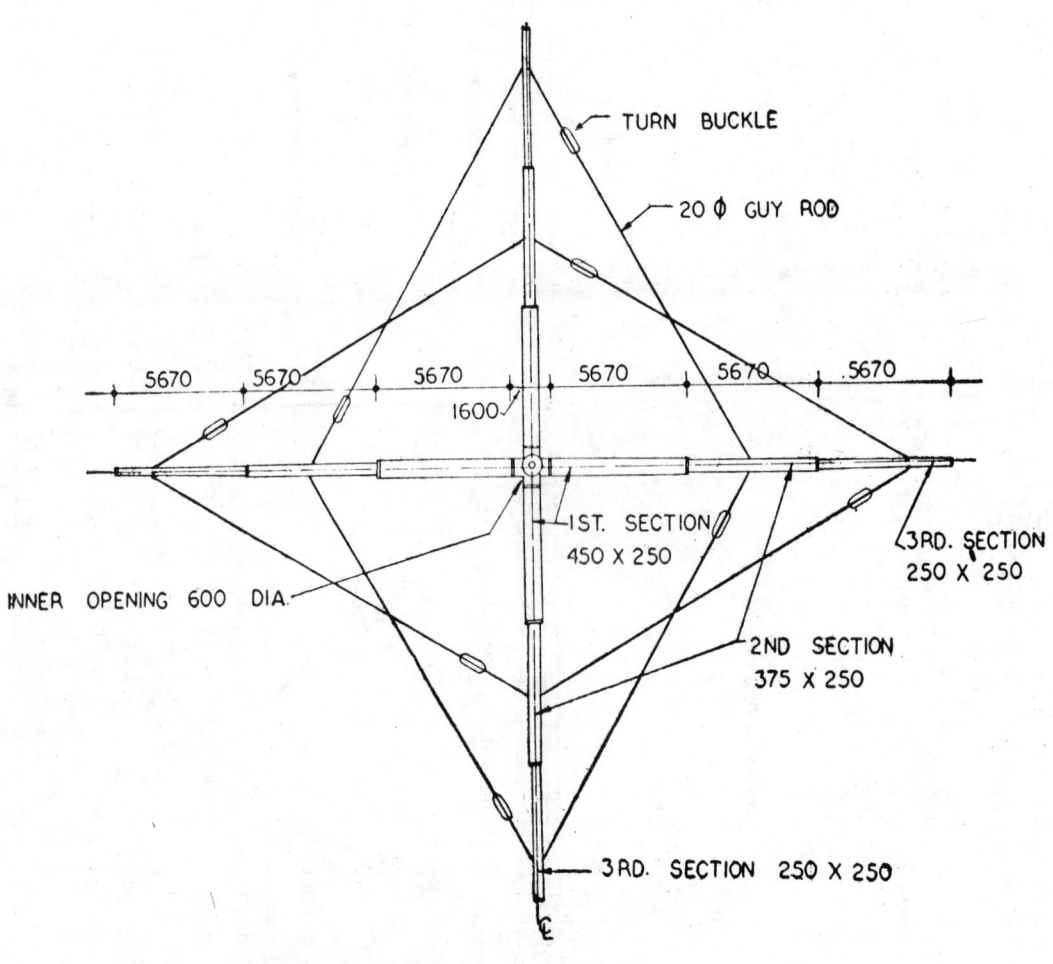

PLAN OF DISTRIBUTOR ARMS

Fig. 27·28.

Fig 27·29 illustrates the sectional view of a bio-filter turntable. This is required to support the moving manifold to which the distributor arms are attached. The turntable assembly has a mercury seal at its base to prevent leakage of sewage.

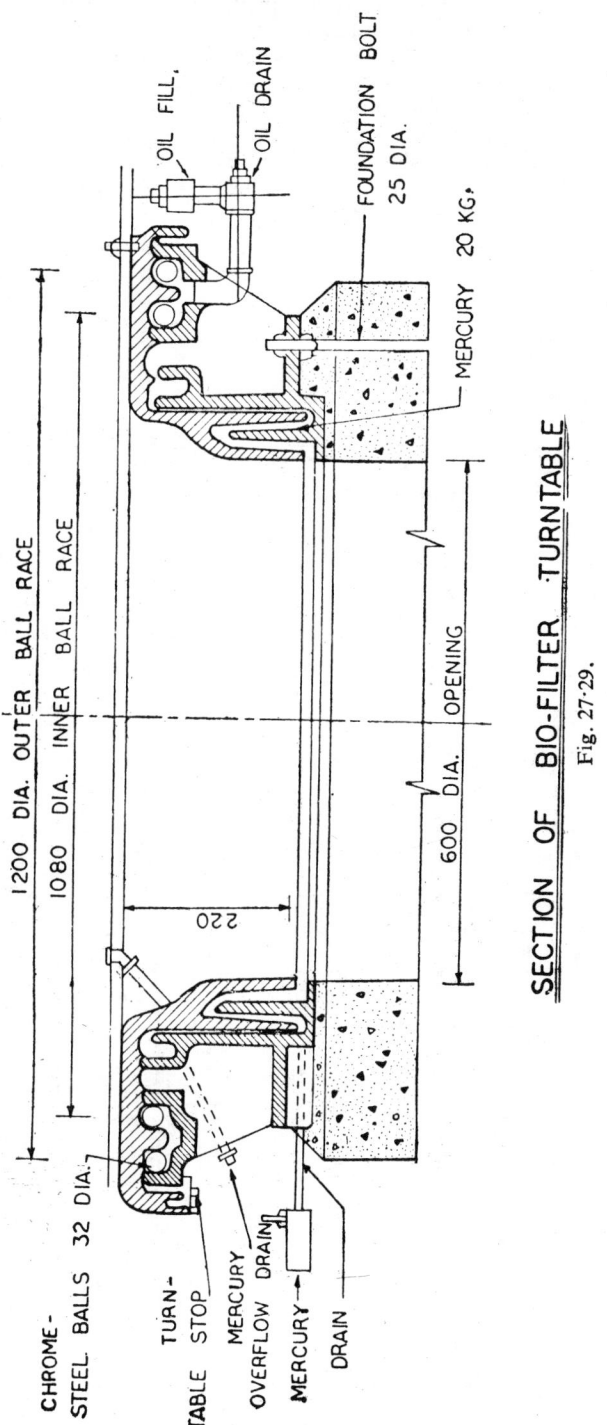

SECTION OF BIO-FILTER TURNTABLE

Fig. 27·29.

The underdrainage system is to collect the filtered sewage and soughed solids and to discharge the same to the main collecting channel. The undedrains are directly laid on the filter floor and they cover the entire floor area forming a false bottom. The underdrains are formed of precast concrete vitrified clay blocks with perforated concrete covers. The cross drains are with round inverts. Fig. 27·30 shows details of cross drains (plan and sectional view). The sizes of drains should be such that the flow occupies less than 50% of the cross sectional area. The total area of the inlet openings in the drain cover should not be less than 15% of the surface of the filter. The main collecting channel may be located along the diameter of the filter. Fig 27·31 shows the details of main drain (collecting channel). At the junctions of underdrains with the main collecting channel the levels should be such that there is a free fall from underdrains to the collecting channel.

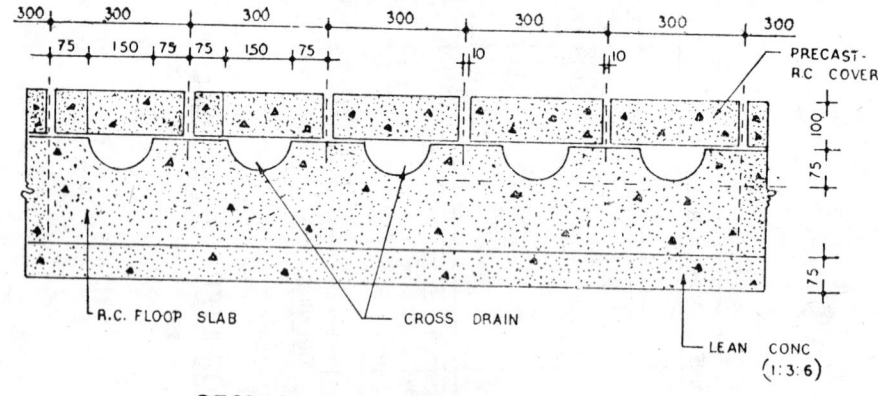

SECTION AT MM

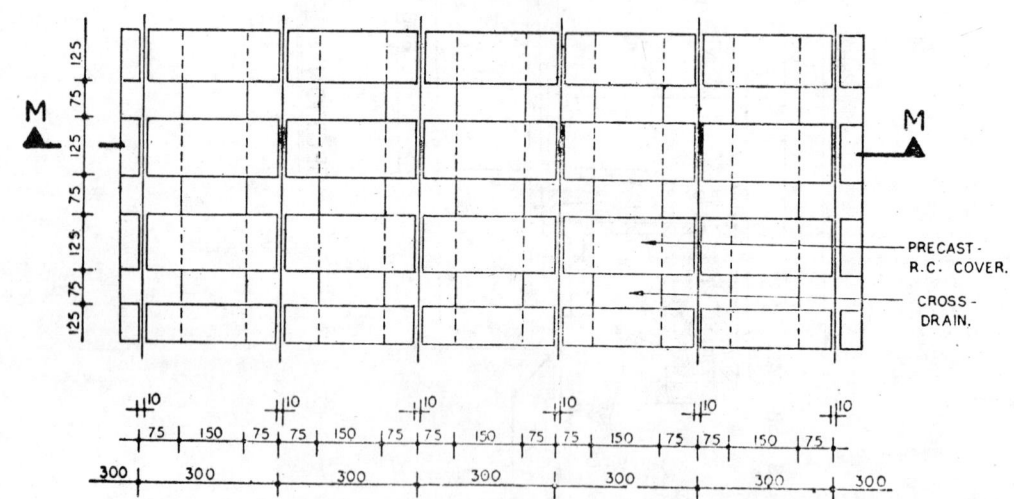

DETAILS OF CROSS DRAINS

Fig. 27·30.

Ventilation is essential for a trickling filter. For improved ventilation, a peripheral head channel on the inside of the filter with vertical vents is desirable. The vertical vents can also be used for flushing the underdrains. 1 m² of open grate ventilating manholes and vent stacks should be provided for every 250 m² of filter area. Fig. 27·32 shows sectional details of ventilation duct.

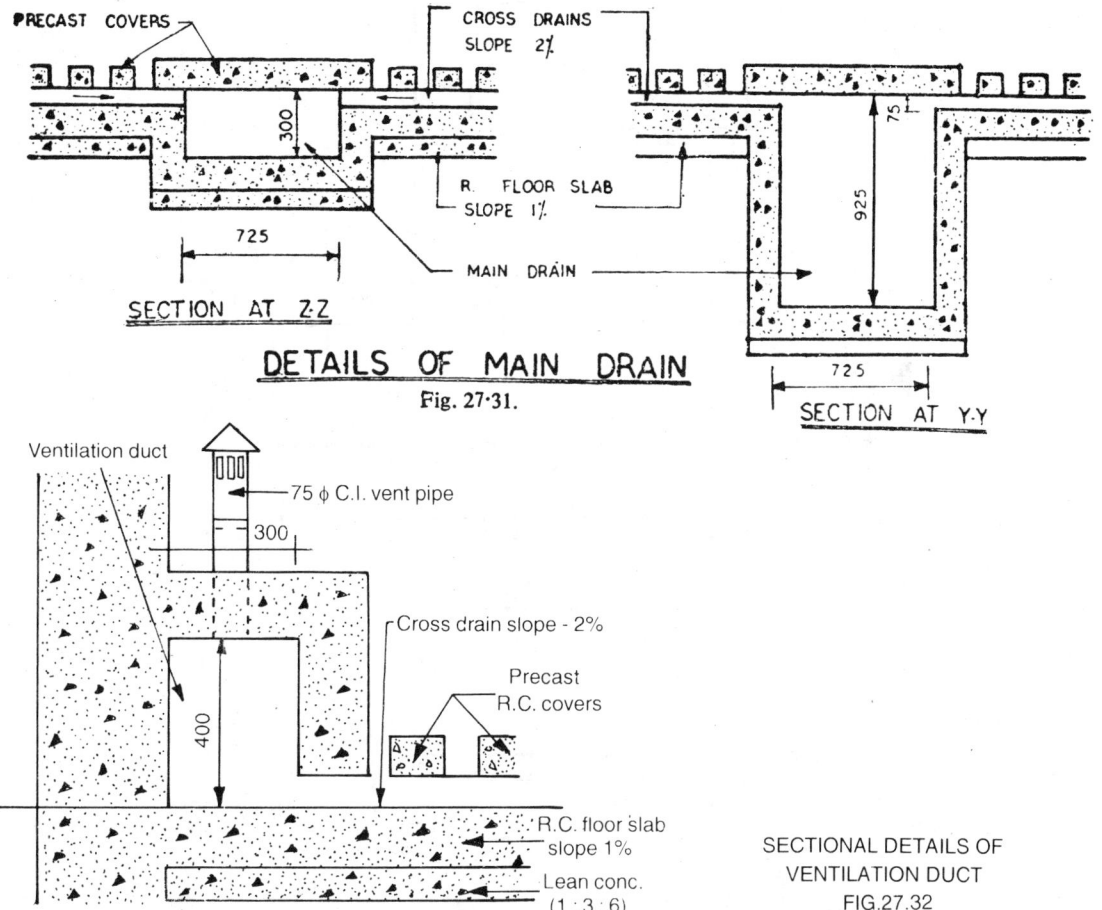

DETAILS OF MAIN DRAIN
Fig. 27·31.

SECTIONAL DETAILS OF
VENTILATION DUCT
FIG.27.32

SECONDARY CLARIFIER :

The sedimentation tank following the trickling filter is the secondary clarifier. The construction is similar to that of a primary clarifier. The side water depth of a secondary clarifier is 3 to 3·5 m. and the detention time is 1·5 to 2 hours for optimum result. The overflow rate for average sewage flow should be 10-25 m³/day/m² of surface area and the peripheral weir loading should not be greater than 100 m³/day/m. The peripheral weir should be provided with shallow V-notches, 50 mm. deep spaced at 0·15 to 0·3 m. apart. The sludge scraping mechanism is similar to that of a primary clarifier. The sludge conveying mains ...ould not be less than 200 mm. diameter. The entry ports in the central feed shaft are submerged 0·3 to 0·6 m. below water surface. The inlet baffle is placed concentric to the central feed shaft, with a diameter of 10-20% of the tank diameter and extending 1 to 2 m below the liquid surface. The mild steel inlet well is hung from the scrapper bridge. Fig. 27·33 shows plan and sectional view of a secondary clarifier. The details of inlet chamber are presented in Fig. 27·34. The sectional details of central feed shaft are shown in Fig. 27·35 The details of sludge chamber and outlet chamber are illustrated in Fig. 27·36 and 27·37 respectively.

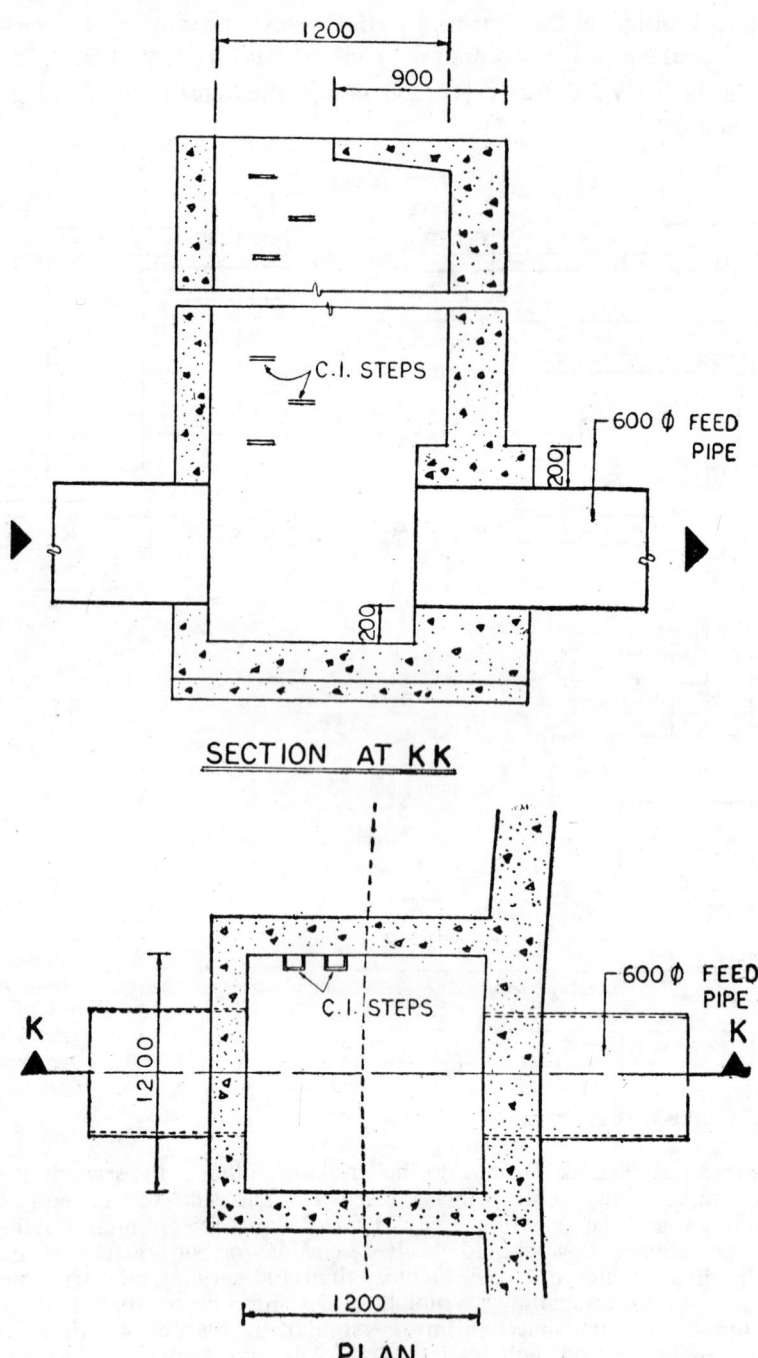

SECTION AT KK

PLAN

DETAILS OF INLET CHAMBER
Fig. 27·34.

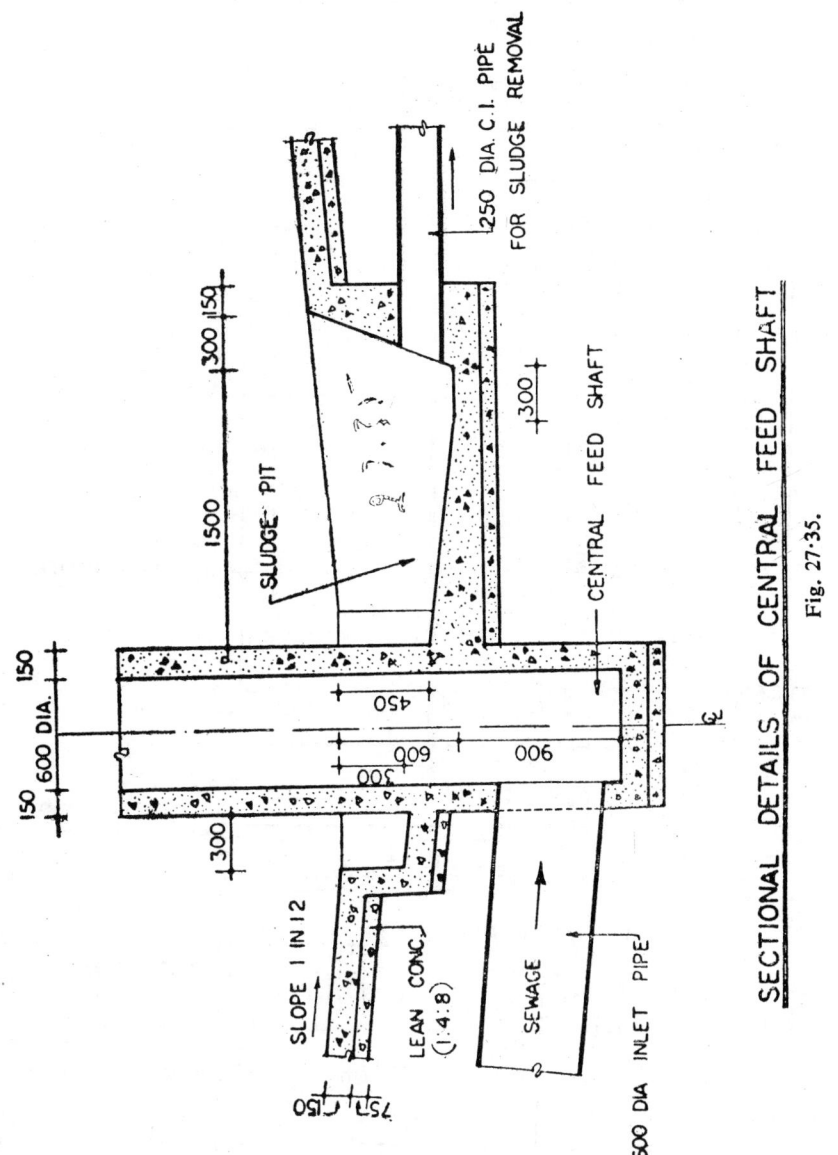

SECTIONAL DETAILS OF CENTRAL FEED SHAFT

Fig. 27·35.

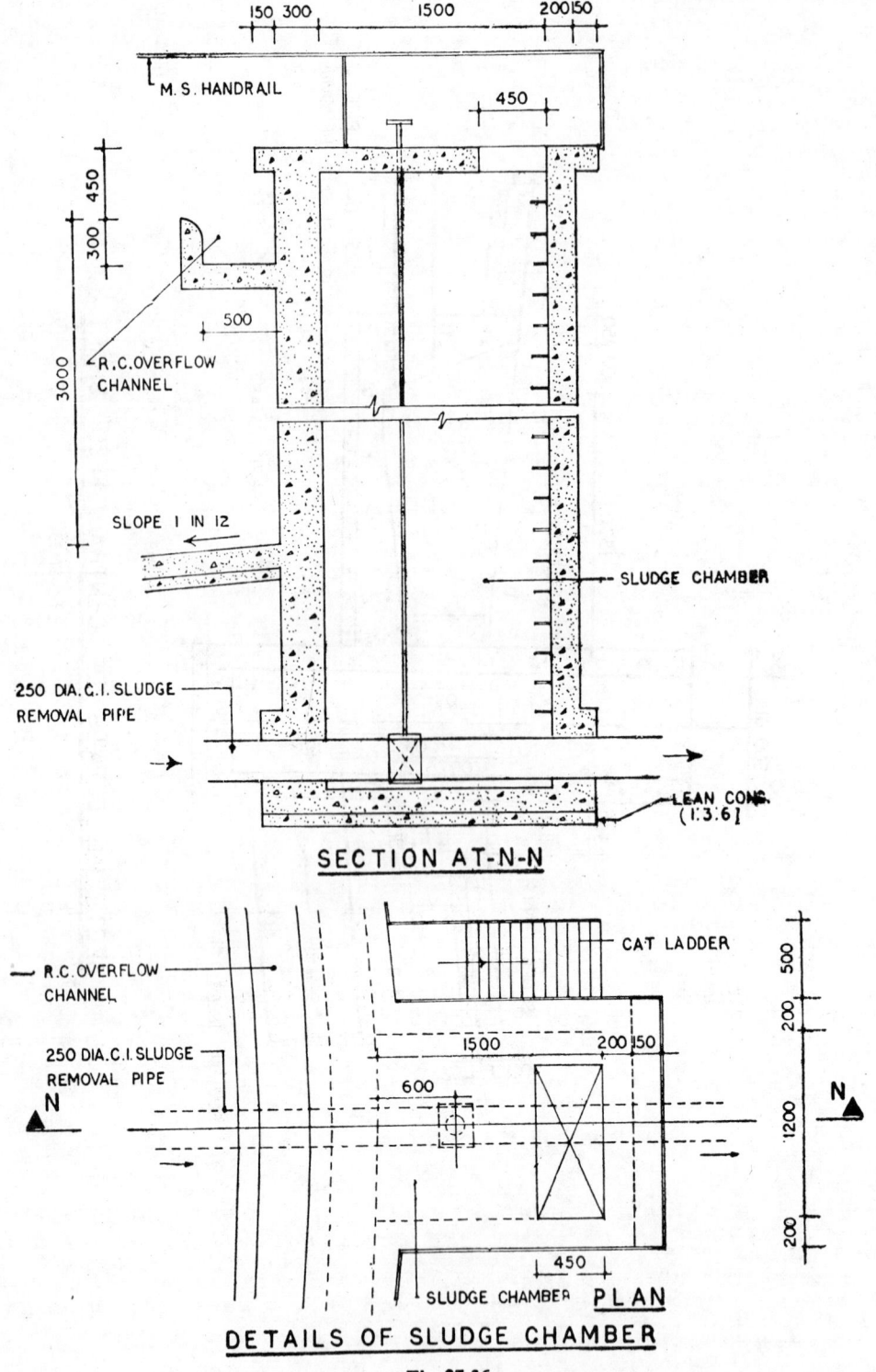

SECTION AT-N-N

DETAILS OF SLUDGE CHAMBER

Fig. 27·36.

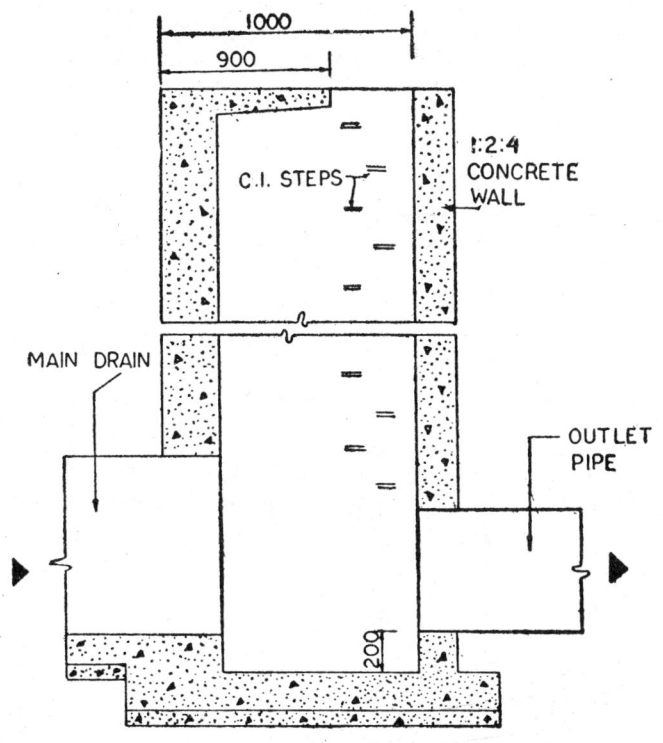

SECTION AT L-L

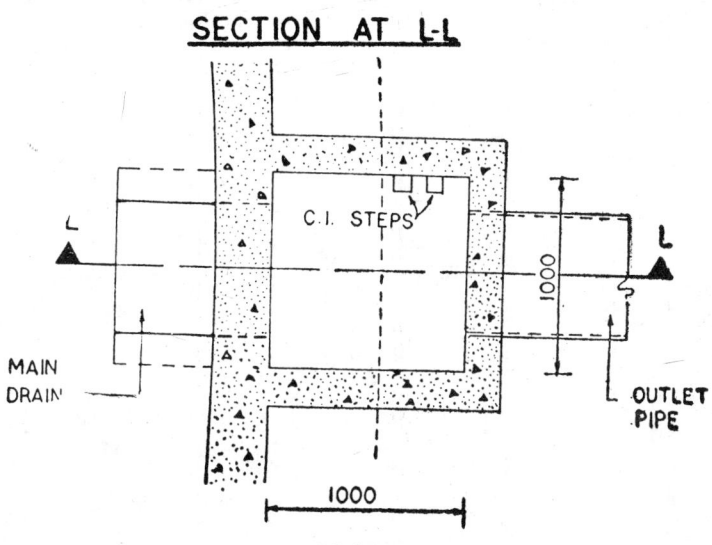

PLAN
DETAILS OF OUTLET CHAMBER
Fig. 27·37.

SLUDGE DRYING BED :

Fig. 27·38 shows the plan of sludge drying beds and the sectional view of a sludge bed is shown in Fig. 27·39.

The area required for sludge drying beds ranges from 0·1 to 0·15 m²/capita with dry solids loading of 80 to 120 kg/m² of bed per year. Commonly, the size of a drying bed is 6 to 8 m. wide and 10 to 30 m. long with a single point of sludge slurry discharge, the bed

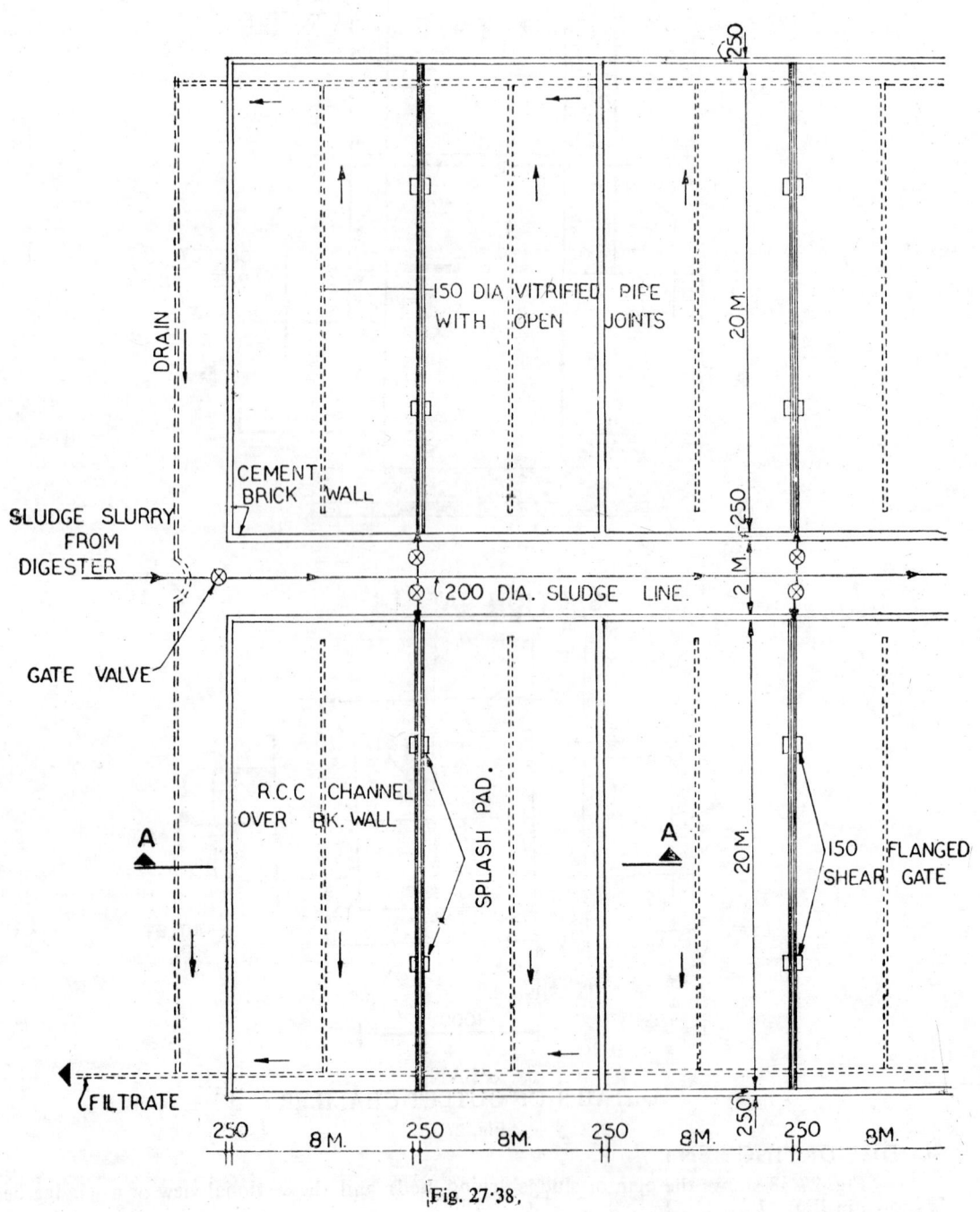

Fig. 27·38,

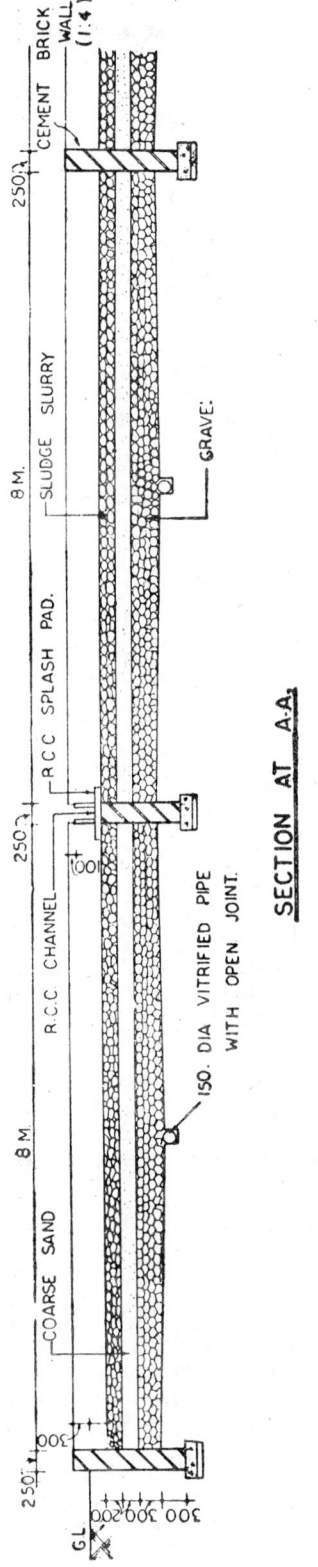

SECTION AT A·A.

Fig. 27·39.

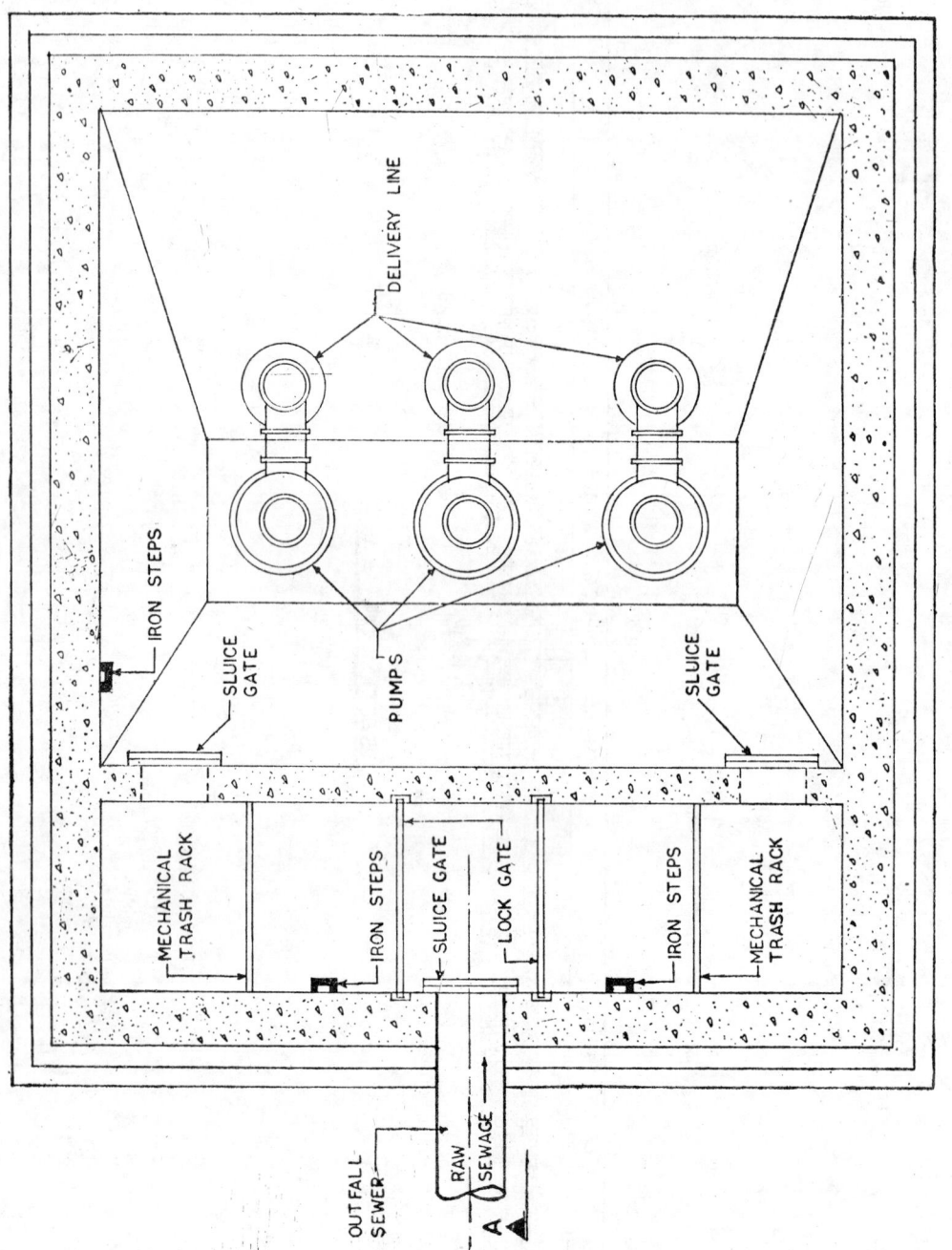

PLAN OF SUMP

Fig. 27·41.

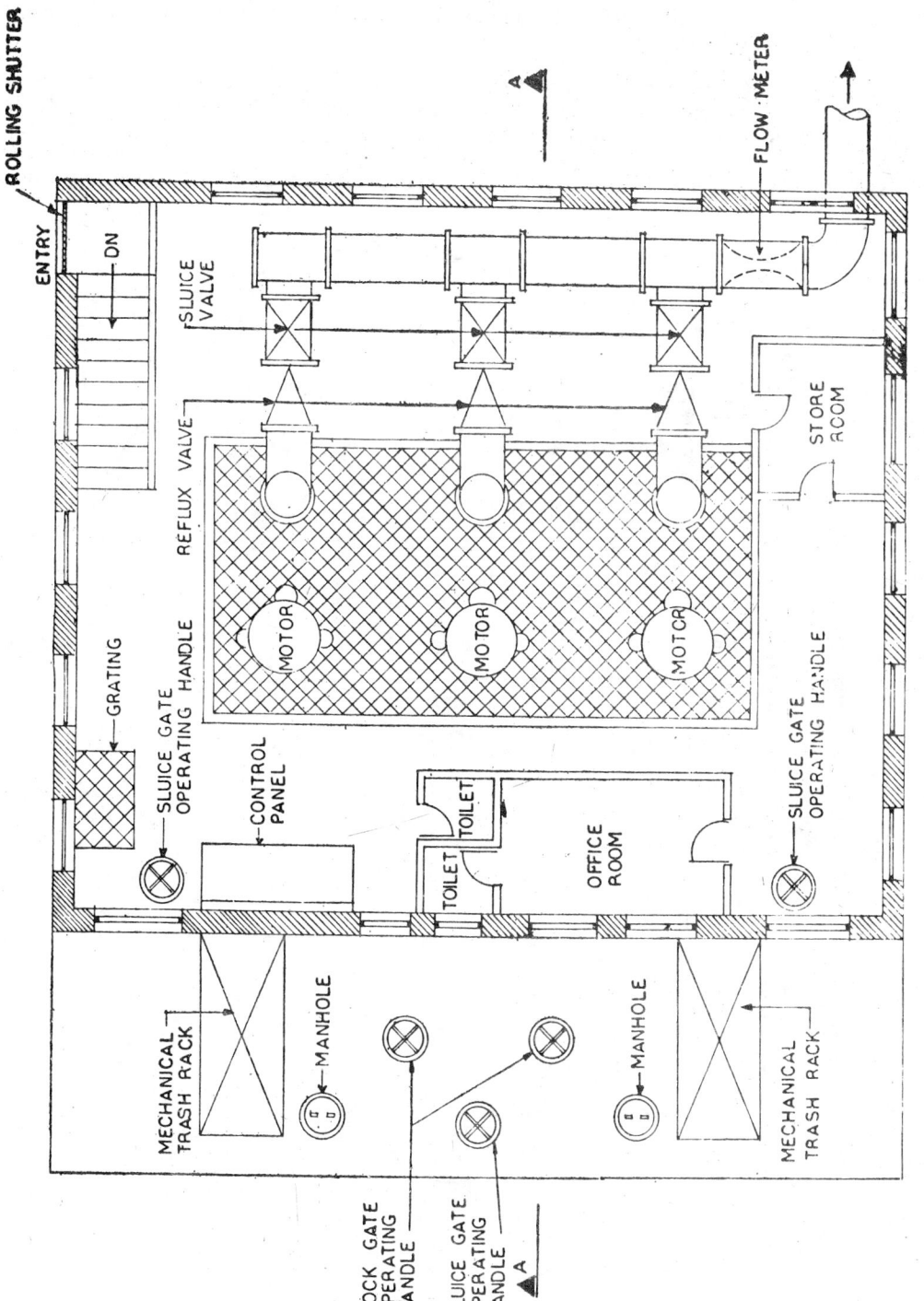

MOTOR ROOM FLOOR LEVEL

Fig. 27·42.

slope being 0·5% away from the inlet. The side wall of a sludge drying bed should extend at least 40 cm above and 15 cm below the sand surface.

A sludge drying bed consists of a bottom layer of graded gravel over which a layer of coarse sand is laid. For underdrains open-jointed tiles are used. Graded gravel is placed around the underdrains in layers upto 30 cm. depth with a minimum of 15 cm. above the top of the undergrains. At least 8 cm. of the top layer shall consist of 3 to 6 mm size gravel. The depth of sand may vary from 15 to 30 cm. The vertical clay pipes laid with open joints and

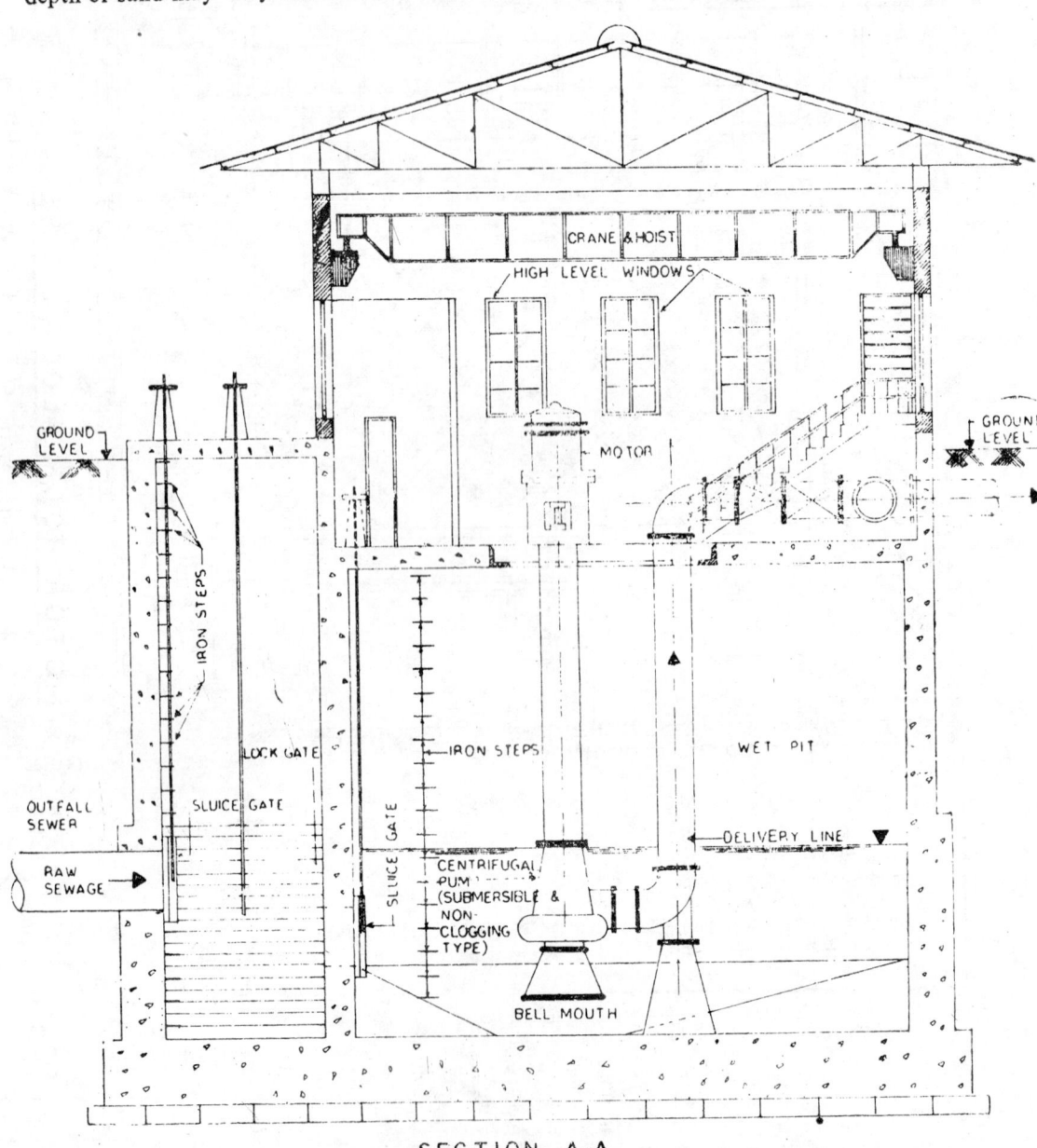

SECTION A-A
Fig. 27·43.

placed not more than minimum 200 mm. diameter terminating at least 30 cm. above the sand surface. Splash pads made of concrete should be used at every discharge point. The sludge slurryshould be deposited evenly to a depth of not greater than 20 cm.

Fig. 27·40 illustrates the details of biofilter rotary arm fitted to the manifold.

The plan of a sewage sump and pump house is shown in Fig. 27·41. The plan of the pump house at motor room floor level is presented in Fig. 27·42. The sectional view of the pump house by taking section through *A A* is shown in Fig. 27·43.

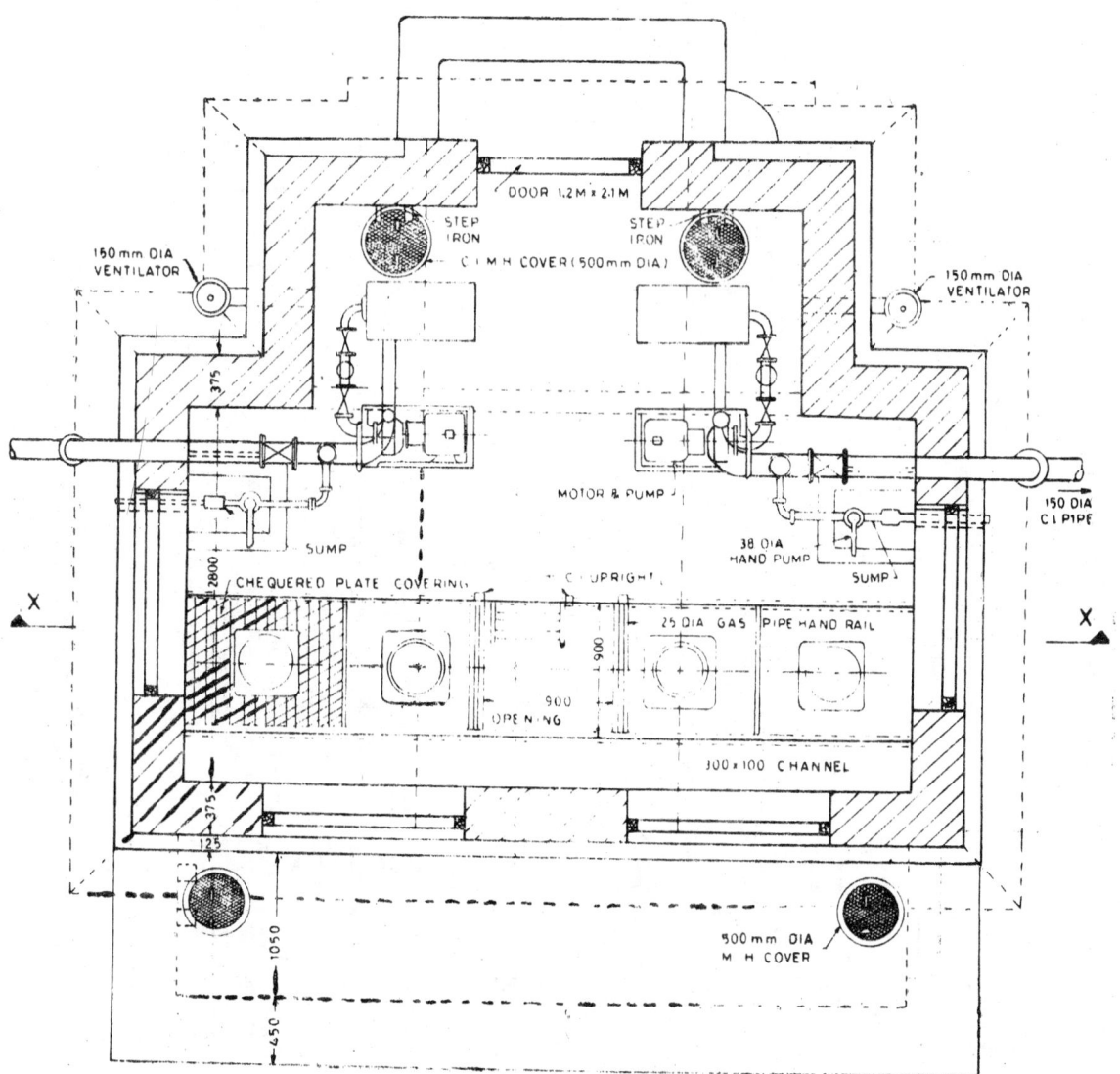

PLAN ON L L

Fig. 27·44.

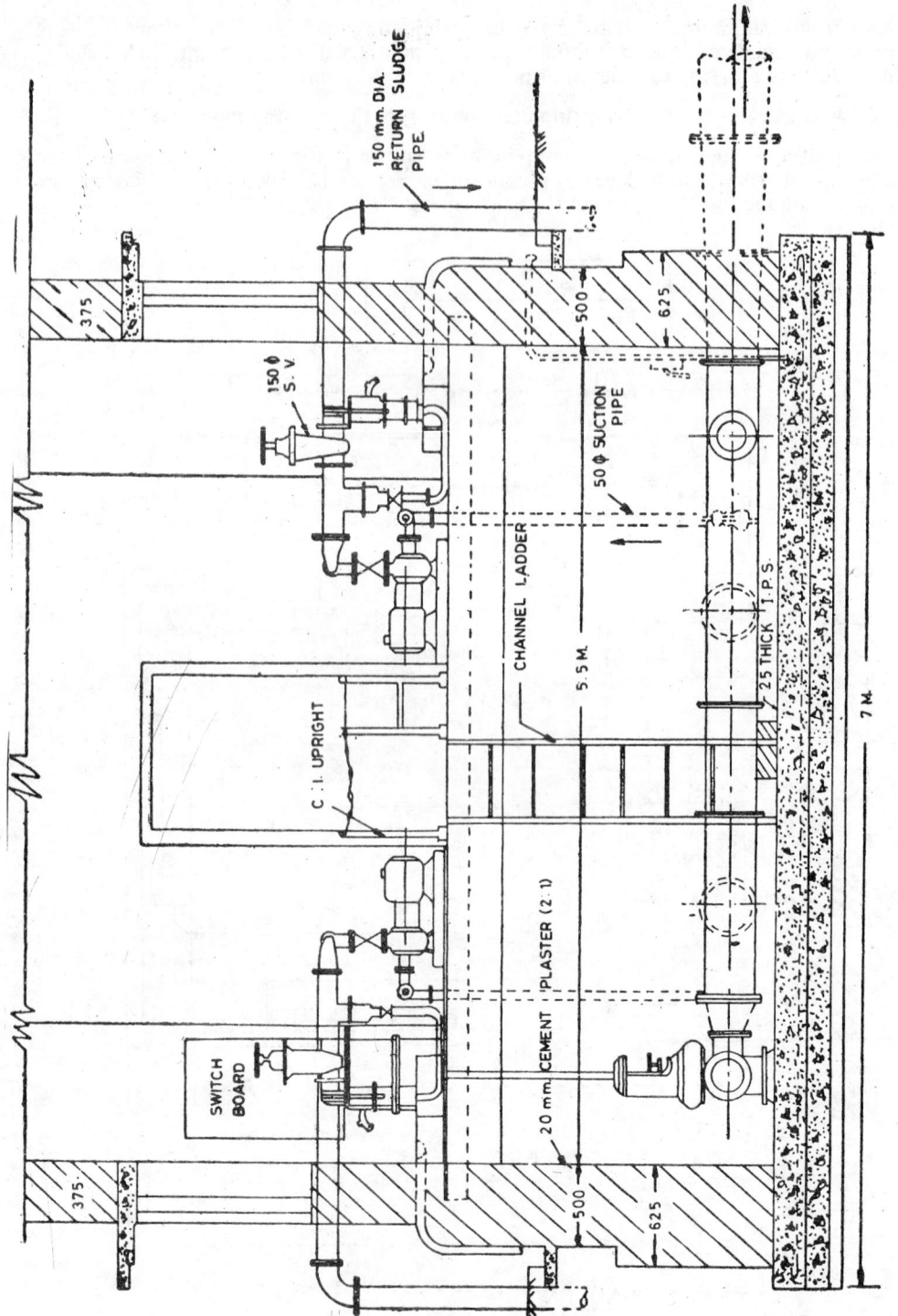

150 mm. DIA. RETURN SLUDGE PIPE

150 φ S. V.

50 φ SUCTION PIPE

500

625

CHANNEL LADDER

5.5 M.

25 THICK I.P.S.

C. I. UPRIGHT

7 M.

SWITCH BOARD

20 mm. CEMENT PLASTER (2:1)

500

625

375

375

SECTION ON X-X

Fig. 27·46.

The plan of another type of sewage pump house is presented in Fig. 27·44, the view being taken from the level *L-L* as shown in Fig. 27·47. The sectional view of the pump house is shown in Fig. 27·45, the section being taken on *X-X*.

The plan of the pump house viewed from the level *K-K* (Ref. 27·47) is shown in Fig. 27·46. Fig. 27·47 is the sectional view of the pump house by taking section on *Y-Y*.

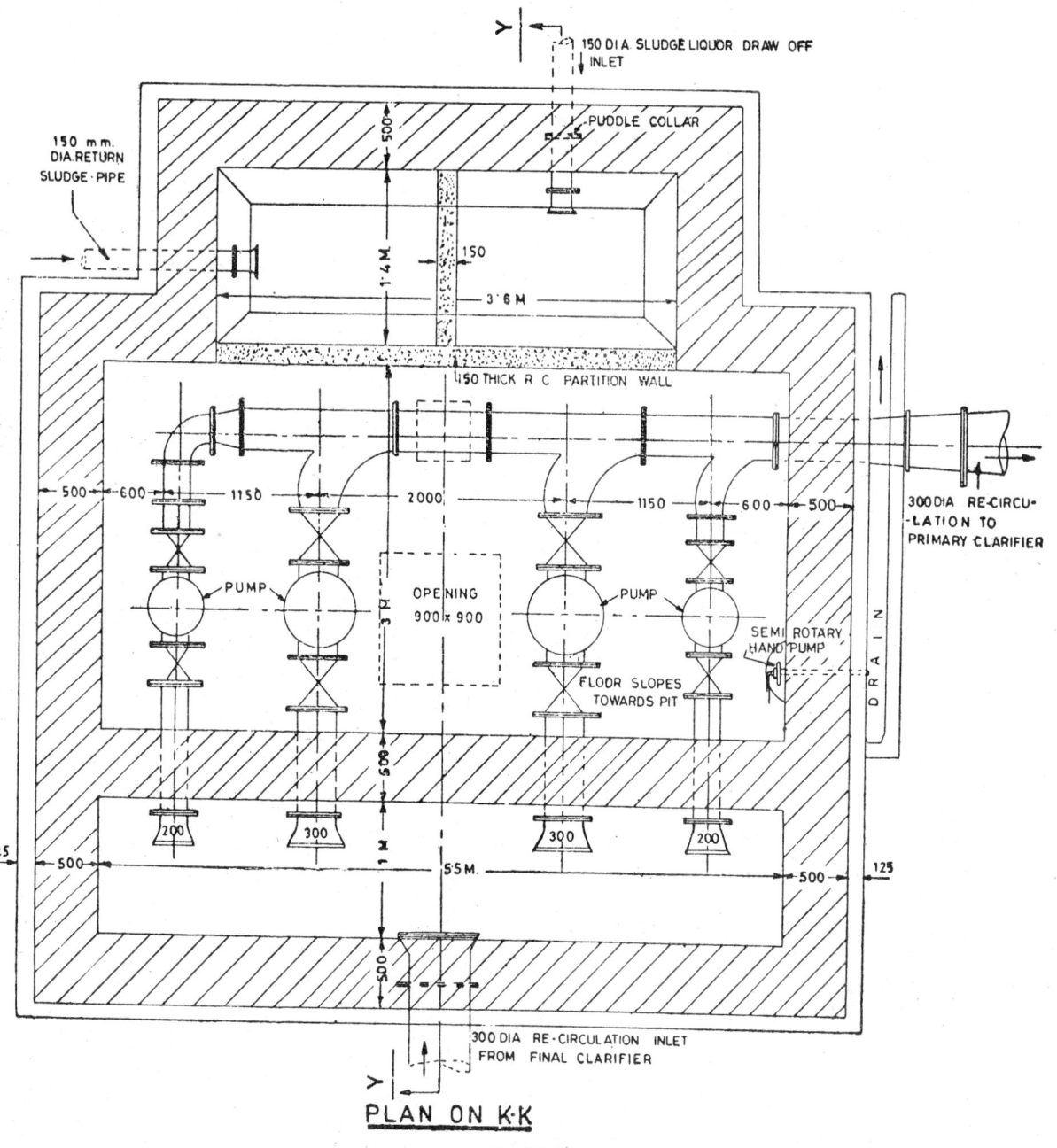

PLAN ON K·K

Fig. 27·46

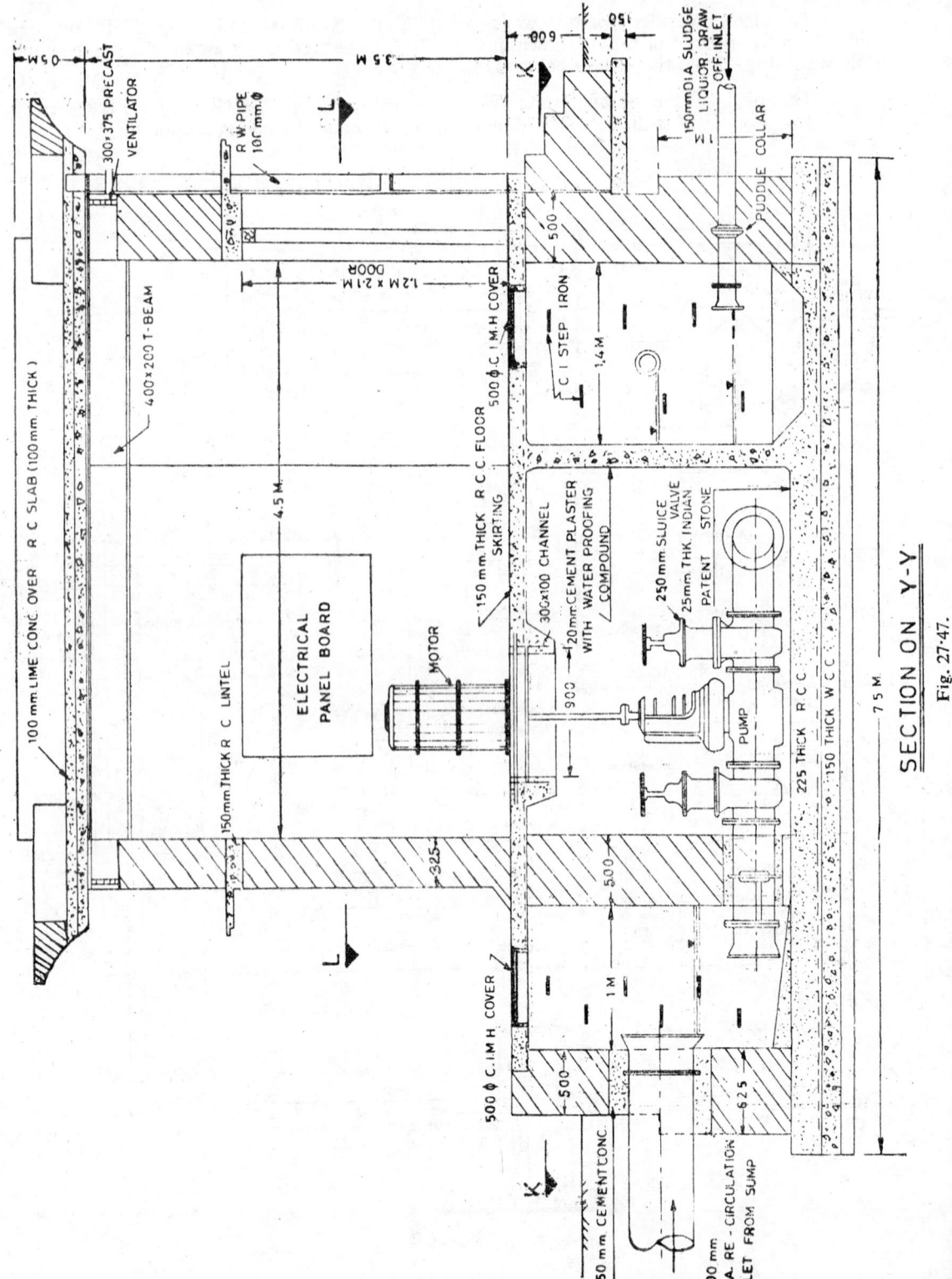

SECTION ON Y-Y

Fig. 27·47.

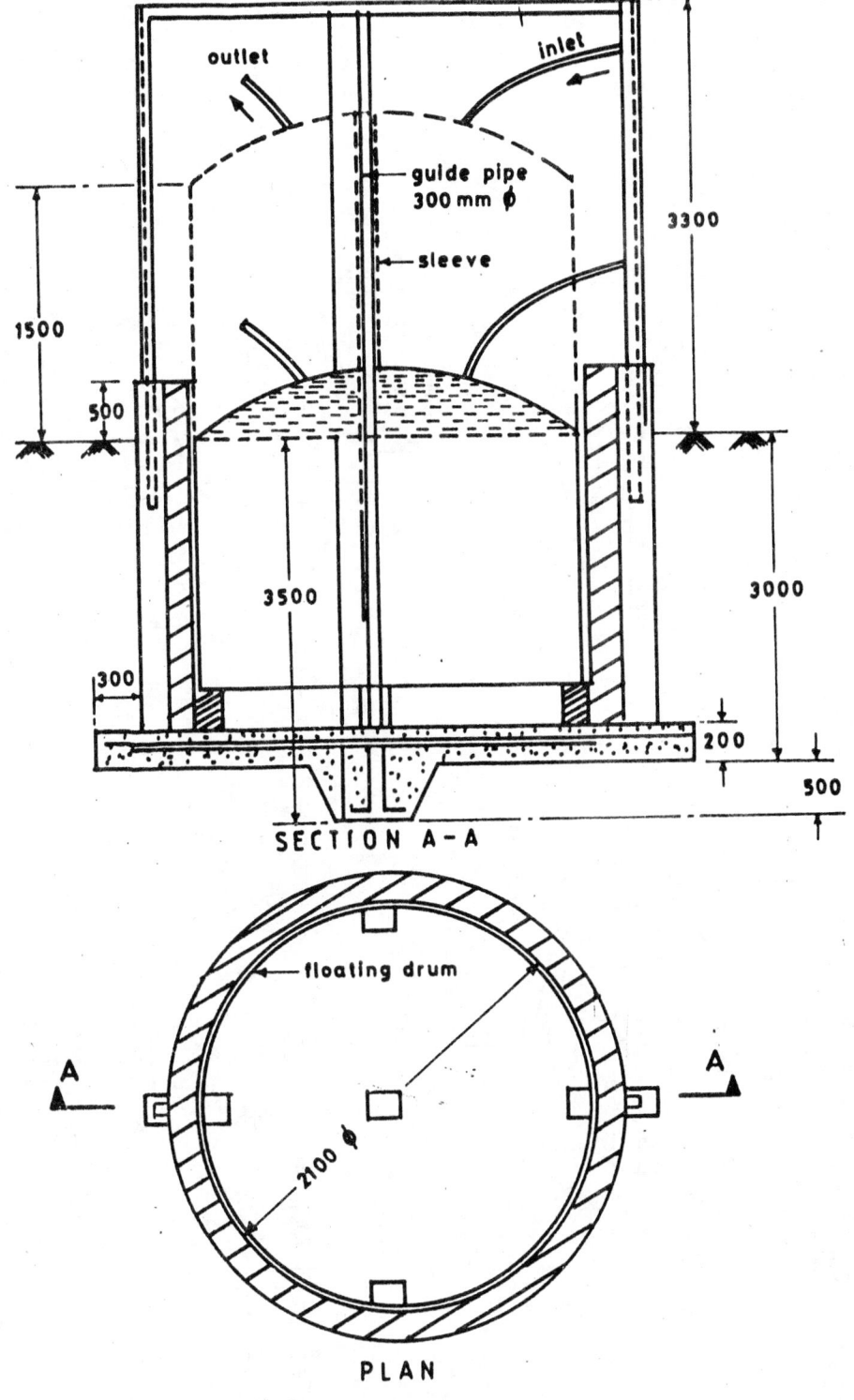

BIO - GAS HOLDER

Fig. 27·4·8.

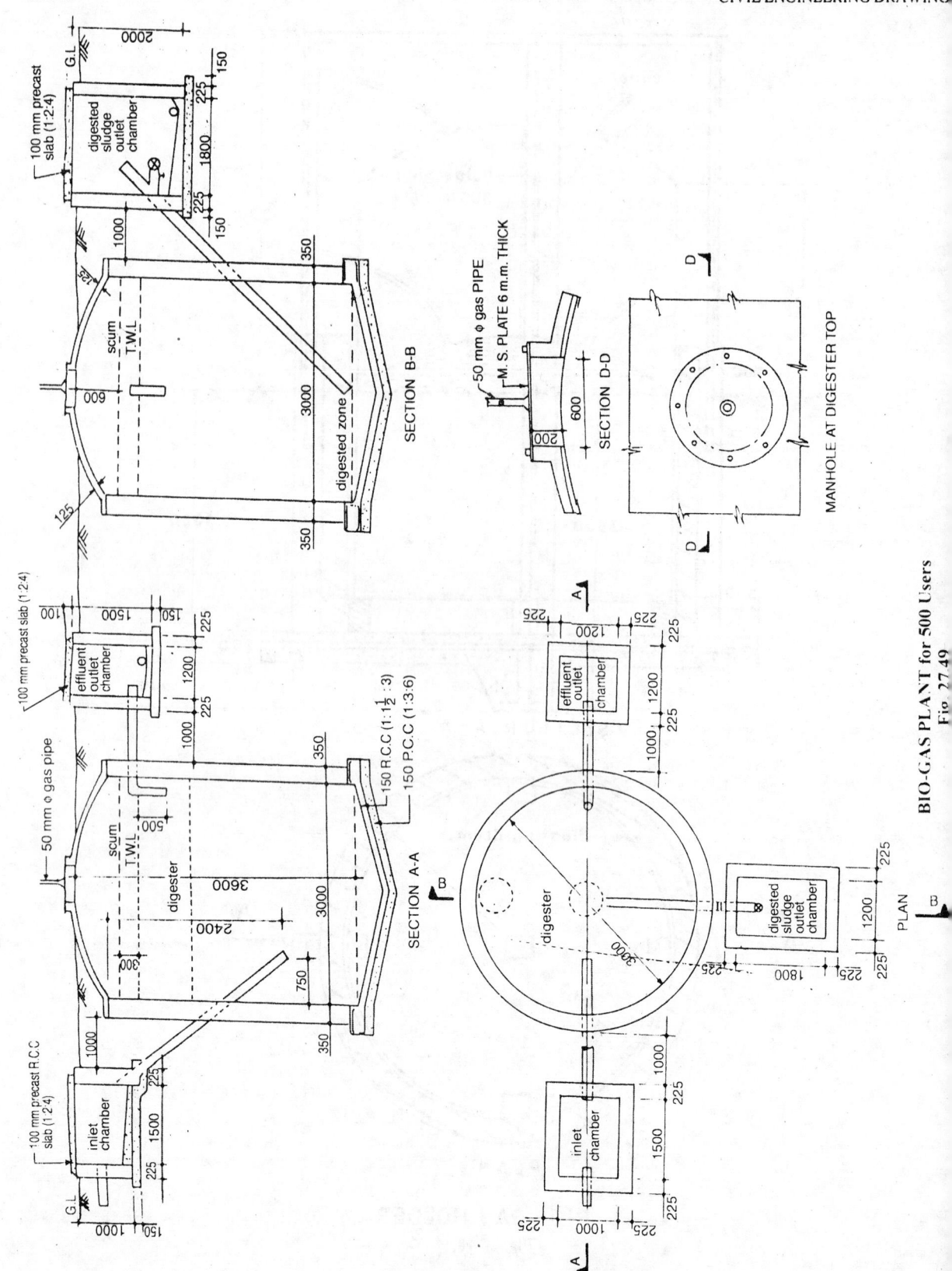

BIO-GAS PLANT for 500 Users
Fig. 27.49

28

Highways and Railway Tracks

This chapter deals with cross-sections of highways and Railway tracks.

The permanent land widths for different classes of roads are given below :

Class of roads	Normal width	Minimum width
National Highway (NH)	60 m	45 m
State Highway (SH)	45 m	30 m
Major District Roads	30 m	20 m
Other District Roads	20—25 m	15 m
Village Roads	10—15 m	10 m

The formation widths (minimun) for different classes of roads are :

Class of road	Formation width (minimum)	
	In plains	In hilly areas
National Highway	12 m	10 m
State Highway	10—12 m	8 m
Major District Roads	10 m	7 m
Other District Roads	8 m	6 m
Village Roads	6 m	5 m

The width of single-lane carriage way for NH, SH and district roads is about 4 m. For two-lane carriage way, the width is 7 to 7·5 m for both NH and SH.

Fig. 28·1 shows the cro s-section of a busy city road. The cross-section of a busy road with mixed traffic is presented in Fig. 28·2. Fig. 28·3 shows the cross-section of two types of hilly roads.

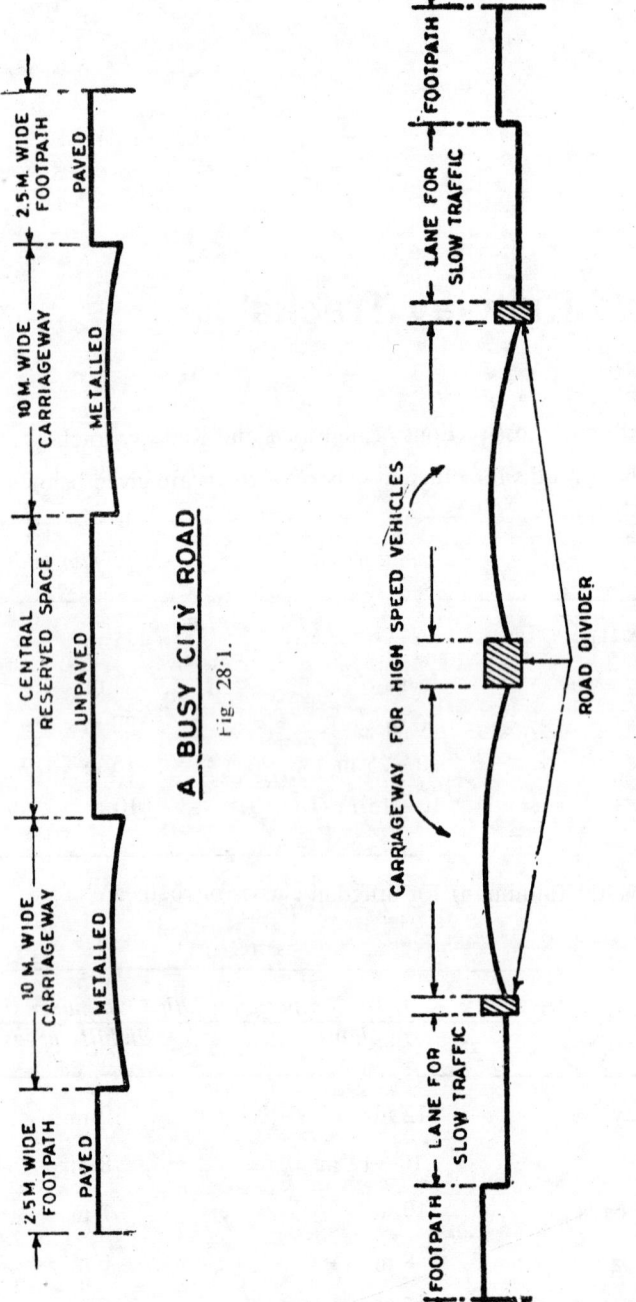

A BUSY CITY ROAD

Fig. 28·1.

A BUSY ROAD WITH MIXED TRAFFIC

Fig. 28·2.

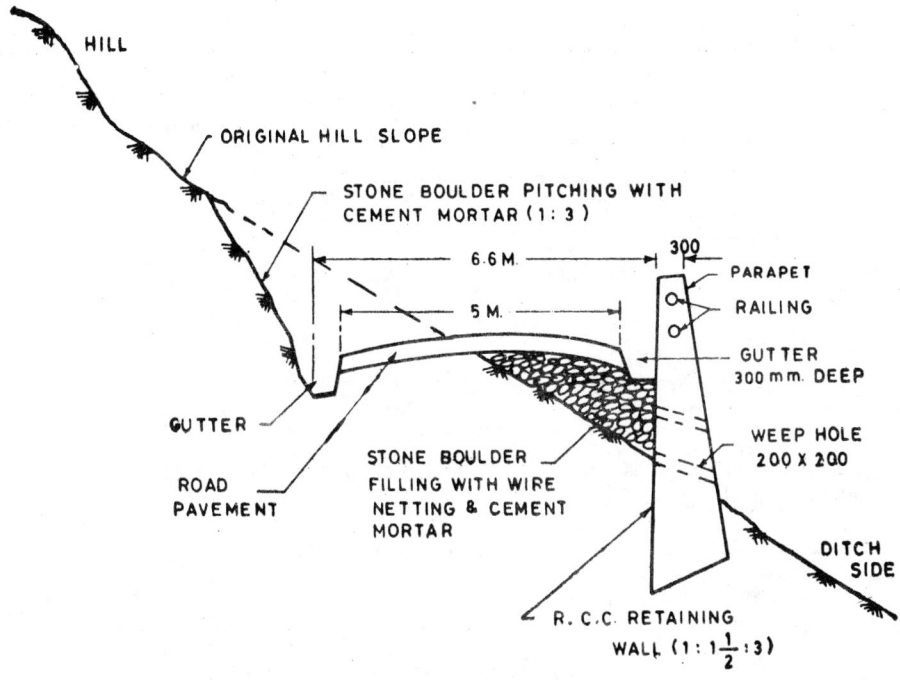

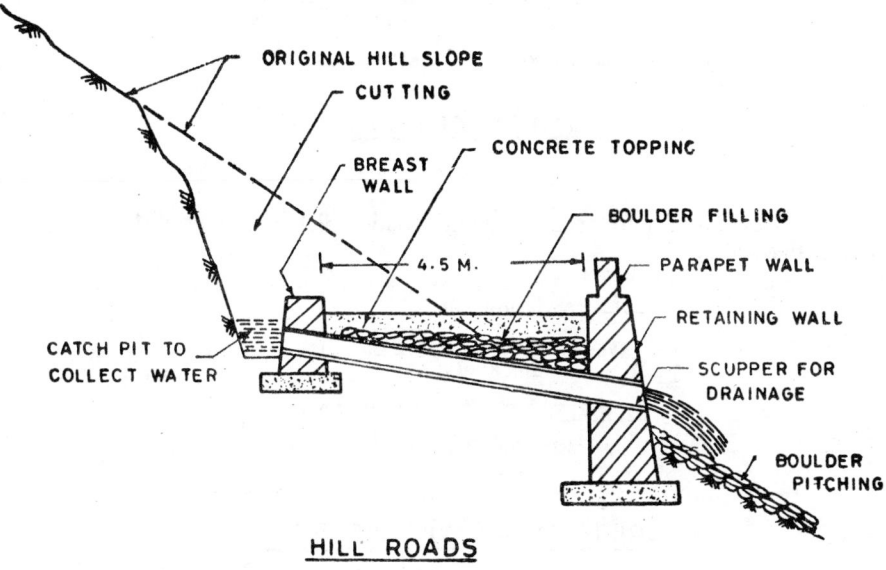

HILL ROADS

Fig. 28·3.

Fig. 28·4 shows the cross-section of a road in embankment. The dimensions will vary depending upon the road class.

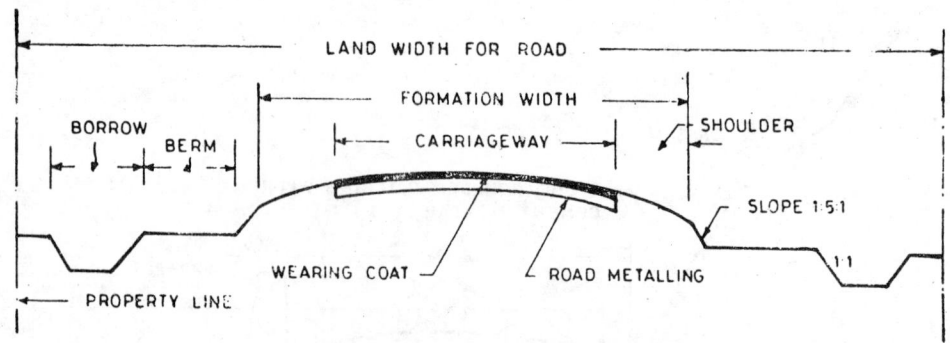

ROAD IN EMBANKMENT

Fig. 28·4.

Fig. 28·5 shows the cross-section of a road in cutting. The slope in cutting is generally kept 1 : 1 and that in filling 1·5 : 1 or 2 : 1.

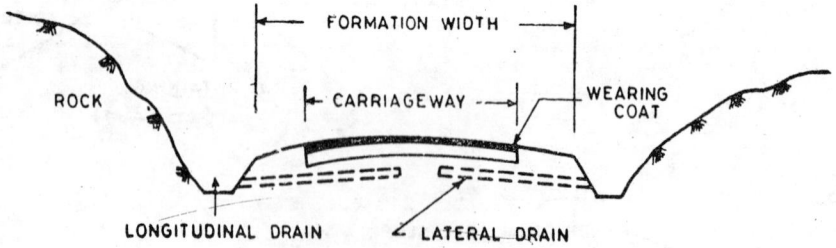

ROAD IN CUTTING

Fig. 28·5.

The cross-section of a creteway—a village road, which can be used for vehicular traffic is presented in Fig. 28·6.

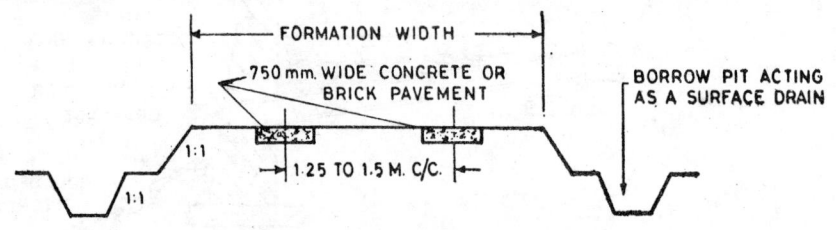

CRETEWAY – A VILLAGE ROAD

Fig. 28·6.

The cross-section of a single line railway track in embankment with the permanent land width required, dimensions of formation level, berm and borrow pit is shown in Fig. 28·7. The same for a single line in cutting is presented in Fig. 28·8. The sizes of sleepers and other dimensions for broadgauge, metre gauge and narrow gauge are also mentioned in the diagrams.

The formation widths both for single track and double track in cutting and in embankment for all the three gauges are given below in tabular form :

Railway Gauge	Minimum width in embankment		Minimum width in cutting	
	Single track	Double track	Single track	Double track
Broad gauge	6·10 m	10·67 m	5·49 m	10·06 m
Metre gauge	4·88 m	8·53 m	4·27 m	7·92 m
Narrow gauge	3·66 m	7·32 m	3·35 m	7·01 m

For embankment, the slope given is 1 : 1 or 2·5 : 1 and for cutting it is 1·5 : 1 or 1 : 1 in normal soil.

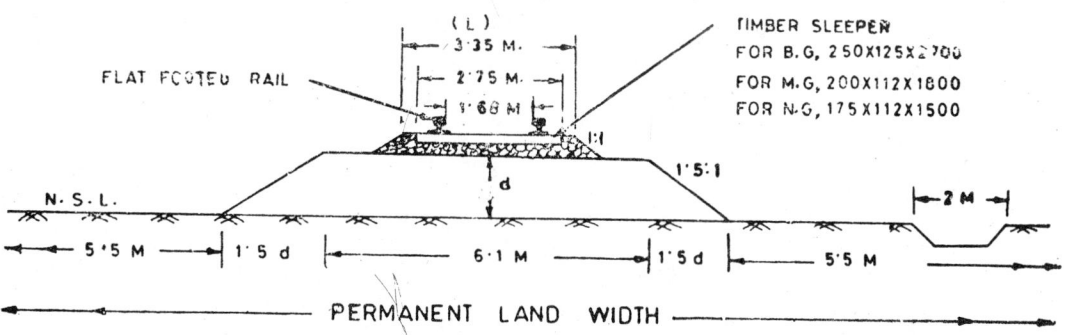

SINGLE LINE IN EMBANKMENT

Fig. 28·7.

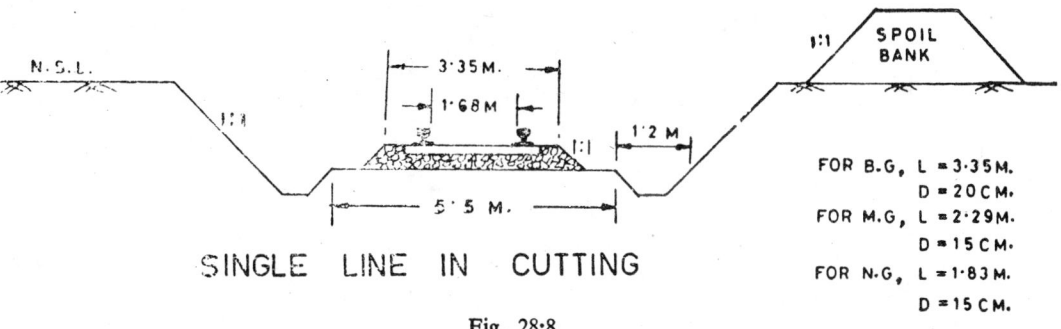

SINGLE LINE IN CUTTING

Fig. 28·8.

29

Culverts and Bridges

Culverts :

When a roadway or railway is to cross a small nullah or drain a culvert is constructed. A culvert is a small form of bridge. There are various types of culvert. These are : Arch culvert, slab culvert, Hume pipe culvert and Box culvert. These are shown in Fig. 29·1.

A culvert may be of single span or double span. The end walls are called 'abutments' and the middle one is called 'pier'. While the pier supports the superstructure only, the abutments support superstructure as well as retain earth on their back.

Pier : For arch culverts, the top width of pier upto 2 m. span is 50 cm. and for

$$2 \text{ m. to } 3 \text{ m. span} = \frac{\text{span}}{4}$$

$$3 \text{ m. to } 4·5 \text{ m. span} = \frac{\text{span}}{5}$$

$$4·5 \text{ m. to } 9 \text{ m. span} = \frac{\text{span}}{6}$$

$$9 \text{ m. to } 18 \text{ m. span} = \frac{\text{span}}{7}$$

For slab culverts, the top width of pier for a span upto 3 m. is 50 cm. and for a span ranging from

3 m. to 6 m. = 60 cm.
6 m. to 9 m. = 70 cm.
9 m. to 12 m. = 80 cm.

Abutment :

For arch culverts, the top width of abutment may be calculated by using the following formula :

$$\text{Top width} = \frac{\text{Radius of soffit}}{\text{Span length}} + \frac{\text{Rise of Arch}}{10} + 0·60 \text{ metre}$$

Alternatively, top width for 60° arch = 0·21 $l + c$
for 75° arch = 0·18 $l + c$
for 90° arch = 0·16 $l + c$
and for 120° arch = 0·15 $l + c$

where l = span length and c is a constant varying with the span.

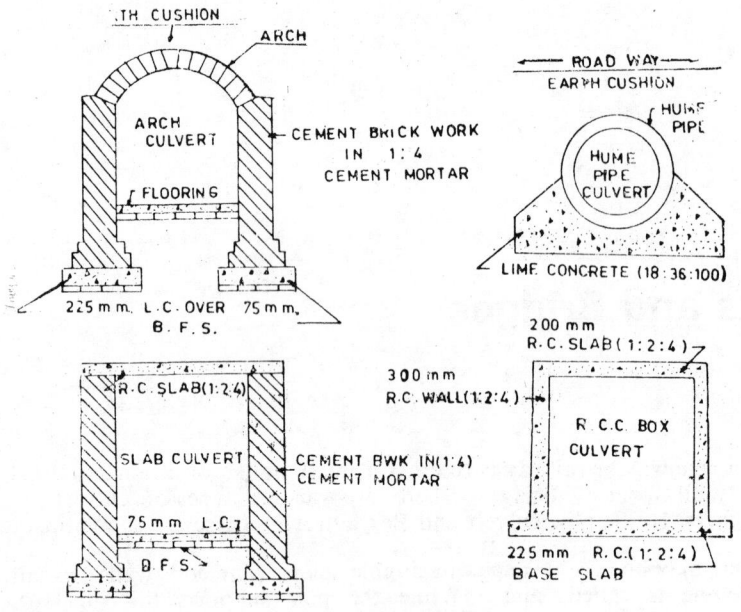

VARIOUS TYPES OF CULVERT

Fig. 29·1.

The values of c are given as :

For a span upto 1·5 m. $c = 50$ cm.

For a span 1·5 m. to 3 m., $c = 60$ cm.

and for 3 m. to 6 m., $c = 70$ cm.

As an example, for a 60° arch upto a span of 1·5 m. the top width of the abutment

$$= 0·21 \times 1·5 + ·5$$
$$= 0·815 \text{ m.} = 81·5 \text{ cm.}$$

The back batter of the abutment wall supporting an arch should be 1 in 2 for 60°, 75° and 90° arch and 1 in 3 for 120° to 180° arch.

Therefore, width of masonry at the top of concrete in foundation *i.e.*, at base is given as :

For 60°, 75° and 90° arch, width = top width $+ \dfrac{h}{2}$

and for 120° to 180° arch, width = top width $+ \dfrac{h}{3}$

where, h is the height from top of concrete in foundation to the springing level.

For slab culverts, the top width of abutment is given as under :

For a span of 3 m. to 6 m., top width = 50 cm.

For a span of 6 m. to 9 m., top width = 60 cm.

For a span of 9 m. to 12 m., top width = 70 cm.

Base width of abutment should be 0·4 to 0·5 h, where h is the height of the road level from the top of concrete in foundation.

Wing Walls :

Wing walls are the walls constructed on both sides of the abutments to retain the soil in embankment. The top width of wing wall is usually 30 to 40 cm, the length of wall being 1·5 to 2 times the height. The width of wing wall at the top of the concrete in foundation should be 0·35 to 0·40 times the height of the road level from the top of concrete in foundation. Wing walls are made either at 90° to the abutment or splayed at an angle of 45° and accordingly they are called 'Return wing wall' and 'splayed wing wall'.

Thickness of Arch :

For arch culverts, mostly segmental arch is used. The thicknses of arch is determined by using the following formulae :

(i) Rankine's formula—

$$T=\sqrt{0·12\ R} \text{ ...for single arch culvert}$$

$$T=\sqrt{0·17\ R} \text{ ...for multi arch culvert.}$$

where, T is thickness of arch ring and R is radius of arch soffit.

(ii) French formula—

Thickness, $T=\dfrac{l}{30}+1·1$; where l is span of arch.

(iii) Hurst's formula—

Thickness, $T=N\sqrt{R}$; where N is a constant and R is radius of arch soffit.
The value of N is 0·40 for single span and 0·45 for multi span.

Of all these three formulae, Hurst's formula is mostly used for ordinary arch culverts. But, for culverts embedded under high fill, the thickness obtained by Hurst's formula should be increased by 50%.

Haunch Filling :

The depth of haunch filling on pier and abutment should be calculated as $\dfrac{r+t}{2}$; where r is the rise of arch and t is thickness of arch ring.

Cover :

The earth filling over the crown of the arch including road metalling should be 60 cm. minimum.

Depth of foundation :

The depth of foundation of a abutment or a pier should be at least 1·2 m. below the floor level of the culvert.

Protection of bed :

For small culverts, pucca floor should be constructed 30 cm. below the bed level of the water course and curtain walls should be provided at either end of the floor.

Culvert Approach :

The approach road on either side of the culvert should have a minimum straight length of 6 metres and preferably 15 metres. The maximum slope of approach should not exceed 1 in 30.

Parapet wall, Railing and wheel guard :

For safety, the height of parapet wall with railing should be 1 m. minimum and wheel guard should be provided on both sides of the roadway crossing the culvert.

The component parts of an arch culvert are presented in Fig. 29·2.

Bridges :

Half-sectional plan, elevation and sectional views of wing wall of a R.C.C. box type bridge are presented in Fig. 29·9. The details of reinforcement for a box type R.C.C. bridge across a channel are shown in Fig. 29·10.

The following views are illustrated in Fig. 29·11.

(a) Elevation of R.C.C. bridge railing ;

(b) Cross-section of safety kerb ;

(c) Cross-section and sectional side elevation of precast handrail ;

(d) Cross-section of railpost ;

(e) Sectional plan and sectional elevation showing details of connection of handrail with railpost ;

(f) Sectional plans of end post and intermediate post.

Fig. 29·12 shows elevation of railway with concrete railpost. The cross-section of concrete railpost is shown in Fig. 29·13.

The following views are illustrated in Fig. 29·14 :

(a) Elevation of railing with and without safety kerb ;

(b) Cross-section of railpost ;

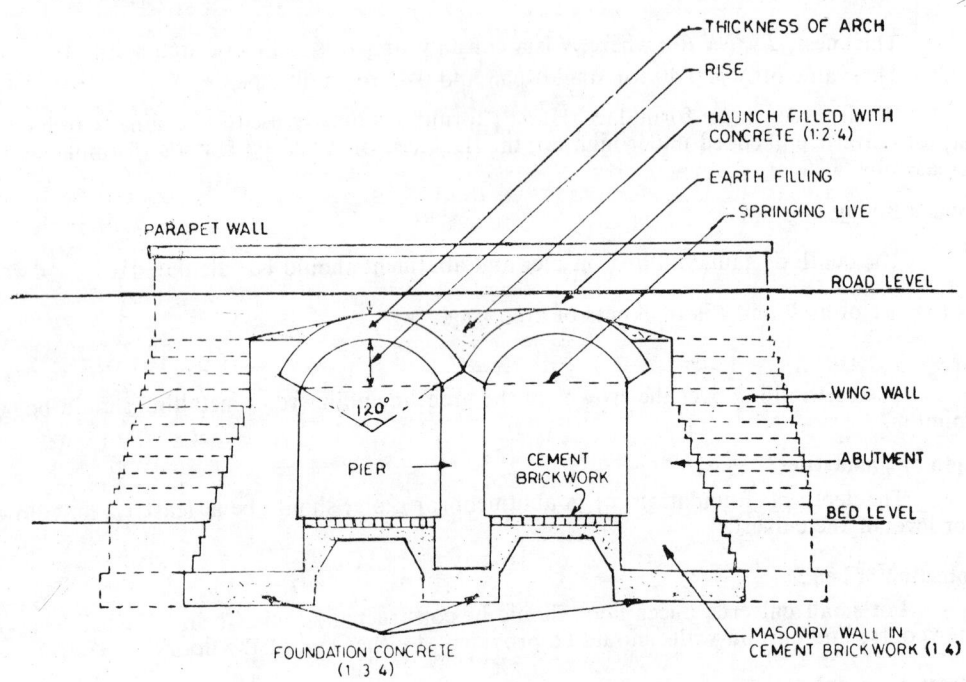

AN ARCH CULVERT

Fig. 29·2.

(c) Cross-section of precast handrail ;

(d) Sectional plan and sectional elevation showing details of fixing of handrail;

(e) Cross-section of safety kerb ;

(f) Sectional plans of end post and intermediate post.

Fig. 29·15 shows the following views :

(a) Elevation of railing ;

(b) Cross-sections of railpost ;

(c) Sectional plans of end post and intermediate post ;

(d) Precast member of railing ;

(e) Cross-section of handrail.

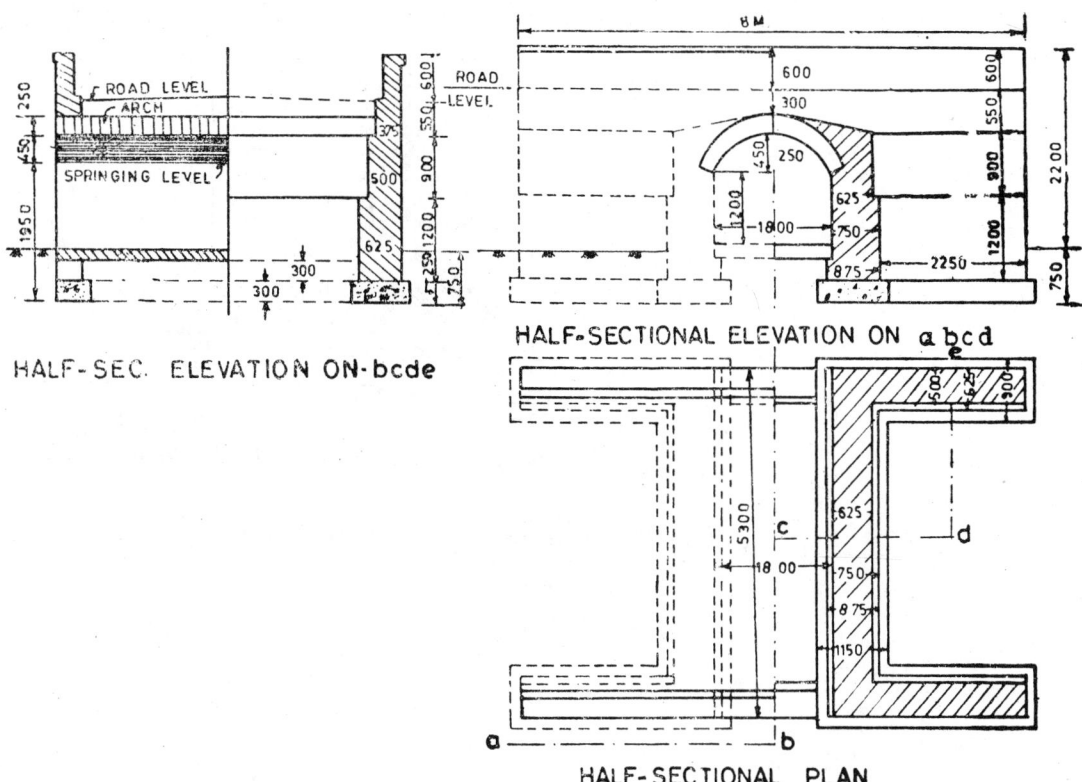

HALF-SEC. ELEVATION ON·bcde

HALF-SECTIONAL ELEVATION ON abcd

HALF-SECTIONAL PLAN

ARCH TYPE CULVERT SHOWING WATER COURSE, ROADWAY, PARAPET WALL, ETC.

Fig. 29 3.

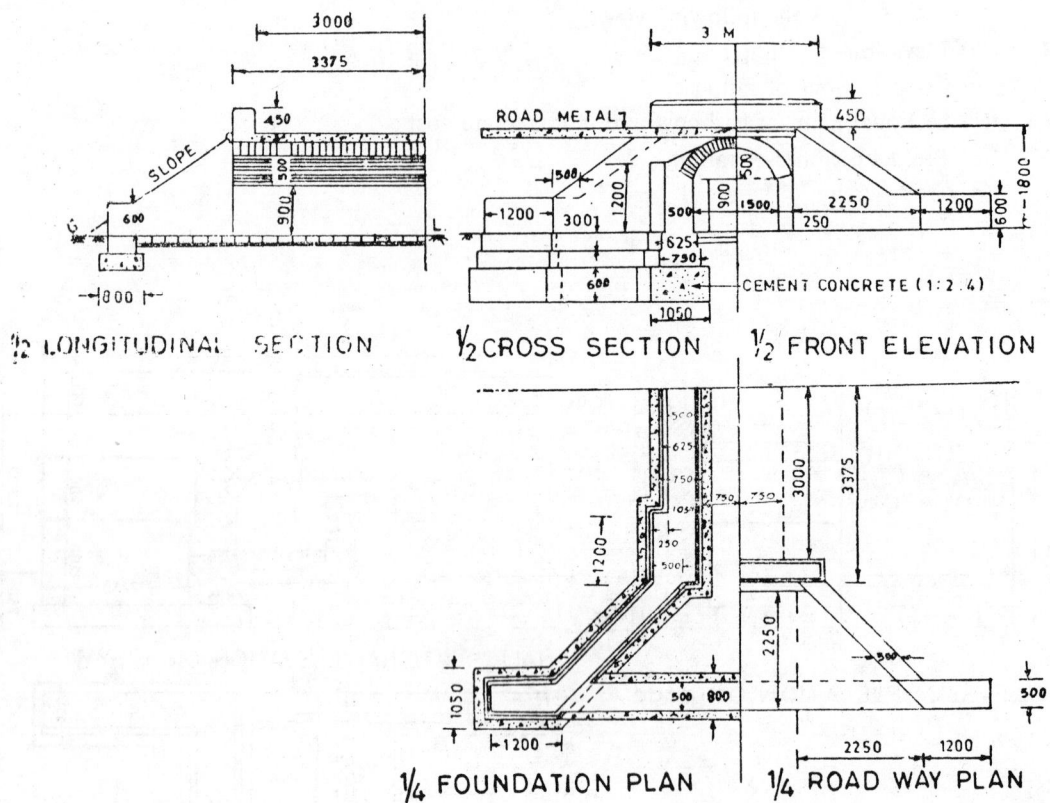

½ LONGITUDINAL SECTION   ½ CROSS SECTION   ½ FRONT ELEVATION

¼ FOUNDATION PLAN   ¼ ROAD WAY PLAN

ARCH TYPE CULVERT WITH SPLAYED WING WALL

Fig. 29·4.

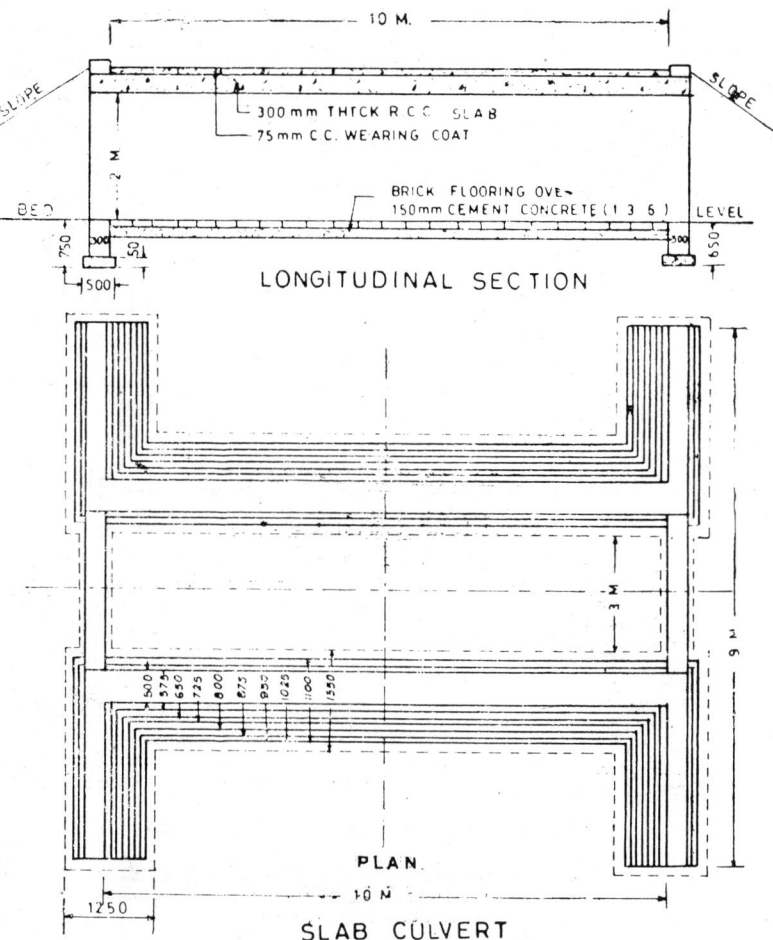

LONGITUDINAL SECTION

PLAN

SLAB CULVERT

Fig. 29·5.

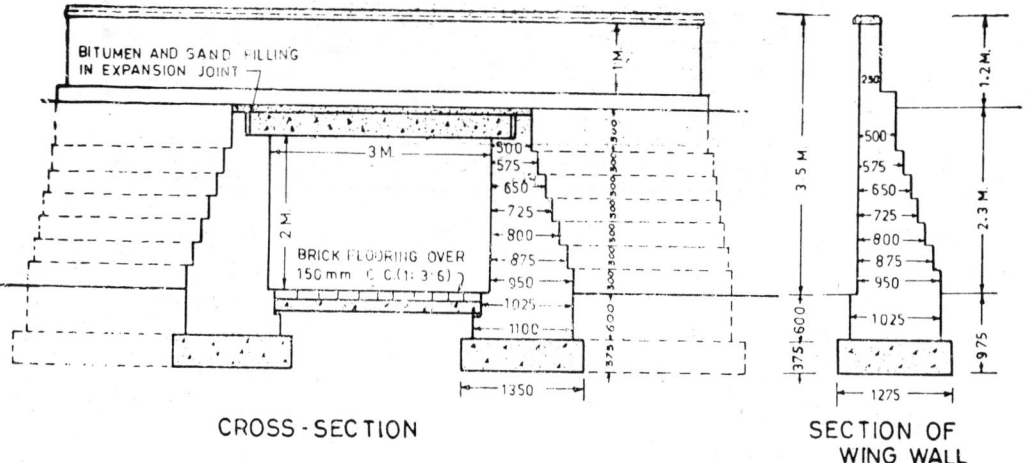

CROSS-SECTION

SECTION OF
WING WALL

Fig. 29·6.

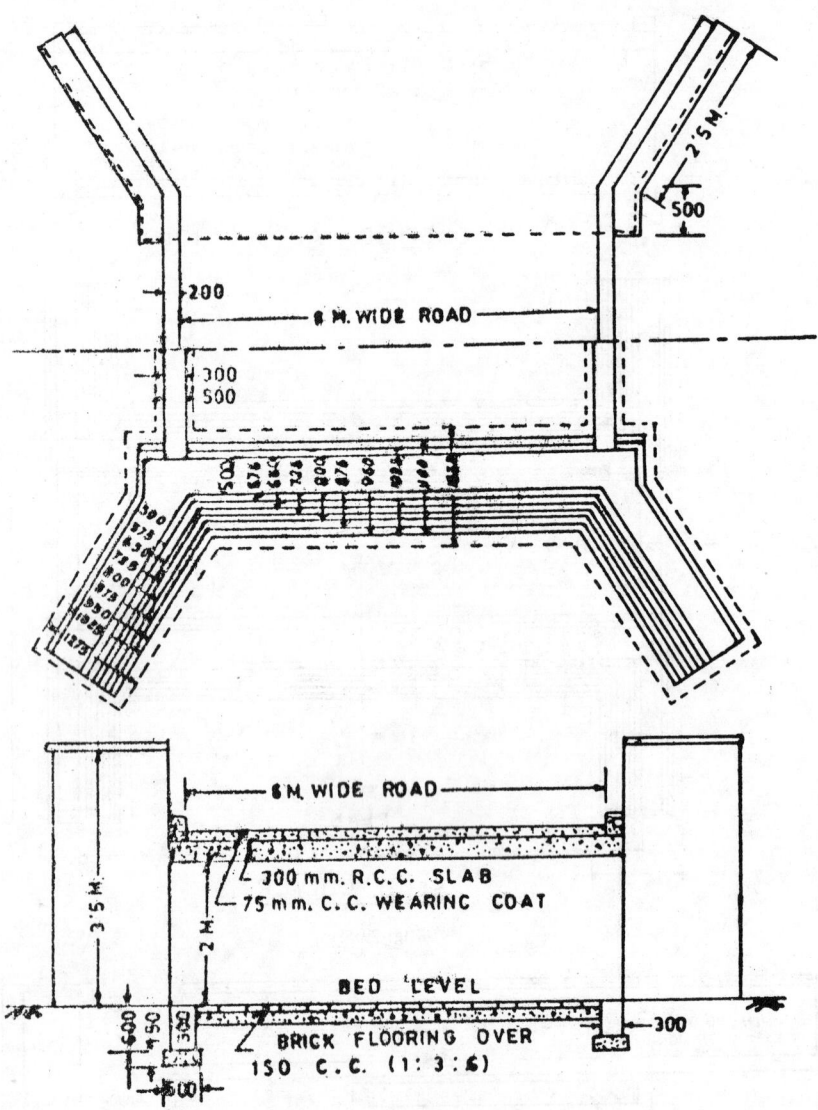

SLAB CULVERT WITH SPLAYED WING WALL

Fig. 29·7.

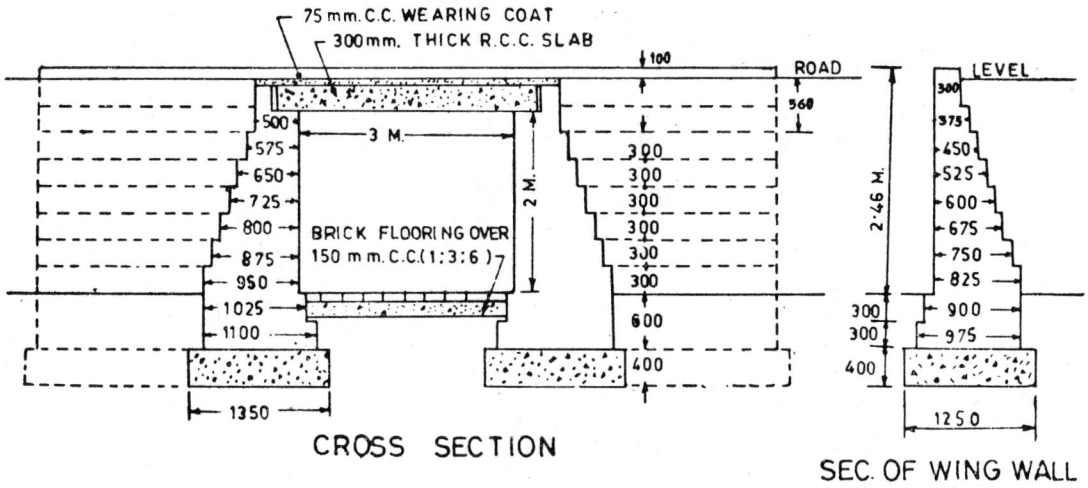

CROSS SECTION

SEC. OF WING WALL

Fig. 29.9

Fig. 29·16 shows the elevation of railing with mild steel angle post. The elevation of mild steel angle rail post is shown in Fig. 29·17.

The span and sectional elevation of a R.C.C. foot bridge are presented in Fig. 29·18. The sectional views of the abutment and the pier are shown in 29·19 The following views are illustrated in Fig. 29·20.

(a) The sectional views are X-X and Y-Y, the sections being taken as shown in Fig. 29·18.

(b) The longitudinal section of the girder.

(c) The details of suspended slab, section being taken the Z-Z as shown in Fig. 29·18.

(d) The sectional view at junction of slab and girder.

(e) The cross-sectional views of the girder (sections on KK and LL).

Fig. 29·21 shows the details of abutment, return wall and wing wall. It also presents the sectional view of the wing wall.

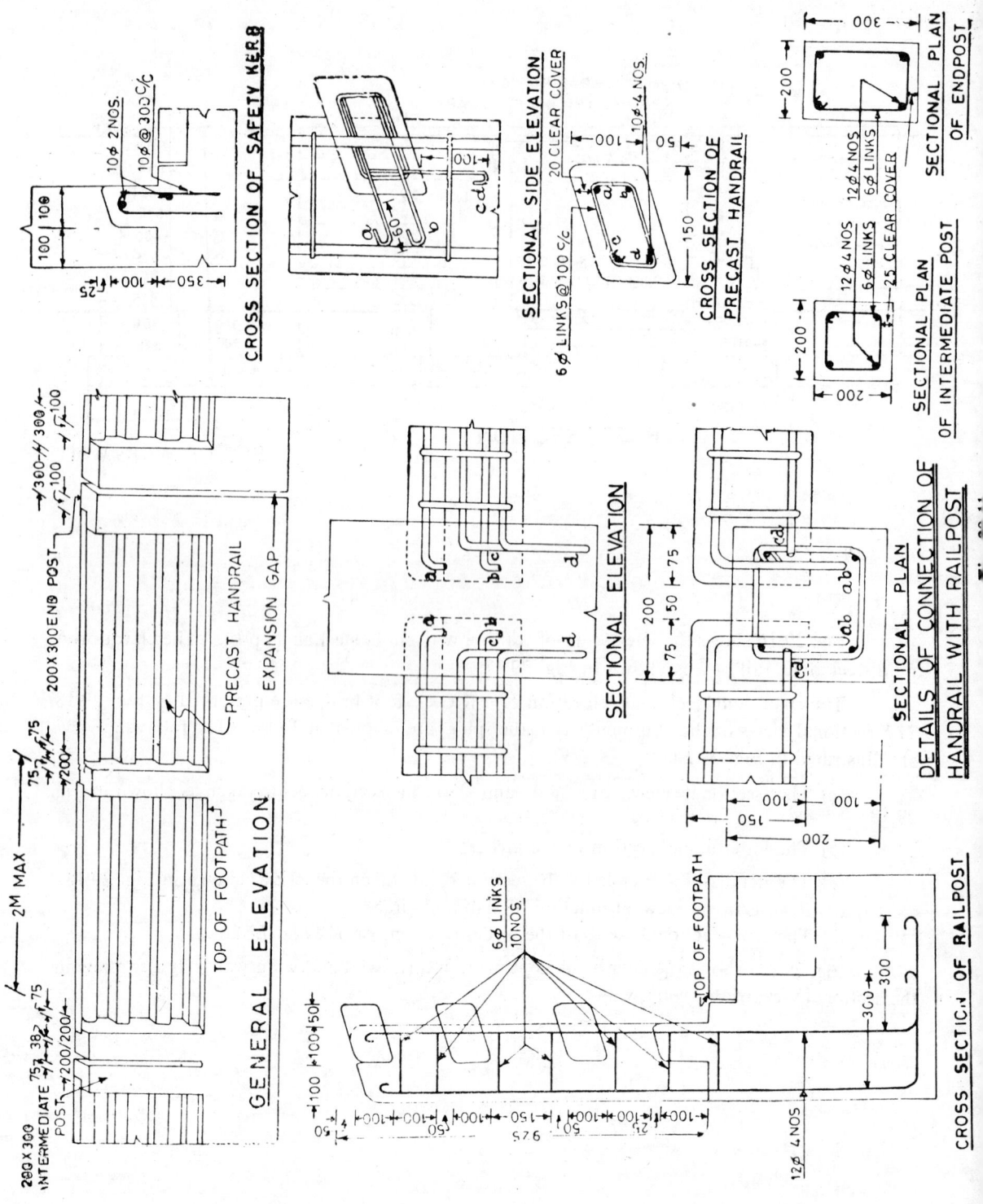

CROSS SECTION OF SAFETY KERB

SECTIONAL SIDE ELEVATION

CROSS SECTION OF PRECAST HANDRAIL

SECTIONAL PLAN OF ENDPOST

SECTIONAL PLAN OF INTERMEDIATE POST

GENERAL ELEVATION

SECTIONAL ELEVATION

SECTIONAL PLAN
DETAILS OF CONNECTION OF
HANDRAIL WITH RAILPOST

CROSS SECTION OF RAILPOST

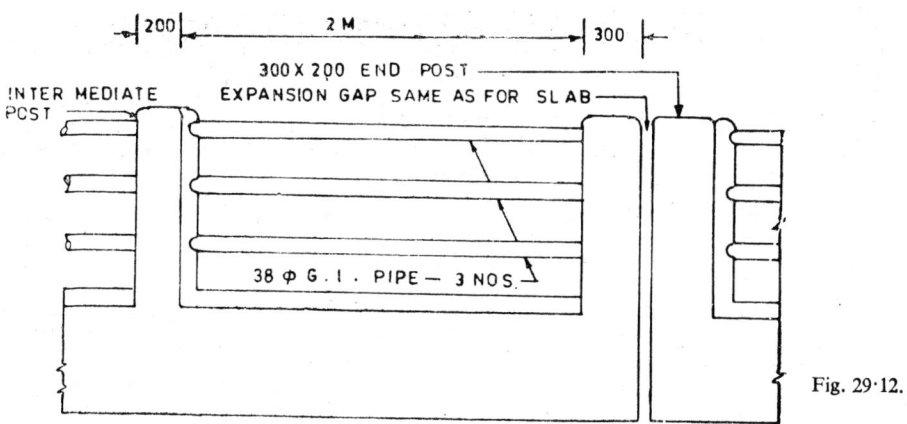

Fig. 29·12.

ELEVATION OF RAILING WITH CONCRETE RAILPOST

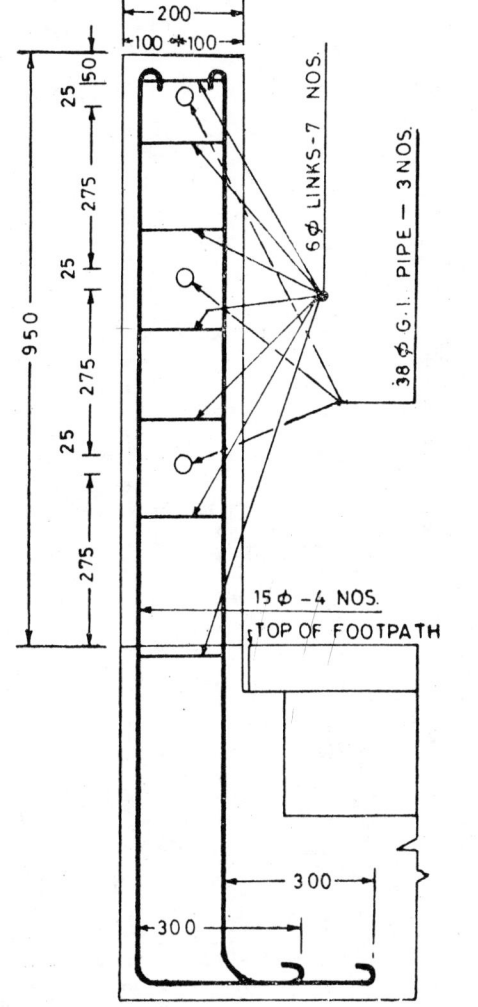

Fig. 29·13.

CROSS SECTION OF CONCRETE RAILPOST

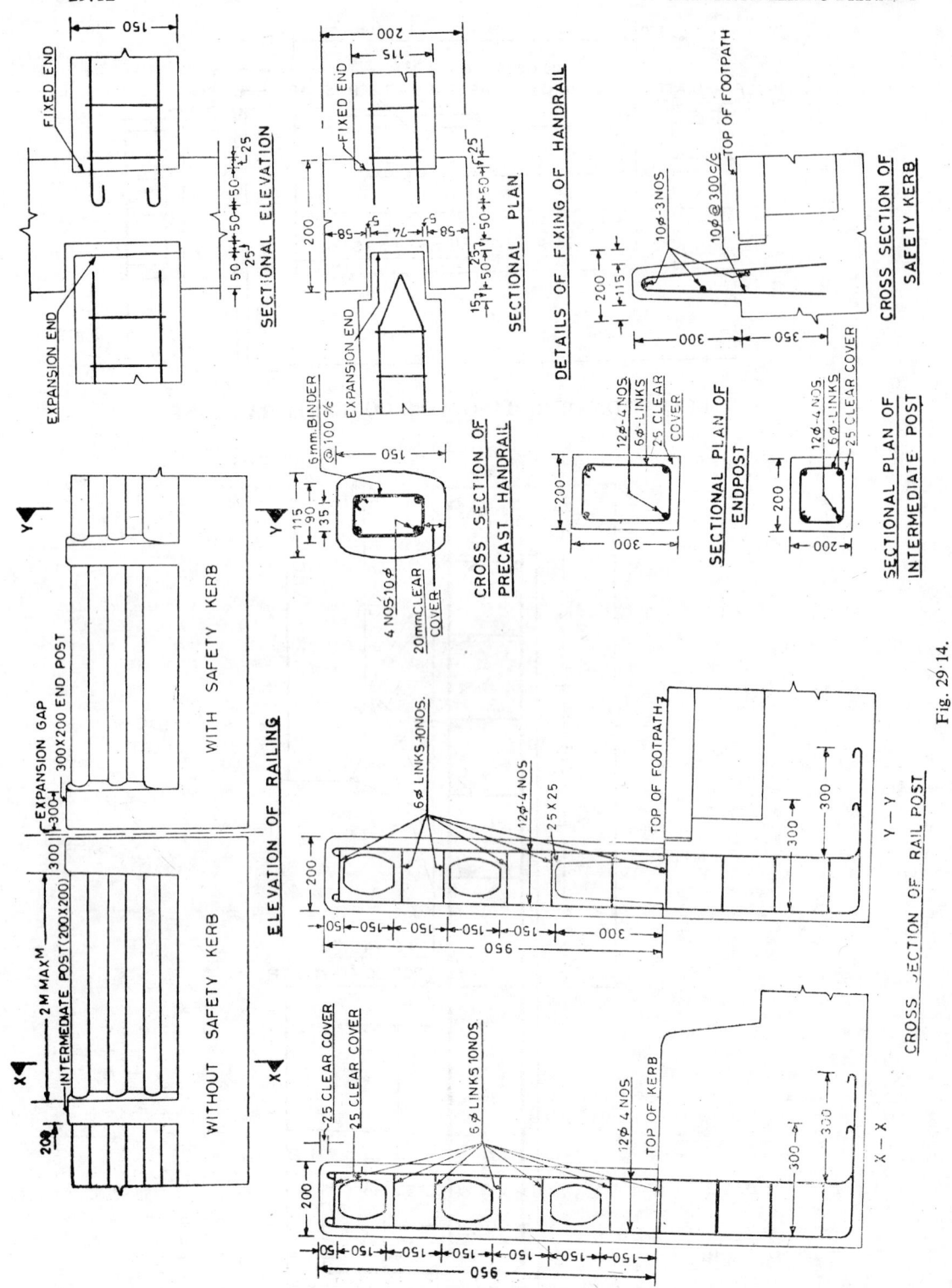

Fig. 29·14.

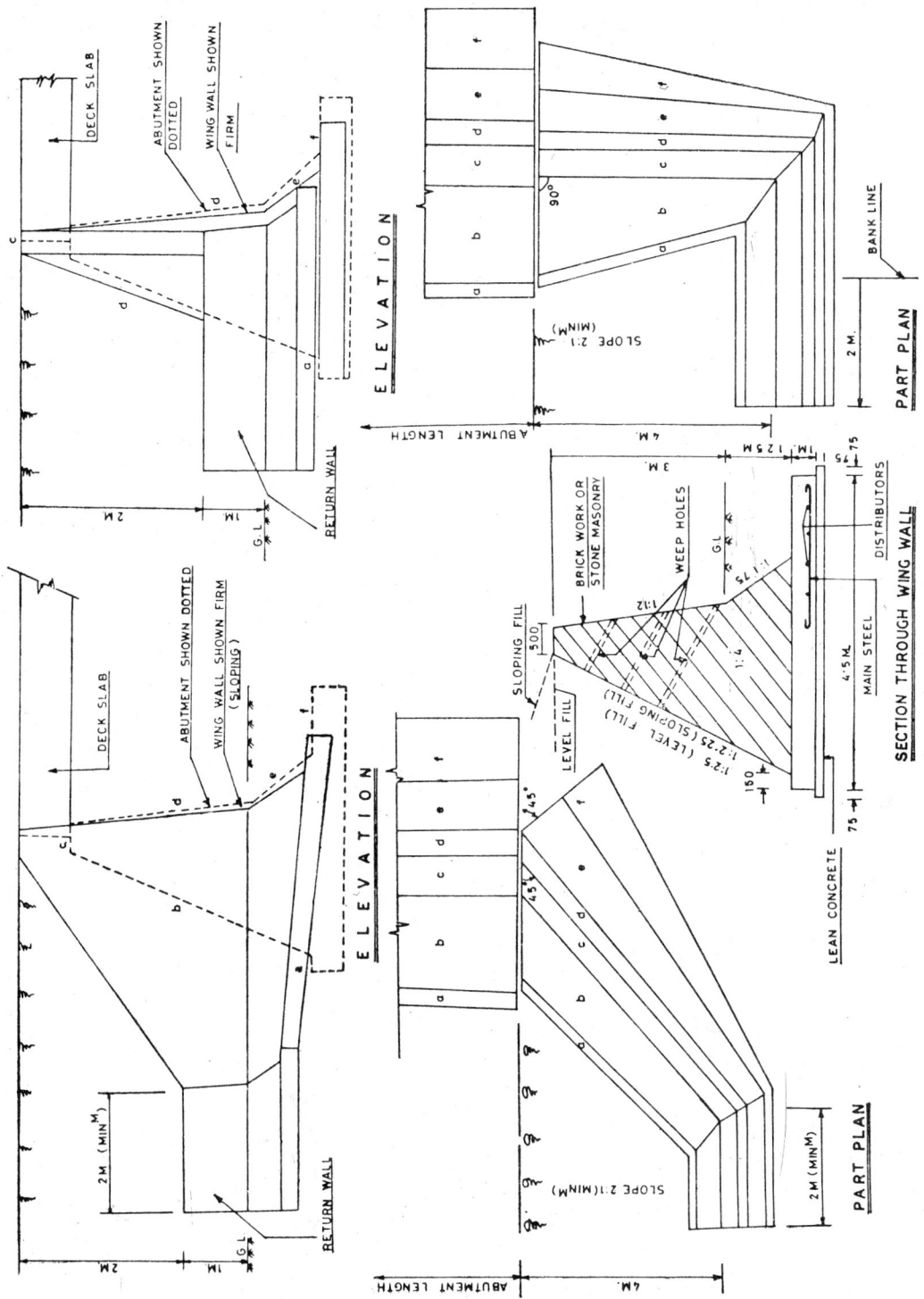

Fig. 29·21.

30

Irrigation Structures

This chapter deals with a few irrigation structures which are commonly employed in an irrigation system.

Fig. 30.1 shows typical cross-sections of a distributory when
(i) Natural surface level is above full supply level ;
(ii) Natural surface level lies between full supply level and bed level ; and
(iii) Natural surface level is below bed level.

Typical cross-section of an Earthen Dycke is shown in Fig. 30.2

Fig. 30.3 shows plan and longitudinal sectional view with details of a canal fall. Such a structure is required when the ground slope (natural surface level) falls very rapidly compared to the bed slope of the canal. Thus, a canal fall is used to lower the full supply level of the canal abruptly from one grade to the other grade, otherwise the canal bed will need high filling.

Fig. 30.4 shows details of installation of a sluice gate. The following views are presented in Fig. 30.4 .

(a) Half Plan.
(b) Longitudinal Section.
(c) End Elevation (up stream).
(d) Cross-section of down stream face wall.
(e) Cross-section of wing wall at junction with face wall (u/s).
(f) Cross-section of wing wall at junction with return wall (u/s).
(g) Section at collared joint.

The details of a lock gate and a head regulator (plan and sectional elevation) are illustrated in Fig. 30.5. The sectional views by taking sections on XX, YY and ZZ are shown in Fig 30.6.

Fig. 30.7 shows the sectional view of a bank protective work.

The plan and sectional views of a Superpassage are presented in Fig. 30.9.

The details of a Syphoned Aqueduct are shown in Fig. 30.8. The following views are given :

(a) Top plan and bottom plan.
(b) Longitudinal section.
(c) Section on XYZ.
(d) Section on AA.
(e) Section on BB.

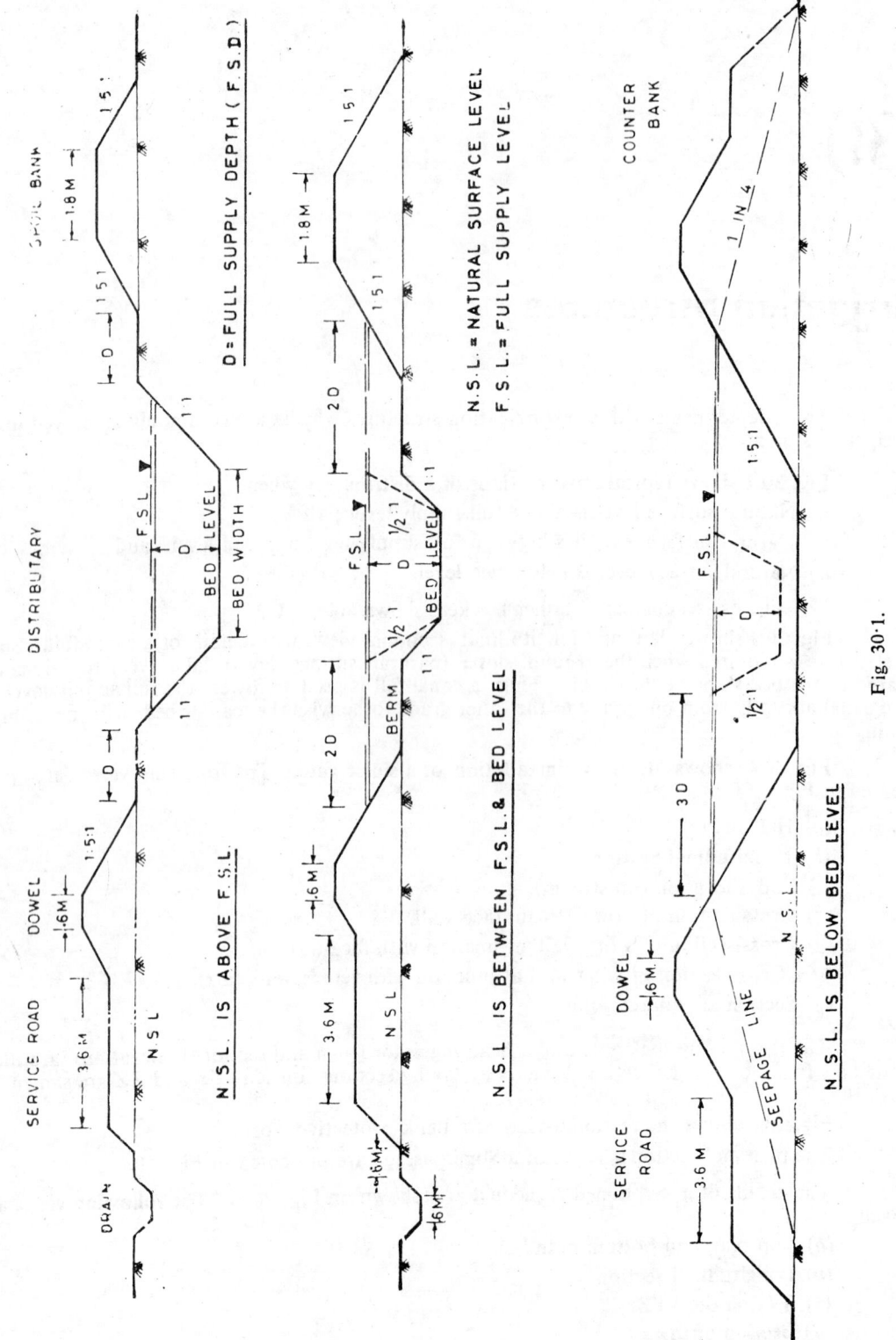

Fig. 30·1.

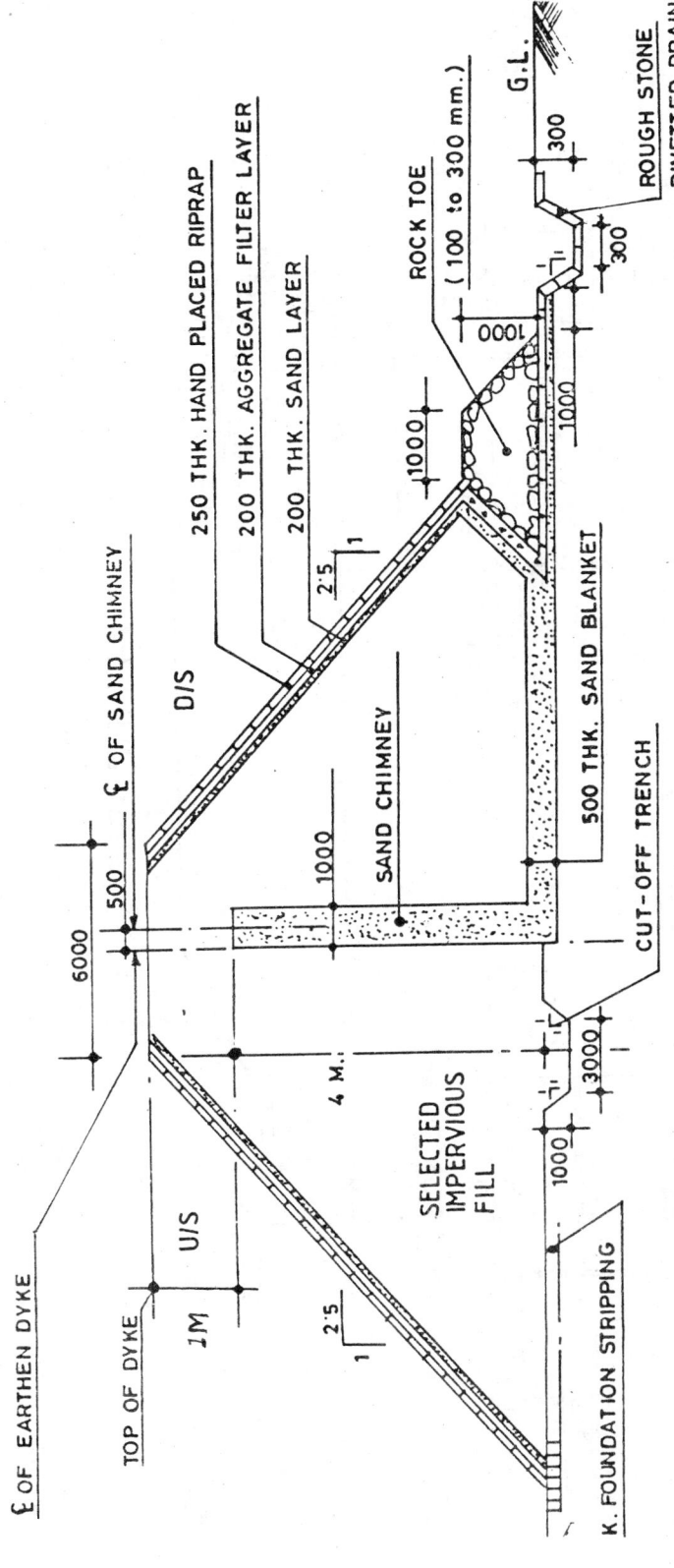

TYPICAL CROSS SECTION OF AN EARTHEN DYKE

Fig. 30.2

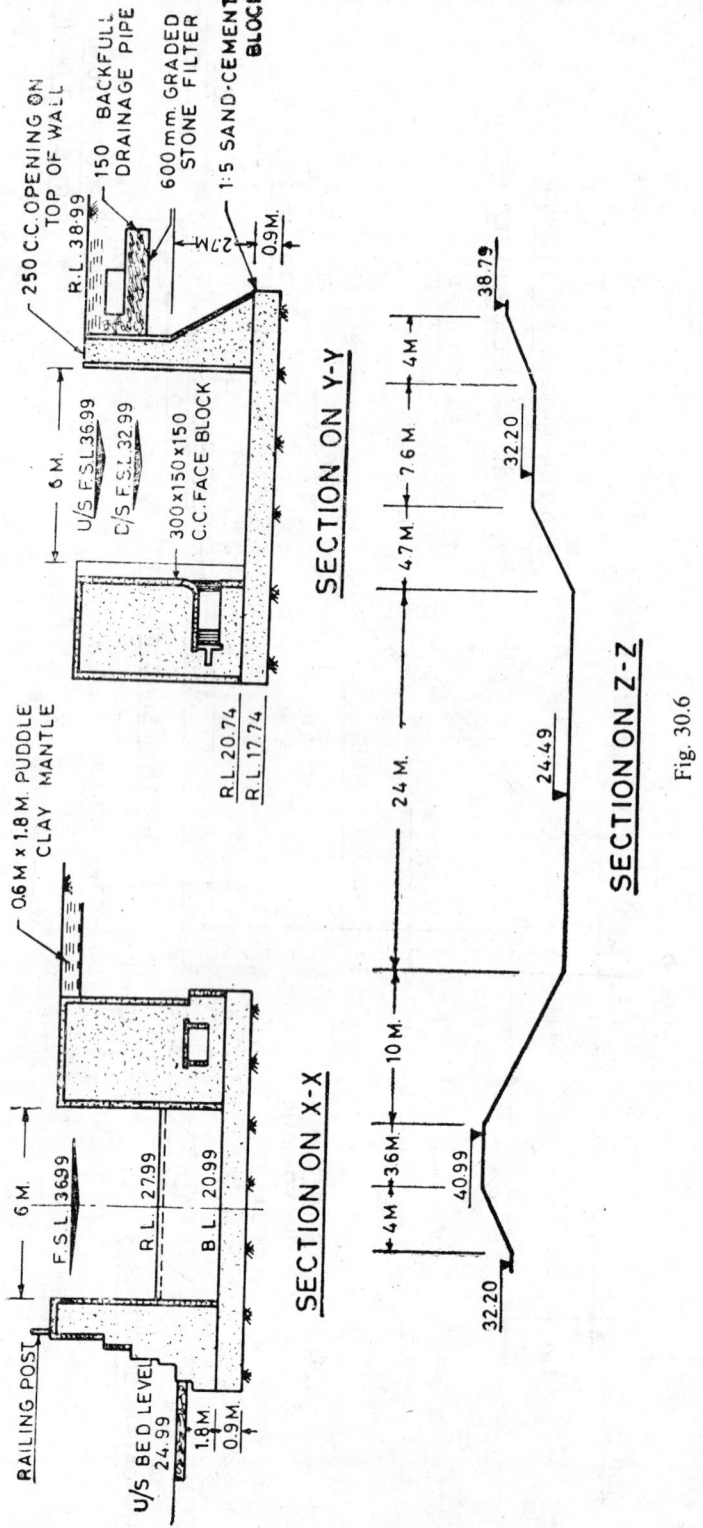

Fig. 30.6

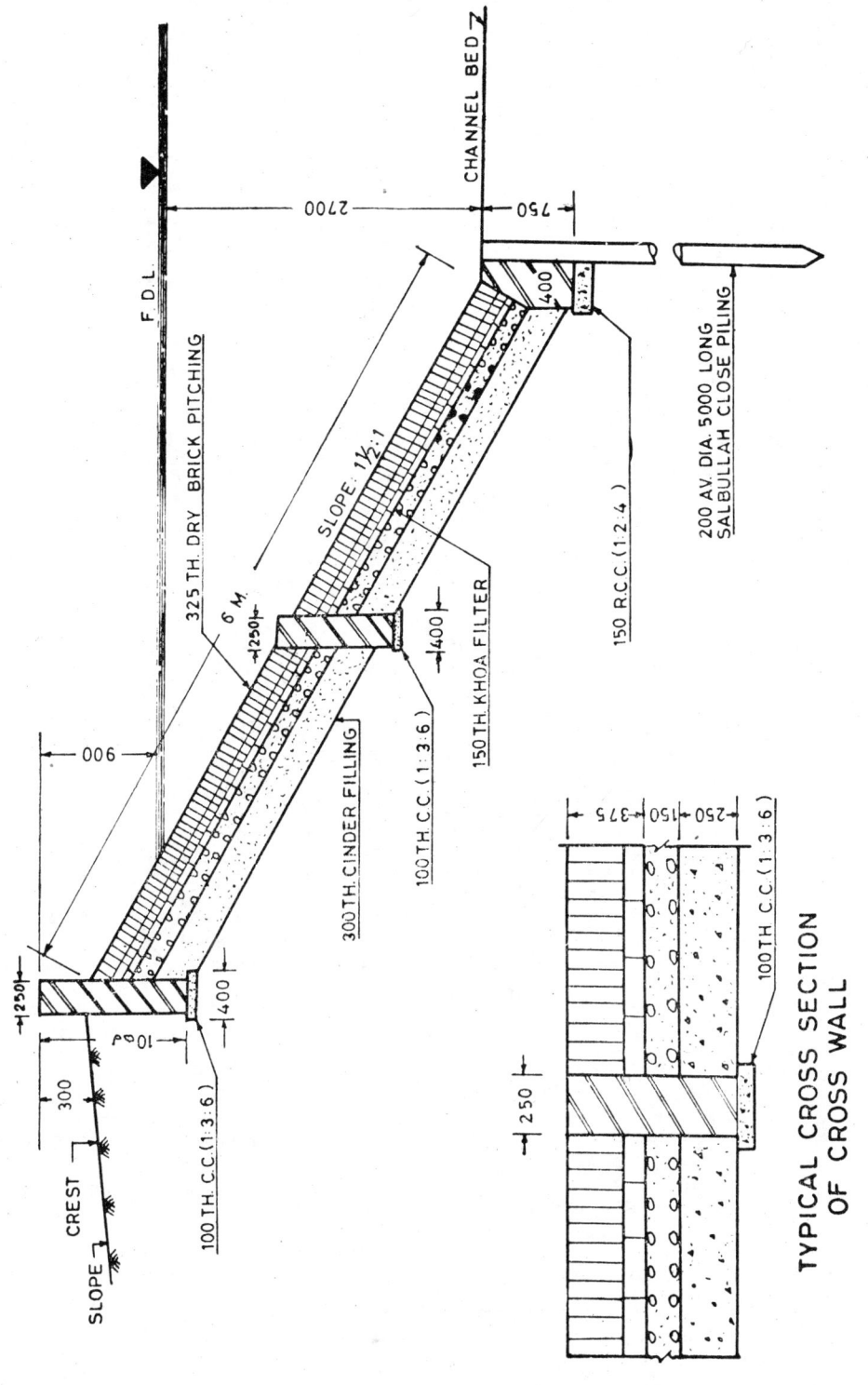

F.D.L.

CHANNEL BED

325 TH. DRY BRICK PITCHING

SLOPE: 1½:1

6 M

250

300 TH. CINDER FILLING

150 TH KHOA FILTER

100 TH C.C (1:3:6)

150 R.C.C (1:2:4)

200 AV. DIA. 5000 LONG
SALBULLAH CLOSE PILING

400

400

2700

750

900

250

100⌀

400

300

CREST

SLOPE

100 TH. C.C (1:3:6)

TYPICAL CROSS SECTION
OF CROSS WALL

375 | 150 | 250

250

100 TH C.C (1:3:6)

SECTIONAL VIEW OF A BANK PROTECTIVE WORK

Fig. 30.7

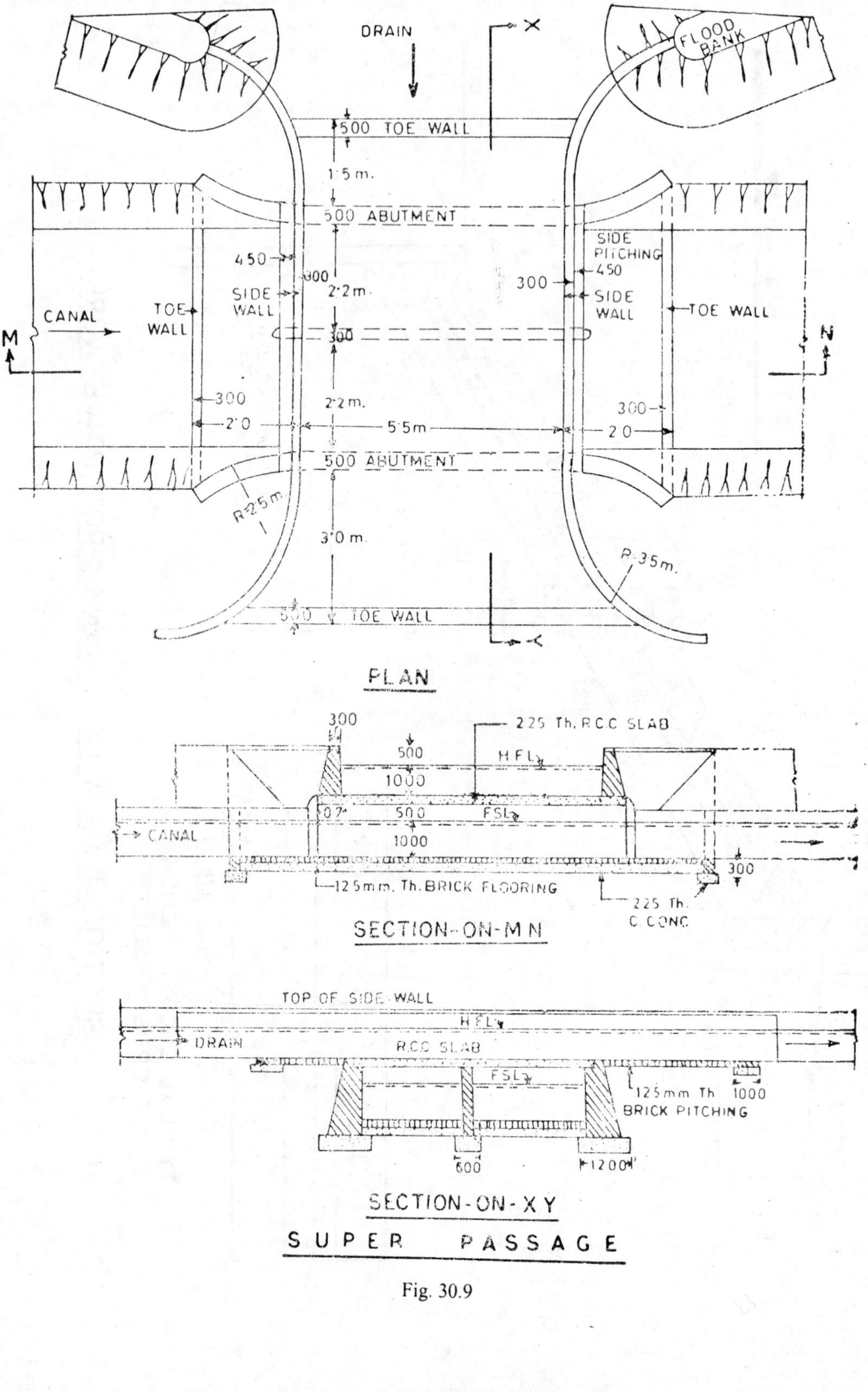

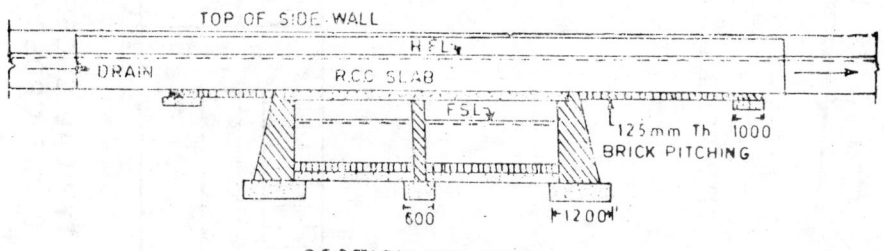

DRAIN

500 TOE WALL

1·5 m.

500 ABUTMENT

450
300
SIDE
WALL
2·2 m.

300

SIDE
PITCHING
450
300
SIDE
WALL

CANAL

TOE
WALL

TOE WALL

M   N

300
2·0

300
5·5m.

300
2·0

2·2 m.

500 ABUTMENT

R·2·5 m.

3·0 m.

500 TOE WALL

R·35 m.

PLAN

300

225 Th. RCC SLAB

500

H.F.L.

1000

0·7
500
FSL

CANAL

1000

125mm. Th. BRICK FLOORING

225 Th.
C. CONC.

300

SECTION-ON-M N

TOP OF SIDE WALL

H.F.L.

DRAIN

RCC SLAB

FSL

125mm Th.
BRICK PITCHING

1000

600

1200

SECTION-ON-X Y

S U P E R   P A S S A G E

Fig. 30.9

31

Miscellaneous Civil Enginering Drawings

      This chapter deals with self explanatory miscellaneous civil Engineering drawings which are not covered in previous chapters. Again, the drawings given here are not comprehensive. Various civil Engineering works are yet to be covered which are considered to be beyond the scope of this book.

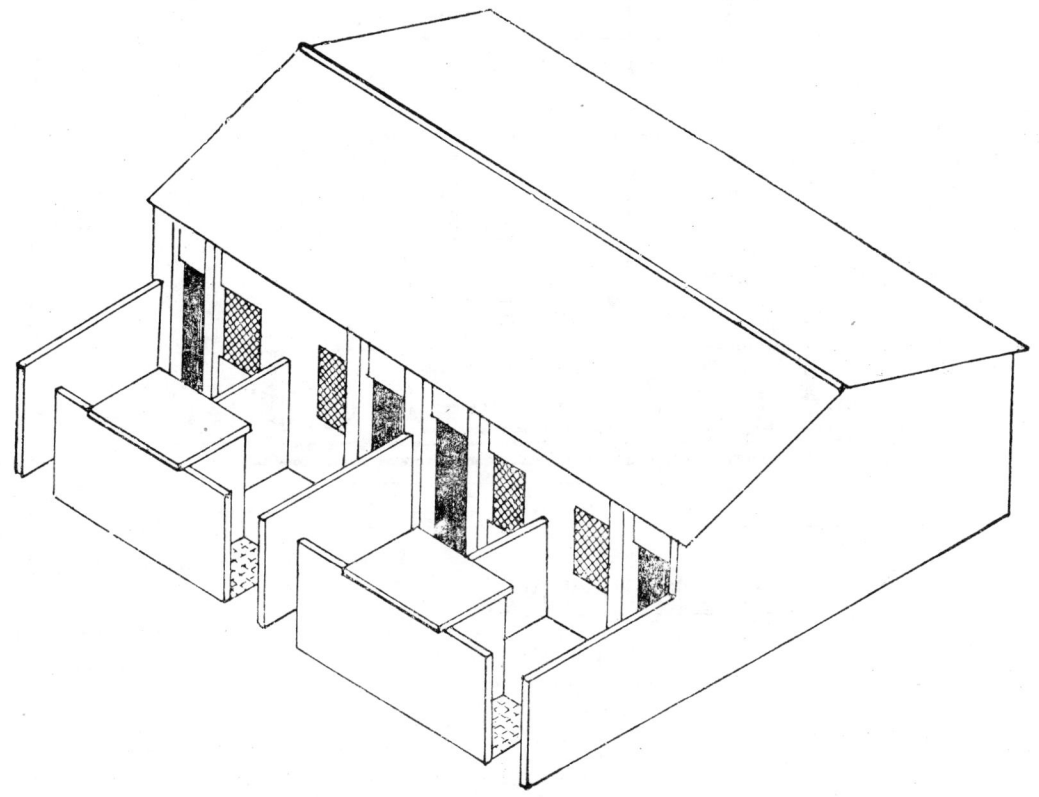

PICTORIAL VIEW OF A DORMITORY

Fig. 31·1.

A MONUMENT

Fig. 31·2.

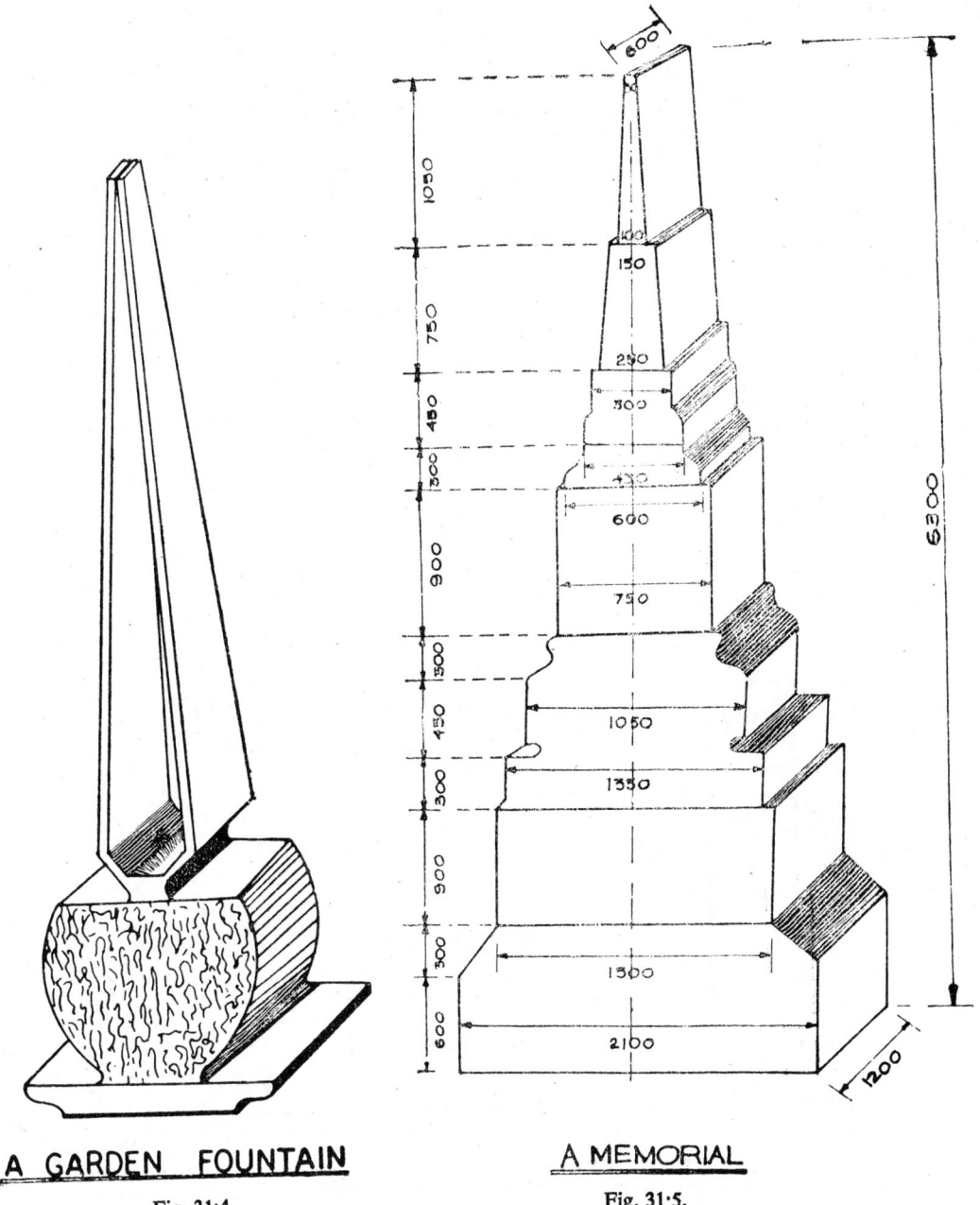

A GARDEN FOUNTAIN

Fig. 31·4.

A MEMORIAL

Fig. 31·5.

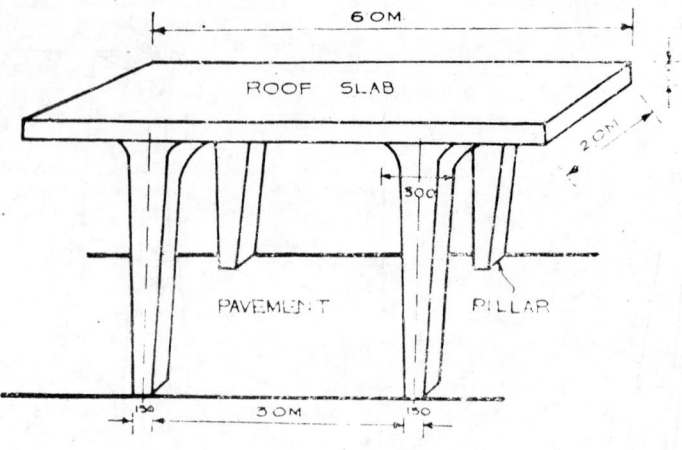

A BUS STOP SHELTER

Fig. 31·5.

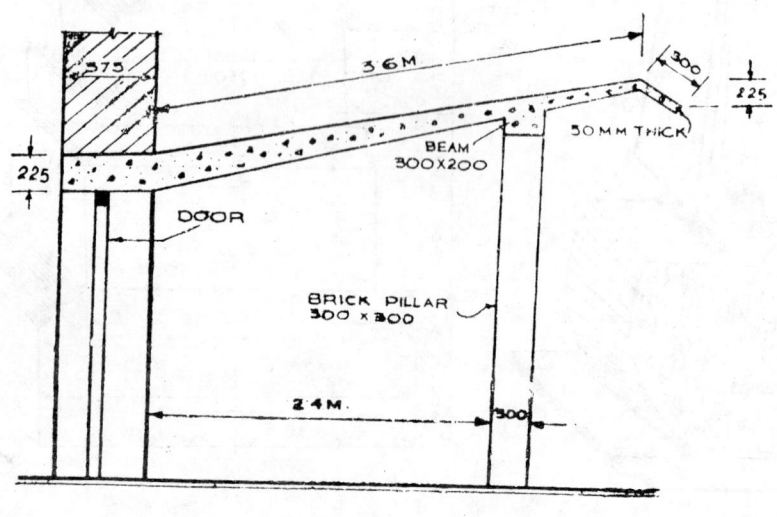

SECTION THROUGH A PORTICO

Fig. 31·6.

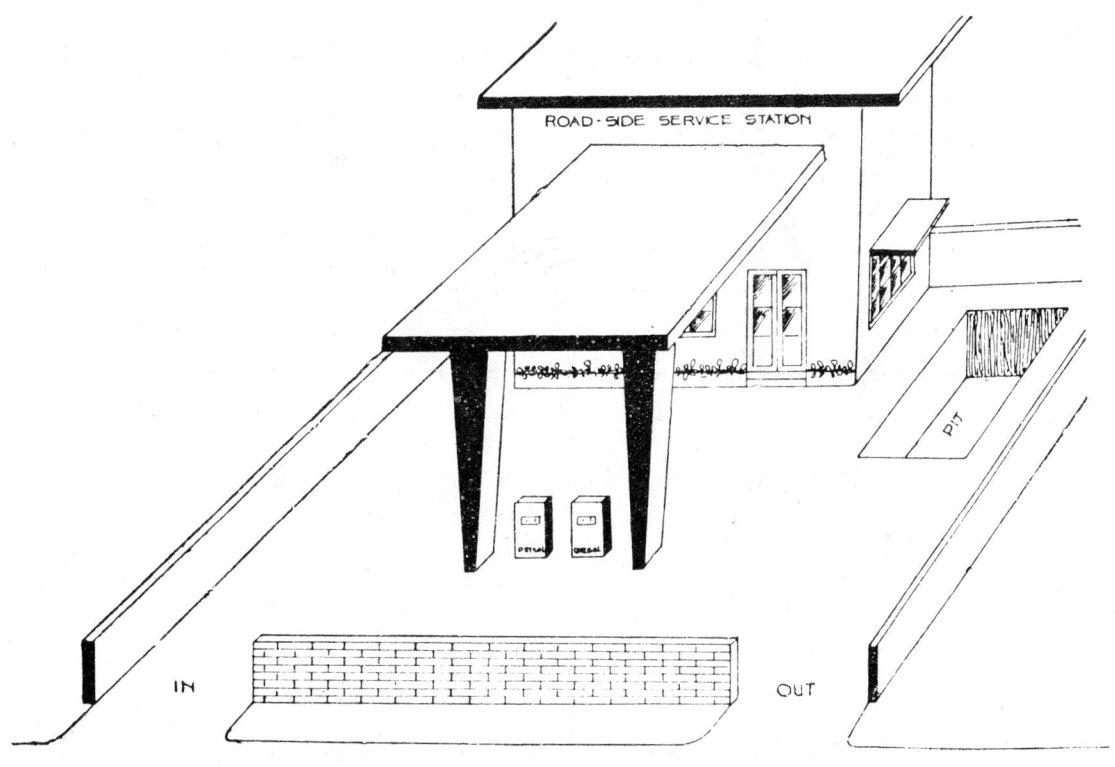

R O A D W A Y

Pictorial view of a Roadside Vehicle Service Station
Fig. 31·7.

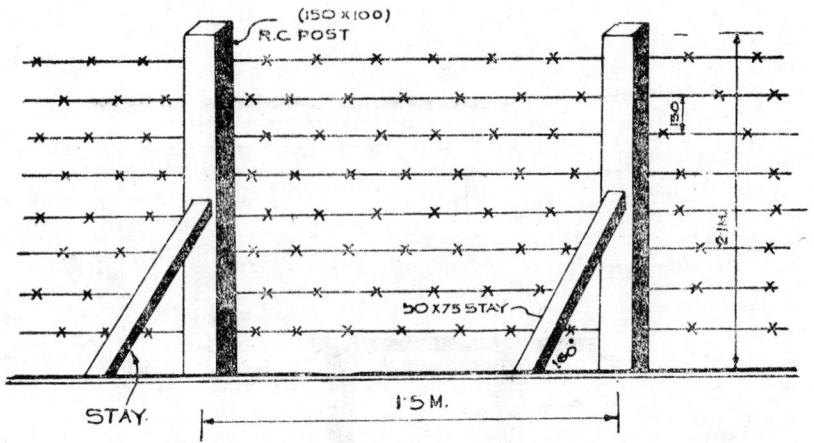

BARBED WIRE FENCING WITH R.C. POST & STAY.

Fig. 31·8.

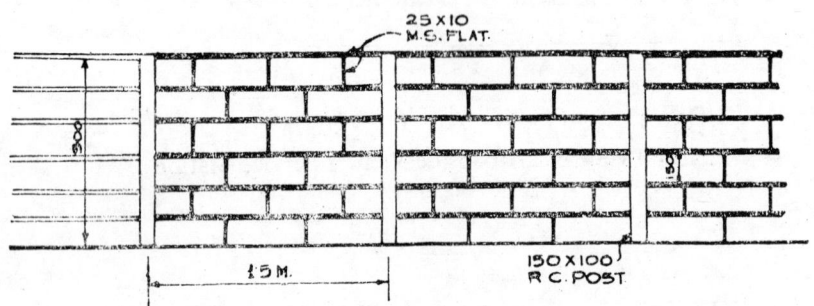

FENCING WITH R.C. POSTS & M.S. FLATS

Fig. 31·9.

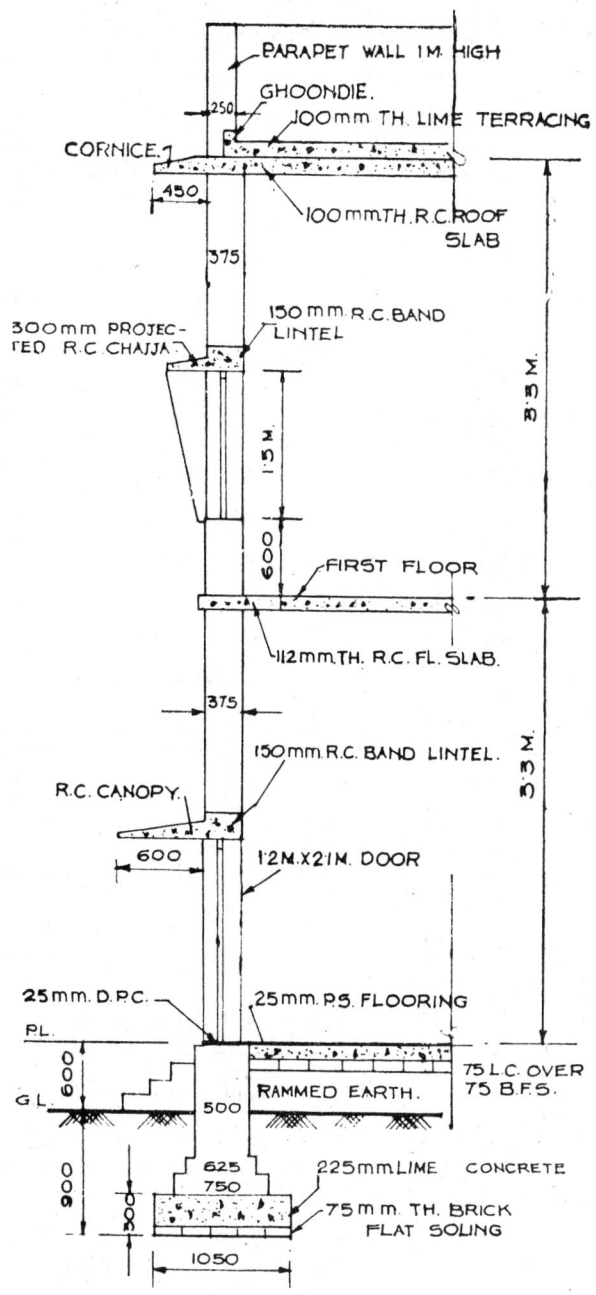

SECTION THROUGH A WALL (FOR A TWO STOREYED DOMESTIC BUILDING) SHOWING DETAILS FROM FOUNDATION TO PARAPET

Fig. 31·10.

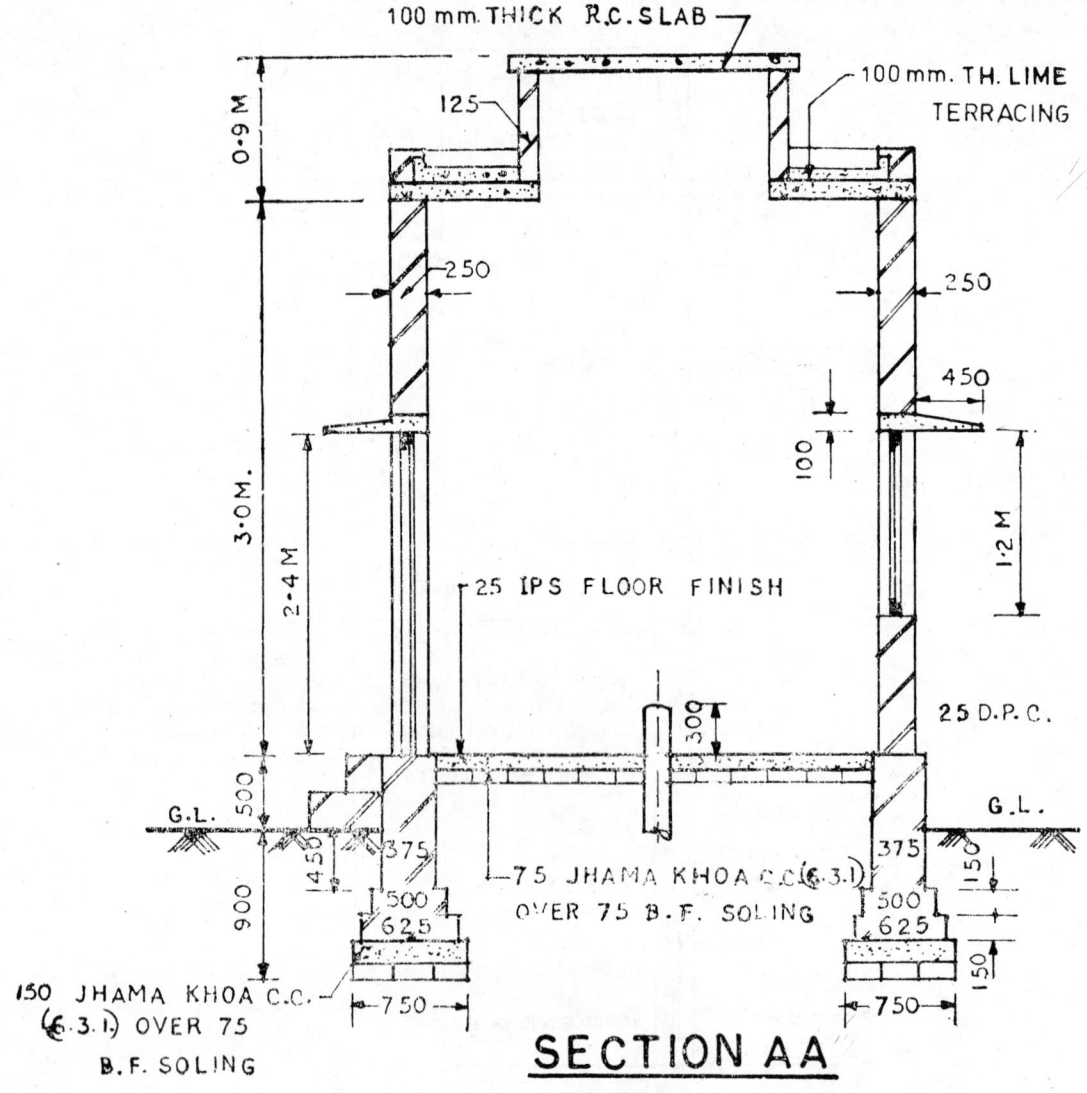

SECTION AA

Fig. 31·11.   (contiued)

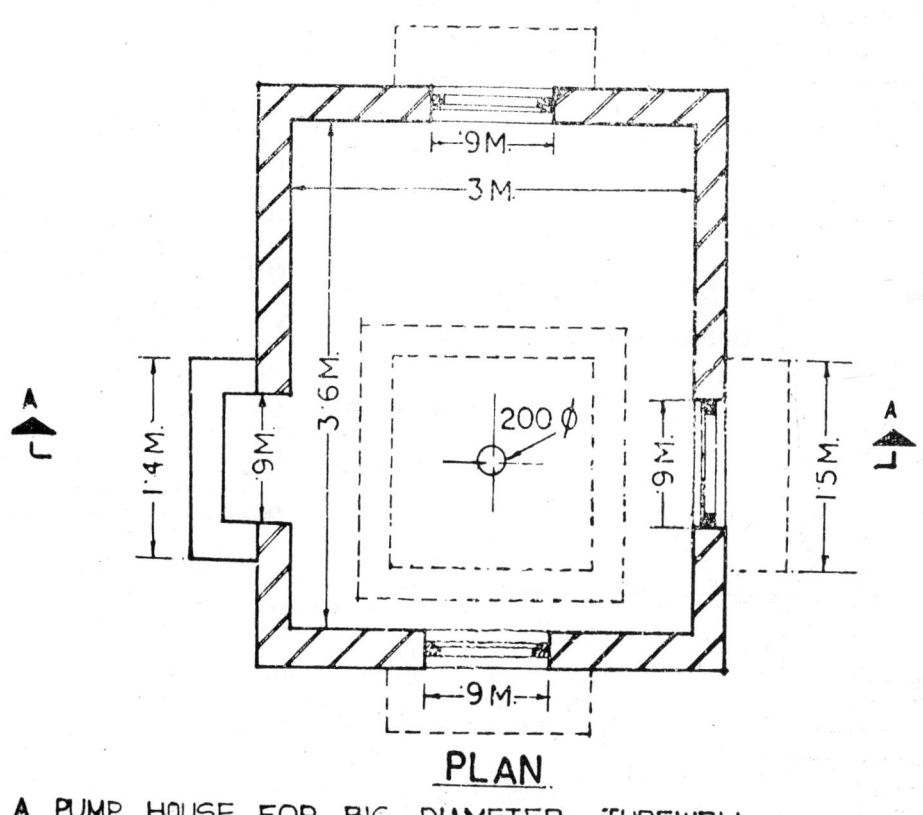

PLAN

A PUMP HOUSE FOR BIG DIAMETER TUBEWELL

Fig. 31·11. (continued)

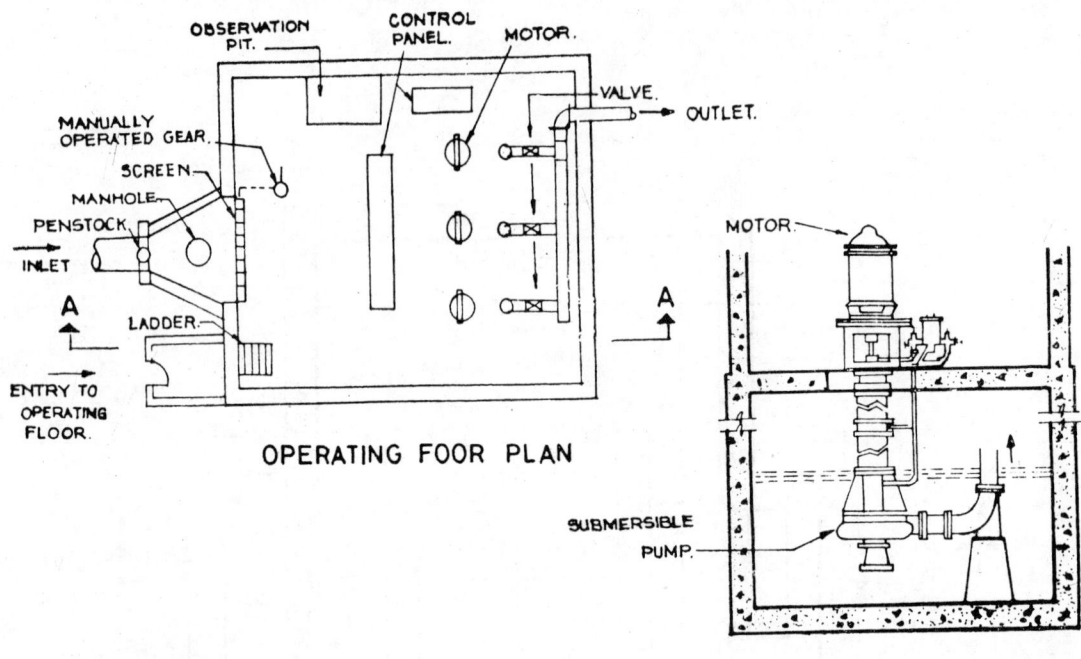

OPERATING FOOR PLAN

INSTALLATION IN WET PIT

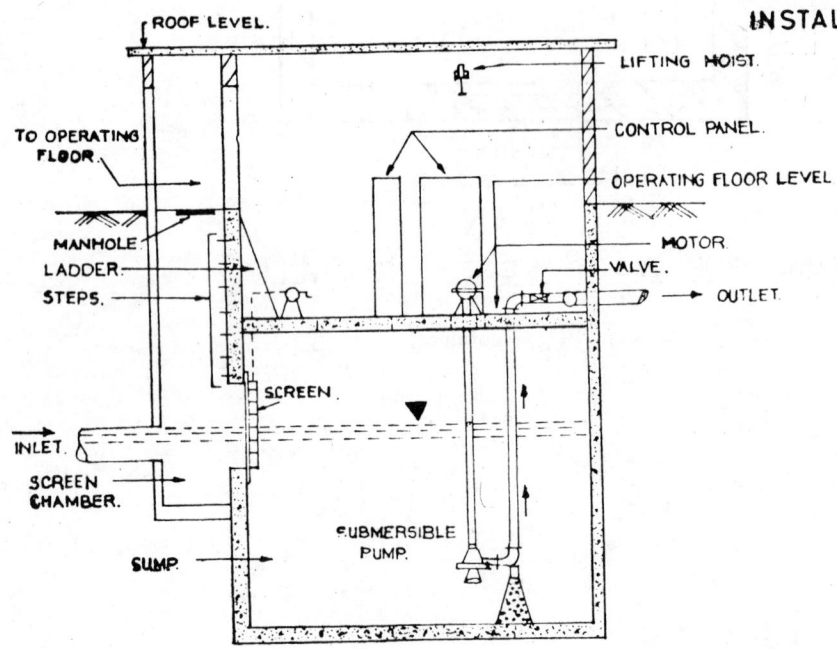

SECTION AA

PLAN AND SECTIONAL VIEW OF A SEWAGE LIFT STATION

Fig. 31·12.

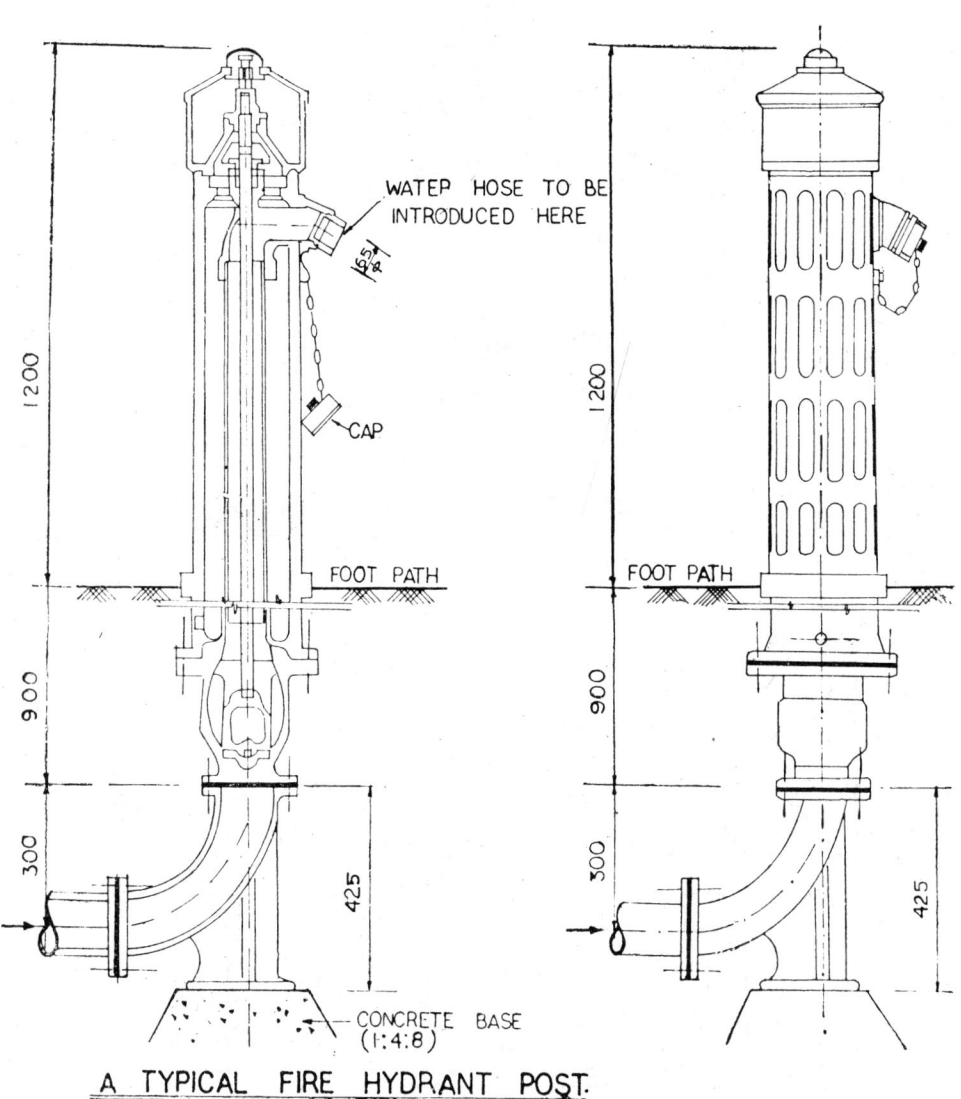

WATER HOSE TO BE
INTRODUCED HERE

CAP

FOOT PATH

FOOT PATH

1200

900

300

425

1200

900

300

425

CONCRETE BASE
(1:4:8)

A TYPICAL FIRE HYDRANT POST.

Fig. 31·13.

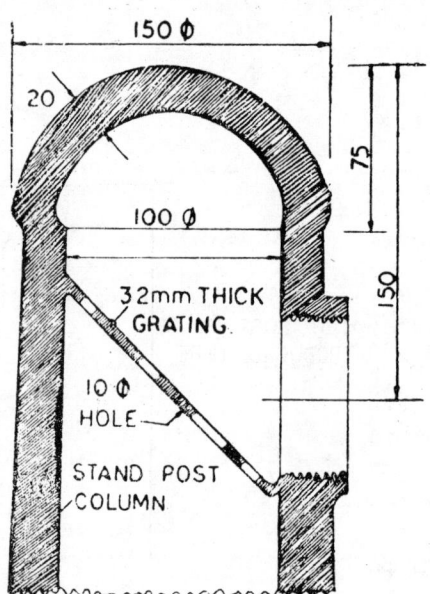

DETAILS OF STAND POST COLUMN

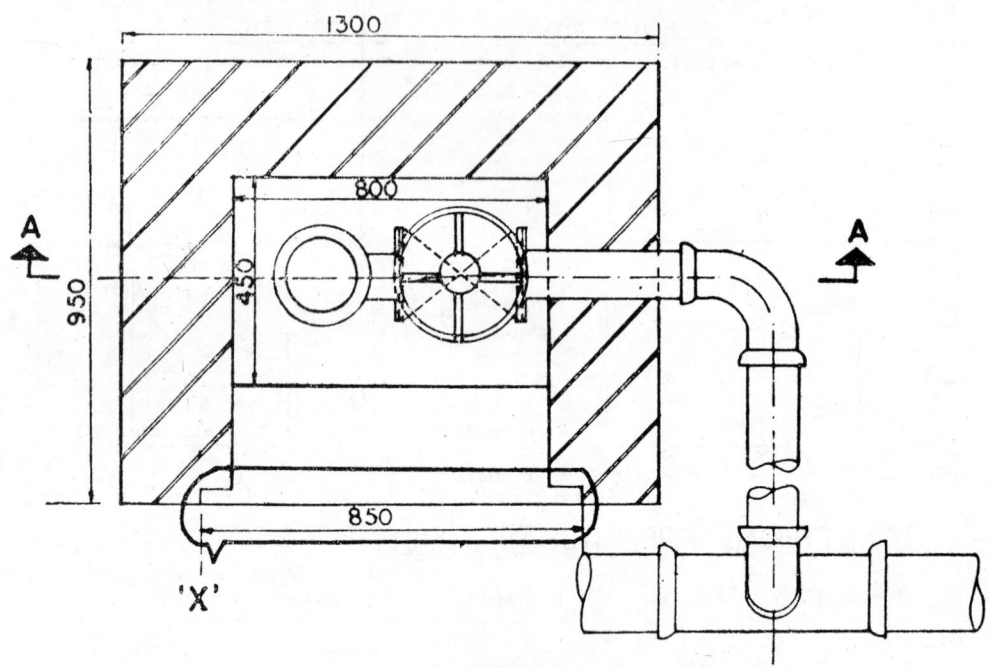

Fig. 31·14.

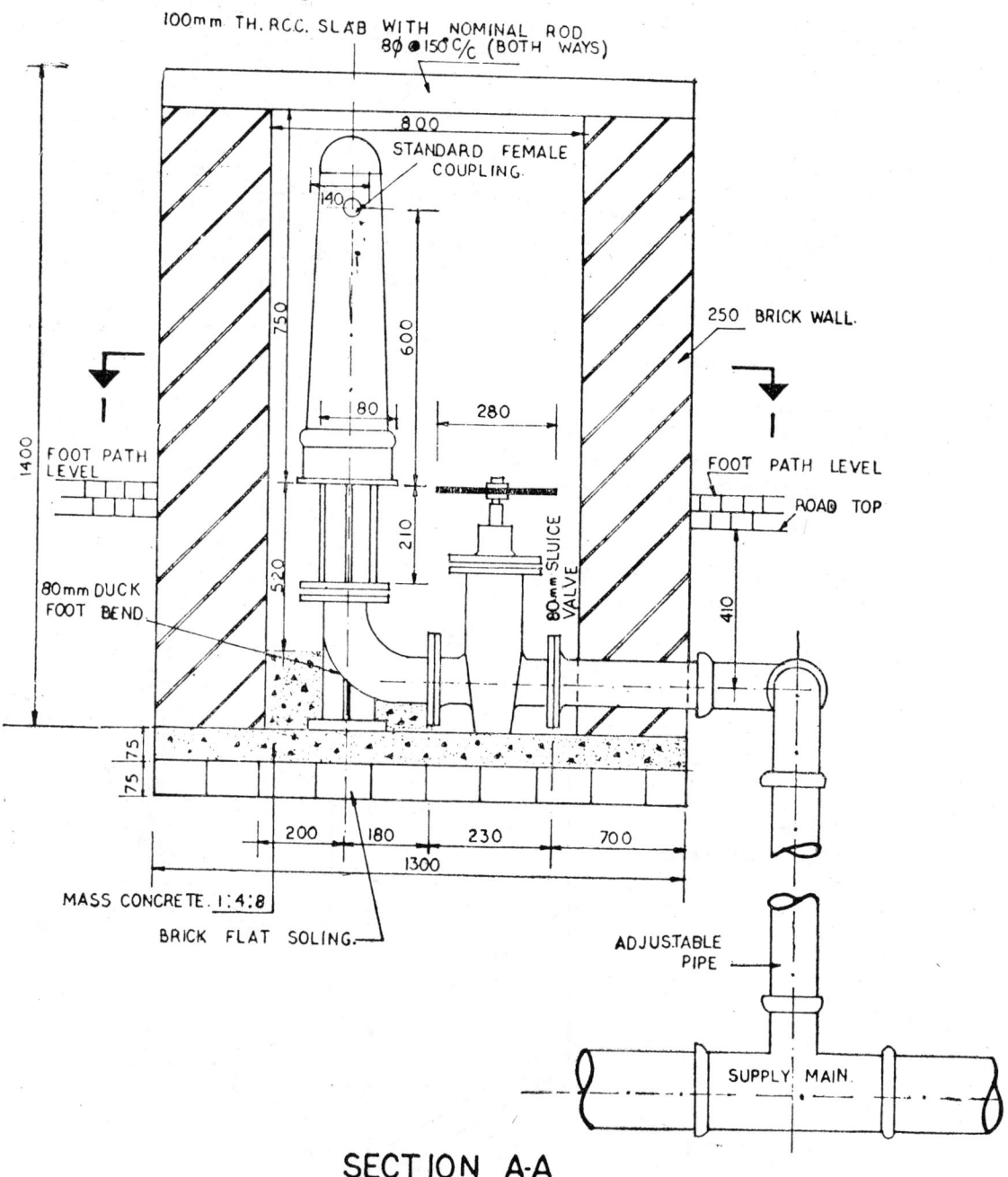

SECTION A-A

Fig. 31·15.

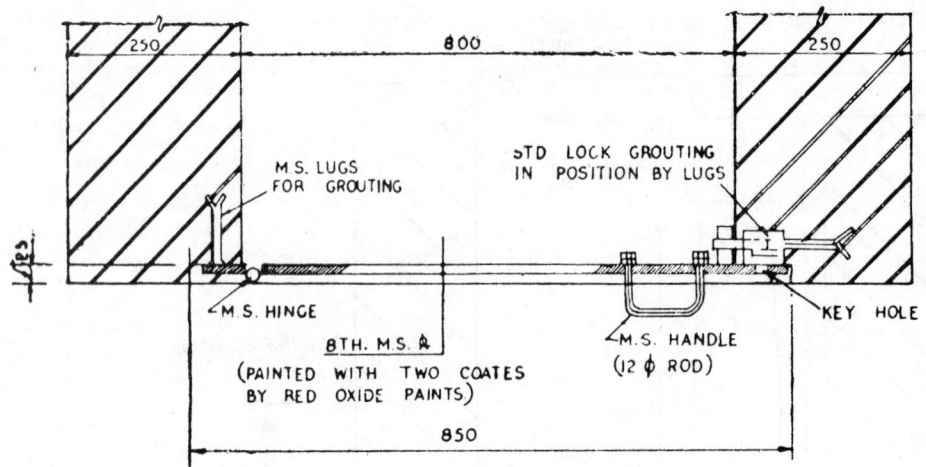

ARRANGEMENT OF DOOR ('X')

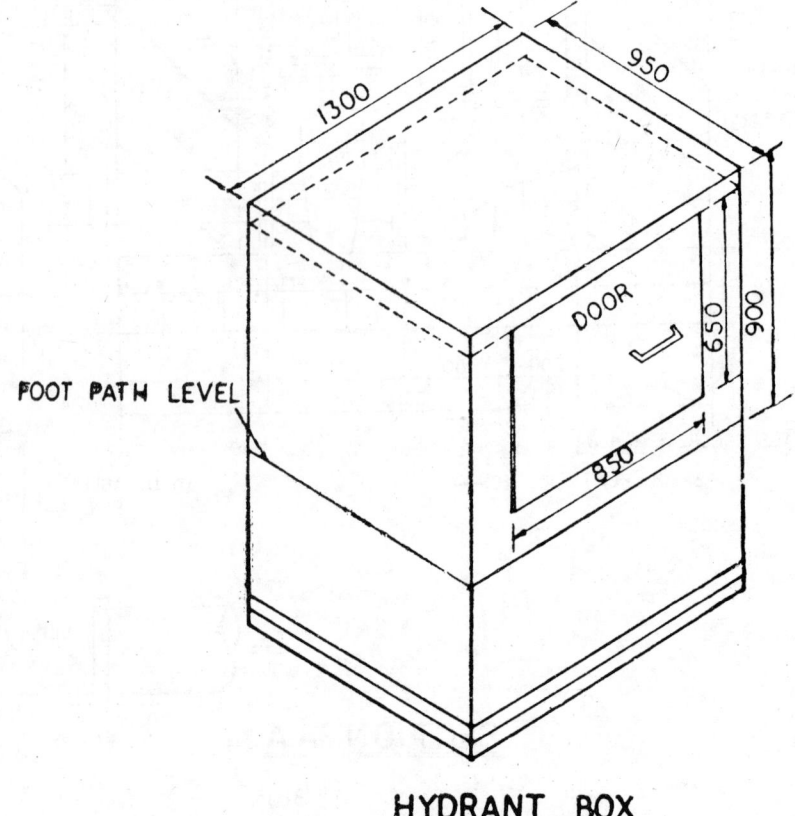

HYDRANT BOX

Fig. 31·16.

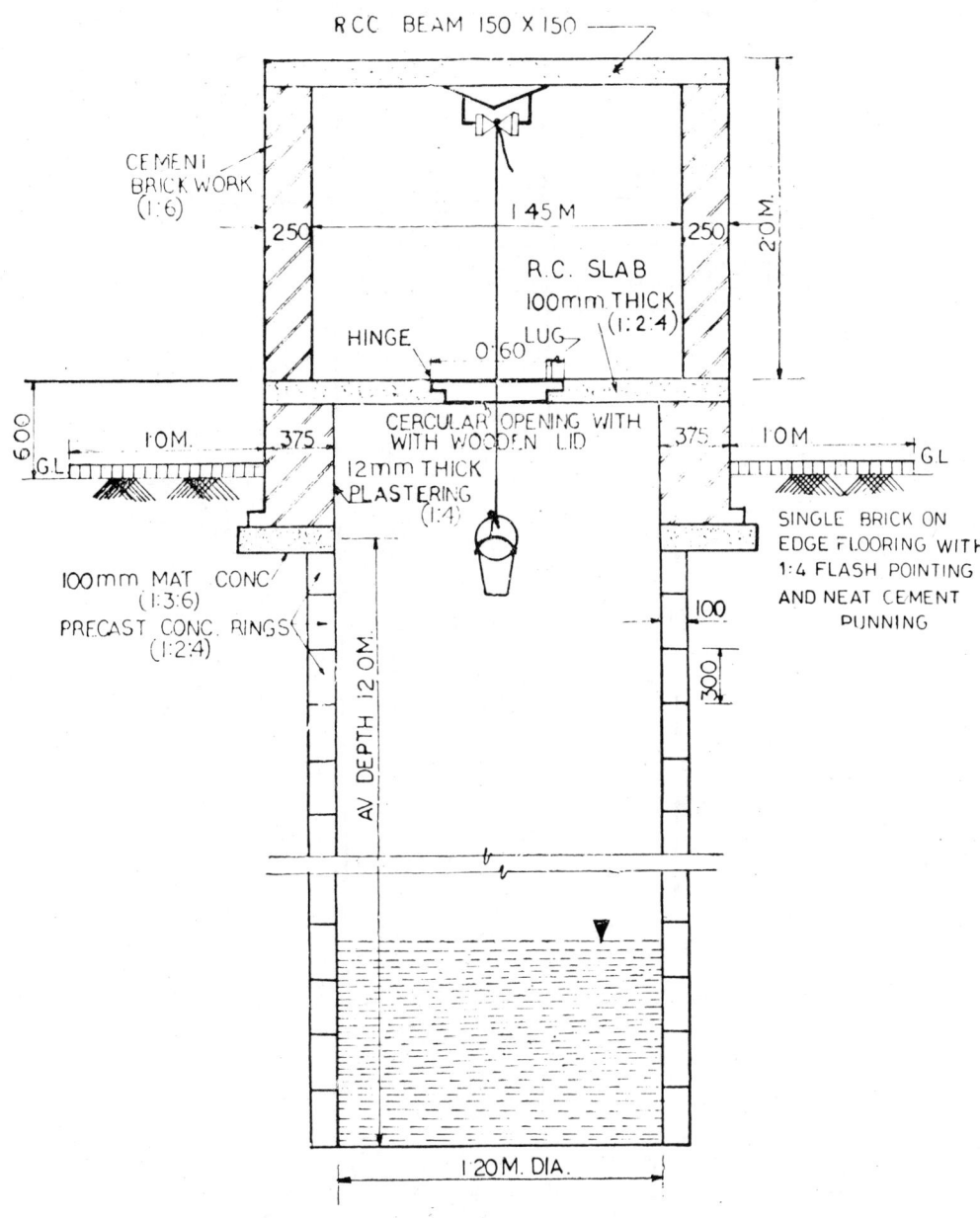

SECTIONAL VIEW OF AN OPEN WEL.

Fig. 31·17.

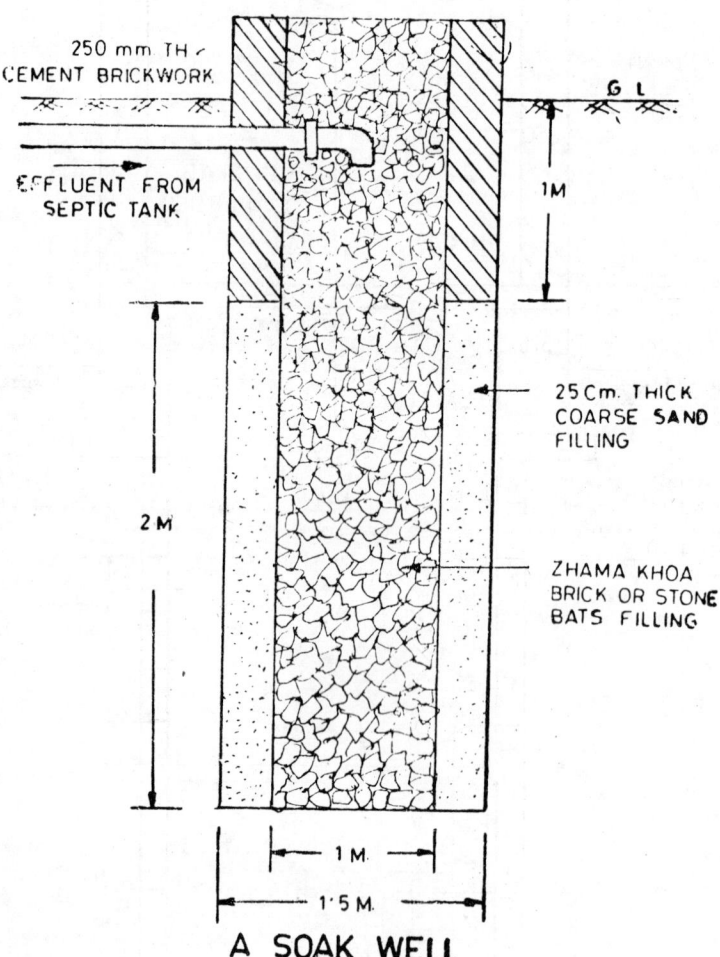

A SOAK WELL

Fig. 31·18.

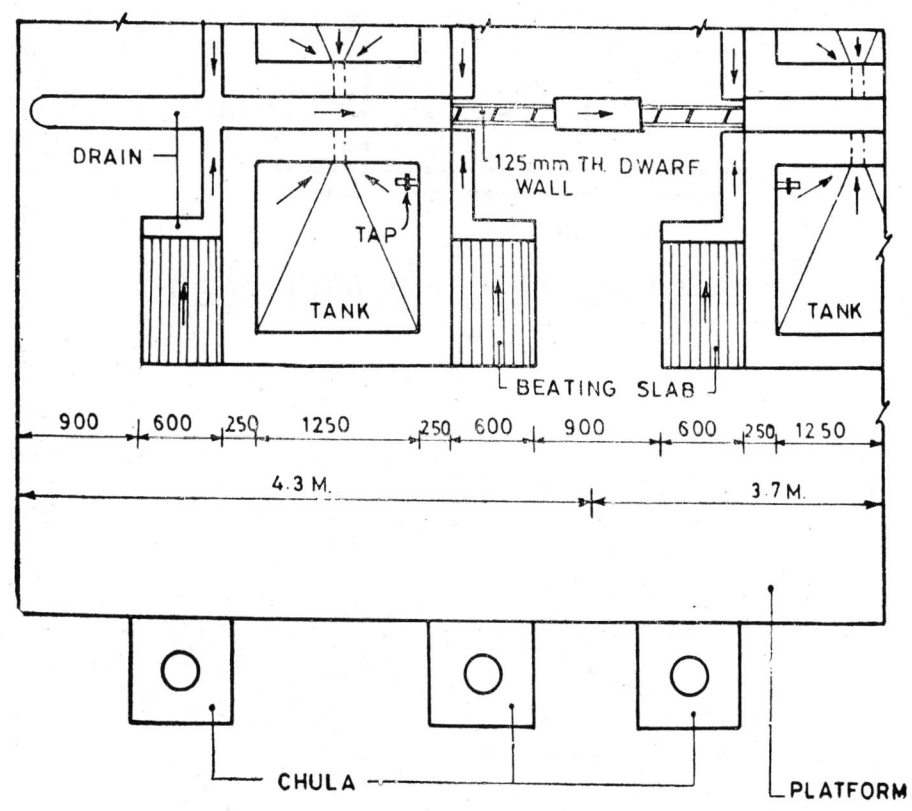

PLAN
DETAILS OF WASHING YARD FOR WASHERMEN
Fig. 31·19.

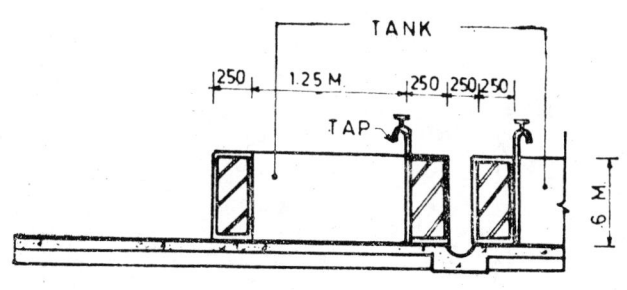

SECTION THROUGH TANK

Fig. 31·20.

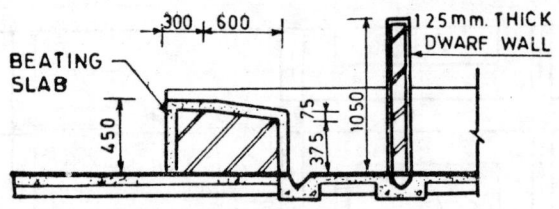

SECTION THROUGH BEATING SLAB

Fig. 31·21.

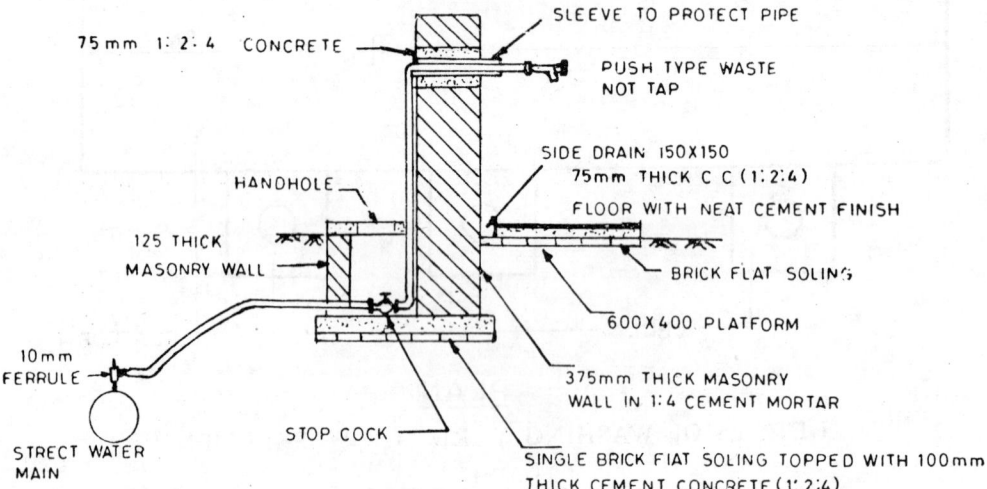

SECTIONAL VIEW OF A STANDPOST

Fig. 31·22.

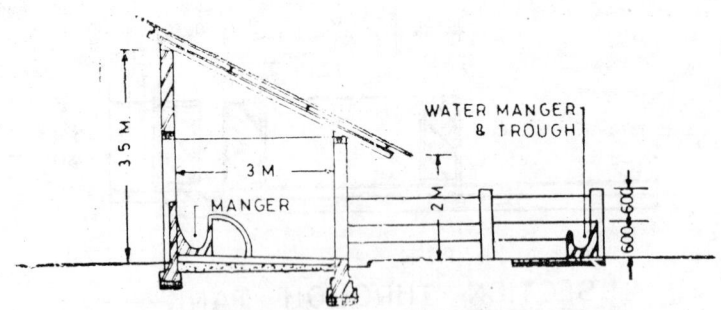

SECTIGNAL VIEW OF A CATTLE SHED
Fig. 31·23.

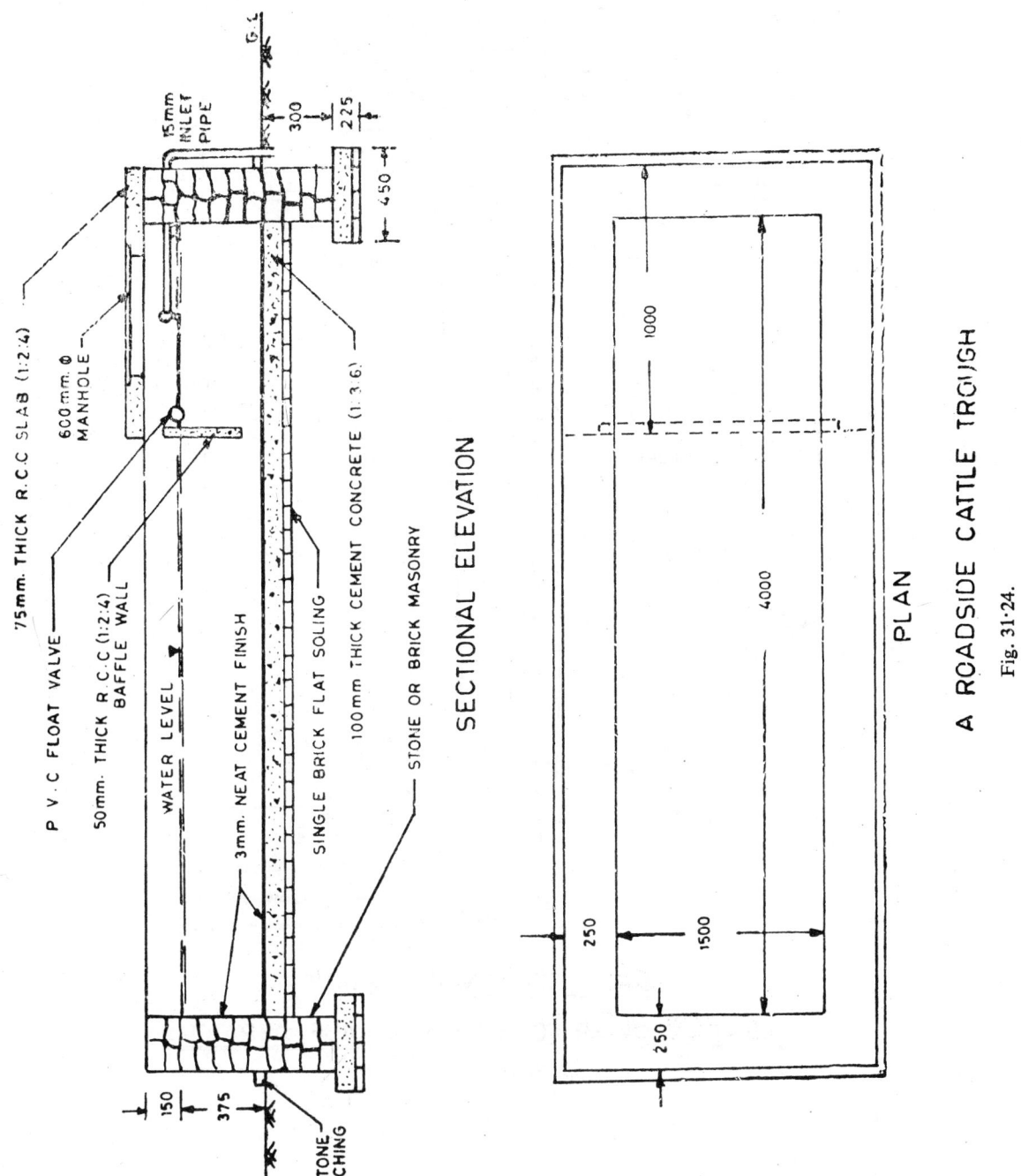

SECTIONAL ELEVATION

PLAN

A ROADSIDE CATTLE TROUGH

Fig. 31·24.

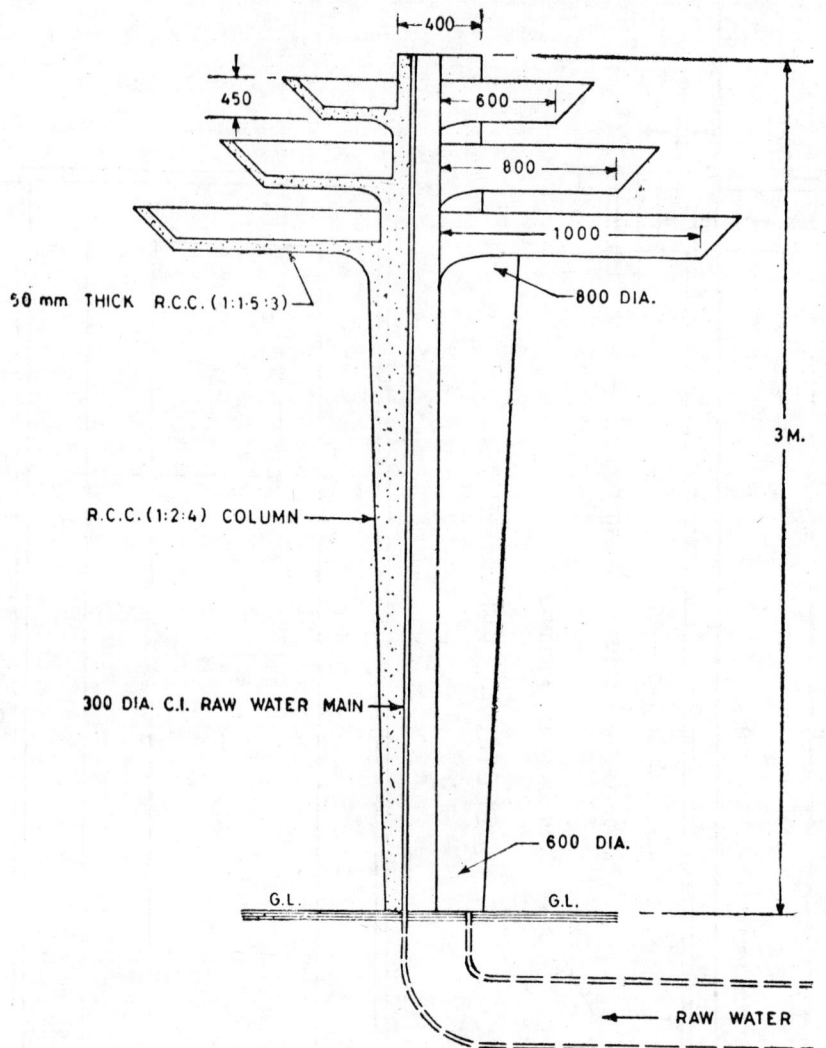

|←—400—→|

450

600

800

1000

50 mm THICK R.C.C. (1:1·5:3)

800 DIA.

R.C.C. (1:2·4) COLUMN

3 M.

300 DIA. C.I. RAW WATER MAIN

600 DIA.

G.L. G.L.

RAW WATER

HALF SECTIONAL ELEVATION
OF RAW WATER FOUNTAIN (CASCADE AERATOR)

Fig. 31·25.

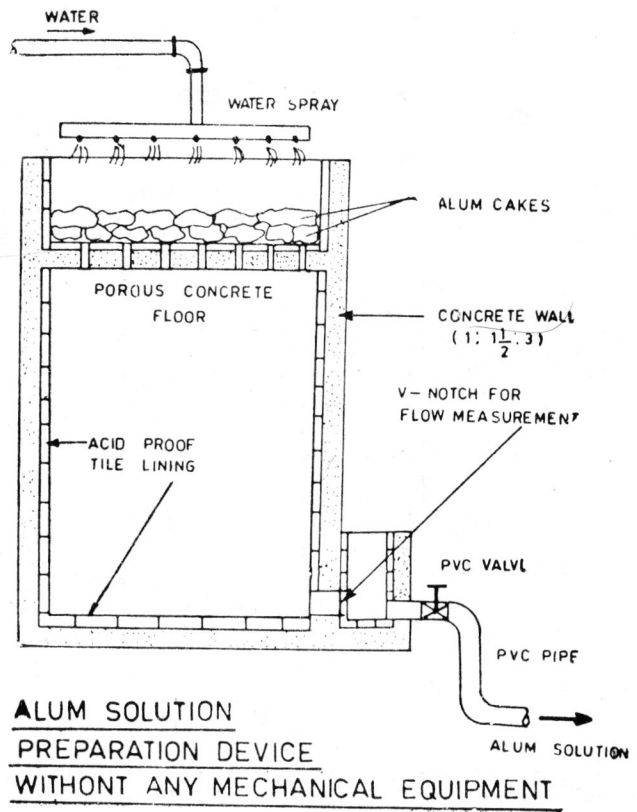

WATER

WATER SPRAY

ALUM CAKES

POROUS CONCRETE
FLOOR

CONCRETE WALL
$(1: 1\frac{1}{2}: 3)$

V – NOTCH FOR
FLOW MEASUREMENT

ACID PROOF
TILE LINING

PVC VALVE

PVC PIPE

ALUM SOLUTION

ALUM SOLUTION
PREPARATION DEVICE
WITHONT ANY MECHANICAL EQUIPMENT

Fig. 31·26.

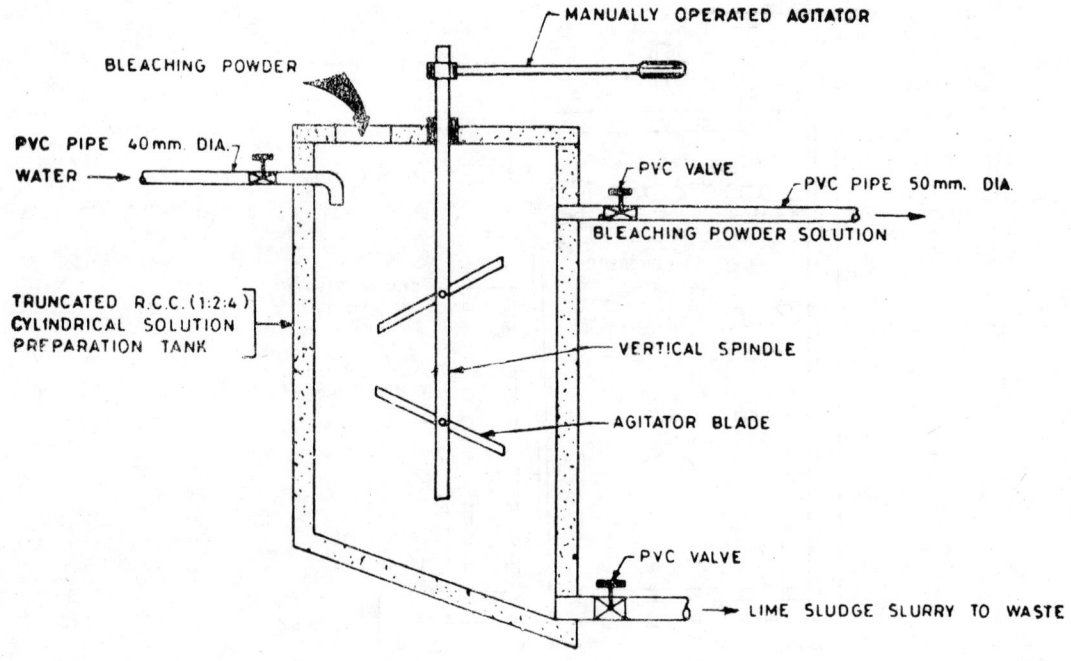

BLEACHING POWDER SOLUTION PREPARATION TANK WITHOUT ANY
MECHANICAL APPARATUS AND ELECTRICAL POWER

Fig. 31·27.

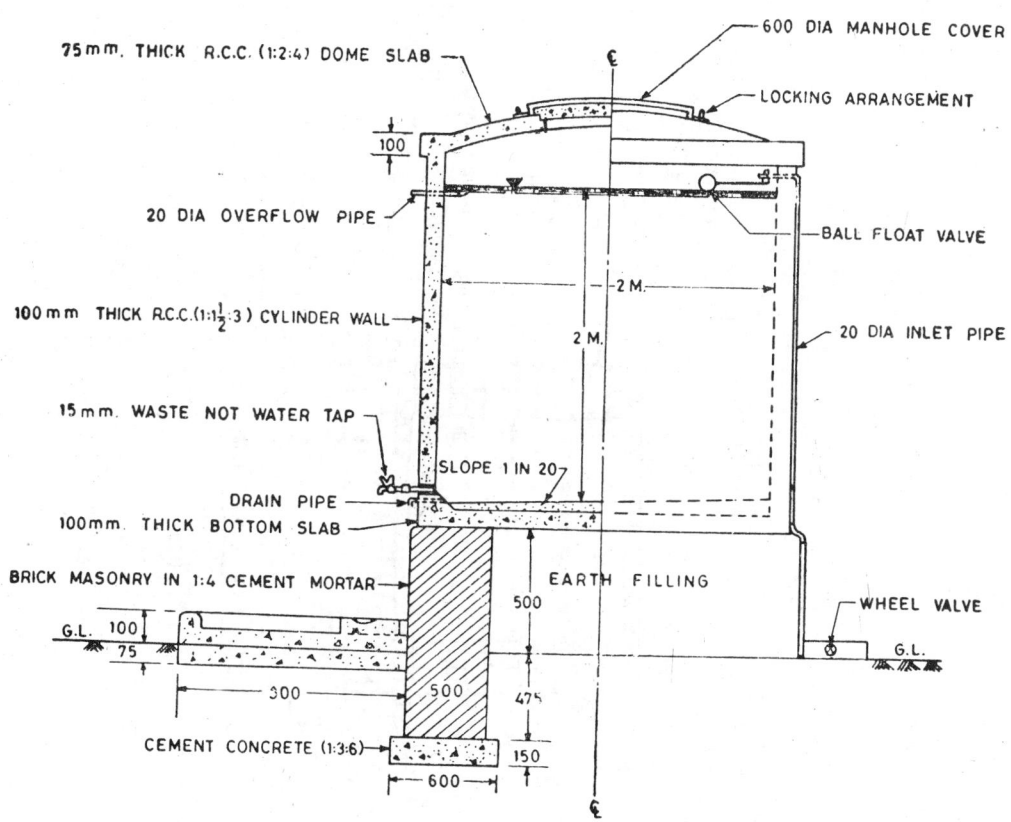

75mm. THICK R.C.C. (1:2:4) DOME SLAB

600 DIA MANHOLE COVER

LOCKING ARRANGEMENT

100

20 DIA OVERFLOW PIPE

BALL FLOAT VALVE

2 M.

100mm THICK R.C.C.(1:1½:3) CYLINDER WALL

2 M.

20 DIA INLET PIPE

15mm. WASTE NOT WATER TAP

SLOPE 1 IN 20

DRAIN PIPE

100mm. THICK BOTTOM SLAB

BRICK MASONRY IN 1:4 CEMENT MORTAR

EARTH FILLING

WHEEL VALVE

G.L. 100

75

500

G.L.

300

500

475

CEMENT CONCRETE (1:3:6)

150

600

SECTIONAL VIEW OF A R.C.C. STAND POST FOR PUBLIC FOUNTAIN

Fig. 31·28.

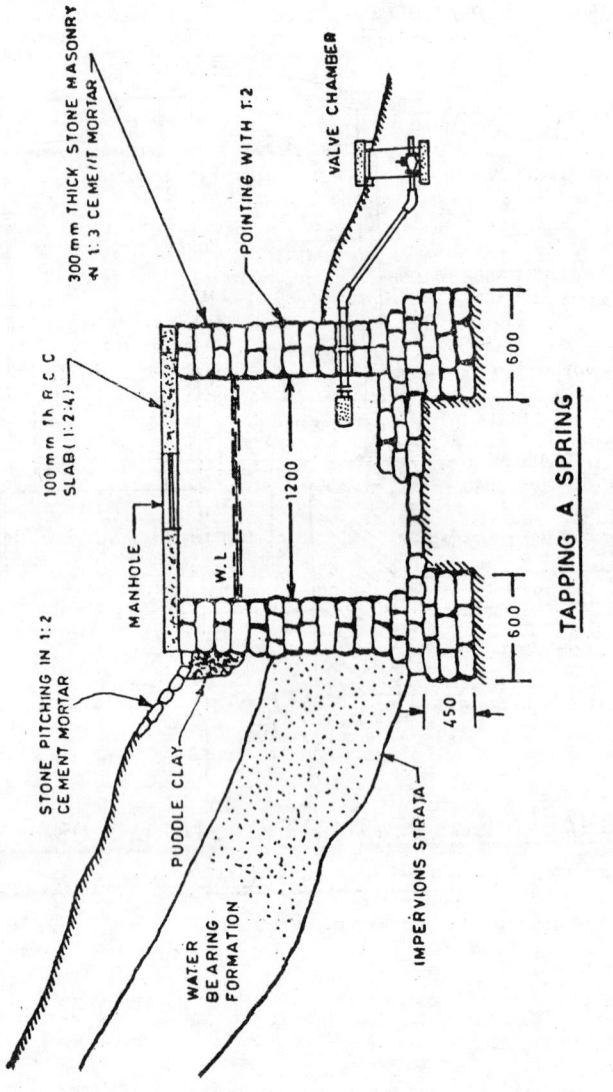

300 mm THICK STONE MASONRY IN 1:3 CEMENT MORTAR

POINTING WITH 1:2

VALVE CHAMBER

100 mm Th R C C SLAB (1:2:4)

MANHOLE

W. L.

1200

600

600

450

STONE PITCHING IN 1:2 CEMENT MORTAR

PUDDLE CLAY

WATER BEARING FORMATION

IMPERVIOUS STRATA

TAPPING A SPRING

Fig. 31·29.

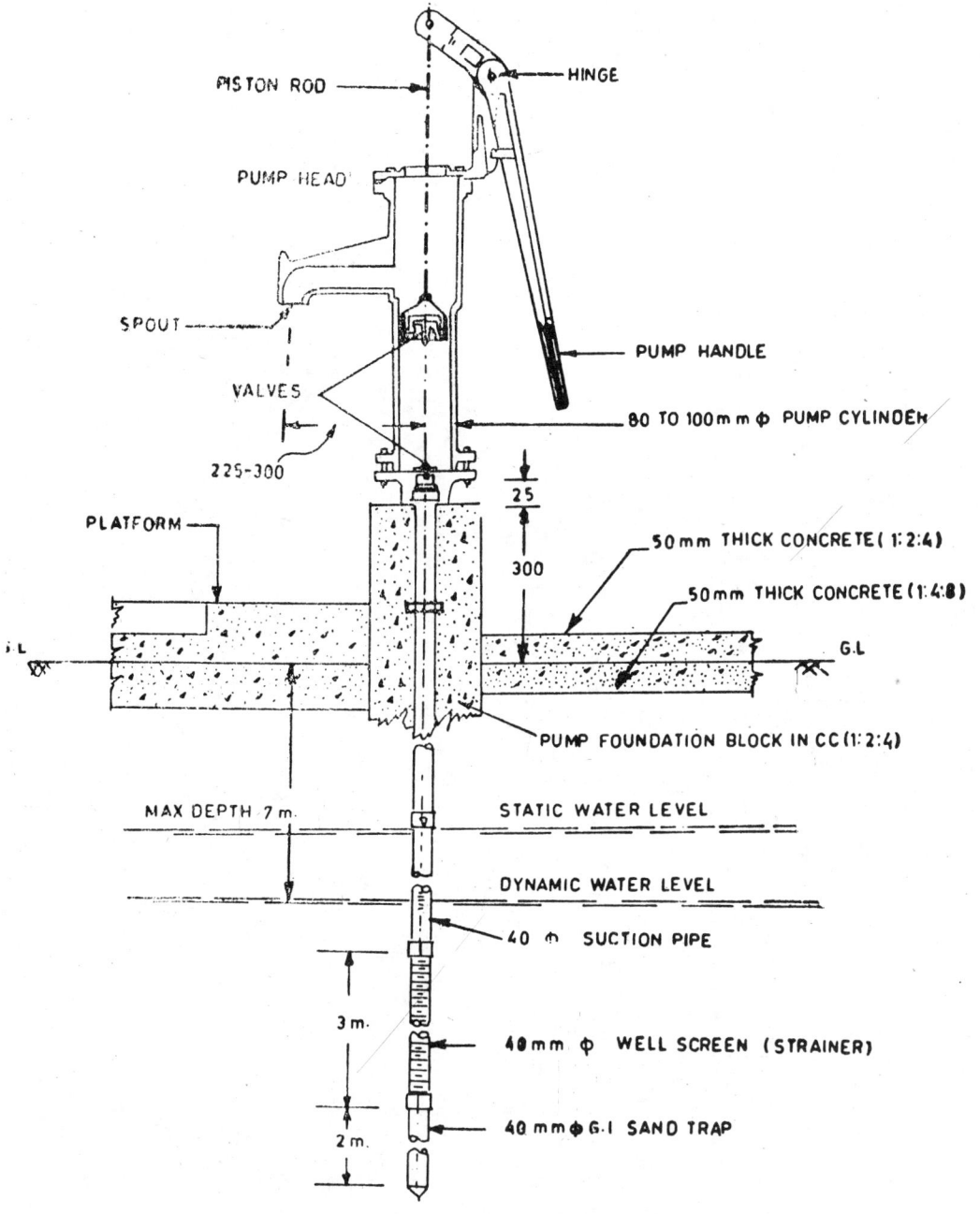

A SHALLOW HAND TUBE WELL

Fig. 31·30.

32

Simple Machine Drawings

This chapter deals with simple machine drawings, which are essentially required by civil engineers. Some of the components presented here are directly related to civil engineering structures. These are thread forms, bolts and nuts, studs, rivets, screws, keys and cotters, joints made with bolts and nuts, rivets, keys and cotters, pump and motor foundation with foundation bolts, adjustable joints, knuckle joint etc. In addition, drawings of lock nuts, pulleys, wall bracket, sole plate, footstep bearing, pedestal bearing, different types of couplings, stuffing box, etc. are given here.

Screw Threads

A screw thread is a continuous spiral or helical ridge and groove of uniform section on the cylindrical surface of a round rod or on a circular hole. A screw thread is a functional element used on bolts, nuts, screws and machine parts for temporary fastening. These can easily be taken out from the parts joined together, as and when required.

A Bolt with a nut is called a 'screwed pair'. The larger diameter of a thread part is called 'outside diameter', 'nominal diameter' or 'major diameter'. The smaller diameter of a threaded part is called 'minor diameter' or 'core diameter' or 'root diameter'. The radial distance between major and minor cylinders is called 'depth of thread'. 'Crest' is the peak point of a thread and 'Root' is the bottom or base of a thread. 'Pitch' of a thread is the crest to crest or root to root distance or is the distance between corresponding points on adjacent thread forms, measured parallel to the axis 'Lead' is the distance which a screw thread advances in one turn. In a single-start thread, pitch=lead and in a double-start thread, Lead=2×pitch.

Screw Thread Forms

In general, there are two main forms of screw threads : Triangular or Vee threads and square threads. Various forms of threads which are modified forms of Vee threads and square threads have come in practice depending upon the strength needed in transmitting motion and power. These are shown in Fig. 32·1.

Acme thread is a modified form of a square thread. Knuckle thread is also a modified form of a square thread. Buteress thread is a combined form of a Vee thread and a square thread.

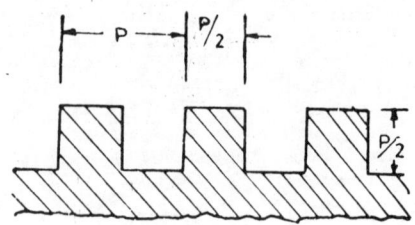

SQUARE THREAD

Square thread. Faces of the thread are normal to the axis of the spindle. The pitch of a square thread is taken as twice that of a B.S.W. thread of the same diameter. The thickness and depth of thread are taken as half the pitch. Square threads offer least resistance compared to the Vee-thread.

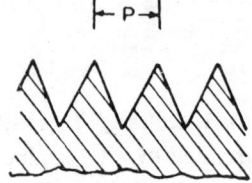

VEE THREAD

Vee thread. These threads have more friction and these are stronger than square threads. Various forms of vee threads are shown in Fig. 32.2.

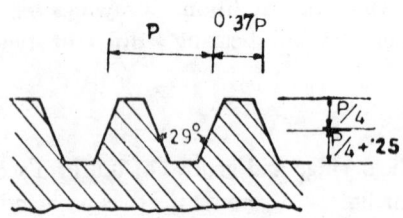

ACME THREAD

Acme thread. Acme thread is stronger than square thread and it can easily be engaged and disengaged. Square threads are replaced by acme threads. The thread angle is 29° and the depth of thread is $0.5 P + 0.25$ mm, P being the pitch of the thread.

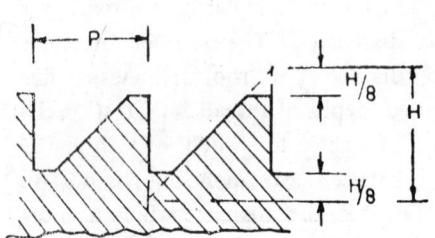

BUTTRESS THREAD

Buttress thread. It transmits power in one direction only. One flank of the thread is normal and the other is inclined to an angle of 45° with the axis. The root and crest are made flat. Pitch of thread is equal to the theoretical depth and the actual depth is 3/4th of theoretical depth.

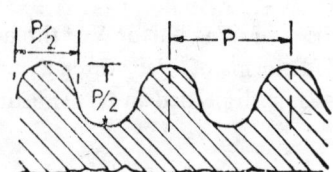

KNUCKLE THREAD

Fig. 32.1.

Knuckle thread. It is a crude form of thread, either rolled or cast. The working surface of the thread is large and it is used for rough work. The depth of thread is half the pitch. The radius of circles forming roots and crests is one-fourth the pitch.

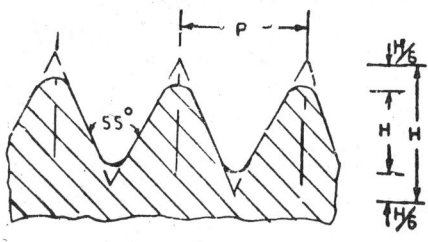

**BRITISH STANDARD
WHITWORTH THREAD**

British standard withworth thread is commonly used for fastenings. One-sixth of the depth of the 'V' is rounded off at top and at bottom of the thread. $H = 0.96\ P$ and $d = 0.64\ P$; Angle of thread is 55°.

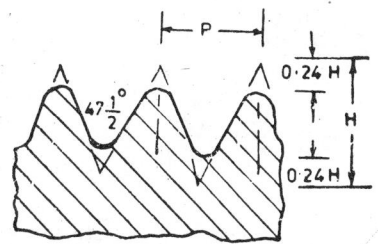

BRITISH ASSOCIATION THREAD

British association thread is used for small screws. The crests and roots are rounded off with equal radii. The thread angle is $47\frac{1}{2}°$. More numbers of threads are availabls per unit length compared to B.S.W. thread.

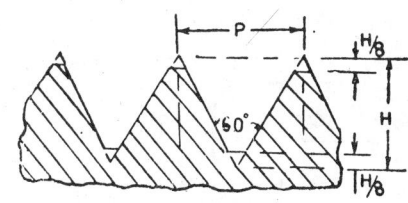

SELLER THREAD

Seller thread is used for making adjustment. This is also known as American thread. The thread angle is 60°. One-eighth of the theoretical depth is cut off at creast and at root. Theoretical depth $= 0.866$ pitch. Actual depth $= 3/4$th of theoretical depth.

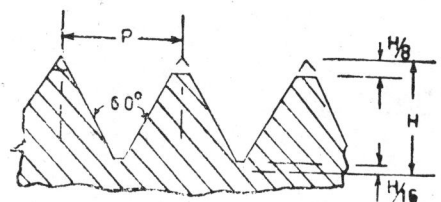

SYSTEM INTERNATIONAL THREAD

System international thread. It is similar to seller thread, having greater depth. The thread angle is 60°. The crest and root are flattened as shown.

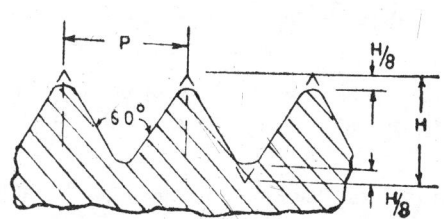

METRIC THREAD
Fig. 32·2.

Metric thread. It is similar to British standard whiteworth thread, only the thread angle is 60° in place of 55°.

Left Hand and Right Hand Threads

A spindle having left hand thread advances into engagement in a direction away from the observer, when turned in anti-clockwise direction. The clockwise rotation of a nut screws it off a bolt having left hand thread.

SECTIONAL VIEW OF A CATTLE SHED

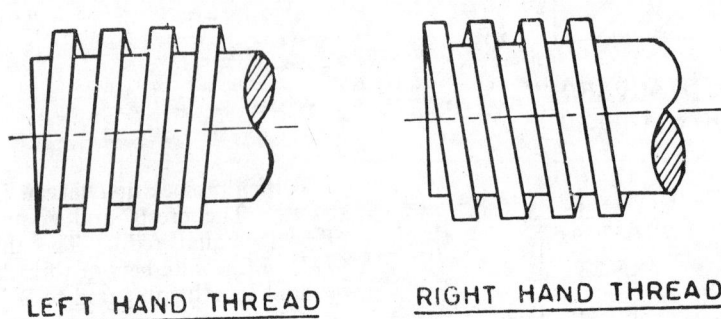

LEFT HAND THREAD   RIGHT HAND THREAD

Fig. 32·3.

A spindle having right hand thread advances into engagement in a direction away from the observer, when turned in clockwise direction. The clockwise rotation of a nut screws it on a bolt. Left Hand and Right Hand threads are shown in Fig. 32·3.

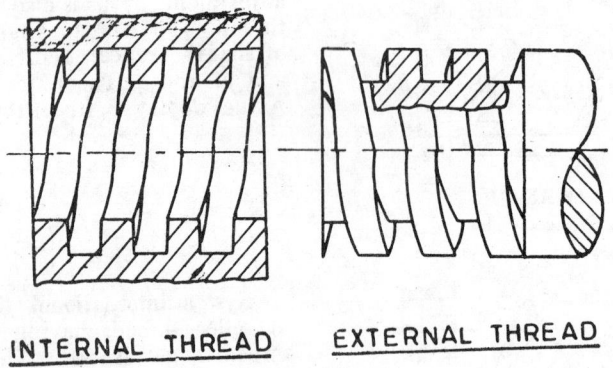

INTERNAL THREAD   EXTERNAL THREAD

Fig. 32·4.

Internal and External Threads

A thread formed on the external surface of a cylinder is called external thread and a thread cut on the internal surface of a hollow cylinder is known as internal tnread. External threads engage into the internal threads. Internal and external threads are also known as Female and Male threads. These are shown in Fig. 32·4.

A Screwed Pair

A bolt-nut is called a screwed pair. Such a pair is used for temporary fastening of two or more parts together. The parts so joined can be separated by screwing off the nut. A screwed pair is shown in Fig. 32·5.

Bolts. A bolt is a threaded spindle with an integral head at one end. The threaded part or the spindle part of the bolt is called stem or shank. There are various types of bolts used to serve different purposes. Most commonly used bolts are either square headed or hexagonal. Fig. 32·5 shows two views of a square headed bolt fitted with a square nut. The plan, elevation and end view of a hexagonal bolt and nut are shown in Fig. 32·6.

A square headed bolt is usually found to be in use, where the bolt head is to be accomodated in a recess of square form. This helps to prevent the head from turning when the nut is screwed off or on. The upper surface of the square head is chamferred to 30° to the base. Sometimes, a square headed bolt is provided with a square neck, when the bolt head projects outside the parts to be joined together. This square neck prevents the turning of the bolt head.

Hexagonal headed bolt is the most common form of bolt widely used in engineering practice. The usual dimensions of a hexagonal headed bolt are given in Fig. 32·6. However,

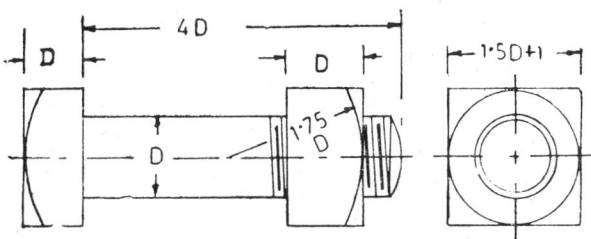

SQUARE BOLT AND NUT

Fig. 32·5.

the diameter and length of the bolt depend upon the thickness or depth of the parts to be joined together. The upper surface of the bolt head is chamferred at 30° to its base.

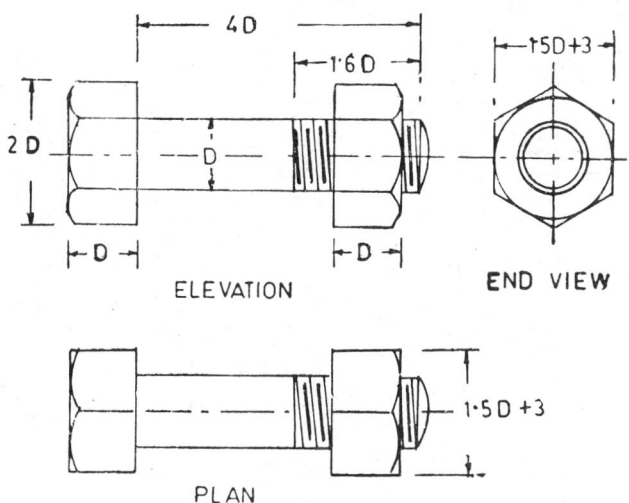

HEXAGONAL BOLT AND NUT

Fig. 32·6.

Other types of bolts in use are : cheese head, cup head and countersunk head bolts, hook bolt, taper or headless bolt, lifting eye bolt, tap or stud bolt and foundation bolts.

Cheese headed bolt is the simplest form of a bolt manufactured by forging the head which is cylindrical. This bolt is used in engine parts. The rotation of the bolt is prevented by inserting a pin (sung) into the shank just below the head.

Cup headed or round headed bolt is similar to cheese headed bolts. The sung is forged with the bolt.

A countersunk headed bolt is used where projection of bolt head is not wanted. Sometimes, a sung is provided with bolt head to prevent turning of the bolt.

The head of a Taper bolt is in form of a tapered cylinder, the tapper being 1 in 30. It is also called a headless bolt. The bolt end of smaller diameter is threaded to receive the nut as shown in Fig. 32.7. For use of such a bolt, the hole is also made tapered which is identical to the bolt head. Its use is similar to that of a countersunk headed bolt.

Hook bolt as shown in Fig. 32.7 is a spherical form of a bolt having a square neck with a hook like head which remains projected. The square neck prevents its turning.

Usually, eye bolt is screwed in a blind hole in the casting of a machine body for lifting purpose. The head of the bolt is a circular ring of square or circular in section. The use of a lifting eye bolt is shown in Fig. 32.7.

Tap bolt or stud bolt is a bolt having thread all throughout its shank. For joining parts together, this bolt does not require any nut. Its use is shown in Fig. 32.7.

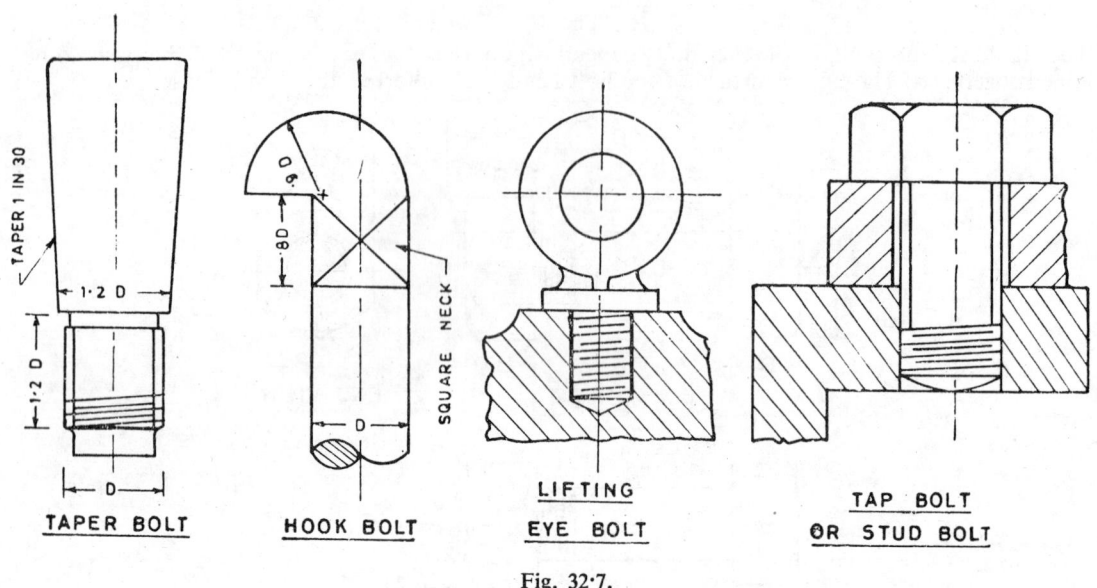

Fig. 32.7.

Foundation Bolts :

These are special bolts required for fixing heavy machines to the concrete foundation. The bolt head remains in concrete under floor and a nut is screwed to the bolt to hold the machine base firmly into the floor. There are various forms of foundation bolts of which most commonly used are :

T-headed bolt, Rag bolt, Lewis bolt and Eye or Loop bolt.

T-headed bolt is the simplest form of foundation bolt, the head being just like a 'Tee'. Usually, a square neck is provided with the head. The shank may be square or circular in section.

Rag bolt is a tapered bolt having square section at upper part and rectangular section at lower part. The sides and edges of the bolt head is serrated as shown in Fig. 32·8.

A lewis bolt is used for a temporary foundation of a machine. One side of the bolt head is straight and other side tapered. A key is inserted to the straight side of the bolt in foundation to hold the machine base. As and when required, the bolt can be taken out by removing the key from foundation. It is shown in Fig. 32·8.

Loop bolt or eye bolt as shown in Fig. 32·8 is a simple form of foundation bolt. The head of the bolt is of eye or loop form which can easily be forged from a mild steel bar. For its use in foundation, a M.S. bar is introduced through the eye of the bolt at right angles to its axis and concreting is done to hold the bolt head in this position.

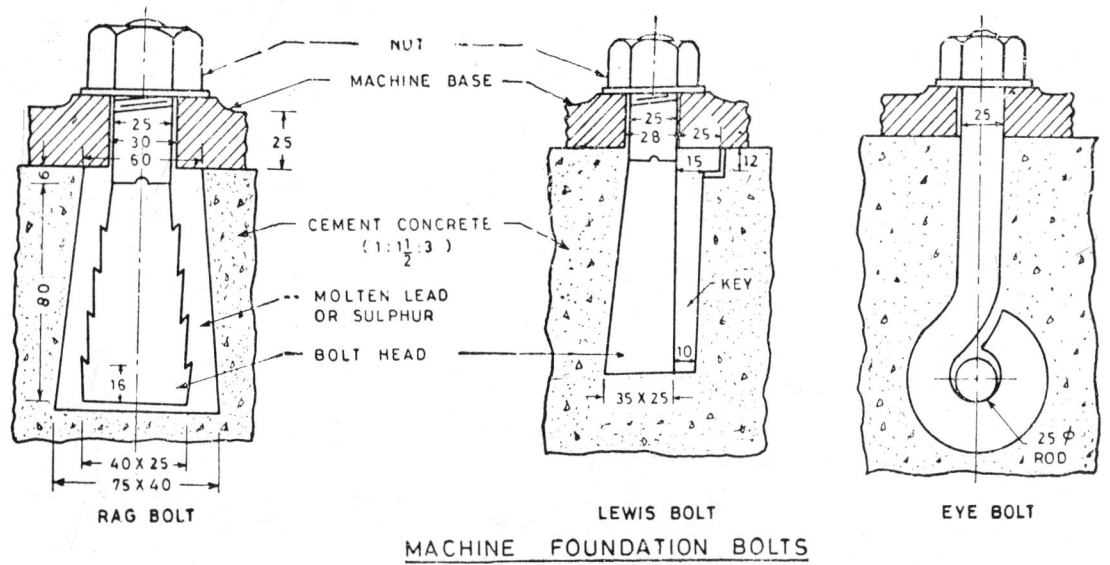

MACHINE   FOUNDATION   BOLTS

Fig. 32·8.

Nuts :

A nut is a component used with a bolt or a threaded spindle to fasten two parts together. A nut screws on the threaded end of a bolt or a stud and draws the parts together, when tightened. Usually, a nut is a hollow hexagonal or square prism with top corners chamfered and threads in the hole. The corners arechamfered at 30° to the base of the nut. A hexagonal nut which is commonly used in engineering works is shown in Fig. 32.9. The height or thickness of the nut is equal to diameter of the bolt. Width across flat is $1.5\ D+3$ mm and width across corners is $2\ D$. Radius of front chamber is $1.2\ D$.

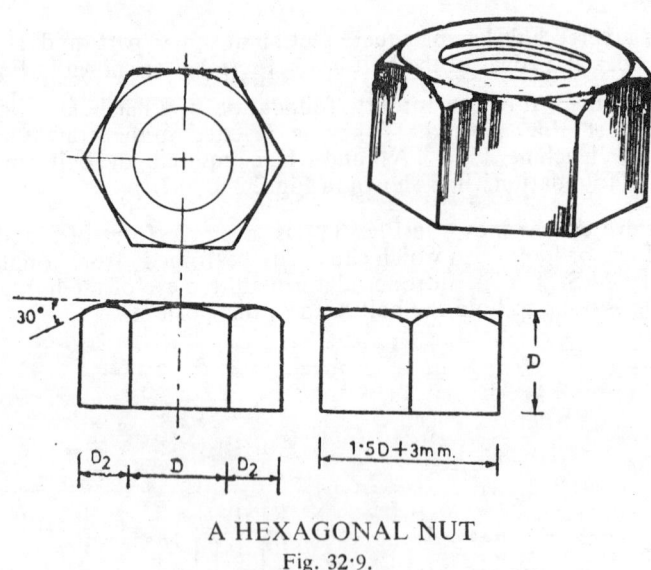

A HEXAGONAL NUT

Fig. 32.9.

For a square nut.

Size across flat $= 1.5\ D+1$ mm ; where d is the diameter of the bolt.

Size across corners $= \sqrt{2} \times (1.5\ D+1$ mm) ;

Height or thickness $= D$ or $0.8\ D$.

Radius of front chamfer $= 1.75\ D$ to $2\ D$.

Angle of chamfer $= 30°$.

The angle of chamfer may vary from 30° to 45°.

Hexagonal and square are the two main forms of nuts. There are also a good many number of other forms of nuts to serve specific funtions.

A flanged nut has a flat circular disc at the base of the nut. The disc thickness is $D/4$ and its diameter is $2.25\ D$. The flange or disc serves the purpose of a washer. A flanged nut is therefore a hexagonal nut with an integral washer.

A cap nut is also a hexagonal nut. The top of the nut is provided with an integral cylindrical cap with a view to protecting the bolt end from corrosion. The cap keeps the bolt end closed in it. The outer thickness of the cap is $D/2$ and the inner thickness is $D/4$. The diameter of the cap is $1.5\ D+1$ mm.

A dome nut is similar to a cap nut, the cap being spherical.   The depth of cap is $D/2$.

A ring nut is a circular nut with slots on the curved surface.   The thickness of nut is $D/2$.   The outer diameter of the nut excluding the slots is $1.5\,D$ and including the slots $1.75\,D$.

A capstan nut is provided with circular holes on the curved surface which facilitates in turning the nut with a tommy bar.

A spherical seated nut has a spherical seat at its bottom.   The diameter of the flat surface on the spherical seat is equal to the diameter of the bolt.   The total depth or height of the nut is D to $1.2\,D$.

All these nuts are shown in Fig. 32.10.

A wing nut is a conical nut having two wings attached to the slant surface of the nut. The nut can easily be turned by hand with the help of wings.   It is shown in Fig. 32.11.

A thumb nut is similar to a wing nut, but the body of the nut is cylindrical and not conical.   The wings are replaced by two conical rod which facilitates in turning the nut.

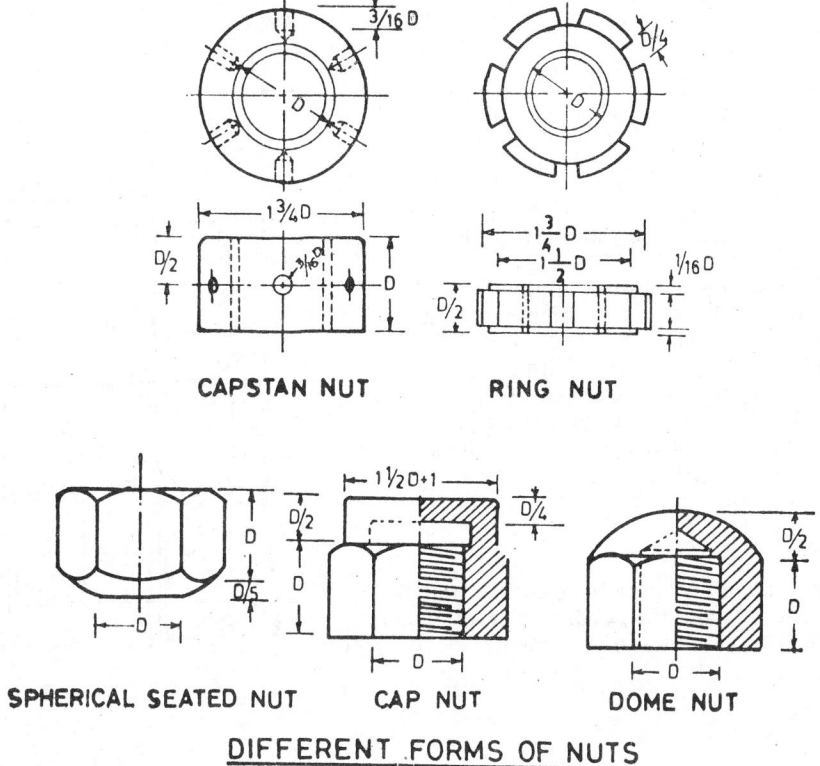

CAPSTAN NUT   RING NUT

SPHERICAL SEATED NUT   CAP NUT   DOME NUT

DIFFERENT FORMS OF NUTS

Fig. 32.10.

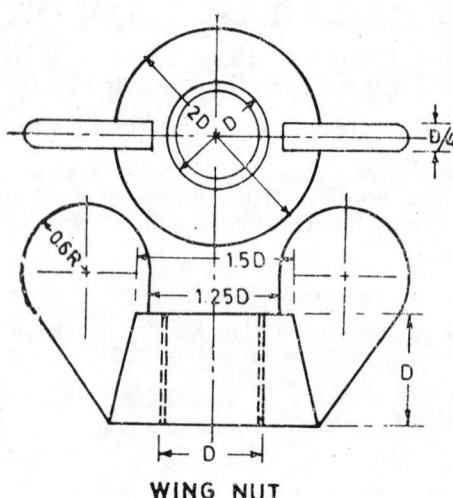

WING NUT

Fig. 32·11.

Locking Arrangements :

When bolts and nuts used in a machine part are subjected to vibration, there is a tendency of the nuts to screw off the bolts by slackening which may lead to a serious breakdown and accident. In such cases, locking arrangement is made to prevent the nuts from slackening and to keep them tight in position. There are various devices for locking arrangement, a few of which are described here along with sketches.

Lock nut : This nut is also known as 'check nut' or 'Jam nut'. Fig. 32·12 shows such a lock nut. This type of nut is most commonly used for locking arrangement. It is simply a nut having thickness of about 0·5 to 0·7 times the thickness of a standard nut and both top and bottom surfaces of this nut are chamfered at an angle of 30°. This lock nut is to be used with an ordinary nut for the purpose of locking.

The lock nut is first screwed on the bolt and tightened. Then, the ordinary nut is placed above the lock nut and tightened.

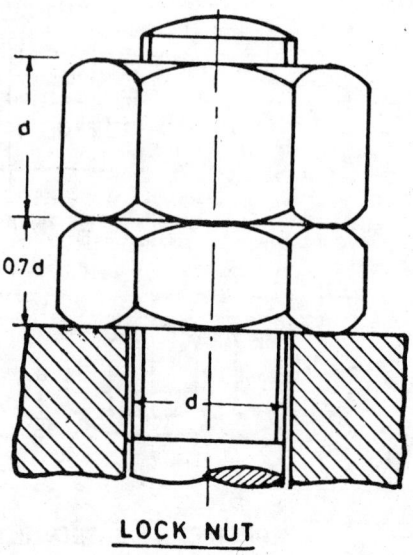

LOCK NUT

Fig. 32·12.

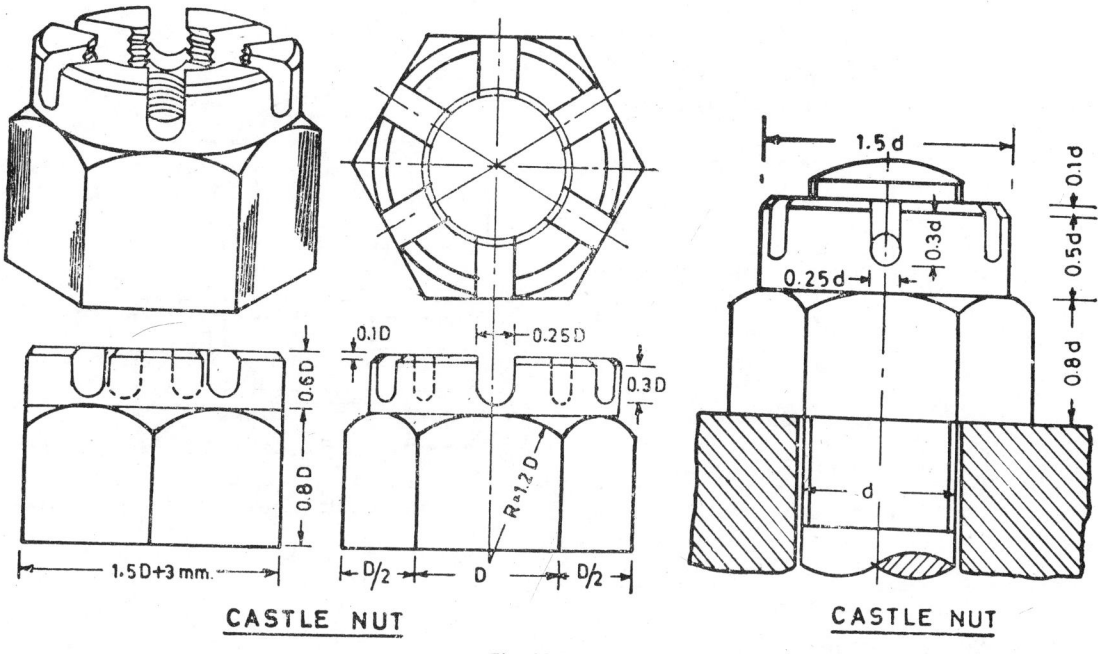

CASTLE NUT

CASTLE NUT

Fig. 32·13.

Castle Nut :

It is a form of a slotted nut. This nut is provided with a cylindrical collar on its upper surface, the collar being slotted in lines with the centre of each face of the nut Fig. 32·13 shows four views of a centre nut and its use. Standard dimensions of the nut, collar. and slots are also given. For locking the nut, a split pin is inserted through the slot and the hole in the bolt. This keeps the nut in position.

Wile's Lock Nut :

This nut is also called 'sawn nut'. It is an ordinary hexagonal nut having a saw-cut half-way across it. The upper part of the cut is provided with hole and the lower part with a tapped hole. This is shown in Fig. 32·14. For locking the nut, the nut after screwing on the bolt, is tightened and a set screw is introduced into the hole of the sawn nut and tightened.

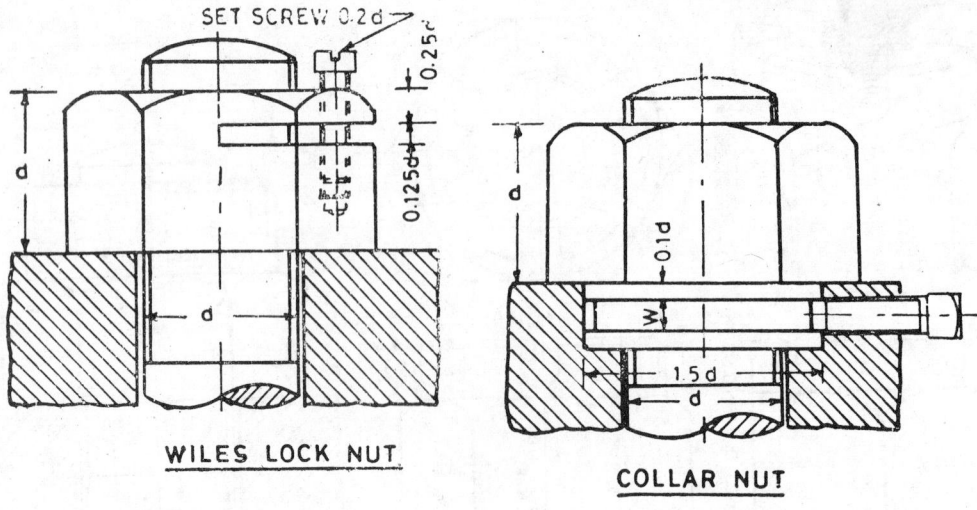

Fig. 32·14.

Fig. 32·15.

Collar Nut or Grooved Nut :

A cylindrical grooved collar is made integral with the nut at its bottom. After fitting it into a counter bored hole in the adjoining piece a set screw is introduced through the side hole of the piece and tightened. This prevents the nut from movement during rotation or vibration. Sucharrangement is shown in Fig. 32·15.

Pin Locking : Two forms of pin locking arrangement are shown in Fig. 32·16. This device is very simple.

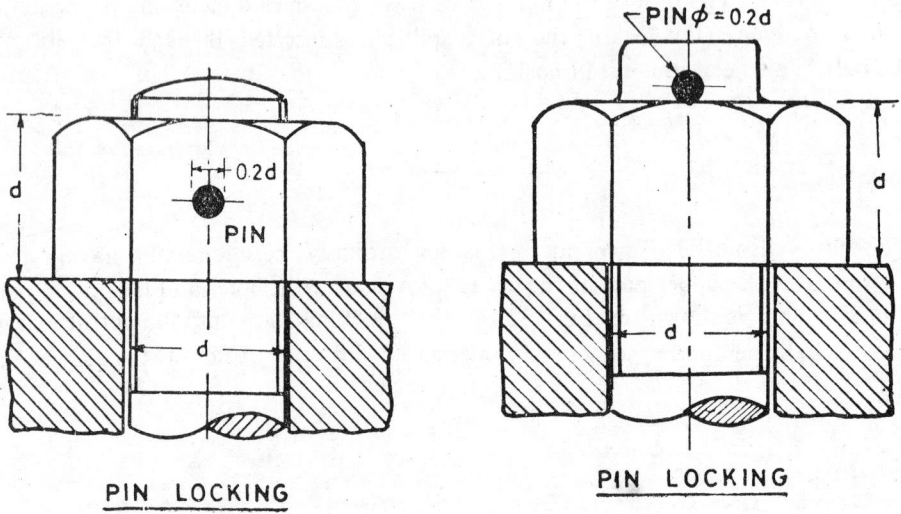

Fig. 32·16.

Set Screws :

Set screws are screws of small size and these are just like small bolts having heads of various shapes and no nut is required for fastening. Where the projection of screw head is unwanted, countersunk head or headless screws are used. Various types of set screws that are commonly in use are shown in Fig. 32.17.

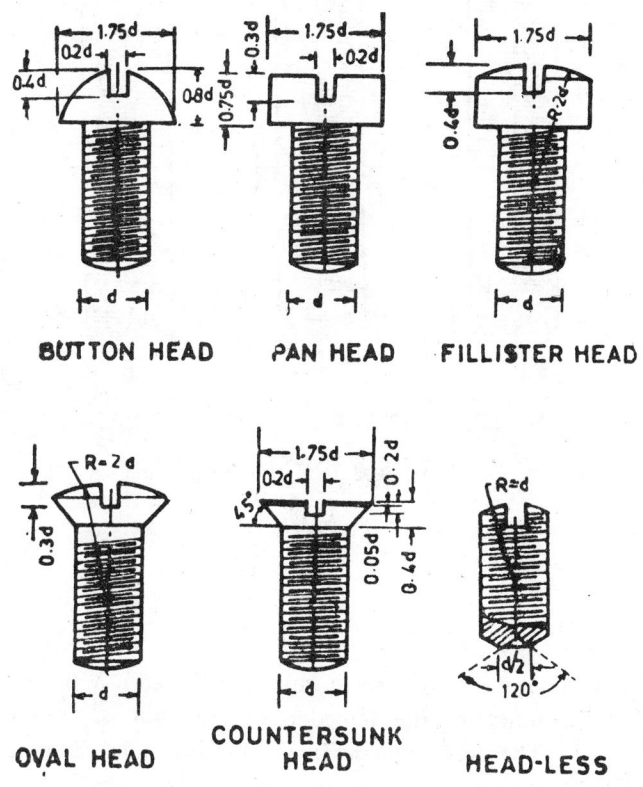

VARIOUS TYPES OF SET SCREWS

Fig. 32.17.

Studs :

   A stud may be called a headless bolt. It is a spindle with threads at both the ends, the central part being left plain or provided with a collar or square neck. A stud is used when the adjoining parts do not have adequate space for bolt head. For joining two parts together,

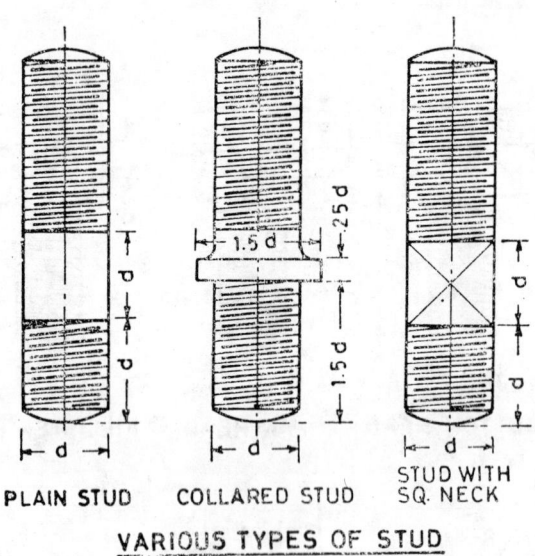

PLAIN STUD   COLLARED STUD   STUD WITH SQ. NECK

VARIOUS TYPES OF STUD

Fig. 32·18.

into the tapped hole of one part and then a nut is screwed on the top of the second part and tightened. For disconnecting the joined parts, the nut is screwed off the stud and the upper part is taken out. The stud remains in the tapped hole. Different types of studs are illustrated in Fig. 32·18.

   Rivets : Rivets are used for permanent fastening of structural parts, pressure vessels and machine parts. A rivet is a round rod with a head at one end. After placing a rivet into the holes of two parts to be joined together, the tail end of the rivet is forged into a head by preessing the parts, with the help of a die. For this purpose, riveting machines are employed. A rivet essentially consists of a head, a central cylindrical portion called shank and a tapered tail which is 1·25 times the dia. of the rivet.

Rivet heads are of various forms. All types of rivets have cylinderical body, but their heads may differ from each other. The most common type of head used is snap head of cup shape. In general, two classes of rivets are used in engineering works. They are structural rivets and boiler rivets. Different types of structural and boiler rivets are shown in Fig. 32.19. The dimensions for each type of rivet head are given in figure. For very small jobs, small diameter rivets of brass, copper and aluminium are also available in market.

Riveting and Riveted Joints :

Riveting is the process by which the tail of the rivet is formed into a head for joining two parts permanently. A riveted joint is a permanent fastening and may be either a lap joint or a butt joint. In a lap joint the parts to be joined together overlap one above another and are

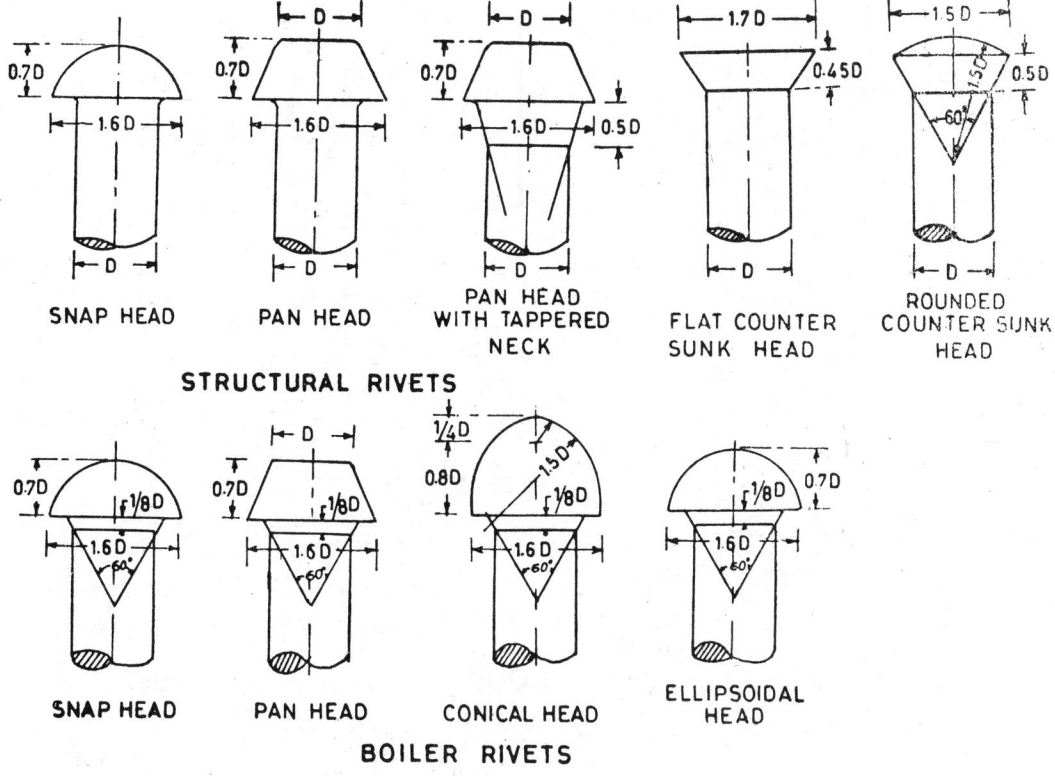

VARIOUS TYPES OF RIVET

Fig. 32·19.

riveted.   In a butt joint, the parts butt up against each other and they are joined together with the help of a single cover strap or double cover straps.   Again, a riveted joint may be a single riveted or a double riveted joint.   Fig. 32·20 shows plan and sectional elevation of a single riveted lap joint.   The plan and sectional elevatian of a double riveted lap joint are presented in Fig. 32.21.   Ref. to these figures P=pitch=3×diameter of rivet ; marginal distance is half the pitch *i.e.*, 1·5×dia. of rivet and backpitch P_B=2×dia, of rivet+6 mm.

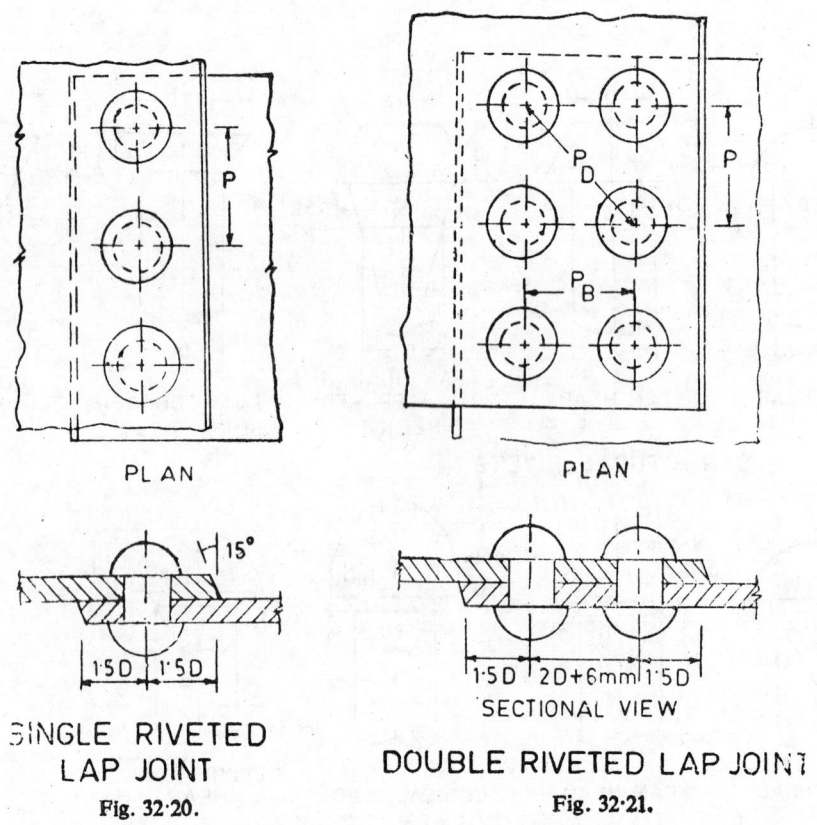

PLAN   PLAN

SECTIONAL VIEW

SINGLE RIVETED
LAP JOINT
Fig. 32·20.

DOUBLE RIVETED LAP JOINT
Fig. 32·21.

    In a butt joint, the plates to be joined together butt against each other and riveting is done by covering the joint with one or two cover straps or cover plates. A butt joint may be single riveted or double riveted. In a butt joint when only one cover plate is used and two rows of rivets, one on either end of the butting end are provided, it is called a single cover single riveted butt joint (shown in Fig. 32.22). A double cover single riveted butt joint is shown in Fig. 32.23. The cover plate end is usually chamfered at an angle of 10°.

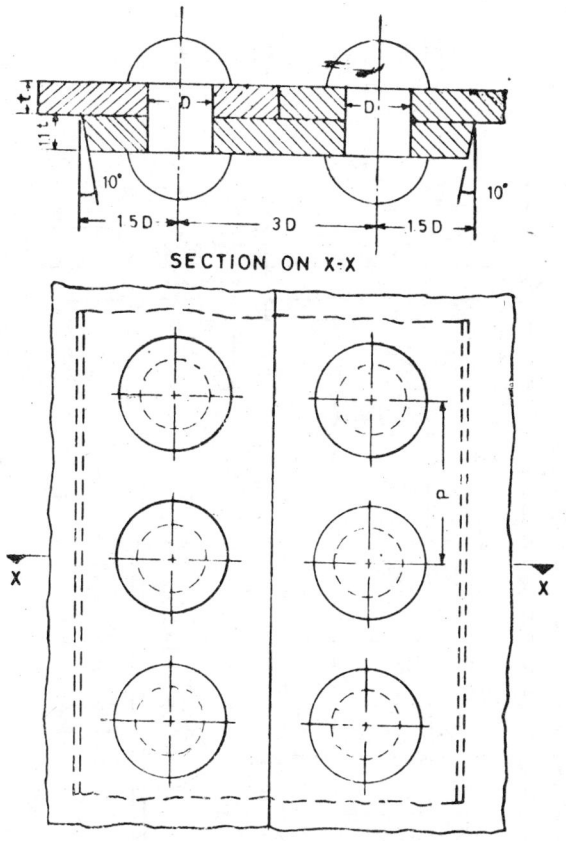

SECTION ON X-X

SINGLE COVER SINGLE RIVETED BUTT JOINT

Fig. 32.22.

When only one cover plate is used, its thickness should be equal to or 1·125 times the thickness of each plate to be joined together. When two cover plates are used, the thickness of each cover plate should be 0·6 to 0·8 times the thickness of plates to be joined together.

If d is the diameter of rivet and t is the thickness of plate, then according to unwin's formula

$$d = 1.2\sqrt{t} \ldots\ldots\ldots \text{in F.P.S. units} \quad \text{(dimensions iu inchs)}$$

$$d = 6\sqrt{t} \ldots\ldots\ldots \text{in M.K.S. units} \quad \text{(dimensions in mm.)}$$

$$= 1.9\sqrt{t} \ldots\ldots\ldots \text{in M.K.S. units (dimensions in cm.)}$$

Maximum Pitch, $P_{max} = 3d$

Minimum Pitch, $P_{min} = d + 30$ mm.

Marginal distance $= 1.5d$

Back Pltch, $P_B = 2d$

Length of rivet $=$ Plate grip $+ 1.25$ to $1.7d$.

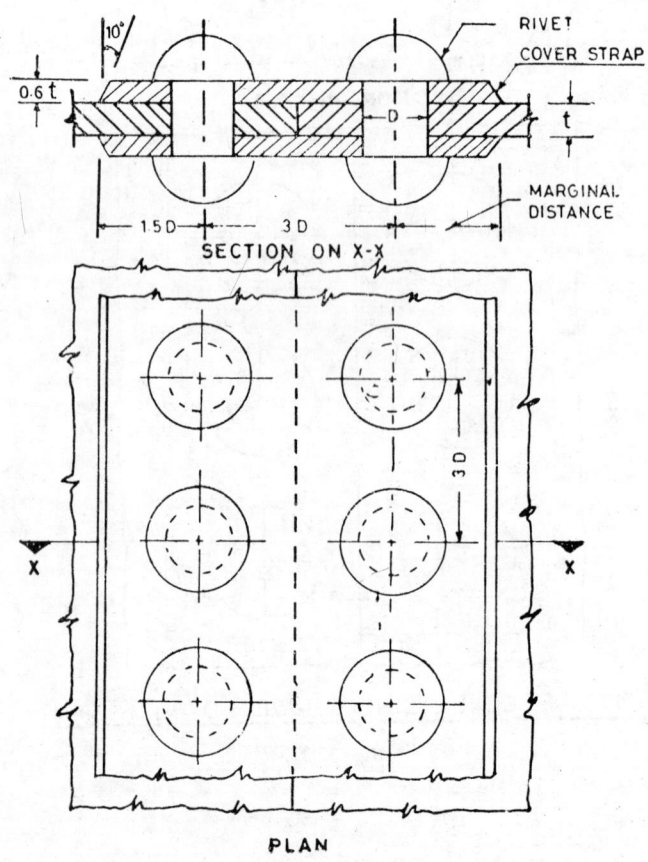

SECTION ON X-X

PLAN

DOUBLE COVER, SINGLE RIVETED BUTT JOINT
Fig. 32·23.

A double cover double riveted butt joint with zig-zag riveting is shown in Fig.32·24.

Here,   Pitch=3 d.

Back Pitch=2 d.

Diagonal Pitch=2·5 d.

Marginal distance=1·5 d.

Thickness of cover strap=0·6 t.

where, d=dia. of rivet and t=thickness of each plate to be joined.

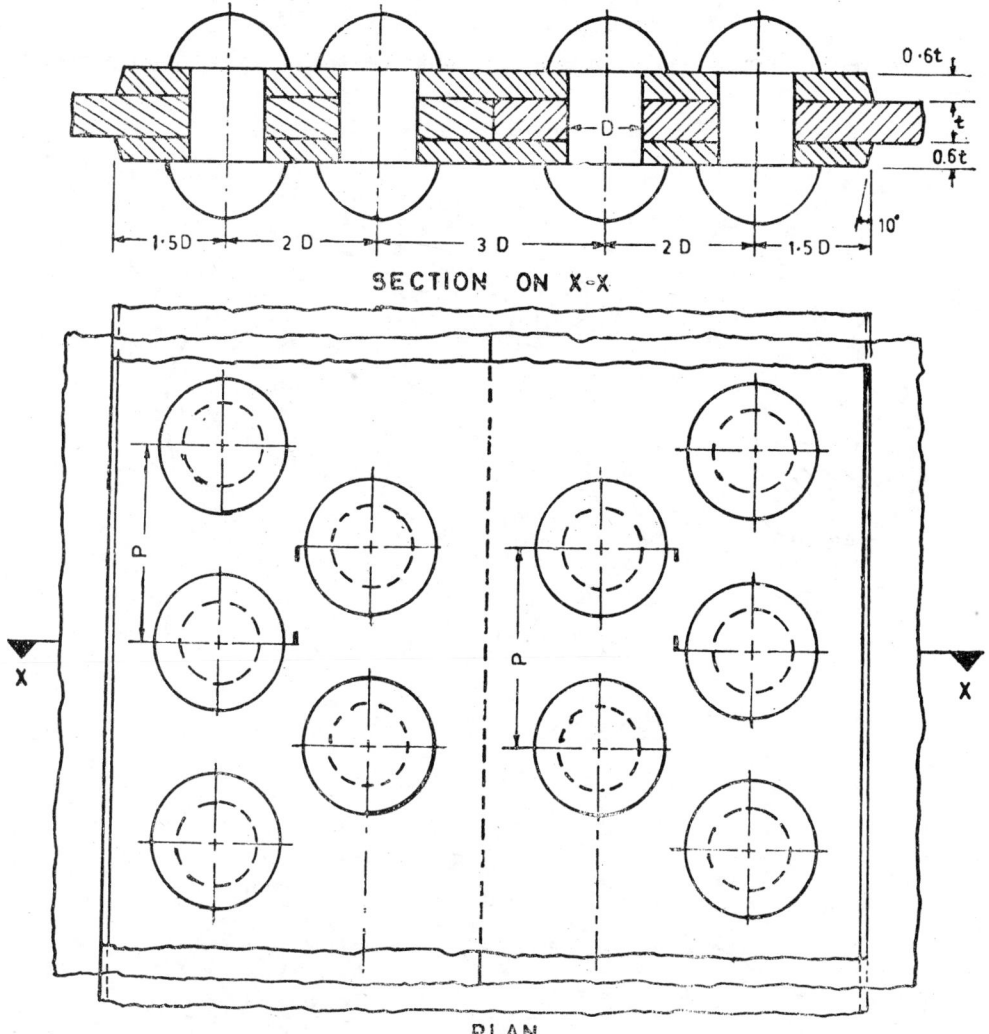

SECTION ON X-X

PLAN

DOUBLE COVER DOUBLE RIVETED BUTT JOINT-ZIGZAG FORMATION

Fig. 32·24.

Keys

A key may be defined as a wedge inserted in an axial direction in a joint of two circular parts. Thus, a key is a fastening device of temporary nature which prevents axial movement and relative rotary motion between the circular parts joined together. Flywheels, pulleys, cranks, etc., are screwed to shafts with the help of keys. A keyway of uniform depth and required length is formed in the shaft as well as in the hub of the flywheel or pulley to be secured to the shaft.

Keys are usually rectangular in cross-section. They may be parallel or tapered. A tapered key is uniform in width, but tapered in thickness. The usual taper is 1 in 100. Keys are proportioned with relation to the shaft diameter. If d is diameter of the shaft, the width of key$=0.25\ d$, thickness of key$=0.17\ d$ and minimum length of key$=1.5\ d$.

There are various types of keys in use, *e.g.*, sunk key, saddle key, key on flat, round or pin key, etc.

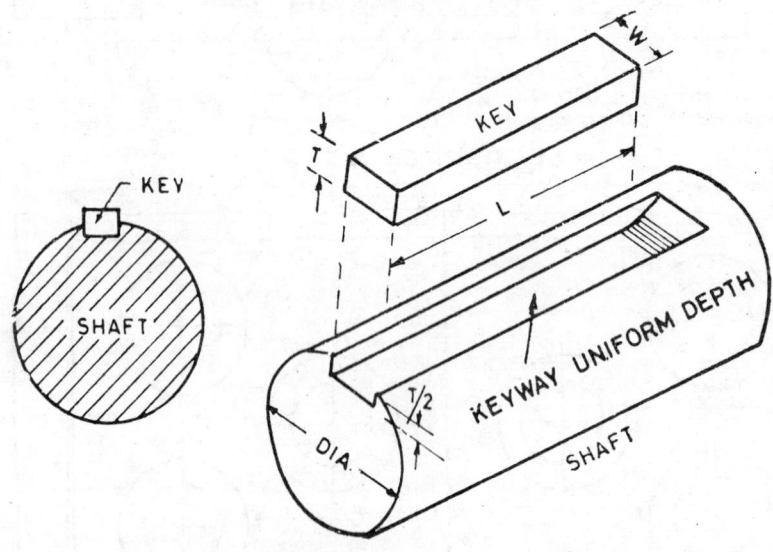

SUNK KEY

Fig. 32·25.

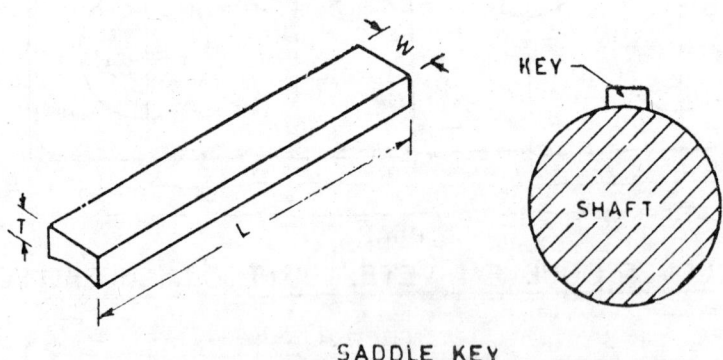

SADDLE KEY

Fig. 32·26.

When a key is inserted in the keyway formed in shaft and hub, *i.e.*, the key is sunk in the keyway of the shaft upto the half thickness of the key, it is called a sunk key.

Fig. 32·25 shows the insertion of a sunk key of rectangular cross-section.

A saddle key is just attached to the surface of the shaft and projects into the boss. It is not used for heavy duty jobs. Fig. 32·26 shows a saddle key.

A key on flat or a flat saddle key has its lower surface flat on the shaft and it projects into the boss. It is shown in Fig. 32·27.

A Round or Pin key is nothing but a circular pin, the diameter being $\frac{1}{8}$th to $\frac{1}{4}$th of shaft diameter. This is introduced just like a sunk key. For insertion of a pin key, a hole

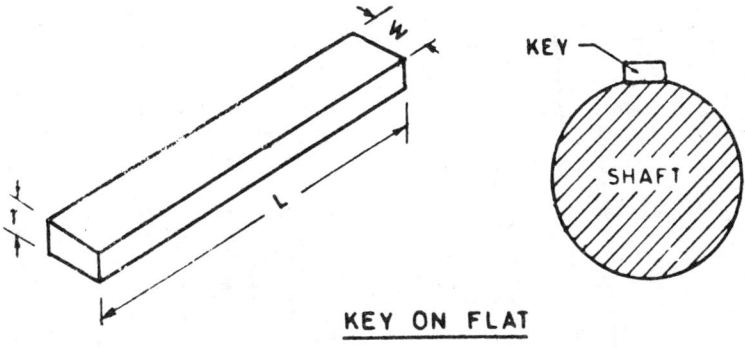

KEY ON FLAT

Fig. 32·27.

of key diameter is drilled parallel to the axis of the shaft, half in the shaft and half in the hub. A taper pin is sometimes used in place of a round key. The taper is generally 1 in 48 on diameter. The taper pin is driven in the mating hole by hammering on the larger end of the pin.

Rectangular parallel or taper sunk keys are sometimes provided with a 'Gib-head' and they are called gib headed keys. The gib head facilitates in removing the key easily. The proportions of the gib head are :

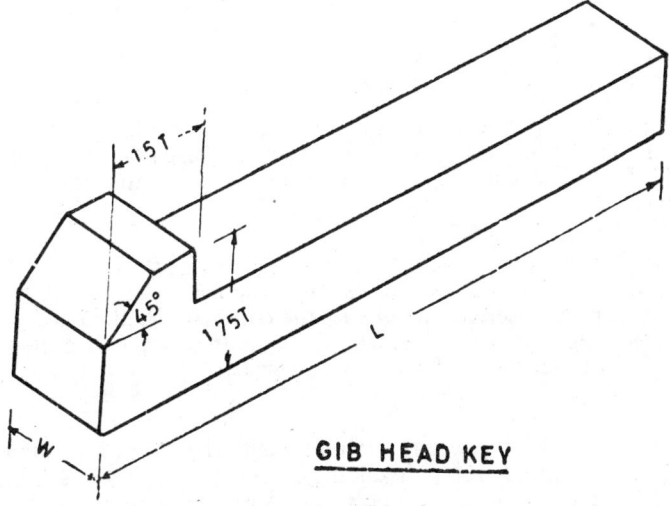

GIB HEAD KEY

Fig. 32·28.

Height of Gib head $= 1\cdot75\ T$.

Length of Gib head $= 1\cdot5\ T$.

Width of Gib head $=$ Width of key.

Angle of chamfer $= 45°$, T is the thickness of the key.

Cotter :

A cotter is a flat wedge of rectangular cross-section. It looks like a key but it is always fitted at right angles to the axis of the parts joined. A cotter is uniform in thickness,

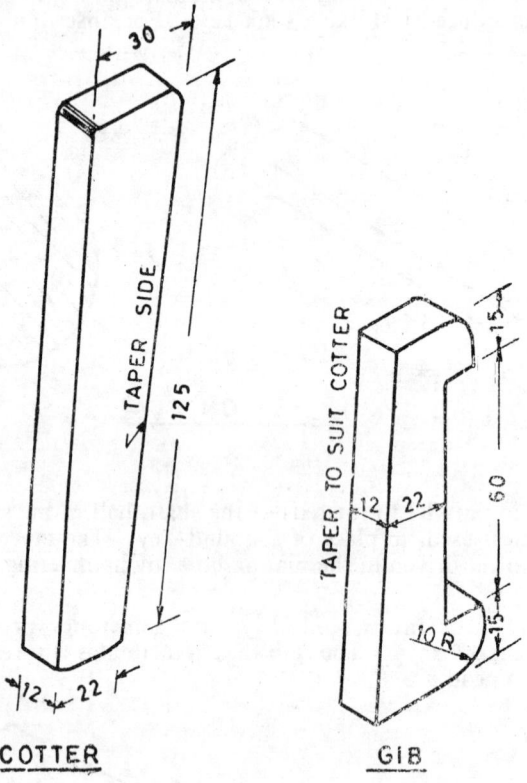

Fig. 32·29.

but tapered in width throughout its length. A cotter having taper one side is shown in Fig. 32·29. The tapering may be on both sides also. They are usually made of mild steel. The taper usually varies from 2 to 4 in 100. The cotter is driven in the slots made in the two parts to be joined together.

Gib :

Gibs as shown in Fig. 32·29 are used to increase the contact area between the mating surfaces. A gib is inserted alongwith the cotter where the parts joined are subjected to pull. The tapered edge of the cotter should come in contact with the tapered edge of the gib. The outer edges of the cotter and the gib are vertical.

Cottered Joint :

Plan and half-sectional view of a simple cottered joint are shown in Fig. 32·30. Such a cottered joint essentially consists of a sleeve with holes corresponding to the holes in the enlarged portions of the shafts and two cotters. A clearance of 1·5 to 3 mm is kept in the holes. This is called 'draw'.

A cottered joint may be spigot and socketed.   A gib is generally used in a cottered joint for square bars.

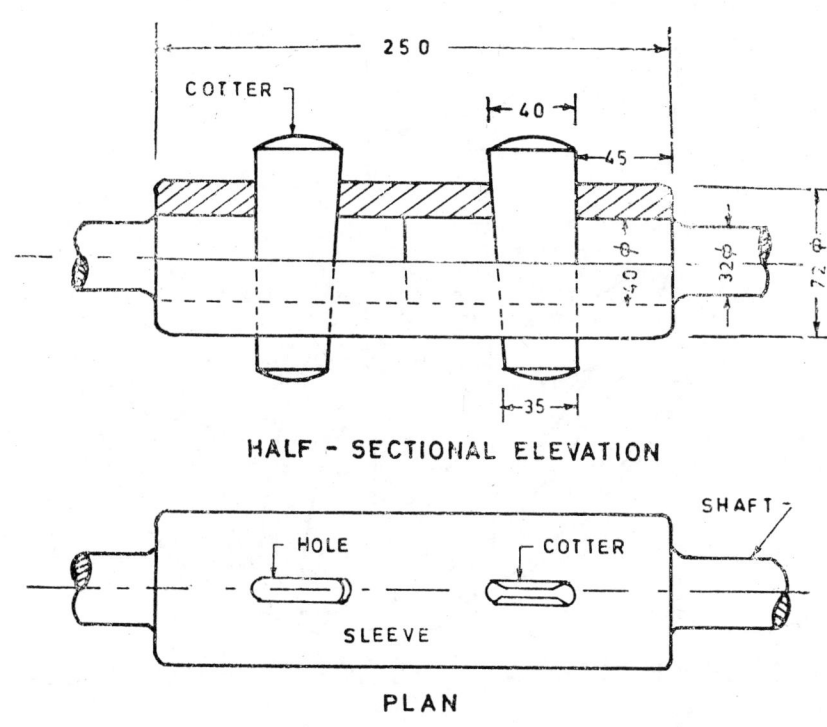

HALF - SECTIONAL ELEVATION

PLAN

COTTERED JOINT

Fig. 32·30.

Pulleys :

A pulley is a wheel fixed on to a shaft with the help of a key. It rotates with the shaft for transmitting motion and power. Pulleys are usually made of cast iron, mild steel, wrought iron and hard timber. In workshops, mostly cast iron pulleys are used. A pulley essentially consists of a rim, a hub or boss and arms or spokes. There are various types of pulleys such as, belt pulley (cast in one piece with straight or curved arms), split pulley or built-up pulley (large pulley made in two halves and joined together ; the spokes are fitted separately), stepped pulley or speed cone (several pulleys of different diameters cast integrally to fit onto a single shaft). Groove pulley (semi-circular groove for rope pulley and V-groove for V-belt pulley), etc.).

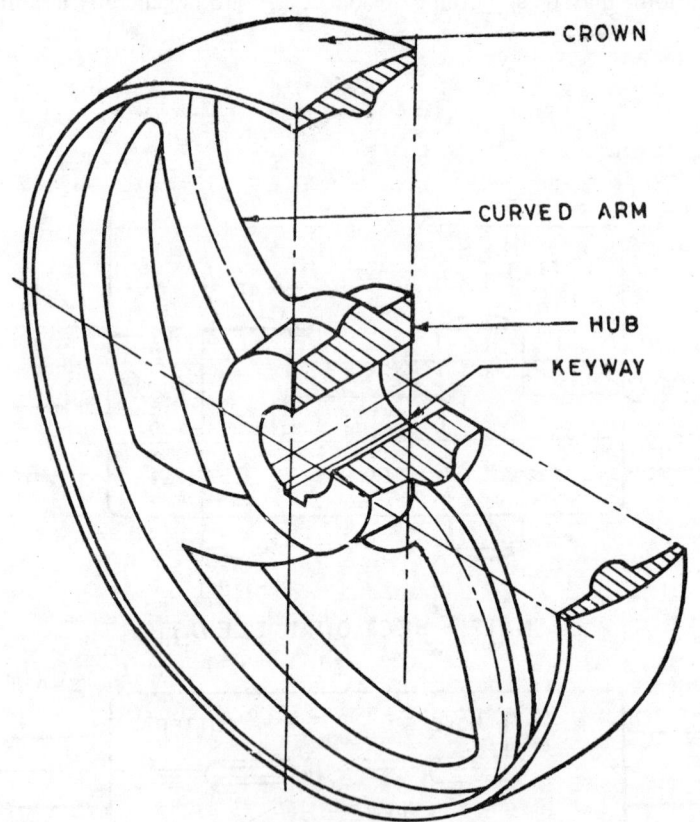

ISOMETRIC VIEW OF A CAST IRON PULLEY IN HALF-SECTION
Fig. 32·31.

Fig. 32·31 shows an isometric view of a cast iron pulley in half section. The arms shown here are curved ordinarily. The arms are made straight. Whether curved or straight, the arms are usually made elliptical in section. For greater strength of pulley, a solid cast iron pulley of solid radial arm is used.

Built up or **split pulleys** are large in diameter and they are made of two halves which are joined together with countersunk bolts and rivets. Such pulleys are stronger, more durable and are preferred for high speed. These are made of mild steel and wrought iron.

The height of the crown of flat pulleys varies with the diameter of the pulleys and the crown values are calculated by the formula :

$$h = 0.003 \ D.$$

where h = height of the crown

and D = diameter of the pulley.

The number of arms to be providedin a pulley also varies with the diameter of the pulley. For pulleys of diameter

> upto 200 mm, use web ;
> 200 mm to 450 mm, use 4 arms ;
> above 450 mm, use 6 arms ;

The arms are usually of elliptical cross section.

$$\text{Thickness of arm near boss} = 2\text{·}94 \sqrt[3]{\frac{WD}{2N}} \quad \text{for single belt}$$

$$= 2\text{·}94 \sqrt[3]{\frac{WN}{4N}} \quad \text{for double belt}$$

where, W = width of pulley
D = diameter of arms
N = number of pulley

Thickness of arm near rim is arrived at by giving a taper of 1 in 25.

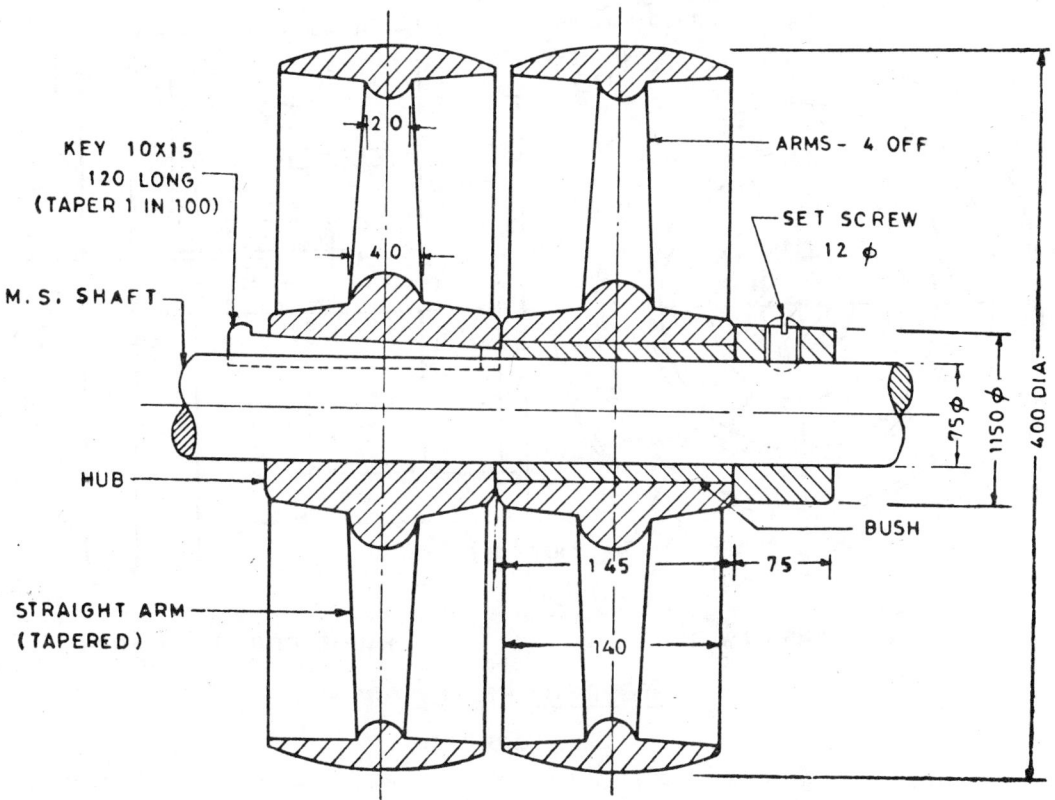

FAST PULLEY & LOOSE PULLEY

Fig. 32·32.

Fast Pulley and Loose Pulley arrangement is required to stop or start a machine without disturbing the rotation of the main shaft. The fast pulley is keyed onto the shaft and the loose pulley is free to move on the shaft. For transmitting motion to a machine *i.e.*, to start a machine, the machine pulley belt is put over the fast pulley fitted to the shaft. For stopping the machine as and when required the machine pulley belt is simply switched over the loose pulley. Fig. 32·32 shows the full sectional view of fast pulley and loose pulley arrangement.

Stepped Pulley or speed cone is a combination of several pulleys of different diameters cast integrally in form of steps. This type of pulley is used when variable speed is required. The speed can be varied just by shifting the belt from one step to the other. Half-sectional elevation and left end view of a stepped pulley are illustrated in Fig. 32·33. Sometimes, these pulleys have flanges on one or both sides so that the belt does not come out of the pulley during rotation. Such a pulley is fitted to a shaft either by means of a set screw or with the help of a key.

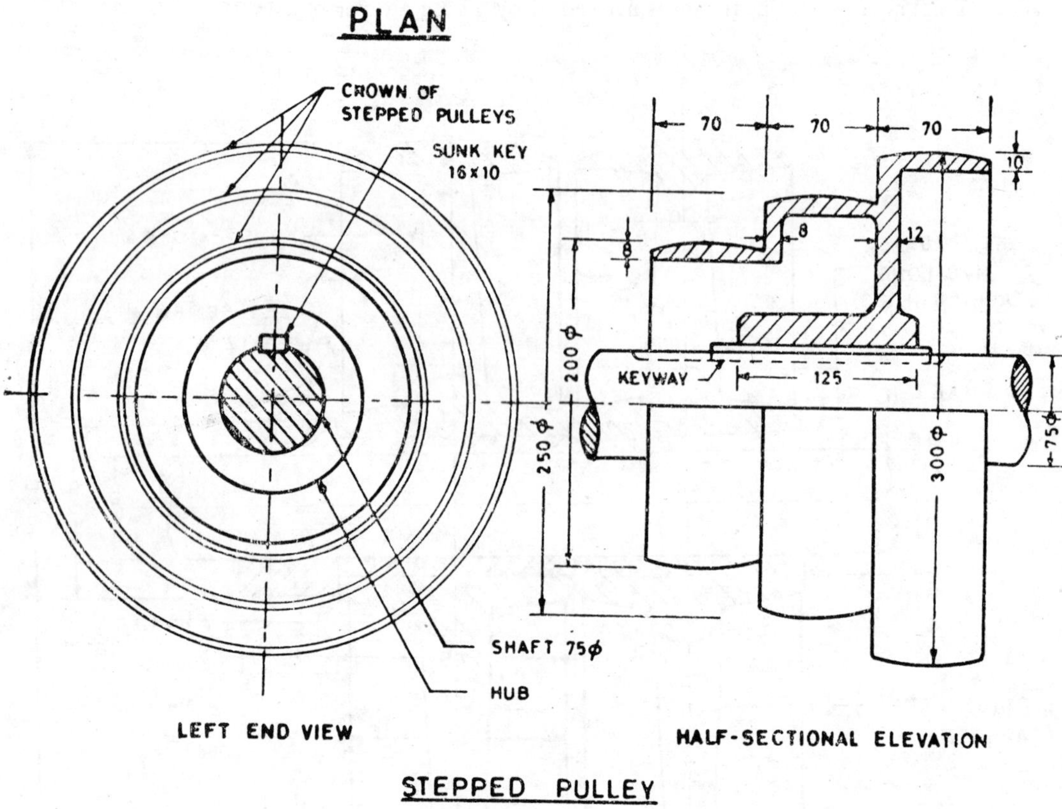

PLAN

STEPPED PULLEY

Fig. 32·33.

Wall Brackets are used to support a bearing which in turn supports a rotating shaft. Two views of a wall bracket are shown in Fig. 32·34. This type of bracket is required to support a pedestal bearing (plummer block). The bracket is fixed either to a wall or to a stanchion. These are usually made of cast iron.

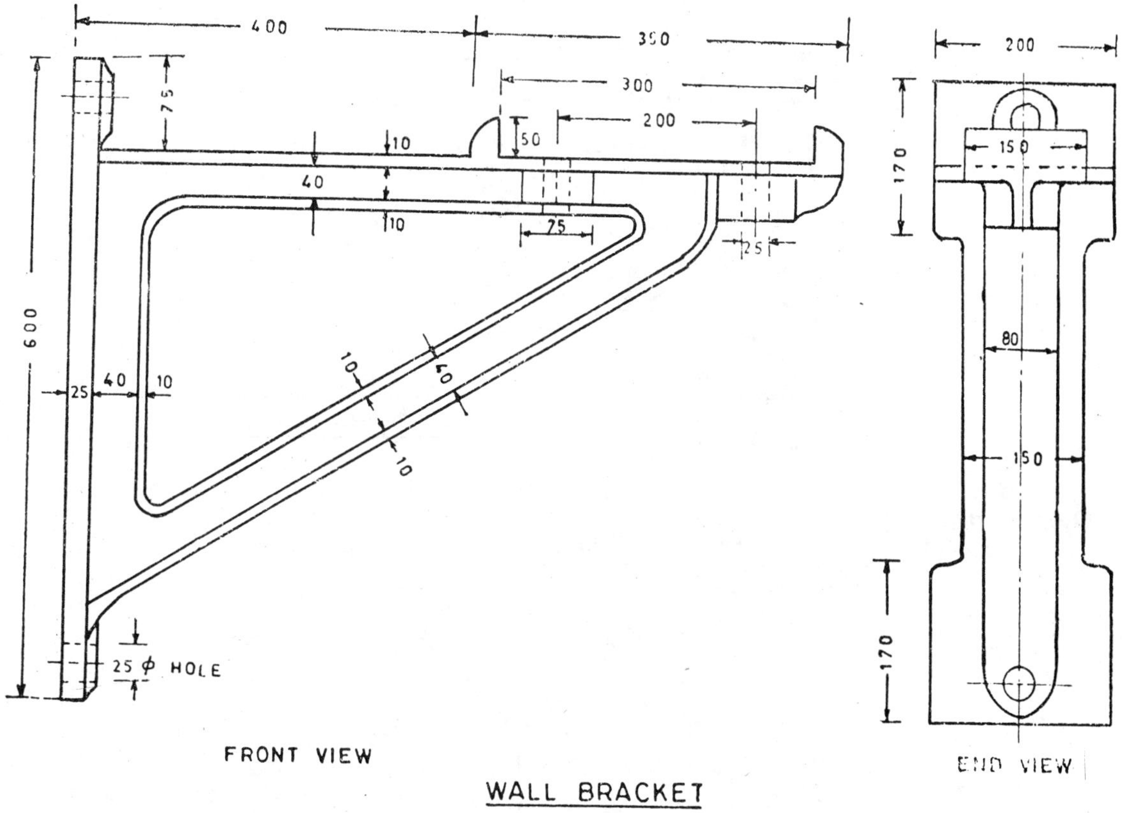

FRONT VIEW

END VIEW

WALL BRACKET

Fig. 32·34.

Sole Plate

Plan and half sectional elevation of a sole plate used to hold a bearing or a machine are shown in Fig. 32·35. Its pictorical view is presented in Fig. 32·36.

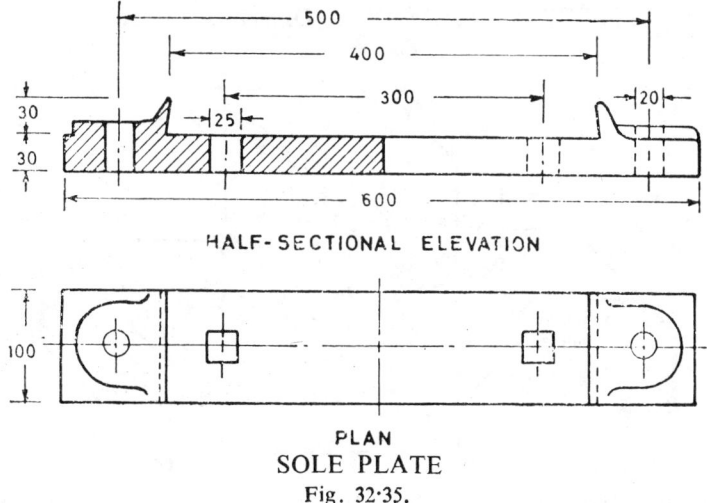

HALF-SECTIONAL ELEVATION

PLAN

SOLE PLATE

Fig. 32·35.

SOLE PLATE

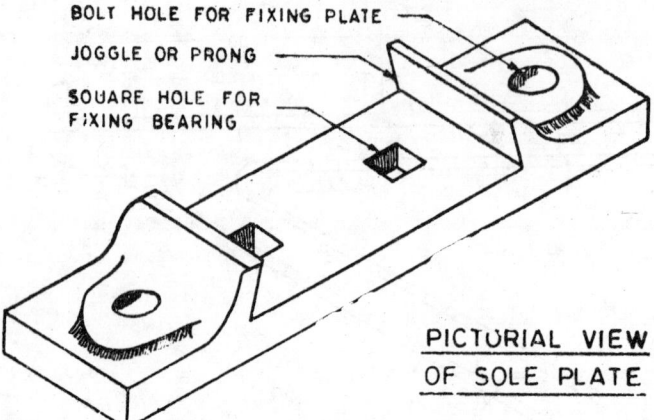

BOLT HOLE FOR FIXING PLATE

JOGGLE OR PRONG

SQUARE HOLE FOR
FIXING BEARING

PICTORIAL VIEW
OF SOLE PLATE

Fig. 32·36.

Open Bearing
      A pictorial view of a open bearing *i.e.,* without a cap at top is shown in Fig. 32·37.

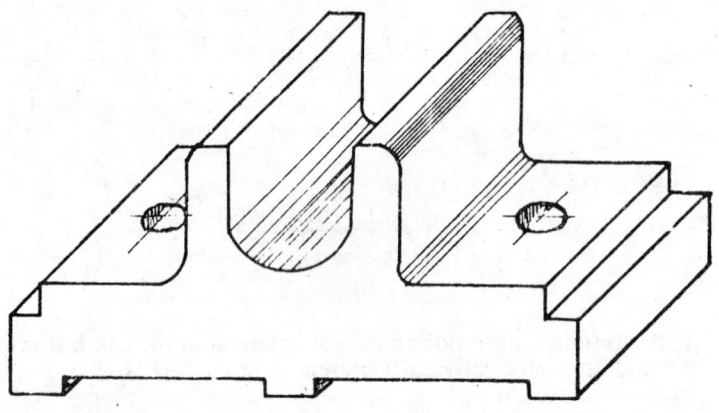

PICTORIAL VIEW OF A OPEN BEARING
Fig. 32·37.

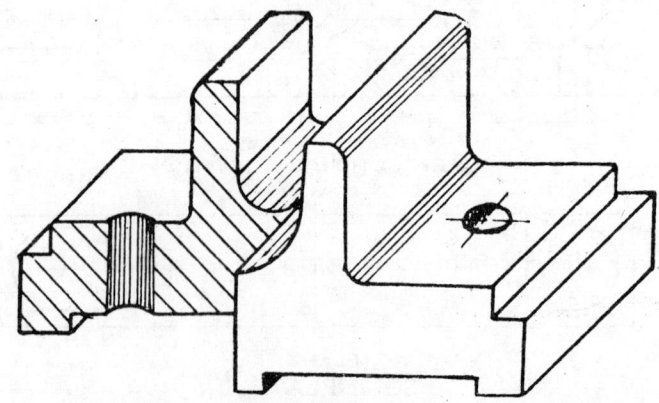

PICTORIAL VIEW OF A OPEN BEARING IN HALF-SECTION
Fig. 32·38.

Fig. 32·38 presents the pictorial view of a open bearing in half-section *i.e.*, ¼th part of the bearing is cut and removed.

The top view and front view of the same open bearing are shown in Fig. 32·39 with dimensions. Its half-sectional elevation and end view are presented in Fig. 32·40.

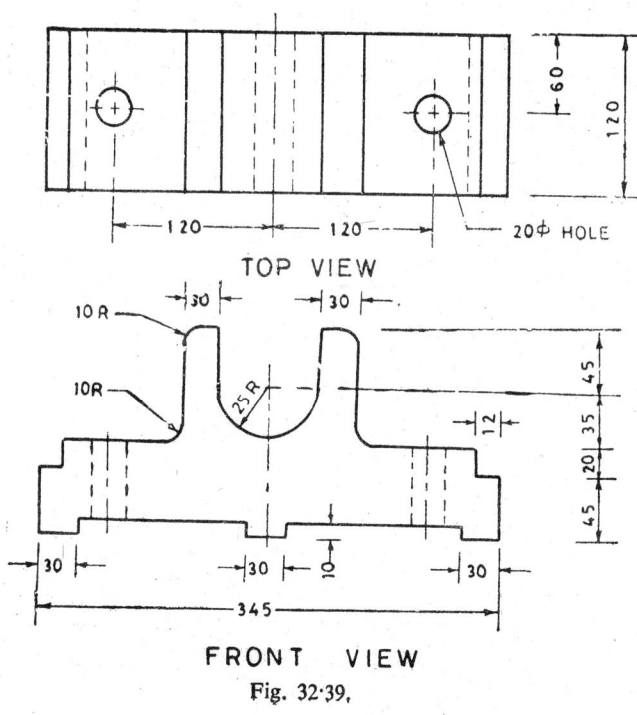

TOP VIEW

FRONT VIEW

Fig. 32·39,

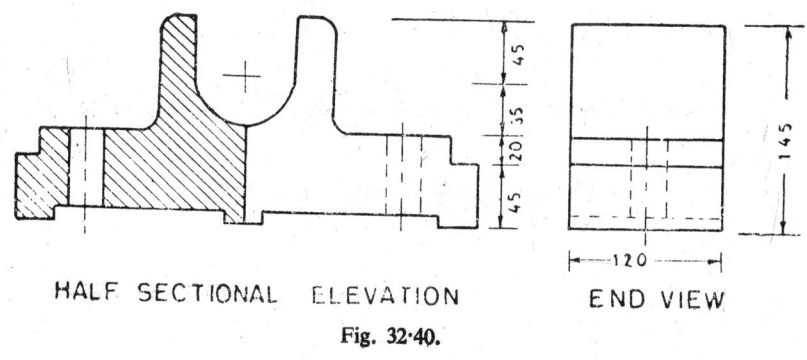

HALF SECTIONAL ELEVATION   END VIEW

Fig. 32·40.

Footstep Bearing is used to hold a vertical shaft. It essentially consists of a pedestal, a bush and a pad. The pedestal is made of cast iron while the bush and the pad are made of gun metal. The plan and elevation of a footstep bearing are shown in Fig. 32·41.

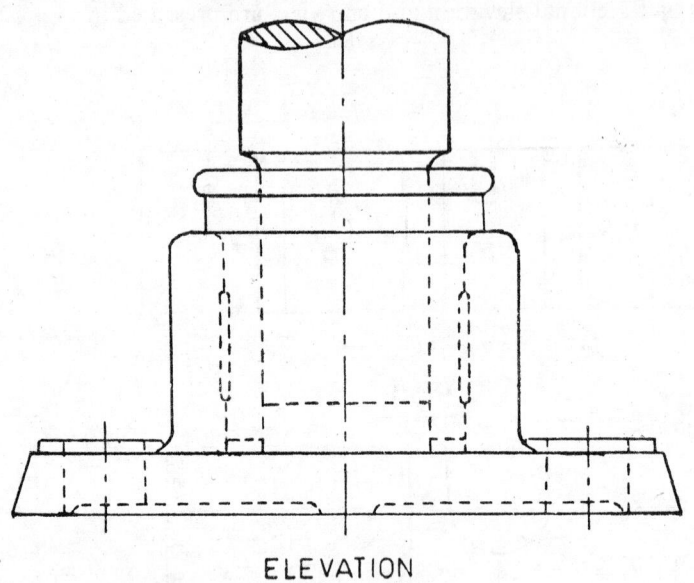

ELEVATION

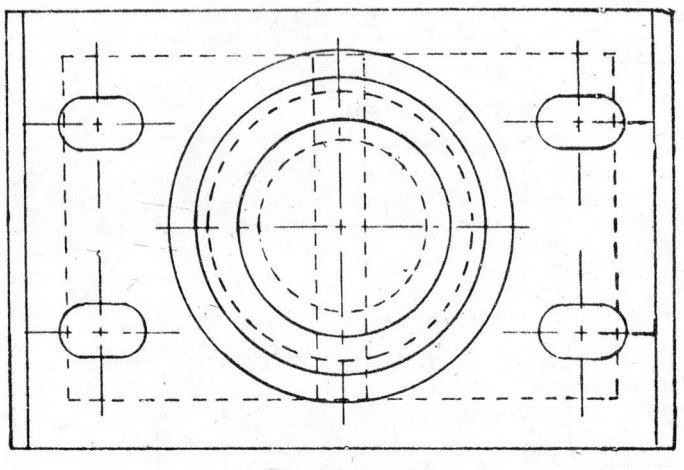

PLAN

FOOTSTEP BEARING

Fig. 32·41.

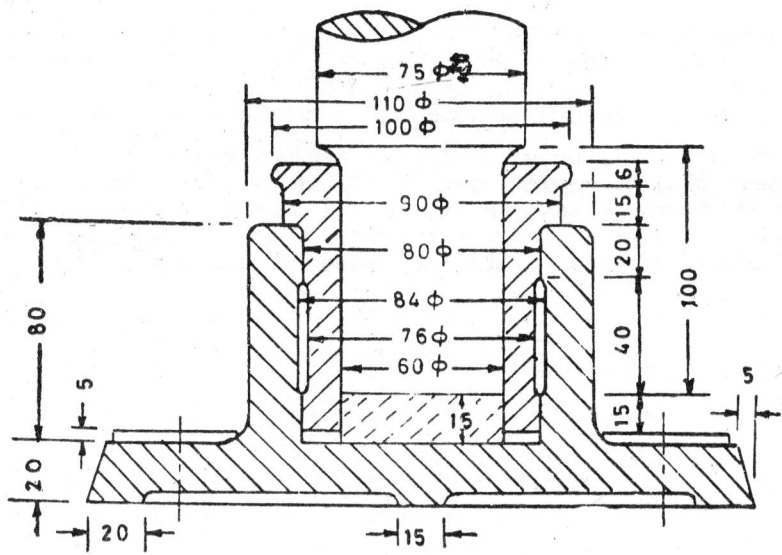

SECTIONAL   ELEVATION

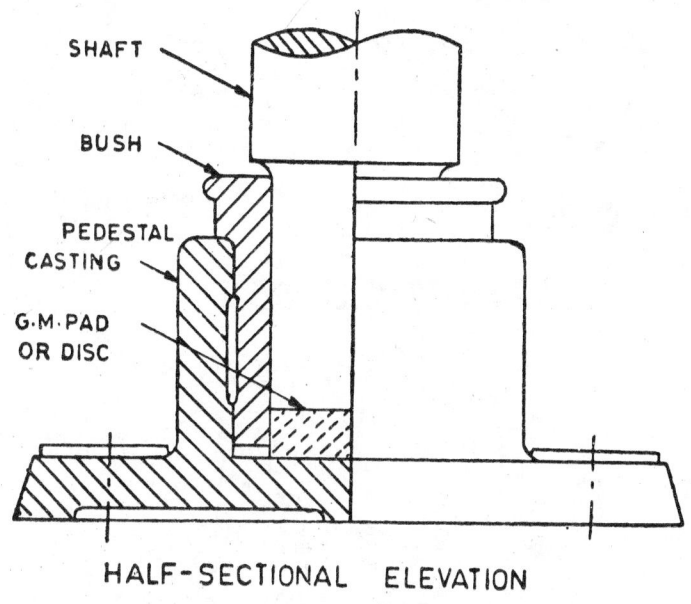

SHAFT

BUSH

PEDESTAL
CASTING

G·M·PAD
OR DISC

HALF-SECTIONAL   ELEVATION

FOOTSTEP BEARING
Fig. 32·42.

Full-sectional elevation and half-sectional elevation of a Footstep bearing with
dimensions are illustrated in Fig. 32·42.

Plummer Block

Plummer block is a pedestal bushed bearing required to support shaft. It essentially consists of a cast iron base or pedestal, gun metal brasses (bush) made in two halves, a cast irou cap provided with a oil hole at top and bolts, nub etc. Fig. 32·43 shows the different parts of the plummer block in pictorial view and its assembled view is given in Fig. 32·44.

There are various forms of plummer blocks required to serve spccific function. An angle plummer block is generally used as crank shaft bearing. The inclined construction of casting does not provide any space for bolt hole and therefore studs are ured with lock nuts at top to hold the shaft. Its construction facilitates power transmission at an angle and easy removal of the shaft by unscrewing the lock nut and stud from top.

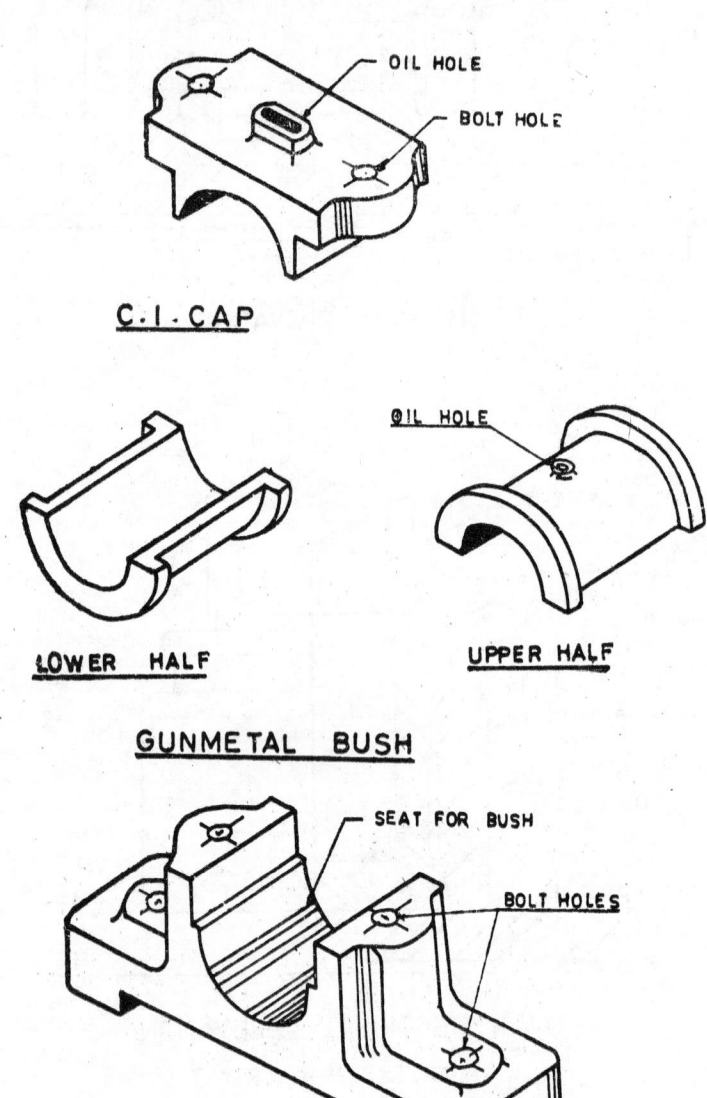

OIL HOLE

BOLT HOLE

C·I·CAP

OIL HOLE

LOWER HALF

UPPER HALF

GUNMETAL BUSH

SEAT FOR BUSH

BOLT HOLES

C·I·BASE

Fig. 32·43.

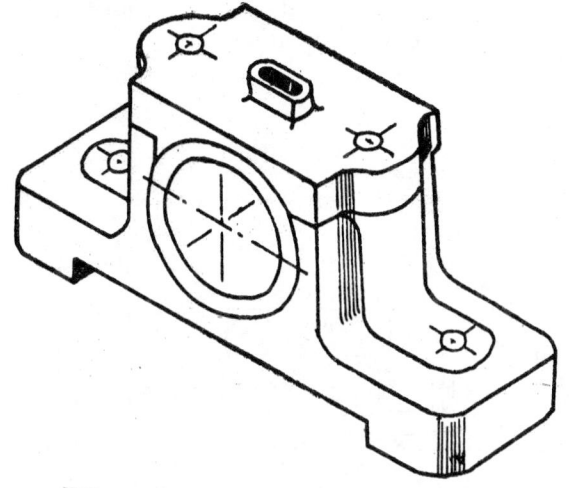

ASSEMBLED VIEW OF
A PEDESTAL BEARING

Fig. 32·44.

Shaft Couplings

Shafts are usually available in a length varying from 6 to 10 metres. When lengthy shafts are required, two or more shafts are connected together to form one shaft. The joining of two shafts is called a shaft coupling. A shaft coupling is classified into two major groups *viz.* Rigid coupling and flexible coupling. The various types of shaft couplings used are : Muff or Box coupling, split Muff coupling, Flanged coupling, Flexible coupling, Friction clutch claw coupling, etc. etc.

A solid flange coupling with forged flanges integral with the shafts is shown in Fig. 32·45. The flanges are held together with the help of taper bolts. This type of coupling is used in marine works.

Fig. 32·45 shows half-sectional elevation of a Pin-type flexible coupling. A hub with a flange is fitted to either shaft with the help of key and the flanges are firmly held by means of pins. Compressible washers of leather, rubber, etc. are used as shown in figure to join the shafts which alleviate shocks and vibrations occuring during transmission of power.

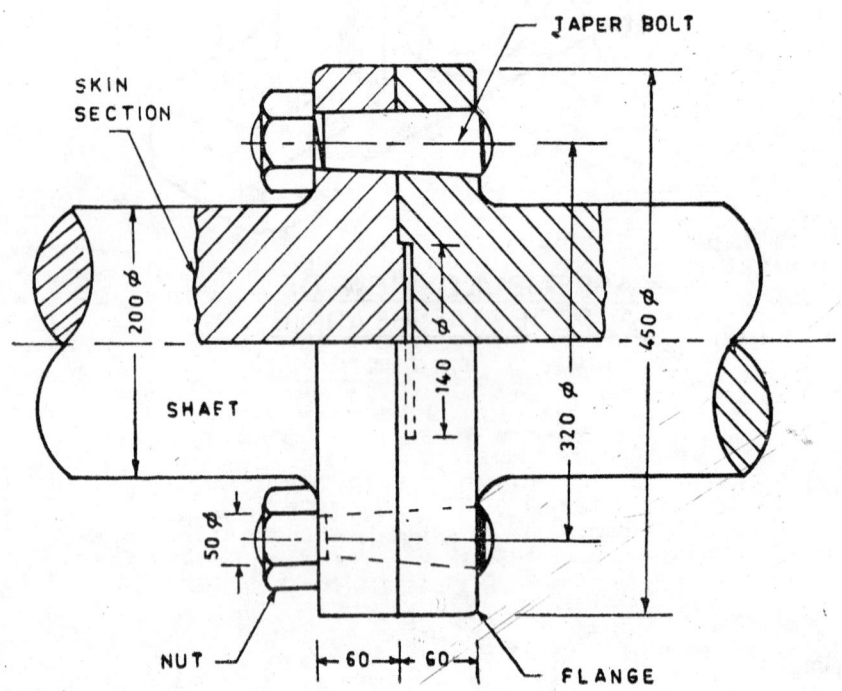

SOLID FLANGED COUPLING

Fig. 32·45.

Flange couplings are of various types. We have already discussed two types of flange couplings. Fig. 32·47 shows another type of flange coupling which is very commonly used. It may be seen that the bolt heads and nuts do not project beyond the flanges and they are guarded by giving projection at the flanges. Fig. 32.47 shows the full-sectional elevation and end view of this coupling. The half-sectional view and a pictorial view of the left half of a protected type flange coupling in half section are presented in Fig. 32·48. This is similar to the flange coupling shown in Fig. 32·47.

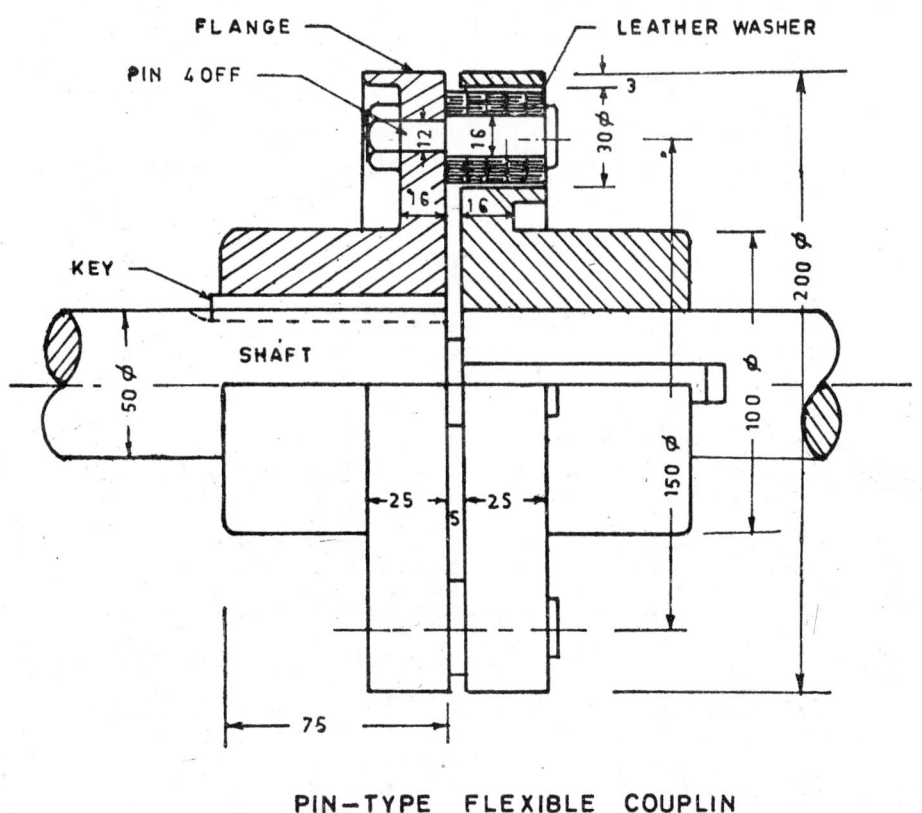

Fig. 32·46.

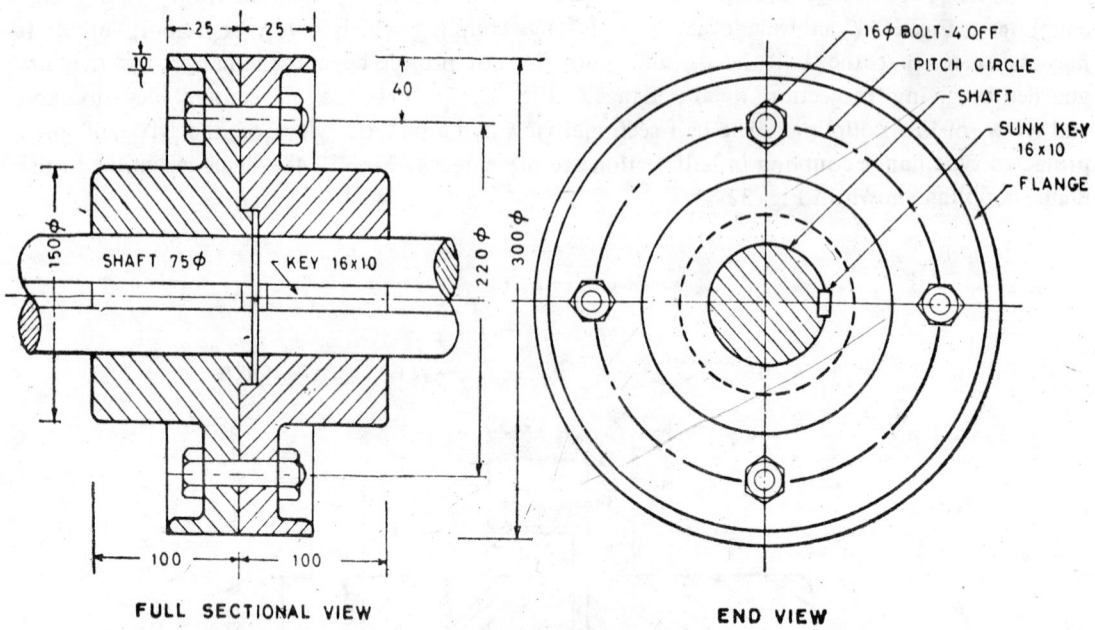

FULL SECTIONAL VIEW   END VIEW

FLANGE COUPLING

Fig. 32·47.

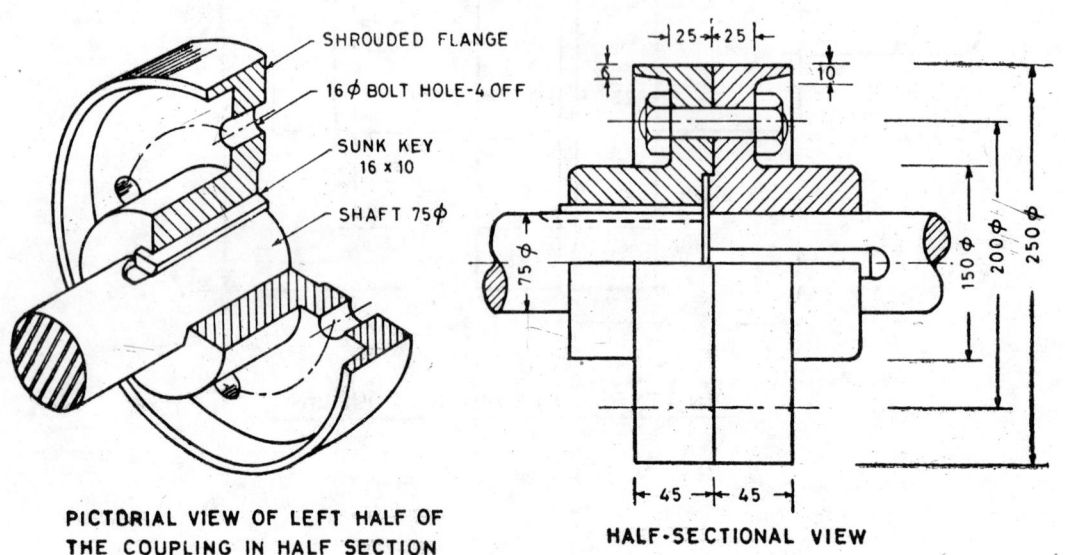

PICTORIAL VIEW OF LEFT HALF OF
THE COUPLING IN HALF SECTION

HALF-SECTIONAL VIEW

PROTECTED TYPE FLANGED COUPLING

Fig. 32·48.

Adjustable Joint :

An adjustable joint is required to join two circular bars or rods which will be subjecteds to tension. The end of the rod to be jointed have threads in opposite direction and they are screwed to a hexagonal screwed coupler. When the hexagonal coupler is screwed on the rod ends, it pulls them inside the hole. Plan and sectional view of such a joint are shown in Fig. 32.49.

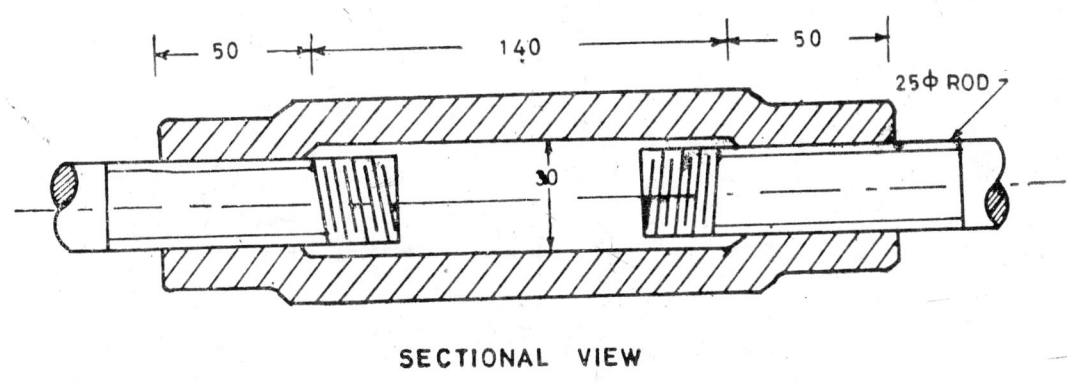

SECTIONAL VIEW

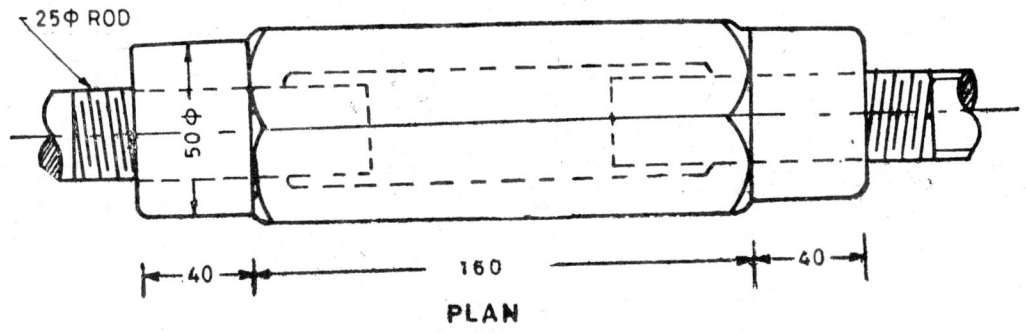

PLAN

ADJUSTABLE JOINT

Fig. 32.49.

Knukckle Joint : This is also known as pin joint, since two or more rods are connected to gether by means of pins. One end of a rod is forged in form of a fork having a hole in both the arms of the fork. The end of the other rod is forged into a single eye end (having a hole). This end is placed within the fork and a cylindrical pin is inserted through the holes in them. Then, they are held together by means of a collar and a taper pin. The rods are

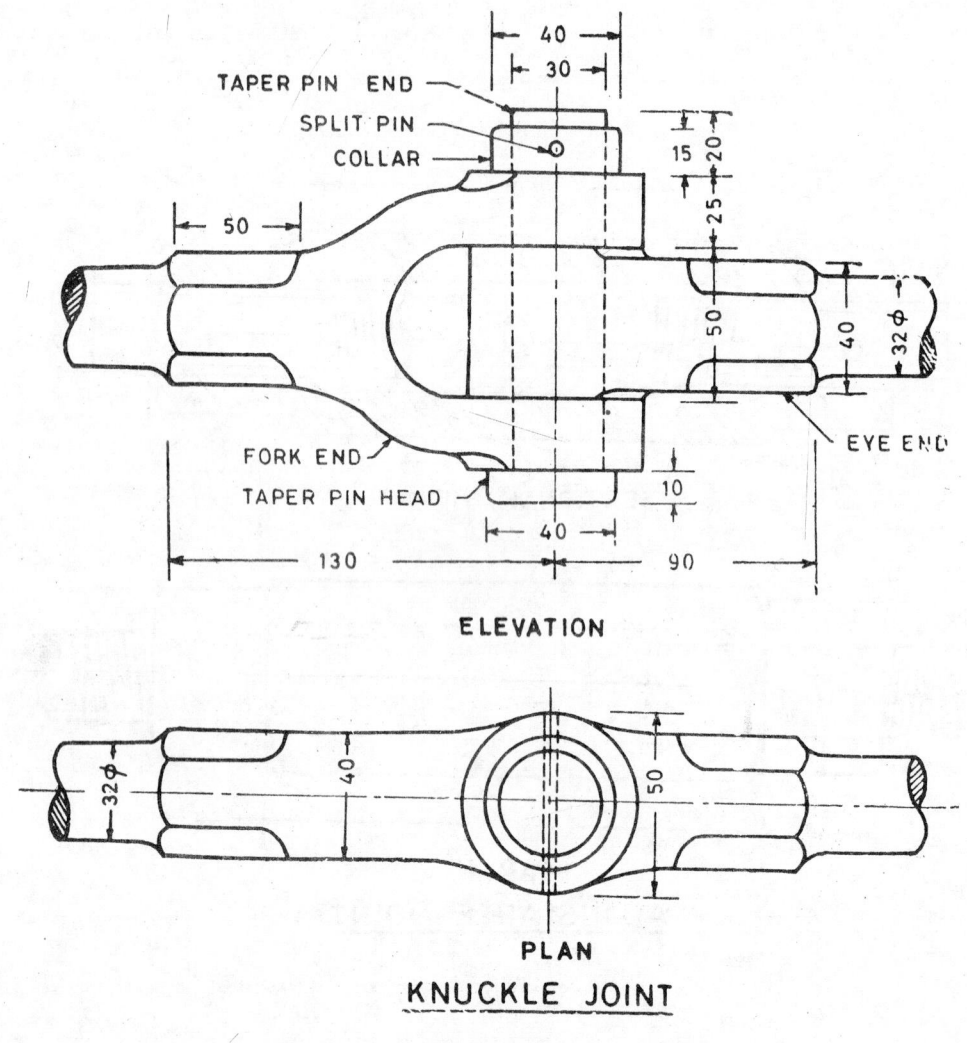

ELEVATION

PLAN

KNUCKLE JOINT

Fig. 32·50.

quite free to rotate in axial direction on the pin. This type of joint is widely used in machine parts where angular movement of one rod about the other is required. Fig. 32·50 shows plan and elevation of a knuckle joint for two rods. It pictorial view is shown in Fig. 32·51.

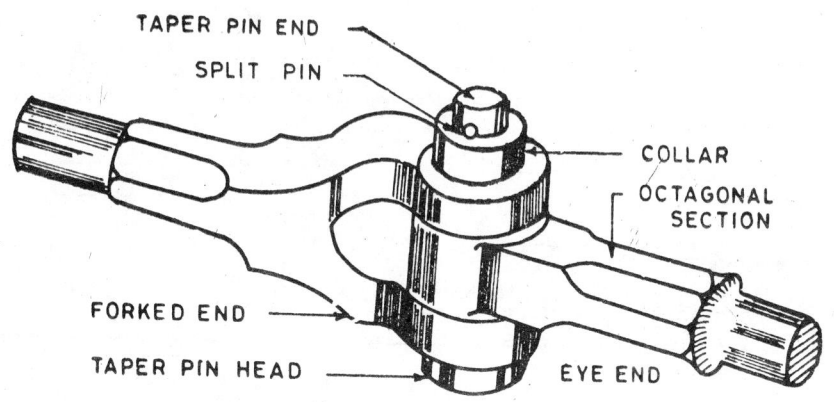

TAPER PIN END
SPLIT PIN
COLLAR
OCTAGONAL SECTION
FORKED END
TAPER PIN HEAD
EYE END

PICTORIAL VIEW OF KNUCKLE JOINT
(FOR 2 RODS)

Fig. 32·51.

A pictorial view of a knuckle joint for three rods is presented in Fig. 32·52.

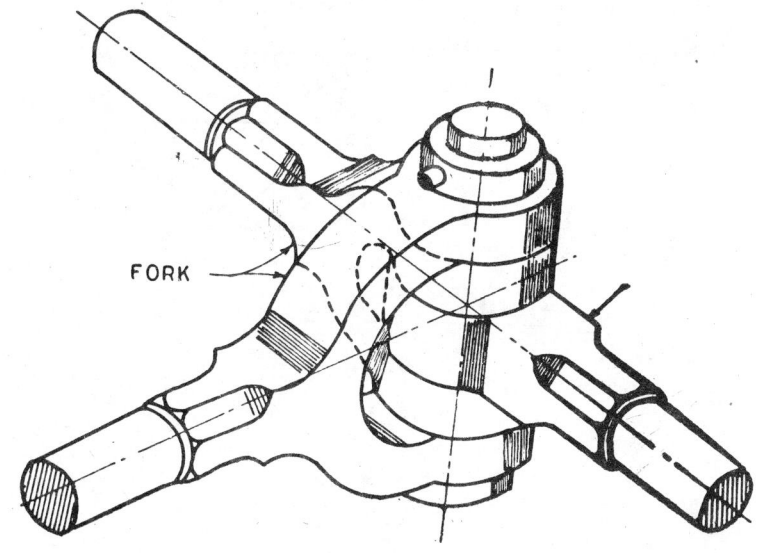

FORK

PICTORIAL VIEW OF KNUCKLE JOINT
(FOR 3 RODS)

Fig. 32·52.

Stuffing Box : A stuffing box is a device used to give free sliding or rotating movement to a shaft without causing any leakage of fluid from the vessel containing fluid under pressure to impart motion to the shaft. A sectional view of a stuffing box is shown in Fig. 32·53. A stuffing box essentially consists of a cast iron body, cast iron gland, neckbush and gland bush made of gunmetal, packing material and studs with nuts.

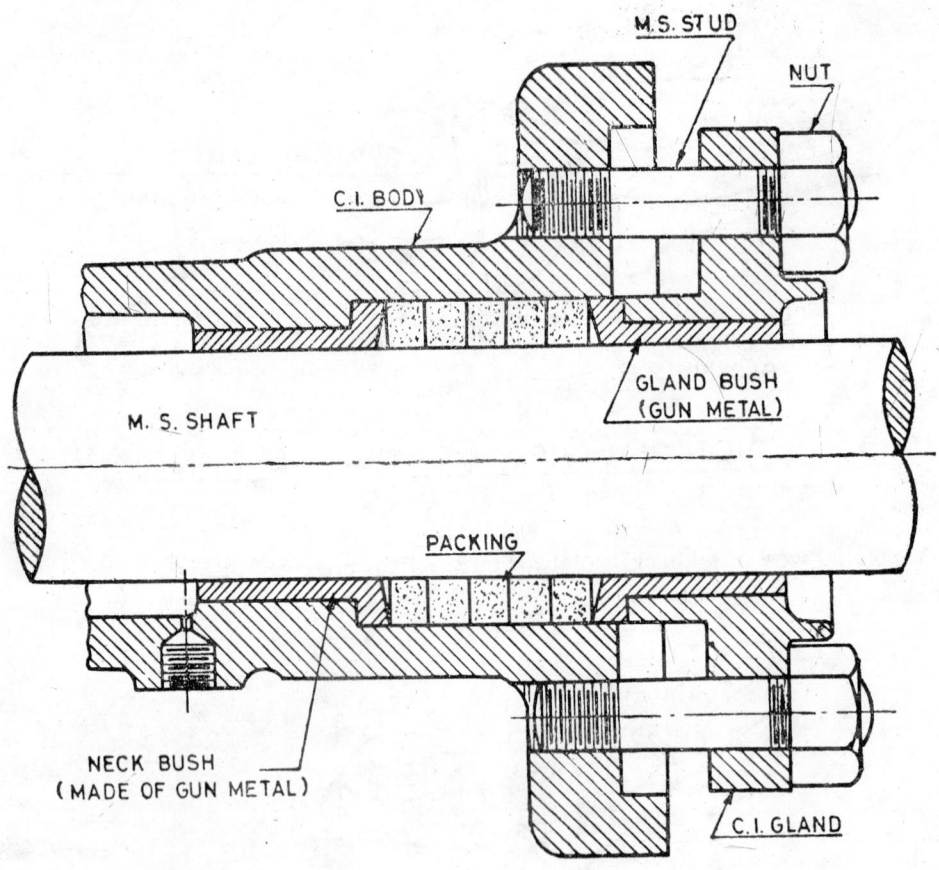

STUFFING BOX

Fig. 32·53,